D1154925

Methods in Enzymology

Volume 325
REGULATORS AND EFFECTORS OF SMALL GTPases
Part D
Rho Family

METHODS IN ENZYMOLOGY

EDITORS-IN-CHIEF

John N. Abelson   Melvin I. Simon

DIVISION OF BIOLOGY
CALIFORNIA INSTITUTE OF TECHNOLOGY
PASADENA, CALIFORNIA

FOUNDING EDITORS

Sidney P. Colowick and Nathan O. Kaplan

Methods in Enzymology

Volume 325

Regulators and Effectors of Small GTPases

Part D
Rho Family

EDITED BY

W. E. Balch

THE SCRIPPS RESEARCH INSTITUTE
LA JOLLA, CALIFORNIA

Channing J. Der

LINEBERGER COMPREHENSIVE CANCER CENTER
THE UNIVERSITY OF NORTH CAROLINA AT CHAPEL HILL
CHAPEL HILL, NORTH CAROLINA

Alan Hall

UNIVERSITY COLLEGE LONDON, LONDON, ENGLAND

QP
601
,M49
V.325
Pt D

ACADEMIC PRESS

San Diego   London   Boston   New York   Sydney   Tokyo   Toronto

Academic Press
A Harcourt Science and Technology Company
525 B Street, Suite 1900, San Diego, California 92101-4495, USA

http://www.academicpress.com

Academic Press Limited
32 Jamestown Road, London NN1 7BY, UK

International Standard Book Number: 0-12-182226-5

PRINTED IN THE UNITED STATES OF AMERICA
00 01 02 03 04 05 06 MM 9 8 7 6 5 4 3 2 1

Table of Contents

Section I. Purification, Posttranslational Modification, and *in Vitro* Regulation

Section II. Purification and Activity of GTPase Targets

Section III. Analysis of Rho GTPase Function

Section IV. Biological Assays of Rho GTPase Function

Contributors to Volume 325

Article numbers are in parentheses following the names of contributors.
Affiliations listed are current.

KARON ABE (38), *Lineberger Comprehensive Cancer Center, University of North Carolina, Chapel Hill, North Carolina 27599*

KLAUS AKTORIES (12), *Institut für Pharmakologie und Toxikologie, Albert-Ludwigs-Universität Freiburg, D-79104 Freiburg, Germany*

MUTSUKI AMANO (14), *Division of Signal Transduction, Nara Institute of Science and Technology, Ikoma, Nara 630-0101, Japan*

ANSER C. AZIM (22), *Division of Hematology, Brigham and Women's Hospital, Harvard Medical School, Boston, Massachusetts 02115*

DIANE L. BARBER (30), *Departments of Stomatology and Surgery, University of California, San Francisco, California 94143*

KURT L. BARKALOW (22, 31), *Division of Hematology, Brigham and Women's Hospital, Harvard Medical School, Boston, Massachusetts 02115*

DAFNA BAR-SAGI (29), *Department of Molecular Genetics and Microbiology, State University of New York, Stony Brook, New York 11794-5222*

GARY M. BOKOCH (28), *Departments of Immunology and Cell Biology, The Scripps Research Institute, La Jolla, California 92037*

GIDEON BOLLAG (5, 6), *Onyx Pharmaceuticals, Richmond, California 94806*

DANIEL BROEK (4), *Department of Biochemistry and Molecular Biology, Keck School of Medicine, University of Southern California, Los Angeles, California 90033*

SHARON L. CAMPBELL (3), *Department of Biochemistry and Biophysics, University of North Carolina, Chapel Hill, North Carolina 27599-7260*

EMMANUELLE CARON (41), *MRC Laboratory for Molecular Cell Biology, University College London, London WC1E 6BT, England, United Kingdom*

CHRISTOPHER L. CARPENTER (18), *Division of Signal Transduction, Beth Israel Deaconess Medical Center, Boston, Massachusetts 02215*

FLAVIA CASTELLANO (25), *Institut Curie-Recherche, CNRS UMR 144, 75248 Paris Cedex 05, France*

CHESTER E. CHAMBERLAIN (35), *Department of Cell Biology, The Scripps Research Institute, La Jolla, California 92037*

PHILIPPE CHAVRIER (25), *Institut Curie-Recherche, CNRS UMR 144, 75248 Paris Cedex 05, France*

EDWIN CHOY (10), *Department of Medicine, Massachusetts General Hospital, Boston, Massachusetts 02114*

JOHN G. COLLARD (26, 36), *Division of Cell Biology, The Netherlands Cancer Institute, 1066 CX Amsterdam, The Netherlands*

ANNE M. CROMPTON (5), *Onyx Pharmaceuticals, Richmond, California 94806*

GIOVANNA M. D'ABACO (37), *Cancer Research Campaign for Cell and Molecular Biology, Chester Beatty Laboratories, Institute of Cancer Research, London SW3 6JB, England, United Kingdom*

BALAKA DAS (4), *Department of Biochemistry and Molecular Biology, Keck School of Medicine, University of Southern California, Los Angeles, California 90033*

SHERYL P. DENKER (30), *Department of Stomatology, University of California, San Francisco, California 94143*

CHANNING J. DER (38), *Lineberger Comprehensive Cancer Center, The University of North Carolina, Chapel Hill, North Carolina 27599*

JOHN F. ECCLESTON (7), *Division of Physical Biochemistry, National Institute for Medical Research, London NW7 1AA, England, United Kingdom*

EVA E. EVERS (36), *Division of Cell Biology, The Netherlands Cancer Institute, 1066 CX Amsterdam, The Netherlands*

TOREN FINKEL (27), *National Heart, Lung, and Blood Institute, Laboratory of Molecular Biology, National Institutes of Health, Bethesda, Maryland 20892-1650*

ALYSON E. FOURNIER (42), *Department of Neurology, Yale University School of Medicine, New Haven, Connecticut 06520*

ANDREA FRIEBEL (8), *Max von Pettenkofer-Institut, Ludwig Maximilians Universität, 80336 Munich, Germany*

MICHAEL A. FROHMAN (17), *Department of Pharmacology and Institute for Cell and Developmental Biology, State University of New York, Stony Brook, New York 11794-8651*

YIXIN FU (44), *Section of Microbial Pathogenesis, Boyer Center for Molecular Medicine, Yale University School of Medicine, New Haven, Connecticut 06536-0812*

KEIGI FUJIWARA (33), *Department of Structural Analysis, National Cardiovascular Center Research Institute, Osaka 565-8565, Japan*

YUKO FUKATA (14), *Department of Cell Pharmacology, Nagoya University School of Medicine, Nagoya AICHI 466-8550, Japan*

JORGE E. GALÁN (44), *Section of Microbial Pathogenesis, Boyer Center for Molecular Medicine, Yale University School of Medicine, New Haven, Connecticut 06536-0812*

PETER GIERSCHIK (16), *Department of Pharmacology and Toxicology, University of Ulm, D-89081 Ulm, Germany*

GASTON G. M. HABETS (5), *Onyx Pharmaceuticals, Richmond, California 94806*

KLAUS M. HAHN (35), *Department of Cell Biology, The Scripps Research Institute, La Jolla, California 92037*

ALAN HALL (41), *MRC Laboratory for Molecular Cell Biology, University College London, London WC1E 6BT, England, United Kingdom*

ANDREW D. HAMILTON (34), *Department of Chemistry, Yale University, New Haven, Connecticut 06511*

JAEWON HAN (4), *Department of Vascular Biology, The Scripps Research Institute, La Jolla, California 92037*

WOLF-DIETRICH HARDT (8), *Max von Pettenkofer-Institut, Ludwig Maximilians Universität, 80336 Munich, Germany*

MATTHEW J. HART (6), *Onyx Pharmaceuticals, Richmond, California 94806*

JOHN H. HARTWIG (22, 31), *Division of Hematology, Brigham and Women's Hospital, Harvard Medical School, Boston, Massachusetts 02115*

PATRICK HEARING (29), *Department of Molecular Genetics and Microbiology, State University of New York, Stony Brook, New York 11794-5222*

MARK R. HOLT (32), *Physiology Department, University College London, London WC1E 6JJ, England, United Kingdom*

JON P. HUTCHINSON (7), *Division of Physical Biochemistry, National Institute for Medical Research, London NW7 1AA, England, United Kingdom*

DARIA ILLENBERGER (16), *Department of Pharmacology and Toxicology, University of Ulm, D-89081 Ulm, Germany*

TOSHIMASA ISHIZAKI (24), *Department of Pharmacology, Kyoto University Faculty of Medicine, Kyoto 606-8315, Japan*

LENNERT JANSSEN (26), *Division of Cell Biology, The Netherlands Cancer Institute, 1066 CX Amsterdam, The Netherlands*

DANIEL G. JAY (43), *Department of Physiology, Tufts University School of Medicine, Boston, Massachusetts 02111*

GARETH E. JONES (40), *Randall Centre for Molecular Mechanisms of Cell Function, King's College London, London SE1 1UL, England, United Kingdom*

KOZO KAIBUCHI (14), *Department of Cell Pharmacology, Nagoya University School of Medicine, Nagoya AICHI 466-8550, Japan and Division of Signal Transduction, Nara Institute of Science and Technology, Ikoma, Nara 630-0101, Japan*

ROBERT G. KALB (42), *Department of Neurology, Yale University School of Medicine, New Haven, Connecticut 06520*

YASUNORI KANAHO (17), *Department of Pharmacology, Tokyo Metropolitan Institute of Medical Science, Tokyo 113-8613, Japan*

YUMIKO KANO (33), *Department of Structural Analysis, National Cardiovascular Center Research Institute, Osaka 565-8565, Japan*

KAZUO KATOH (33), *Department of Structural Analysis, National Cardiovascular Center Research Institute, Osaka 565-8565, Japan*

CHARLES C. KING (15, 28), *Department of Immunology, The Scripps Research Institute, La Jolla, California 92037*

ULLA G. KNAUS (15), *Department of Immunology, The Scripps Research Institute, La Jolla, California 92037*

ANNA KOFFER (32), *Physiology Department, University College London, London WC1E 6JJ, England, United Kingdom*

VADIM S. KRAYNOV (35), *Department of Cell Biology, The Scripps Research Institute, La Jolla, California 92037*

IAN G. MACARA (1), *The Markey Center for Cell Signaling, University of Virginia, Charlottesville, Virginia 22908*

LAURA M. MACHESKY (20), *Division of Molecular Cell Biology, School of Biosciences, University of Birmingham, Birmingham B15 2TT, England, United Kingdom*

AKIKO MAMMOTO (9), *Department of Molecular Biology and Biochemistry, Osaka University Graduate School of Medicine, Faculty of Medicine, Osaka 565-0871, Japan*

DANNY MANOR (13), *Division of Nutritional Sciences, Cornell University, Ithaca, New York 14853*

FRITS MICHIELS (26), *Galapagos Genomics, 2333 AL Leiden, The Netherlands*

MICHAEL MOOS (11), *Institut für Medizinische Mikrobiologie, Universität Mainz, D-55101 Mainz, Germany*

ANDREW J. MORRIS (17), *Department of Pharmacology and Institute for Cell and Developmental Biology, State University of New York, Stony Brook, New York 11794-8651*

RAYMOND MOSTELLER (4), *Department of Biochemistry and Molecular Biology, Keck School of Medicine, University of Southern California, Los Angeles, California 90033*

R. DYCHE MULLINS (20), *Department of Cellular and Molecular Pharmacology, University of California School of Medicine, San Francisco, California 94143*

ROBERT K. NAKAMOTO (2), *Department of Molecular Physiology and Biological Physics, University of Virginia, Charlottesville, Virginia 22908-0736*

SHUH NARUMIYA (24), *Department of Pharmacology, Kyoto University Faculty of Medicine, Kyoto 606-8315, Japan*

CHERYL L. NEUDAUER (1), *The Markey Center for Cell Signaling, University of Virginia, Charlottesville, Virginia 22908*

MARGARETA NIKOLIC (19), *Molecular Neurobiology Group, King's College, London, England, United Kingdom*

CATHERINE D. NOBES (39), *MRC Laboratory for Molecular Cell Biology and Department of Anatomy and Developmental Biology, University College London, London WC1E 6BT, England, United Kingdom*

GARRY NOLAN (26), *Stanford University School of Medicine, Stanford, California 94305*

MICHAEL F. OLSON (37), *Cancer Research Campaign for Cell and Molecular Biology, Chester Beatty Laboratories, Institute of Cancer Research, London SW3 6JB, England, United Kingdom*

JAYESH C. PATEL (41), *MRC Laboratory for Molecular Cell Biology, University College London, London WC1E 6BT, England, United Kingdom*

DANIELLE PEVERLY-MITCHELL (5), *Onyx Pharmaceuticals, Richmond, California 94806*

MARK PHILIPS (10), *Departments of Medicine and Cell Biology, New York University School of Medicine, New York, New York 10016*

PAUL W. READ (2), *Department of Molecular Physiology and Biological Physics, University of Virginia, Charlottesville, Virginia 22908-0736*

ABINA M. REILLY (15), *Department of Immunology, The Scripps Research Institute, La Jolla, California 92037*

XIANG-DONG REN (23), *State University of New York, Stony Brook, New York 11794-8165*

ANNE J. RIDLEY (40), *Ludwig Institute for Cancer Research, London W1P 8BT, England, United Kingdom*

KATRIN RITTINGER (7), *Division of Protein Structure, National Institute for Medical Research, London NW7 1AA, England, United Kingdom*

WILLIAM ROSCOE (6), *Onyx Pharmaceuticals, Richmond, California 94806*

KENT L. ROSSMAN (3), *Department of Biochemistry and Biophysics, University of North Carolina, Chapel Hill, North Carolina 27599-7260*

LURAYNNE C. SANDERS (28), *Department of Immunology, The Scripps Research Institute, La Jolla, California 92037*

TAKUYA SASAKI (9), *Department of Biochemistry, The University of Tokushima, School of Medicine, Kuramoto, Japan*

GUDULA SCHMIDT (12), *Institut für Pharmakologie und Toxikologie, Albert-Ludwigs-Universität Freiburg, D-79104 Freiburg, Germany*

FRIEDER SCHWALD (16), *Department of Pharmacology and Toxicology, University of Ulm, D-89081 Ulm, Germany*

MARTIN ALEXANDER SCHWARTZ (23), *The Scripps Research Institute, La Jolla, California 92037*

SAÏD M. SEBTI (34), *Drug Discovery Program, H. Lee Moffitt Cancer Center and Research Institute, University of South Florida, Tampa, Florida 33612*

HIROAKI SHIMOKAWA (14), *Research Institute of Angiocardiology and Cardiovascular Clinic, Kyushu University School of Medicine, Fukuoka 812-8582, Japan*

PATRICIA A. SOLSKI (38), *Lineberger Comprehensive Cancer Center, University of North Carolina, Chapel Hill, North Carolina 27599*

ILONA STEPHAN (16), *Department of Pharmacology and Toxicology, University of Ulm, D-89081 Ulm, Germany*

STEPHEN M. STRITTMATTER (42), *Department of Neurology, Yale University School of Medicine, New Haven, Connecticut 06520*

DANIEL M. SULLIVAN (27), *National Heart, Lung, and Blood Institute, Laboratory of Molecular Biology, National Institutes of Health, Bethesda, Maryland 20892-1650*

MARC SYMONS (5), *Onyx Pharmaceuticals, Richmond, California 94806*

KAZUO TAKAHASHI (9), *Second Department of Internal Medicine, Chiba University Medical School, Chiba 260-0856, Japan*

YOSHIMI TAKAI (9), *Department of Molecular Biology and Biochemistry, Osaka University Graduate School of Medicine/Faculty of Medicine, Osaka 565-0871, Japan*

LAURA J. TAYLOR (29), *Department of Molecular Genetics and Microbiology, State University of New York, Stony Brook, New York 11794-5222*

JEAN P. TEN KLOOSTER (36), *Division of Cell Biology, The Netherlands Cancer Institute, 1066 CX Amsterdam, The Netherlands*

KIMBERLEY TOLIAS (18), *Division of Signal Transduction, Beth Israel Deaconess Medical Center, Boston, Massachusetts 02215*

Li-Huei Tsai (19), *Howard Hughes Medical Institute, Department of Pathology, Harvard Medical School, Boston, Massachusetts 02115*

Masayoshi Uehata (24), *Drug Discovery Laboratories, WelFide (Yoshitomi) Corporation, Osaka 573-1153, Japan*

Rob A. van der Kammen (26, 36), *Division of Cell Biology, The Netherlands Cancer Institute, 1066 CX Amsterdam, The Netherlands*

Christoph von Eichel-Streiber (11), *Verfügungsgebäude für Forschung und Entwicklung, Institut für Medizinisch Mikrobiologie und Hygiene, Johannes Gutenberg-Universität, 55101 Mainz, Germany*

Amy B. Walsh (29), *Department of Molecular Genetics and Microbiology, State University of New York, Stony Brook, New York 11794-5222*

Eric V. Wong (43), *Department of Physiology, Tufts University School of Medicine, Boston, Massachusetts 02111*

Weihong Yan (30), *Department of Stomatology, University of California, San Francisco, California 94143*

Yue Zhang (17), *Department of Pharmacology and Institute for Cell and Developmental Biology, State University of New York, Stony Brook, New York 11794-8651*

Daniel Zicha (40), *Imperial Cancer Research Fund, London WC2A 3PX, England, United Kingdom*

Sally H. Zigmond (21), *Biology Department, University of Pennsylvania, Philadelphia, Pennsylvania 19104-6018*

Preface

In 1955 we edited three volumes of *Methods in Enzymology* (255, 256, 257) dedicated to small GTPases. Since then this field has exploded, and these monomeric, regulatory proteins are now firmly established as a common focus of interest in a wide variety of research areas including cell and developmental biology, immunology, neurobiology, and, more recently, microbiology. After talking with colleagues, it became apparent that all three volumes needed to be significantly updated. We have, therefore, attempted to identify the major new areas and themes that have emerged.

This volume covers the Rho GTPase family. These proteins are key regulators of the actin cytoskeleton, and since the last volume on the subject there has been significant progress in identifying and characterizing the biochemical pathways associated with the three best characterized members of this family, Rho, Rac, and Cdc42. In the past five years, interest has also widened to a much broader community, as it has become clear that Rho GTPases also participate in the regulation of many other signaling pathways, notably activation of the JNK and p38 MAP kinase pathways and of transcription factors such as SRF and NF-κB. This ability to coordinately regulate changes in the actin cytoskeleton with changes in gene transcription and other associated activities appears to be conserved from yeast to mammals.

When the last volumes were published, the large diversity of both downstream targets and upstream guanine nucleotide exchange factors that interact with Rho GTPases was not fully appreciated. Not surprisingly, therefore, these figure more prominantly this time around. Also, although it was thought likely that Rho GTPases might participate in many processes dependent on the organization of filamentous actin, it has now been directly shown that these proteins control cell movement, phagocytosis, growth cone guidance, and cytokinesis. An additional exciting new development has been the identification and characterization of numerous proteins encoded by pathogenic bacteria that directly affect the activity of mammalian Rho GTPases.

We very much hope that this and the accompanying volumes covering the Ras family (Volumes 332 and 333) and the small GTPases involved in membrane trafficking (Volume 329) will provide a useful source of practical information for anyone entering the field. None of this would have been possible without the talents and commitment of all our colleagues who have contributed to these volumes. We are indebted to them.

ALAN HALL
WILLIAM E. BALCH
CHANNING J. DER

METHODS IN ENZYMOLOGY

VOLUME LV. Biomembranes (Part F: Bioenergetics)
Edited by SIDNEY FLEISCHER AND LESTER PACKER

VOLUME LVI. Biomembranes (Part G: Bioenergetics)
Edited by SIDNEY FLEISCHER AND LESTER PACKER

VOLUME LVII. Bioluminescence and Chemiluminescence
Edited by MARLENE A. DELUCA

VOLUME LVIII. Cell Culture
Edited by WILLIAM B. JAKOBY AND IRA PASTAN

VOLUME LIX. Nucleic Acids and Protein Synthesis (Part G)
Edited by KIVIE MOLDAVE AND LAWRENCE GROSSMAN

VOLUME LX. Nucleic Acids and Protein Synthesis (Part H)
Edited by KIVIE MOLDAVE AND LAWRENCE GROSSMAN

VOLUME 61. Enzyme Structure (Part H)
Edited by C. H. W. HIRS AND SERGE N. TIMASHEFF

VOLUME 62. Vitamins and Coenzymes (Part D)
Edited by DONALD B. MCCORMICK AND LEMUEL D. WRIGHT

VOLUME 63. Enzyme Kinetics and Mechanism (Part A: Initial Rate and Inhibitor Methods)
Edited by DANIEL L. PURICH

VOLUME 64. Enzyme Kinetics and Mechanism (Part B: Isotopic Probes and Complex Enzyme Systems)
Edited by DANIEL L. PURICH

VOLUME 65. Nucleic Acids (Part I)
Edited by LAWRENCE GROSSMAN AND KIVIE MOLDAVE

VOLUME 66. Vitamins and Coenzymes (Part E)
Edited by DONALD B. MCCORMICK AND LEMUEL D. WRIGHT

VOLUME 67. Vitamins and Coenzymes (Part F)
Edited by DONALD B. MCCORMICK AND LEMUEL D. WRIGHT

VOLUME 68. Recombinant DNA
Edited by RAY WU

VOLUME 69. Photosynthesis and Nitrogen Fixation (Part C)
Edited by ANTHONY SAN PIETRO

VOLUME 70. Immunochemical Techniques (Part A)
Edited by HELEN VAN VUNAKIS AND JOHN J. LANGONE

VOLUME 71. Lipids (Part C)
Edited by JOHN M. LOWENSTEIN

VOLUME 72. Lipids (Part D)
Edited by JOHN M. LOWENSTEIN

Section I

Purification, Posttranslational Modification, and *in Vitro* Regulation

[1] Purification and Biochemical Characterization of TC10

By Cheryl L. Neudauer and Ian G. Macara

Introduction

TC10 is a member of the Rho family of small GTPases. We have previously characterized the biochemistry, cellular effects, and effector interactions of TC10.[1] This study established TC10 as a distinct member of the Rho family most closely related to Cdc42. In NIH 3T3 cells, the ectopic expression of hemagglutinin (HA)-tagged, gain-of-function TC10 induces long filopodia and loss of stress fibers. TC10 interacts with a subset of those effector proteins that bind to Cdc42. TC10 also interacts with several distinct proteins that are specific for TC10.[2] This article describes the methods used for the mutagenesis of TC10 and the set of vectors used for the purification of recombinant proteins and mammalian expression of TC10. It also describes the biochemical characterization of TC10 and various methods used to study the interaction of TC10 with putative effector proteins.

Mutagenesis and Subcloning of TC10

We have tested several methods for the introduction of point mutations and have found megaprimer polymerase chain reaction (PCR) to be a relatively consistent, cost effective, and reliable technique.[3] Briefly, an internal primer is designed that contains the nucleotide substitution(s). An initial round of PCR is performed with this primer and a primer to either the 5′ or 3′ end of the sequence. In general, we include a *Bam*HI site in the 5′ primer and an *Eco*RI site in the 3′ primer to facilitate subcloning. Restriction enzymes usually cut the ends of PCR products inefficiently and are therefore digested with high concentrations of *Bam*HI and *Eco*RI at 37° for greater than 4 hr.

To facilitate the expression of TC10 and other proteins in bacteria, yeast, and mammalian cells, we have designed a set of vectors with similar

---

[1] C. L. Neudauer, G. Joberty, N. Tatsis, and I. G. Macara, *Curr. Biol.* **8**, 1151 (1998).

[2] G. Joberty, unpublished results (1999).

[3] S. Barik, *in* "PCR Protocols: Current Methods and Applications" (B. A. White, ed.), p. 277. Humana Press, Totowa, NJ, 1993.

0076-6879/00 $30.00

cloning sites (Table I). The majority of these vectors produce N-terminally tagged fusion proteins; C-terminal tagging of the small GTPases is usually avoided as most of these proteins undergo posttranslational modification (e.g., prenylation, carboxymethylation) at their C termini. Each vector contains a *Bam*HI site in the same reading frame as pGEX-2T (Amersham Pharmacia, Piscataway, NJ; Fig. 1).

The pK series of vectors derives expression from a cytomegalovirus (CMV) promoter and contains splice donor and acceptor sites upstream of the initiation codon to increase the efficiency of mRNA export from the nucleus. The vectors contain a simian virus 40 (SV40) origin, so they will replicate in COS-7 cells (which contain the SV40 large T antigen). They are designed for high-level expression in transient transfections and do not contain a eukaryotic selectable marker. This set of vectors allows for the rapid characterization of TC10 or other proteins by prokaryotic expression and purification and by mammalian expression and immunoprecipitation, immunoblotting, or immunofluorescence. The purification methods are listed in Table I. The antibodies used and their concentrations for immunoblots or immunofluorescence are listed in Table II.

TABLE I
VECTOR SUMMARY

Parent vector	Vector	Tag	Purification	Expression
pQE70 Qiagen	pQNzz	ZZ	IgG Sepharose	Prokaryotic
	pRK7	—	—	Mammalian
pRK7	pKH3	Triple HA	12CA5 with Sigma protein A–Sepharose	Mammalian
pRK7	pKMyc	Myc	9E10 with Amersham Pharmacia GammaBind Plus Sepharose	Mammalian
pRK7	pKFLAG	FLAG	Sigma anti-FLAG M2-agarose	Mammalian
pRK7	pRK7-GFP	GFP	Santa Cruz anti-GFP with Sigma protein A–Sepharose	Mammalian
pRK7	pKNzz	ZZ	IgG Sepharose	Mammalian
pGBT9 Clontech	pGBT10	GAL4 DNA-binding domain		Yeast
pVP16	pVP16-CP	GAL4 activation domain		Yeast

pGEX-2T
```
                Smal
    BamHI ┌──────────┐ EcoRI
    GGA TCC CCG GGA ATT C
```

pQNzz
```
    NcoI
    ┌──────────┐BamHI        NotI         EcoRI      BglII
    CC ATG GGA TCC GCG CGG CCG CGA ATT CAG ATC T
```

pRK7
```
    HindIII      PstI        SalI        XbaI        BamHI          EcoRI       ClaI
    AAG CTT CTG CAG GTC GAC TCT AGA GGA TCC CCG GGG AAT TCA ATC GAT
```

PKH3
```
    BamHI      EcoRI        ClaI
    GGA TCC GAA TTC AAT CGA T
```

PKMyc
```
                Smal
    BamHI ┌──────────┐ NheI          NotI         EcoRI      ClaI
    GGA TCC CGG GCT AGC GGG CGG CCG CTT GAA TTC ATC GAT
```

pKFLAG
```
    BamHI      EcoRI        ClaI            SfiI
    GGA TCC GAA TTC AAT CGA TGG CCG CCA TGG CCA
```

PRK7-GFP
```
    BamHI      EcoRI          ClaI
    GGA TCC TGA GAA TTC AAT CGA T
```

pKNzz
```
    BamHI          NotI         EcoRI        ClaI
    GGA TCC GCG CGG CCG CGA ATT CAA TCG AT
```

pGBT10
```
    BamHI      AatII       EcoRI
    GGA TCC GAC GTC GGT ACC
```

VP16-CP
```
    BamHI      AatII       EcoRI
    GGA TCC GAC GTC GGT ACC
```

FIG. 1. Multiple cloning sites of expression vector set. These vectors were designed to place the *Bam*HI cloning site in the same reading frame as pGEX-2T (Amersham Pharmacia, shown as reference).

TABLE II

ANTIBODY DILUTIONS FOR IMMUNOBLOTTING AND IMMUNOFLUORESCENCE

Antibody[a]	Company	Immunoblotting concentration (μg/ml)	Immunofluorescence concentration (μg/ml)
Anti-GST	Santa Cruz (Santa Cruz, CA)	0.008	0.2
12CA5 (anti-HA)	—	0.2	3.75
Polyclonal anti-HA	BAbCO (Richmond, CA)	1	2
9E10 (anti-myc)	—	0.4	4
Anti-FLAG M2	Kodak (Rochester, NY)	0.66	10
Polyclonal anti-GFP	Molecular Probes (Eugene, OR)	2	4

[a] All antibodies are monoclonal, except those indicated.

Purification of Glutathione S-Transferase (GST)–TC10 Fusion Proteins

To decrease the loss of plasmid due to destruction of the ampicillin by the β-lactamase product of the ampicillin resistance gene, grow overnight cultures as lawns on four LB/ampicillin plates. Scrape the colonies off the plates into 1 liter of LB/ampicillin and grow at either room temperature or 37° with shaking until the OD_{600} = 0.8. Induce the cultures with 1 mM isopropyl-β-D-thiogalactoside (IPTG) at room temperature for 2–4 hr with shaking. Resuspend the pelleted cells in a lysis buffer containing $MgCl_2$ [50 mM Tris, pH 8, 1 mM $MgCl_2$, 0.1 mM EDTA, 1 mM dithiothreitol (DTT), 1 mM phenylmethylsulfonyl fluoride (PMSF), 25 μg/ml leupeptin, 10 μg/ml DNase I, and 1 mg/ml lysozyme]. $MgCl_2$ is necessary to maintain guanine nucleotide complexed to the TC10 proteins. In the absence of a complexed nucleotide, small GTPases rapidly denature. In general, we have found lysis in a French press to provide a higher fraction of soluble, functional protein than sonication and/or freeze-thawing. Certain point mutations in TC10 (e.g., Q75L) substantially reduce the solubility of the GST–fusion protein, particularly when the cells are lysed by sonication.

To prepare recombinant TC10 lacking the N-terminal GST tag, either cleave directly from the glutathione–Sepharose beads or in solution after elution from the beads. To cleave from the beads, wash the beads with thrombin cleavage buffer (50 mM Tris, pH 7.5, 150 mM NaCl, 2.5 mM $CaCl_2$) and incubate with thrombin at 4° overnight. Remove the thrombin by incubation with p-aminobenzamide–Sepharose (Sigma, St. Louis, MO; washed first with thrombin cleavage buffer) at 4° for 30 min. To cleave in solution, first remove the glutathione by passage over a PD10 column (Amersham Pharmacia, Piscataway, NJ) or a Centricep spin column

(Princeton Separations, Adelphia, NJ), with buffer exchange into thrombin cleavage buffer. Remove the GST by incubation with glutathione–Sepharose at 4° for 30 min and then remove the thrombin with washed *p*-aminobenzamide–Sepharose. Concentrate the proteins and exchange into appropriate buffers using a Centricon-30 or Centricon-10 (Millipore, Bedford, MA). Freeze the proteins prepared in this manner in liquid nitrogen and store at −80°, under which conditions they are stable for several months.

Conversion of [α-^{32}P]GTP to [α-^{32}P]GDP

Reagents

> 100 mM Magnesium acetate
> 1 mg/ml Nucleotide diphosphate kinase (NDPK; Sigma)
> 100 mM Uridine diphosphate (UDP)
> 1 M HEPES, pH 7.4
> 25 mM DTT
> 10 mM EDTA, pH 7.0
> Glycerol
> Distilled water
> [α-^{32}P]GTP (3000–5000 Ci/mmol)
> 1 N NaOH
> 1 N HCl

Procedure

In a microcentrifuge tube, combine 4 μl magnesium acetate, 1.5 μl NDPK, 4 μl UDP, 4 μl HEPES, 2 μl DTT, 2 μl EDTA, 40 μl glycerol, 40 μl distilled water, and 40 μl [α-^{32}P]GTP and incubate at 30° for 30 minutes.[4] Stop the reaction by the addition of 8 μl 1 N NaOH and incubate on ice for 10 min. Neutralize the reaction by the addition of 8 μl HCl. Aliquot and store at −20°. Detect the conversion efficiency by thin-layer chromatography on polyethyleneimine-cellulose plates (Baker, Phillipsburg, NJ) with 0.75 M Tris base, 0.5 M LiCl, and 0.45 M HCl running buffer.[5] Soak the plates in methanol prior to and after use to remove buffer.

Loading Recombinant TC10 with Labeled Nucleotide

Small GTPases are loaded with radiolabeled nucleotide in the presence of EDTA to chelate Mg^{2+} ions. After loading, the nucleotide is trapped on the protein by the addition of excess Mg^{2+}.[6]

---

[4] I. G. Macara and W. H. Brondyk, *Methods Enzymol.* **257**, 117 (1995).
[5] B. R. Bochner and B. N. Ames, *J. Biol. Chem.* **257**, 9759 (1982).
[6] E. S. Burstein and I. G. Macara, *Biochem. J.* **282**, 387 (1992).

Procedure

In a microcentrifuge tube, combine 1–5 μg of recombinant TC10, 5 μl 1% (w/v) bovine serum albumin (BSA), 1 μl [α-^{32}P]GTP or [γ-^{32}P]GTP (3000–5000 Ci/mmol) or 3.8 μl [α-^{32}P]GDP (equivalent to 1 μl GTP), and 25 mM MOPS, pH 7.1, and 1 mM EDTA to 50 μl and incubate on ice for 20 min. Add 1 μl 1 M MgCl$_2$ and incubate on ice for an additional 10 min. Store loaded proteins on ice prior to use.

To quantitate the amount of complexed nucleotide, bind loaded TC10 to nitrocellulose filters (Millipore, Bedford, MA; HAWP02400) in the presence of quench buffer (15 mM sodium phosphate, 10 mM MgCl$_2$, 1 mM ATP). Wash filters twice with quench buffer. Measure the radioactivity bound to the filters by scintillation counting. To remove unincorporated nucleotide, pass loaded TC10 over a PD-10 or Centricep column, equilibrated in appropriate buffer (containing $\geq$1 mM MgCl$_2$).

Biochemical Characterization of TC10

To determine the intrinsic GTPase and exchange activities of TC10,[7,8] load recombinant protein with [γ-^{32}P]GTP for GTPase activity or [α-^{32}P]GTP for exchange activity as described previously. Dilute loaded TC10 in 25 mM MOPS, pH 7.1, 1 mM GTP, 1 mM GDP, 5 mM MgCl$_2$, and incubate at 30°. Remove aliquots at timed intervals, filter bind as described previously, and quantitate by scintillation counting. The k_{off} and k_{cat} values are calculated assuming single-exponential kinetics. However, the rate of loss of [γ-^{32}P]GTP from the TC10 is actually the sum of the release and hydrolysis rates. Therefore, it is necessary to correct the apparent k_{cat} value by subtraction of k_{off}.

To determine whether a GTPase-activating protein (GAP) has activity on TC10, load recombinant, cleaved TC10 with [γ-^{32}P]GTP as described earlier. Serially dilute the GAP protein in an appropriate buffer in threefold steps. In a microcentrifuge tube, combine 452.5 μl 25 mM MOPS, pH 7.1, 5 μl 100 mM GTP, 5 μl 100 mMGDP, and 2.5 μl 1 M MgCl$_2$ and incubate at 30°. Add 25 μl of diluted GAP or buffer and incubate at 30°. Initiate the reaction by the addition of 10 μl [γ-^{32}P]GTP-TC10, vortex, and incubate at 30° for 3 min. At t_o and at various time points, remove 20 μl for filter binding and quantitate by scintillation counting. Intrinsic k_{cat} values are calculated and subtracted from k_{cat} values in the presence of GAP, assuming single-exponential kinetics. The apparent affinity of GAP for TC10 is esti-

[7] J. B. Gibbs, M. D. Schaber, W. J. Allard, I. S. Sigal, and E. M. Scolnick, *Proc. Natl. Acad. Sci. U.S.A.* **85**, 5026 (1988).

[8] J. John, M. Frech, and A. Wittinghofer, *J. Biol. Chem.* **263**, 11792 (1988).

mated as the GAP concentration yielding a half-maximal acceleration of hydrolysis.

Kinase Assays

Rac1 and Cdc42 have been shown to activate Jun N-terminal kinase (JNK)[9-12] and p21-activated kinase (PAK).[13] To determine if TC10 and its effectors can activate these kinases, tagged TC10, effector proteins, and kinase are coexpressed by transient transfection, immunoprecipitated, and assayed *in vitro*. In some cases, coexpression of a kinase with another protein may diminish the expression of one of the proteins. In these cases, the amount of plasmid transfected is modified to obtain similar expression levels or the tagged kinase is immunoprecipitated and activated *in vitro* with recombinant TC10 and/or effector proteins.

JNK assays are performed similar to Derijard *et al.*[14] and Coso *et al.*[9] Cotransfect pKH3-JNK with pKMyc-TC10(Q75L) into NIH 3T3 or COS-7 cells. To determine the basal activation of JNK, transfect one plate of cells with pKH3-JNK (and empty vector to normalize plasmid levels). To test the effect of putative TC10 effectors on JNK activity, cotransfect the effector in pKMyc with pKH3-JNK or with pKH3-JNK and pKMyc-TC10(Q75L). At 24 hr after transfection, transfer cells to serum-free medium and starve overnight. Place cells on ice, wash once with phosphate-buffered saline (PBS), and lyse cells with 400 μl of lysis buffer [25 mM HEPES, pH 7.4, 0.3 M NaCl, 1.5 mM MgCl$_2$, 0.5 mM DTT, 20 mM β-glycerophosphate, 1 mM sodium vanadate, 1 μM okadaic acid, 20 μg/ml aprotinin, 10 μg/ml leupeptin, 1 mM PMSF, and 0.1% (v/v) Triton X-100]. Scrape the cells from the plate and centrifuge at 13,000g at 4° for 5 min. Remove 50 μl of each soluble lysate to determine protein expression by immunoblotting. Immunoprecipitate HA-tagged JNK from the soluble supernatant with 3 μg 12CA5 at 4° for 1 hr, followed by incubation with 30 μl of protein A–Sepharose (washed with lysis buffer) at 4° for 1 hr. Wash the beads three times with 2 mM sodium vanadate, 1% Igepal in PBS, once with 0.1 M MOPS, pH 7.5, 0.5 M LiCl, and once with kinase buffer (12.5 mM MOPS, pH 7.5, 12.5 mM β-glycerophosphate, 7.5 mM MgCl$_2$, 0.5 mM

[9] O. A. Coso, M. Chiariello, J. C. Yu, H. Teramoto, P. Crespo, N. Xu, T. Miki, and J. S. Gutkind, *Cell* **81**, 1137 (1995).

[10] S. Bagrodia, B. Derijard, R. J. Davis, and R. A. Cerione, *J. Biol. Chem.* **270**, 27995 (1995).

[11] A. Minden, A. Lin, F. X. Claret, A. Abo, and M. Karin, *Cell* **81**, 1147 (1995).

[12] M. F. Olson, A. Ashworth, and A. Hall, *Science* **269**, 1270 (1995).

[13] E. Manser, T. Leung, H. Salihuddin, Z. S. Zhao, and L. Lim, *Nature* **367**, 40 (1994).

[14] B. Derijard, M. Hibi, I. H. Wu, T. Barrett, B. Su, T. Deng, M. Karin, and R. J. Davis, *Cell* **76**, 1025 (1994).

EGTA, 0.5 mM NaF, and 0.5 mM sodium vanadate). Resuspend the beads in 300 μl kinase buffer; remove 30 μl of resuspended beads to analyze the amount of immunoprecipitated protein by immunoblotting. Pellet beads and initiate JNK reactions by the addition of 30 μl kinase buffer containing 2 μg recombinant GST-Jun(1-79) and 2 μCi [γ-^{32}P]ATP (6000 Ci/mmol). Incubate reactions at 30° for 20 min and terminate by the addition of 10 μl 4× SDS–PAGE sample buffer. Fractionate phosphorylated substrates with 12% SDS–PAGE and visualize by fluorography. Fractionate expressed and immunoprecipitated proteins with 12% SDS–PAGE and transfer to nitrocellulose for immunoblotting. To avoid detection of the antibody used in the immunoprecipitation by the anti-mouse secondary antibody, use 12CA5 or 9E10 coupled directly to horseradish peroxidase.

Our methods to assay the activation of PAK are based on those of Knaus *et al.*[15] and Lamarche *et al.*[16] Because αPAK expression is often perturbed by its cotransfection with other plasmids, pCMV6M-αPAK (a gift from G. Bokoch, Scripps Research Institute, La Jolla, CA) is transfected alone into NIH 3T3 or COS-7 cells. PAK is immunoprecipitated as described earlier with 1 μg polyclonal anti-PAK antibody (Santa Cruz Biotechnology, Inc., Santa Cruz, CA) and protein A–Sepharose. Alternatively, Myc-αPAK can be immunoprecipitated with 4 μg 9E10 and GammaBind Plus Sepharose. Stimulate the immunoprecipitated αPAK by the addition of 45 μl recombinant TC10, loaded as described earlier in the presence of 2 mM guanylyl imidodiphosphate tetralithium salt (GMP-PNP; Boehringer Mannheim, Indianapolis, IN), and incubate on ice for 5 min. Initiate PAK reactions by the addition of 30 μl kinase buffer containing 5 μg of the substrate, myelin basic protein (Sigma, St. Louis, MO), and 5 μCi [γ-^{32}P]ATP. Incubate reactions at 30° for 20 min. Terminate reactions by the addition of 30 μl 4× SDS–PAGE sample buffer and analyze results as described earlier.

Interaction of TC10 with Putative Effectors

We routinely use five assays to detect the interaction of TC10 with putative effectors. These assays include yeast two-hybrid interactions, overlay assays, coprecipitation assays, coimmunoprecipitation assays, and *in vitro* competition assays. The yeast two-hybrid assay is the most sensitive of these, but it does not determine if the interaction is direct and yields little information about affinity. An interaction can appear to be much higher affinity in the yeast two-hybrid interaction than *in vitro* due to self-

[15] U. G. Knaus, S. Morris, H. J. Dong, J. Chernoff, and G. M. Bokoch, *Science* **269,** 221 (1995).
[16] N. Lamarche, N. Tapon, L. Stowers, P. D. Burbelo, P. Aspenstrom, T. Bridges, J. Chant, and A. Hall, *Cell* **87,** 519 (1996).

activation by the effector. The yeast two-hybrid interactions have been described elsewhere[17] and will not be discussed here. Coimmunoprecipitation also does not necessarily detect a direct interaction. Overlay and coprecipitation assays measure direct interactions but require nanomolar affinities. The *in vitro* competition assay is the most sensitive. It measures direct interactions, and we have determined affinities with K_D values of approximately 20 μM.

Overlay Assays

In the overlay assay, a putative effector protein is immobilized on nitrocellulose and is then overlaid with TC10 that has been complexed with radioactive nucleotide. The guanine nucleotide specificity can be examined by loading recombinant TC10 with either [α-^{32}P]GTP or [α-^{32}P]GDP and overlaying two filters bound to the same putative effector proteins. The specificity of the interaction of the effector with small GTPases can be assessed by overlaying individual filters with various GTPases. To decrease the likelihood of false positives due to the dimerization of GST, it is important to avoid using GST-fusions of both the effector and the GTPase. Our methods for the overlay assay are modified from Manser *et al.*[18]

Procedure. Fractionate recombinant proteins or lysates of cells expressing effector proteins by SDS–PAGE followed by transfer of the proteins to nitrocellulose. Renature the proteins and block the membrane by incubation at 4° overnight in binding buffer [20 mM MOPS, pH 7.1, 100 mM potassium acetate, 5 mM magnesium acetate, 5 mM DTT, 0.5% (w/v) BSA, 0.05% (v/v) Tween 20] containing 0.25% (v/v) Tween 20 and 5% (w/v) milk. Alternatively, recombinant proteins to be tested can be spotted directly onto nitrocellulose. Spot small volumes of putative effectors (up to 2 μg) on small pieces of nitrocellulose and allow to dry at room temperature for 1 hr. Block the membrane in binding buffer containing 5% (w/v) milk at 4° for 1 hr in a small container.

To block nonspecific GTP binding, incubate the membrane in a small volume ($\leq$5 ml) of binding buffer containing 100 μM GTP at 4° for 30 min. Load recombinant GTPases with [α-^{32}P]GTP or [α-^{32}P]GDP, remove unincorporated nucleotide, and quantitate complexed nucleotide as described earlier. Add equal counts per minute (cpms) of loaded GTPases to the blots at 4° and incubate for 10 min with rocking. Wash the blots briefly (5–10 sec) with binding buffer until no further radioactivity is removed. Analyze by fluorography with exposures of 1–2 hr and then over-

[17] P. L. Bartel and S. Fields, *Methods Enzymol.* **254,** 241 (1995).

[18] E. Manser, T. Leung, C. Monfries, M. Teo, C. Hall, and L. Lim, *J. Biol. Chem.* **267,** 16025 (1992).

night. The $[\alpha\text{-}^{32}P]GTP$ or $[\alpha\text{-}^{32}P]GDP$ will diffuse away from the proteins over time, especially at room temperature, so the film exposures should be done immediately after completion of the assay.

Coprecipitation Assay

Coprecipitation assays can be performed with a recombinant GST–fusion protein, and glutathione–Sepharose beads, to study its interaction with another recombinant protein or a protein expressed either ectopically or endogenously in cells.[19] The interaction can be analyzed most easily by SDS–PAGE and Coomassie staining if the proteins are of different sizes. If there is an antibody against the protein to be precipitated or if a tagged protein is precipitated, the interaction can be analyzed by immunoblotting; an anti-GST antibody can be used to quantitate the amount of GST–fusion protein bound to the beads. Either GST–TC10 bound to glutathione–Sepharose can be used to assay its interaction with an effector protein or a GST–fusion protein of the effector protein can be used to assay its interaction TC10. Because there are currently no available antibodies against TC10, either a tagged version of TC10 or $[\alpha\text{-}^{32}P]GTP$-loaded TC10 needs to be used.

Procedure. Exchange recombinant GST–TC10 into binding buffer (see earlier discussion) with a PD10 or Centricep column. Bind 25–50 μg GST–TC10 to 10 μl of glutathione–Sepharose beads (washed in binding buffer) in a microcentrifuge tube at 4° for 1 hr. Wash excess GST–TC10 from the beads once with binding buffer. Add an equimolar concentration of recombinant, cleaved effector protein in a small volume (40–200 μl) of binding buffer and incubate at 4° for 1 hr. (Alternatively, GST–TC10 and a cleaved effector protein can be added to the beads at the same time.) Wash the beads three to five times with binding buffer. Add 10 μl 2× SDS–PAGE sample buffer to the beads, and fractionate the proteins by SDS–PAGE. As controls, add effector to glutathione–Sepharose beads and to glutathione–Sepharose beads bound to GST. For comparison, fractionate the amount of GST–fusion coupled to the beads and effector added to the beads.

To measure the interaction of TC10 with a GST–fusion of an effector protein by scintillation counting, load recombinant, cleaved TC10 with $[\alpha\text{-}^{32}P]GTP$ as described earlier. Incubate the loaded TC10 with glutathione–Sepharose beads bound to a GST–fusion of the effector protein and assay as described earlier. After washing the beads, cut the top of the microcentrifuge tube and place the tube in a scintillation vial; fill the vial

---

[19] P. H. Warne, P. R. Viciana, and J. Downward, *Nature* **364,** 352 (1993).

with scintillation fluid and quantitate. Alternatively, this interaction can be assayed with recombinant or ectopically expressed, tagged TC10.

To affinity precipitate a protein from mammalian cells, lyse cells in 400 μl of a cell lysis buffer with the cells on ice. Preclear the lysate with 0.5 ml of glutathione–Sepharose beads (washed with lysis buffer) and 2.5 mg GST at 4° for 1 hr. Add the cleared lysate to the glutathione–Sepharose beads bound to GST–TC10 and assay as described previously; wash the beads with lysis buffer.

Coimmunoprecipitation

To detect interaction of TC10 with a putative effector protein by co-immunoprecipitation,[20] coexpress HA- or Myc-tagged TC10(Q75L) or TC10(T31N) (as a negative control) with an effector protein fused to another tag (HA or Myc) in NIH 3T3 or COS-7 cells. Two days after transfection, place the cells on ice, wash once with PBS, and add 400 μl lysis buffer [25 mM HEPES, pH 7.4, 300 mM NaCl, 1.5 mM MgCl$_2$, 0.5 mM DTT, 20 mM β-glycerophosphate, 1 mM sodium vanadate, 1 mM PMSF, 20 μg/ml aprotinin, 10 μg/ml leupeptin, and 0.1% (v/v) Triton X-100]. Scrape the cells from the plate and centrifuge at 13,000g at 4° for 5 min. Remove 50 μl of each soluble lysate to determine protein expression by immunoblotting. Incubate the remaining soluble supernatant with 3 μg 12CA5 or 4 μg 9E10 antibody at 4° for 1 hr. Add 30 μl protein A–Sepharose (washed with lysis buffer) to 12CA5 immunoprecipitations or GammaBind Plus Sepharose to 9E10 immunoprecipitations and incubate at 4° for 1 hr. Wash the beads three times with PBS containing 0.1% Triton X-100 and three times with PBS. Add 30 μl 2× SDS–PAGE sample buffer to the beads. Fractionate the expressed and immunoprecipitated proteins with 12% SDS–PAGE and transfer to nitrocellulose for immunoblotting. To avoid detection of the antibody used in the immunoprecipitation by the antimouse secondary antibody, use 12CA5 or 9E10 coupled directly to horseradish peroxidase.

Competition Assays

To detect a low-affinity interaction between a GTPase and an effector protein *in vitro*, a competition assay can be performed.[19] The assay is based on the fact that epitopes on the GTPases to which effector proteins bind overlap with the GAP-binding site. Thus, effector proteins inhibit GAP activity. Because this inhibition is competitive, it can be used to assess the affinity of the effector protein for the GTPase. The concentration of GAP

---

[20] E. Harlow and D. Lane, *in* "Using Antibodies: A Laboratory Manual," p. 223. Cold Spring Harbor Laboratory Press, Cold Spring Harbor, NY, 1999.

used in the assay is determined empirically so that in the absence of effector protein, there is approximately 95% hydrolysis of GTP during the period of the assay. The specificity of binding for an effector can be determined by comparing this assay to various GTPases. The concentration of GAP may have to be adjusted for each GTPase.

Procedure. To determine the lowest concentration of TC10 feasible for this assay, load recombinant, cleaved TC10 with $[\gamma\text{-}^{32}P]GTP$ as described previously. Dilute the loaded TC10 to various concentrations. In a microcentrifuge tube, add 25 μl 2× reaction buffer (50 mM MOPS, pH 7.1, 2 mM GDP, 10 mM MgCl$_2$, 1 mM sodium phosphate, 2 mM 2-mercaptoethanol, and 0.1% (v/v) BSA), 20 μl of the buffer in which the effector is diluted, and 2.5 μl of diluted, loaded TC10 and incubate on ice for 20 min. Add 2.5 μl of the buffer in which the GAP is diluted and incubate at 30° for 3 min. Filter bind 40 μl as described earlier and quantitate with scintillation counting. Select a concentration of TC10 that yields approximately 3000–5000 cpm for the competition assay; a concentration in the picomolar range is ideal.

To determine the concentration of GAP to be used in the assay, add 25 μl 2× reaction buffer, 20 μl of the buffer in which the effector is diluted, and 2.5 μl TC10 and incubate on ice for 20 min. Add 2.5 μl of diluted GAP and incubate at 30° for 3 min. Filter bind 40 μl as described earlier and quantitate with scintillation counting. Plot GAP concentration vs the mean cpm remaining complexed. Select a concentration of GAP that provides 95% hydrolysis of GTP.

The competition assay is performed as for the determination of GAP concentration, using the concentrations of TC10 and GAP determined as described previously. The effector is diluted to various concentrations and incubated with TC10 on ice for 20 min prior to the addition of GAP. Plot effector concentration vs the mean cpm remaining complexed. As controls, the values obtained with TC10 alone, TC10 and GAP, and TC10 and effector are compared to the results of the competition assay.

Acknowledgments

This work was supported by Grant CA40042 (to I.G.M.) and a University of Virginia Pratt Fellowship (to C.L.N.).

[2] Expression and Purification of Rho/Rhogdi Complexes

By PAUL W. READ and ROBERT K. NAKAMOTO

Introduction

Recombinant protein production of Rho proteins and Rho guanine dissociation inhibitor (RhoGDI) has resulted in a tremendous wealth of information. To date, the majority of studies using recombinant Rho proteins has utilized protein expressed in *Escherichia coli*. Relatively high yields can be obtained, but the Rho protein is not posttranslationally modified. Expression in eukaryotic cells provides posttranslational modifications of the carboxyl-terminal CAAX sequence: transfer of a geranylgeranyl (or farnesyl for some Rho proteins) group to the cysteine, proteolytic removal of the final three residues, and carboxy methylation of the new terminal cysteine. It appears that overexpression can overwhelm the processing machinery, resulting in posttranslational modification on only a fraction of the Rho protein. Prenylation appears to be the most important modification and is required for full functionality, including membrane association and binding to RhoGDI.[1-4]

To study Rho/RhoGDI interactions, posttranslationally processed Rho proteins have been purified from native sources[1,5-9] or eukaryotic expression systems such as Sf9 cells[8,10,11] and from *in vitro* translation of Rho

[1] Y. Hori, A. Kikuchi, M. Isomura, M. Katayama, Y. Muira, H. Fujioka, K. Kaibuchi, and Y. Takai, *Oncogene* **6,** 515 (1991).
[2] T. Mizuno, K. Kaibuchi, T. Yamamoto, M. Kawamura, T. Sakoda, H. Fujioka, Y. Matsuura, and Y. Takai, *Proc. Natl. Acad. Sci. U.S.A.* **88,** 6442 (1991).
[3] J. F. Hancock, K. Cadwallader, and C. J. Marshall, *EMBO J.* **10,** 641 (1991).
[4] J. F. Hancock, K. Cadwallader, H. Paterson, and C. J. Marshall, *EMBO J.* **10,** 4033 (1991).
[5] Y. Fukumoto, K. Kaibuchi, Y. Hori, H. Fujioka, S. Araki, T. Ueda, A. Kikuchi, and Y. Takai, *Oncogene* **5,** 1321 (1990).
[6] Y. Matsui, A. Kikuchi, S. Araki, Y. Hata, J. Kondo, Y. Teranishi, and Y. Takai, *Mol. Cell. Biol.* **10,** 4116 (1990).
[7] T. Ueda, A. Kikuchi, N. Ohga, J. Yamamoto, and Y. Takai, *J. Biol. Chem.* **265,** 9373 (1990).
[8] M. J. Hart, Y. Maru, D. Leonard, O. N. Witte, T. Evans, and R. A. Cerione, *Science* **258,** 812 (1992).
[9] D. Leonard, M. J. Hart, J. V. Platko, A. Eva, W. Henzel, T. Evans, and R. A. Cerione, *J. Biol. Chem.* **267,** 22860 (1992).
[10] S. Ando, K. Kaibuchi, T. Sasaki, K. Hiraoka, T. Nishiyama, T. Mizuno, M. Asada, H. Nunoi, I. Matsuda, Y. Matsuura, P. Polakis, F. McCormick, and Y. Takai, *J. Biol. Chem.* **267,** 25709 (1992).
[11] T. K. Nomanbhoy and R. A. Cerione, *J. Biol. Chem.* **271,** 10004 (1996).

proteins in rabbit reticulocyte lysates supplemented with canine pancreatic microsomal membranes.[3,12] In addition, prokaryotic-derived Rho proteins prenylated *in vitro* with recombinant geranylgeranyltransferase[13] or with C-terminal truncation deletions with prenylated peptides added to mimic full-length prenylated Rho proteins have also been used.[14]

We have developed a coexpression system in *Saccharomyces cerevisiae* in which either a wild-type or constitutively active mutant Rho protein and RhoGDI are coexpressed: one with a hexahistidine (His_6) amino-terminal tag and the other with a FLAG (DYKDDDK-) amino-terminal tag.[15] The purification by sequential passage over a metal chelate column followed by the anti-FLAG antibody column effectively prevents contamination by endogenous yeast Rho or RhoGDI proteins. Because the Rho proteins are isolated as a complex with RhoGDI, the Rho protein must be prenylated, as RhoGDI association requires the geranylgeranyl moiety.[1,3,10,16] Expression of both proteins, particularly RhoGDI, is toxic to yeast and therefore requires the use of a tightly regulated promoter such as *GAL1*. Purification of the complex from the yeast cytosol yields 100–300 μg of complex/liter of culture, which is greater than 98% pure. The complexes have a 1:1 stoichiometry of Rho:RhoGDI without contamination from yeast homologs and a 1:1 nucleotide:protein molar ratio. Unlike crude preparations from tissue, the purified complex can be stoichiometrically ADP-ribosylated by the *Clostridium botulinum* C3 exoenzyme and immunoprecipitated by the 26-C4 monoclonal antibody (Santa Cruz Biotechnology, Santa Cruz, CA) made against the Rho insert helix (amino acids 124–136).[15,17] The FLAG-RhoA/His_6-RhoGDI complex has been used for structure determination by X-ray crystallography, signal transduction studies, and Rho protein activation studies.[15,17] As expected, while bound to RhoGDI, RhoA has kinetics of nucleotide exchange (3.3×10^{-4} sec^{-1} $\pm$ 0.3×10^{-4} at 22°) and hydrolysis (0.45×10^{-4} sec^{-1} $\pm$ 0.02×10^{-4} at 22°) significantly slower than for free RhoA and consistent with the inhibitory properties of RhoGDI.[15] Thus far, we have not observed any significant differences in

[12] P. Lang, F. Gesbert, M. Delespine-Carmagnat, R. Stancou, M. Pouchelet, and J. Bertoglio, *EMBO J.* **15**, 510 (1996).

[13] S. A. Armstrong, V. C. Hannah, J. L. Goldstein, and M. S. Brown, *J. Biol. Chem.* **270**, 7864 (1995).

[14] A. R. Newcombe, R. W. Stockley, J. L. Hunter, and M. R. Webb, *Biochemistry* **38**, 6879 (1999).

[15] P. W. Read, X. Liu, K. Longenecker, C. G. DiPierro, L. A. Walker, A. V. Somlyo, A. P. Somlyo, and R. K. Nakamoto, *Protein Sci.* **9**, 376 (2000).

[16] J. F. Hancock and A. Hall, *EMBO J.* **12**, 1915 (1993).

[17] K. Longenecker, P. Read, U. Derewenda, Z. Dauter, X. Liu, S. Garrard, L. Walker, A. V. Somlyo, R. K. Nakamoto, A. P. Somlyo, and Z. S. Derewenda, *Acta Crystallogr.* **D55**, 1503 (1999).

nucleotide binding or GTPase characteristics of the proteins if the amino-terminal affinity tags are switched between the proteins or removed proteolytically. In addition to RhoA, CDC42 and Rac1 and their constitutively active mutants (G14V for RhoA and G12V for CDC42 and Rac1) have been expressed successfully in this system.

Construction of Vectors

Construction of YEpP$_{Gal}$/t$_{PMA1}$ Expression Vector

The *Gal10* promoter region[18] was inserted between the *Eco*RI and the *Bam*HI sites of the yeast 2-μm shuttle vectors YEplac181 (*LEU2*) and YEplac195 (*URA3*).[19] In addition, the 3' half of the yeast *PMA1* gene, including its transcription termination signal, was inserted between *Bam*HI and *Xba*I.[20] The *PMA1* segment included a *Xho*I site, which was created by ligation of a linker in the *Sal*I site 260 bases upstream of the *PMA1* termination codon. This created plasmid YEpP$_{GAL}$/t$_{PMA1}$ (Fig. 1A).

Insertion of Human cDNA into Yeast Expression Vectors

In order to optimize translational efficiency in yeast, a fragment was cloned into the *Bam*HI site of YEpP$_{GAL}$/t$_{PMA1}$, which contained the sequence immediately upstream of the initiation codon and the first five codons from the highly expressed yeast *PMA1* gene followed by His$_6$ or FLAG affinity tags (Fig. 1B). An *Eag*I site following the tags provided the cloning site for the Rho protein[21] or RhoGDI cDNA. When a protease site was desired for removal of the affinity tags, the recognition sequence for the rTEV protease (-DYDIPTTENLYFQG-) was added to the Rho protein and RhoGDI cDNA, 3' of the *Eag*I site, by PCR with extended primers.

Expression and Purification of the RhoA/RhoGDI Complex

Yeast Growth

A leucine and uracil auxotrophic yeast strain is cotransformed with the expression plasmids by standard methods.[22] We used strain SY1 (*MATa,*

[18] M. Johnston and R. W. Davis, *Mol. Cell. Biol.* **4,** 1440 (1984).
[19] R. D. Gietz and A. Sugino, *Gene* **74,** 527 (1988).
[20] R. K. Nakamoto, R. Rao, and C. W. Slayman, *J. Biol. Chem.* **266,** 7940 (1991).
[21] H. F. Paterson, A. J. Self, M. D. Garrett, I. Just, K. Aktories, and A. Hall, *J. Cell Biol.* **111,** 1001 (1990).
[22] D. M. Becker and L. Guarente, *Methods Enzymol.* **194,** 182 (1991).

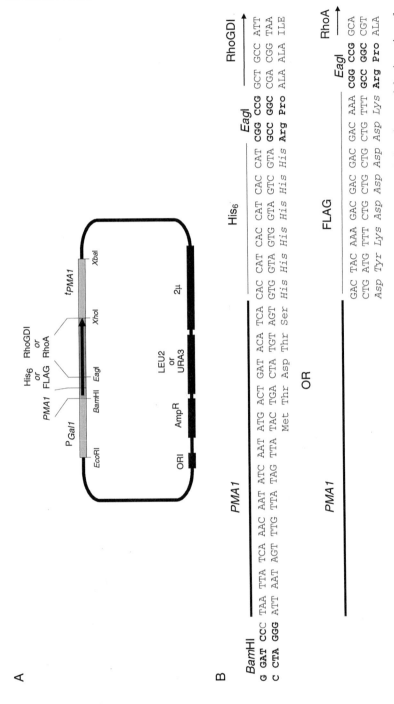

FIG. 1. Yeast expression plasmids for RhoA and RhoGDI. (A) Plasmid map indicating the unique restriction sites used for insertion of Rho protein or RhoGDI cDNA. (B) Nucleotide and protein sequences near the translational initiation sites of YEpPMA1-His₆RhoGDI or YEpPMA1-FLAGRhoA.

ura3-52, leu2-3,112, his4-619, sec6-4, GAL[20]), but others can be used. Initially, an inoculum of a 50-ml culture of SC medium containing 2% glucose, 6.7 gm/liter yeast nitrogen base without amino acids (Difco Laboratories, Detroit, MI), 30 mg/liter L-isoleucine, 150 mg/liter L-valine, 20 mg/liter L-arginine, 20 mg/liter L-histidine hydrochloride, 30 mg/liter L-lysine hydrochloride, 20 mg/liter L-methionine, 50 mg/liter L-phenylalanine, 200 mg/liter L-threonine, 20 mg/liter L-tryptophan, 30 mg/liter L-tyrosine, and 20 mg/liter adenine sulfate (uracil and leucine are omitted to provide auxotrophic selection of the plasmids[23]). The culture is grown for 24 hr at 25° with vigorous shaking. This initial culture is used to inoculate 2.8-liter Fernbach flasks containing 1 liter of the same minimal medium, except 2% glucose is replaced with 2% raffinose. Raffinose is a poorly utilized carbon source that neither induces nor represses the *GAL1* promoter. To induce, galactose is added and, unlike glucose, raffinose need not be removed. Growth is allowed to continue until the optical density at the 650-nm wavelength ($OD_{650\ nm}$) reaches 1.0–1.2. If less than a milligram of purified complex is required, protein expression is induced by the addition of 2% galactose in the Fernbach flasks. If several milligrams are desired, 2 liters of culture is used to inoculate a fermentor containing the same growth media [we use a 19-liter capacity fermentor (Bellco, Vineland, NJ) with a high-speed mixer or large stir bar and aeration by an aquarium pump]. The doubling time is approximately 4.5–5.0 hr. When the $OD_{650\ nm}$ reaches 1.1–1.2, 2% galactose is added, and the culture is allowed to grow for 7–8 hr or until growth reaches late log, which usually occurs at an $OD_{650\ nm}$ of 2.7–3.0.

Yeast are harvested by centrifugation at 1100g for 5 min at 4°, washed in 10 volumes of ice-cold 10 mM NaN_3, and resuspended in 4 ml of 10 mM NaN_3 per gram of cells. Azide helps reduce proteolysis by stopping the synthesis of ATP, which is required to activate cellular proteases. Cells are flash frozen in liquid nitrogen and stored at $-80°$ and can remain for several months prior to protein purification. Routinely, 4–5 g wet cell weight/liter of culture is obtained. We have observed that for strain SY1, the pellets are tan colored when RhoGDI is expressed and white when not, whether or not Rho protein is coexpressed.

Fractionation of Yeast

The cell walls are first digested for efficient lysis by treatment with Zymolyase (ICN Biochemicals, Irvine, CA): 1.25 mg of Zymolase 20T per gram wet cell weight of yeast and 50 μM 2-mercaptoethanol final concentration are added to 40 ml of spheroplasting buffer (2.8 M sorbitol,

[23] F. Sherman, *Methods Enzymol.* **194**, 3 (1991).

100 mM KH_2PO_4, 10 mM NaN_3, pH 7.4) and are incubated at 37° for 1 hr to activate the Zymolyase. The spheroplasting buffer (containing the activated Zymolyase) is added in equal volume to the cell suspension and is gently mixed at room temperature for 1 hr. The spheroplasts are sedimented at 2800g for 10 min at 4° and resuspended in 4 ml of resuspension buffer [50 mM Tris–base, 150 mM NaCl, 5 mM $MgCl_2$, 10% (v/v) glycerol, pH 8.0 at 4°]. A protease cocktail is added consisting of 1 μg/ml leupeptin, 2 μM pepstatin, 2 μg/ml aprotinin, 40 μg/ml benzamidine, and 1 mM Pefabloc (or AEBSF; Pentapharm, Basil, Switzerland) (final concentrations), and the suspension is subjected to two passes through a cooled French press cell at 25,000 psi. Phenylmethylsulfonyl fluoride (1 mM freshly dissolved in dimethyl sulfoxide) is added after the first pass. After a 5-min centrifugation at 13,000g at 4°, the supernatant is centrifuged at 240,000g for 1 hr at 4°. The Rho/RhoGDI complex is purified from the supernatant (cytosolic fraction), and prenylated RhoA can be isolated from the membrane fraction. Both cytosol and membrane fractions can be flash frozen in liquid nitrogen and stored at −80°. Cell lysis and fractionation require 5 hr.

Purification of the Rho/RhoGDI Complex

Purification of the complex via the amino-terminal tags involves sequential utilization of metal affinity resin (e.g., Ni-NTA resin from Qiagen, Valencia, CA) and M2 anti-FLAG antibody (Sigma Chemicals, St. Louis, MO) columns and requires 6 hr for a large-scale preparation. The column material may be regenerated according to manufacturers' instructions and reused several times without loss of effectiveness. The cytosolic fraction is diluted 1 : 4 with Tris buffer (25 mM Tris–base, 150 mM NaCl, 5 mM $MgCl_2$, 10% glycerol, pH 8.0) to decrease protein concentration to less than 2 mg/ml to reduce nonspecific protein interactions. The protein suspension is incubated with 1 ml of equilibrated metal affinity resin per 250 mg of cytosolic protein with gentle mixing for 45 min at room temperature. The resin is sedimented by centrifugation at 700g for 2 minutes, packed into a column, and washed with 10 bed volumes of Tris buffer. Bound protein is eluted with metal elution buffer (50 mM Tris–base, 150 mM NaCl, 5 mM $MgCl_2$, 150 mM imidazole, pH 7.4) and collected in 2-ml fractions. The absorbance at 280 nm of each fraction is measured. The peak fractions, usually eluting in the second and third bed volume, are pooled and then passed five times over an equilibrated M2 anti-FLAG antibody column (0.7 ml bed volume per 250 mg of cytosolic protein). The column is then washed with 10 bed volumes of M2 wash buffer (50 mM Tris–base, 150 mM NaCl, and 5 mM $MgCl_2$, pH 7.4) and eluted with four bed volumes

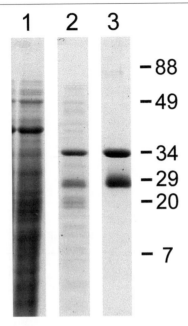

FIG. 2. Purification of the FLAG-RhoA/His$_6$-RhoGDI complex. Samples (15 μg of protein each) were separated by 15% SDS–PAGE and stained with Coomassie blue. Lane 1, cytosolic fraction from yeast expressing the RhoA/RhoGDI complex; lane 2, imidazole elution from the metal affinity resin; and lane 3, FLAG peptide elution from the M2 anti-FLAG antibody column. Numbers on the right indicate the positions of molecular weight standards multiplied by 10^{-3}.

TABLE I

PROTEIN YIELDS DURING PURIFICATION OF FLAG–RhoA/His$_6$-RhoGDI[a]

Protein	Protein in cytosolic fraction	Elution from Talon column	Elution from anti-FLAG M2 column
Total	2000–2400 mg[b]	30–40 mg	4–5 mg
In cytosolic fraction	100%	1.5%	0.2%

[a] From an 18-liter yeast culture.
[b] Values represent the range of quantities observed over a multitude of preparations.

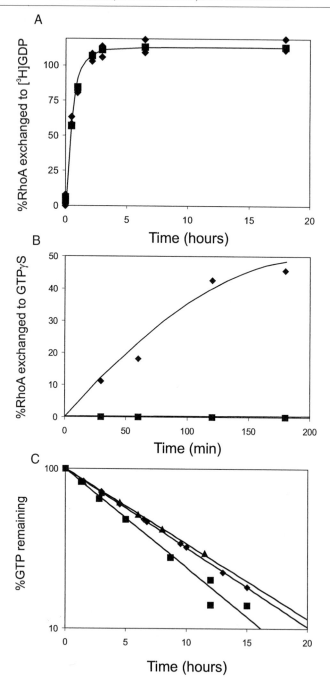

of M2 wash buffer plus 100 μg/ml FLAG octapeptide (DYKDDDDK). The protein again usually elutes during the second and third bed volumes. The protein concentration in the peak fractions is determined by absorbance at 280 nm. RhoA/RhoGDI has an extinction coefficient of 1.3 for 1 mg/ml protein. The degree of purification is shown in Fig. 2 and the normal yields from each purification step are shown in Table I. The concentration of the purified complex can be increased to over 20 mg/ml with minimal loss in centrifuge concentrators (10,000 molecular weight cutoff). To remove the affinity tags, 30–50 units of rTEV protease (GIBCO-BRL, Bethesda, MD) per milligram of protein are added and incubated for 4 hr at room temperature. After proteolysis, the protein is passed over the metal affinity column to separate the purified complex from uncleaved material and rTEV protease, which also contains a His$_6$ tag.

ADP-Ribosylation of RhoA

ADP-ribosylation of the RhoA/RhoGDI complex is carried out with the C3 exoenzyme from *C. botulinum*. A volume of protein sample containing 100 ng of RhoA is incubated in a buffer containing 50 mM Tris–HCl, 140 mM NaCl, 2 mM thymidine, 200 mM GTP, 10 mM DTT with 1 μg of C3, and 100 μM [^{32}P]NAD$^+$ (25 μCi from Amersham Pharmacia Biotech, Piscataway, NJ) for 30 min at room temperature. Protein-associated radioactivity is determined by liquid scintillation counting to quantify the ribosylation reaction. For visualization of the incorporation of radioactivity, the proteins can be separated by SDS–PAGE, electrotransferred to polyvinylidene difluoride (PVDF) membranes, and exposed to film.

FIG. 3. Enzymatic characteristics of the yeast expressed RhoA/RhoGDI complex. (A) Exchange of [^{3}H] GDP/GDP. At time 0, 1 mM [^{3}H]GDP was added to 20 μM FLAG-RhoA/His$_6$-RhoGDI in nucleotide exchange buffer and incubated at 22°. Aliquots were taken at the specified times, and MgCl$_2$ was added to 13 mM final concentration to stop the exchange reaction. The amount of protein-associated [^{3}H]GDP was determined by removing the unbound nucleotide with a centrifuge desalting column and measuring the radioactivity remaining in the filtrate. The line is a fit to a single exponential rise to a maximum. (B) Exchange to GTPγS. Complex was exchanged with 1 mM GTPγS final concentration. Aliquots were taken at the specified times, and the reaction was stopped by the addition of MgCl$_2$ to 13 mM final concentration. The samples were passed over a desalting column, and the amount of protein-associated nucleotide was determined by HPLC anion-exchange chromatography. Diamonds, exchange in the presence of EDTA, squares, exchange in the absence of EDTA. (C) GTP hydrolysis by the complex. GTP-loaded complexes were incubated at 22°, aliquots were taken at specified time points, the reaction was quenched, the protein was precipitated by the addition of perchloric acid, and the nucleotide content was determined. Diamonds, FLAG-RhoA/His$_6$-RhoGDI; triangles, His$_6$-RhoA/FLAG-RhoGDI; and squares, RhoA/RhoGDI with affinity tags removed.

Assays for Nucleotide Content and Exchange

Nucleotide Determination

The guanine nucleotide is measured using high-performance liquid chromatography (HPLC) with an anion exchange (e.g., Waters, Milford, MA, 8PSAX10μ) column equilibrated with 0.7 M ammonium phosphate, pH 4.0. Fifteen microliters of purified complex is added to 7.5 μl of 1% perchloric acid and 7.5 μl of 280 mM sodium acetate to precipitate the protein and release the guanine nucleotide. Thirty microliters of 1.4 M ammonium phosphate, pH 4.0, is added to the sample followed by centrifugation at 10,000g for 5 min at 4° to pellet the precipitated protein. Fifty microliters of supernatant is injected onto the column, and the absorbance is monitored at 254 nm. The integrated area of the nucleotide peak is compared to that of known quantities of each nucleotide. Nucleotide measurement and protein assays for the wild-type RhoA/RhoGDI complex routinely yield a GDP:protein molar ratio of 1.0 ± 0.1 (Fig. 3A), and the ratio remains stable for several days at 22°. The nucleotide content of the G14V–RhoA/RhoGDI complex was also stoichiometric, but a small percentage of the bound nucleotide was GTP.

Nucleotide Exchange of RhoA/RhoGDI

The nucleotide GDP–RhoA/RhoGDI complex can be exchanged to [^{3}H]GDP, GTP, or GTPγS in nucleotide exchange buffer (65 mM Tris–HCl, 100 mM NaCl, 5 mM MgCl$_2$, 10 mM EDTA, pH 7.6) with a nucleotide:protein molar ratio of 30–50:1 or greater.[24] The exchange reaction is stopped by adding MgCl$_2$ to a total of 13 mM. The amount of protein-associated [^{3}H]GDP is determined by removing unbound nucleotide with a centrifuge desalting column[25] equilibrated with buffer A (25 mM Tris–HCl, 100 mM NaCl, 5 mM MgCl$_2$, pH 8.0) and measuring the radioactivity remaining in the filtrate. The rate constant for exchange is 3.3×10^{-4} sec^{-1} ± 0.3×10^{-4} with complete exchange of RhoA/RhoGDI occurring in 3 hr at 22° (Fig. 3A). This rate is two orders of magnitude slower than the rate for free RhoA. The G14V–RhoA mutant complex takes slightly longer, about 5–6 hr for maximal exchange. Because of hydrolysis of GTP on binding, even in complex with RhoGDI or GDP contamination of GTPγS preparations, the maximal level of GTP or GTPγS loaded on the complex is approximately 70% (Fig. 3B).

[24] A. J. Self and A. Hall, *Methods Enzymol.* **256,** 67 (1995).
[25] H. S. Penefsky, *Methods Enzymol.* **56,** 527 (1979).

GTP Hydrolysis Assay

With GTP exchanged onto the complex, the hydrolysis of GTP can be measured by using $[\gamma^{32}\text{-P}]\text{GTP}^{24}$ or by HPLC determination of bound nucleotide described in the previous section. After exchanging to buffer A (containing 5 mM MgCl$_2$, which stops the exchange of nucleotide), the GTP-loaded complex is incubated at 22°, aliquots are taken at specified time points, the reaction is quenched, and the protein is precipitated by the addition of perchloric acid. Nucleotide bound is determined as described earlier. At 22°, the FLAG-RhoA/His$_6$-RhoGDI complex hydrolyzed bound GTP with a rate constant of 0.45×10^{-4} sec^{-1} $\pm$ 0.02×10^{-4} (Fig. 3C). There is little effect if the affinity tags are switched or removed. The heterodimer is stable in solution; however, if a membrane or lipid fraction is present, GTP–RhoA will dissociate from RhoGDI and translocate to the membrane fraction.[15] The GTP–G14V–RhoA/RhoGDI complex is therefore useful for signal transduction studies in which the investigator wishes to deliver highly concentrated, posttranslationally modified RhoA into cellular systems, without the worry of detergents and rapid nucleotide loss or hydrolysis to study downstream signaling effects.

[3] Bacterial Expressed DH and DH/PH Domains

By KENT L. ROSSMAN and SHARON L. CAMPBELL

Introduction

Like Ras, Rho proteins can exist in two distinct structural conformations in response to binding either GDP or GTP. In contrast to the GDP-bound form, Rho–GTP participates actively in intracellular signaling by specifically recognizing downstream target effector molecules to perpetuate an upstream signal. Therefore, Rho family members function as binary molecular switches *in vivo* by cycling between an "inactive" GDP-bound form and an "active" GTP-bound form. Rho GTPases normally exist in the inactive GDP-bound form in unstimulated cells and activation requires the exchange of GDP for GTP. Rho GTPases bind guanine nucleotides with very high affinity and possess an intrinsic rate of GDP for GTP exchange too slow to allow for efficient signal transduction.[1,2] The Dbl family of

---

[1] I. M. Zohn, S. L. Campbell, R. Khosravi-Far, K. L. Rossman, and C. J. Der, *Oncogene* **17,** 1415 (1998).

[2] D. J. Mackay and A. Hall, *J. Biol. Chem.* **273,** 20685 (1998).

proteins act as specific guanine nucleotide exchange factors (GEFs), efficiently catalyzing the GDP/GTP exchange and activation of Rho GTPases *in vivo*. The Dbl family is a rapidly growing family of proteins (over 30 distinct human family members) that share an approximate 300 residue span, which exhibits significant sequence similarity to Dbl, a transforming protein originally isolated from a diffuse B-cell lymphoma. This region of sequence similarity encodes an approximate 200 residue Dbl homology (DH) domain in tandem with an approximate 100 residue pleckstrin homology (PH) domain invariantly carboxy-terminal to the DH domain.[3,4]

The DH domains from several Dbl family members have been shown to be sufficient to act as GEFs for a number of Rho GTPases both *in vitro* and *in vivo*. Dbl family members are further recognized to exhibit a range of substrate specificity toward members of the Rho family. For example, Dbl proteins such as Vav (a GEF for RhoA, Rac1, and Cdc42) and Dbl (a GEF for Cdc42 and RhoA) can catalyze guanine nucleotide exchange on multiple Rho family members, whereas others such as Lsc (a GEF for RhoA), Tiam-1 (a GEF for Rac1), and Fgd1 (a GEF for Ccd42) are highly specific for only one subset of Rho family proteins.[5-9] Other Dbl family members have only been detected to bind Rho members without facilitating exchange. For example, the oncogene Ect2 associates with RhoC and Rac1 but showed no detectable GEF activity toward them *in vitro*.[10] In addition, Lfc catalyzes exchange specifically on RhoA, but can also bind Rac1 *in vitro*.[7] Therefore, the DH domain may also participate in Rho signal transduction via a binding mechanism that is separate from its exchange activity.

The three-dimensional structures of the DH domains of Trio, β-Pix, and Sos1 have been determined.[11-13] The structures revealed that DH

[3] R. A. Cerione and Y. Zheng, *Curr. Opin. Cell Biol.* **8**, 216 (1996).

[4] I. P. Whitehead, S. Campbell, K. L. Rossman, and C. J. Der, *Biochim. Biophys. Acta* **1332**, F1 (1997).

[5] Y. Zheng, M. J. Hart, and R. A. Cerione, *Methods Enzymol.* **256**, 77 (1995).

[6] J. Han, B. Das, W. Wei, L. Van Aelst, R. D. Mosteller, R. Khosravi-Far, J. K. Westwick, C. J. Der, and D. Broek, *Mol. Cell. Biol.* **17**, 1346 (1997).

[7] J. A. Glaven, I. P. Whitehead, T. Nomanbhoy, R. Kay, and R. A. Cerione, *J. Biol. Chem.* **271**, 27374 (1996).

[8] F. Michiels, G. G. Habets, J. C. Stam, R. A. van der Kammen, and J. G. Collard, *Nature* **375**, 338 (1995).

[9] Y. Zheng, D. J. Fischer, M. F. Santos, G. Tigyi, N. G. Pasteris, J. L. Gorski, and Y. Xu, *J. Biol. Chem.* **271**, 33169 (1996).

[10] T. Miki, *Methods Enzymol.* **256**, 90 (1995).

[11] X. Liu, H. Wang, M. Eberstadt, A. Schnuchel, E. T. Olejniczak, R. P. Meadows, J. M. Schkeryantz, D. A. Janowick, J. E. Harlan, E. A. Harris, D. E. Staunton, and S. W. Fesik, *Cell* **95**, 269 (1998).

domains consist of 10 or 11 α helices forming an elongated helical bundle that is structurally unrelated to other GEFs of known structure. The proximity of conserved regions within the DH domain in three-dimensional structures, as well as mutagenesis studies, reveals the potential GTPase-binding surface on the molecule. This surface is composed of residues in and surrounding helices 1, 9, and 10 in the Trio DH domain structure and helices A, I, J, and K in the structure of β-Pix, which corresponds to the conserved regions 1 and 3 (CR1 and CR3) and regions carboxy-terminal to CR3 in the primary sequence.[4] Whereas the structures of the DH domains have revealed the putative Rho GTPase-binding surface, it is still not known what elements within DH domains determine specificity toward their GTPase substrates. Reciprocally, although it has been determined that Rho GTPases interact with Dbl members in part through their switch I and switch II regions, how the specificity of this interaction is derived is unknown.[14,15]

While light has been shed on the role of the DH domain in stimulating guanine nucleotide exchange, what functional role PH domains are supplying to the Dbl family members awaits to be illuminated. Three main themes are emerging, however, which indicate a multifunctional role for this domain. First, the PH domain appears to be important for proper cellular membrane localization in some Dbl members. Transformation studies with Lfc show that removal of the PH domain results in loss of membrane localization and transforming ability. Transformation is restored when the PH domain is functionally replaced by the plasma membrane targeting H-Ras tetrapeptide isoprenylation signal, CAAX.[16] This phenomenon was also observed for Dbs, although transformation was only restored to approximately 30% with the addition of the CAAX motif.[17] At least for Lfc, localization may occur directly to the cytoskeleton, as Lfc localizes to microtubules *in vivo* and the Lfc PH domain can bind tubulin *in vitro*.[18] Second, PH domains within Dbl-related proteins may act to regulate the GEF activity of the DH domain through binding of phosphatidylinositides

[12] B. Aghazadeh, K. Zhu, T. J. Kubiseski, G. A. Liu, T. Pawson, Y. Zheng, and M. K. Rosen, *Nature Struct. Biol.* **5**, 1098 (1998).

[13] S. M. Soisson, A. S. Nimnual, M. Uy, D. Bar-Sagi, and J. Kuriyan, *Cell* **95**, 259 (1998).

[14] R. Li and Y. Zheng, *J. Biol. Chem.* **272**, 4671 (1997).

[15] K. L. Rossman, S. Snyder, D. Broek, J. Sondek, C. J. Der, and S. L. Campbell, unpublished observations (1998).

[16] I. P. Whitehead, H. Kirk, C. Tognon, G. Trigo-Gonzalez, and R. Kay, *J. Biol. Chem.* **271**, 18388 (1995).

[17] I. P. Whitehead, Q. T. Lambert, J. A. Glaven, K. Abe, K. L. Rossman, G. M. Mahon, J. M. Trzaskos, R. Kay, S. L. Campbell, and C. J. Der, *Mol. Cell Biol.* **19**, 7759 (1999).

[18] J. A. Glaven, I. Whitehead, S. Bagrodia, R. Kay, and R. A. Cerione, *J. Biol. Chem.* **274**, 2279 (1999).

(PIs). *In vitro,* Vav binding of phosphatidylinositol 3,4,5-trisphosphate (PIP_3) enhances the rate of nucleotide exchange on Rac1, whereas phosphatidylinositol 4,5-bisphosphate (PIP_2) diminishes GEF activity by Vav.[19] Furthermore, Sos1 can bind PIP_3, and the Sos1 DH/PH bidomain activates the c-Jun NH_2-terminal kinase (JNK) in a Ras and phosphoinositide kinase (PI-3 kinase)-dependent manner *in vivo.*[20,21] These data indicate that binding of PIs through the PH domain can regulate Dbl members by targeting them to the plasma membrane, where PIs are located, and by regulating the exchange factor activity of the adjacent DH domain. Third, the presence of the PH domain reportedly can increase the catalytic efficiency of Dbl members toward their GTPase substrates. When compared to the GEF activity of the isolated DH domain, the DH/PH bidomains of Dbl and Trio exhibited greatly enhanced catalytic activity toward Cdc42 and Rac1, respectively.[5,11] Furthermore, the structure of the SOS1 DH/PH bidomain shows a possible structural interdependence between the DH and the PH domains for catalysis.[13] Taken together, these observations suggest that although the DH domain alone can catalyze exchange, the DH/PH tandem domains may constitute the complete catalytic unit.

Unquestionably, the ability to express and purify active forms of Dbl-related proteins and their DH and DH/PH domains will greatly facilitate structural and biochemical studies of these proteins. This article presents methods for the expression and purification of active DH and DH/PH domains from the Dbl family proteins, Dbs and Vav2, in *Escherichia coli.* Dbs is an oncoprotein originally identified in a retrovirus-based cDNA expression screen for transforming genes, which in addition to sharing a 50% amino acid sequence identity, exhibits a similar Rho GTPase exchange profile to Dbl.[17,22] The Vav2 gene, originally discovered on human chromosome 9q34, due to its residence within the tuberous sclerosis gene, is the second of three Vav homologs discovered thus far.[23] Vav and Vav2 proteins are 53% identical in sequence and are also predicted to share a similar exchange profile toward the Rho family of GTPases.

[19] J. Han, K. Luby-Phelps, B. Das, X. Shu, Y. Xia, R. D. Mosteller, U. M. Krishna, J. R. Falck, M. A. White, and D. Broek, *Science* **279,** 558 (1998).

[20] A. S. Nimnual, B. A. Yatsula, and D. Bar-Sagi, *Science* **279,** 560 (1998).

[21] L. E. Rameh, Arvidsson Ak, K. L. Carraway III, A. D. Couvillon, G. Rathbun, A. Crompton, B. VanRenterghem, M. P. Czech, K. S. Ravichandran, S. J. Burakoff, D. S. Wang, C. S. Chen, and L. C. Cantley, *J. Biol. Chem.* **272,** 22059 (1997).

[22] I. P. Whitehead, H. Kirk, and R. Kay, *Oncogene* **10,** 713 (1995).

[23] E. P. Henske, M. P. Short, S. Jozwiak, C. M. Bovey, S. Ramlakhan, J. L. Haines, and D. J. Kwiatkowski, *Ann. Hum. Genet.* **59,** 25 (1995).

Expression and Purification of Vav2 DH Domain and Dbs DH and
  DH/PH Domains in *Escherichia coli*

A cDNA fragment encoding the human Vav2 DH domain (residues
191–402), murine Dbs DH domain (residues 623–832), and DH/PH domains (residues 623–967) was generated by polymerase chain reaction and
inserted into the *NcoI/XhoI* sites of the bacterial expression vector pET-
28a (Novagen). The Vav2 and Dbs DH domains were designed to encode
the entire DH domain and extend into the linker region between the DH
and the PH domains (Fig. 1A). This was done to avoid truncation of any
structured elements that may be present at the carboxy terminus of the
DH domain. The Dbs DH/PH domain was extended a few residues beyond
what is predicted to be the PH domain carboxy-terminal helix by the same
reasoning. The Vav2 DH domain contains a nonnative Glu(His$_6$) on the
carboxy terminus of the polypeptide, whereas the Dbs DH and DH/PH
domains contain a nonnative amino-terminal Met and carboxy-terminal
Glu(His$_6$). The bacterial expression constructs were transformed into the
E. coli strain BL21 (DE3), grown in LB/Kan (50 μg/ml) cultures to an OD
of 0.6. Protein expression is induced with 1 m*M* isopropylthiogalactoside
(IPTG) and cultures are grown for 3 to 5 hr at 25°. Cells are pelleted by
centrifugation at 6000*g* for 10 min at 4° and resuspended in lysis buffer
containing 50 m*M* NaH$_2$PO$_4$, pH 8.0, 150 m*M* NaCl, 5 m*M* (Dbs) or no
(Vav2) imidazole, and 0.5 mg/ml Pefabloc and lysed by French pressing
three times. Lysates are next clarified by centrifugation at 25,000*g* at 4° for
20 min. DNA is precipitated from the supernatant solution by adding 0.02%
polyethyleneimine while stirring on ice. The solutions are again centrifuged
at 25,000*g* for 20 min at 4° and filtered through a 0.45-μm filter (Whatman,
Clifton, NJ). The recombinant proteins contain a carboxy-terminal hexahistidine (His$_6$) tag and are initially loaded onto a Ni-NTA agarose column
(Qiagen, Hilden, Germany), followed by washing with 10 column volumes
of lysis buffer containing 20 m*M* (Dbs) or 2 m*M* (Vav2) imidazole. Proteins
are eluted from the column with 3 column volumes of lysis buffer containing
300 m*M* imidazole. Next, the protein solutions are concentrated and loaded
onto an S-200 column (Pharmacia, Piscataway, NJ) equilibrated in 20 m*M*
Tris, pH 8.0, 150 m*M* NaCl, 5 m*M* dithiothreitol (DTT), and 5% glycerol
at a flow rate of 1 ml/min. The Vav2 DH domain is then further purified
by loading onto a Source-Q (Pharmacia) column in 10 m*M* Tris, pH 8.0,
5 m*M* NaCl, 5 m*M* DTT, and 10% glycerol. The Vav2 DH domain is then
eluted with a linear gradient from 5 to 500 m*M* NaCl. The recovered Vav2
DH, Dbs DH, and Dbs DH/PH domain proteins are estimated to be greater
than 95% pure (Fig. 1B). Bacterially expressed glutathione *S*-transferase
(GST)–Rho(F25N), GST–Rac1(wt), and GST–Cdc42(wt) are expressed

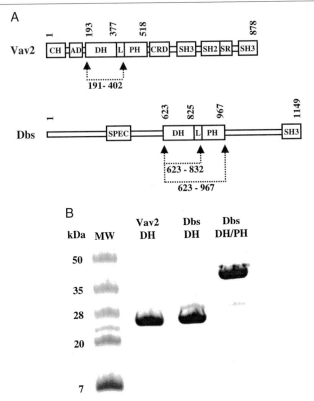

FIG. 1. Bacterial expressed and purified Vav2 DH and Dbs DH and DH/PH domains. (A) The domain structures of full-length Vav2 and Dbs are shown. Also indicated are the DH and DH/PH domain regions from Vav2 and Dbs that were cloned and expressed in *Escherichia coli* to produce the recombinant proteins used in this study: calponin homology domain (CH), acidic domain (AD), Dbl homology domain (DH), linker region (L), pleckstrin homology domain (PH), cysteine-rich domain (CRD), serine-rich region (SR), Src homology 2 domain (SH2), Src homology 3 domain (SH3), and spectrin repeat (SPEC). (B) SDS–PAGE and Coomassie staining of purified recombinant bacterial-expressed Vav2 DH and Dbs DH and DH/PH domains. Lane 1, molecular weight markers; lane 2, Vav2 DH (3 μg); lane 3, Dbs DH (3 μg); and lane 4, Dbs DH/PH (3 μg).

and initially purified on glutathione agarose essentially as described.[24] The GST–GTPases are then further purified on an S-200 column (Pharmacia) equilibrated in 20 mM Tris, pH 8.0, 150 mM NaCl, 5 mM DTT, 5% glycerol, and 50 μM GDP at a flow rate of 1 ml/min.

[24] A. J. Self and A. Hall, *Methods Enzymol.* **256**, 3 (1995).

GDP Dissociation Assay

The GDP dissociation assays are carried out by the filter-binding method essentially as described previously.[5,25] To prepare [^{3}H]GDP-loaded GST–Rho, GST–Rac, or GST–Cdc42, solutions containing 10 mM HEPES, pH 7.5, 100 mM NaCl, 7.5 mM EDTA, 15 μM GDP, 5.5 μM [^{3}H]GDP, and 12.5 μM of the Rho family GTPase are incubated for 25 min at 23°. The [^{3}H]GDP-bound GTPases are stabilized by supplementing the solution with 20 mM MgCl$_2$. Nucleotide exchange reactions are performed at 23° by diluting the [^{3}H]GDP-loaded GST–Rho, GST–Rac, or GST–Cdc42 to 4 μM in 250-μl reaction mixtures. Reaction mixtures contain 0.1 μM (40:1 GTPase:GEF), 0.8 μM (5:1 GTPase:GEF), 4 μM (1:1 GTPase:GEF), or 80 μM (1:20 GTPase:GEF) GEF (Vav2 DH, Dbs DH, Dbs DH/PH) or no GEF, 10 mM HEPES, pH 7.5, 5 mM MgCl$_2$, 100 mM NaCl, 1 mM DTT, 50 μg/ml BSA, and 100 μM GTP (final concentrations). Thirty microliters of each reaction mixture is sampled at 0, 5, 10, 15, and 20 min and quenched in 1 ml of ice-cold dilution buffer containing 20 mM Tris, pH 7.5, 100 mM NaCl, and 20 mM MgCl$_2$. The amount of [^{3}H]GDP remaining bound to the GTPases is measured by filtering 0.9 ml of the quenched samples over BA85 nitrocellulose filters and placing in scintillation fluid, dissolving, and counting. The percentage [^{3}H]GDP remaining bound at each time point for the GEF catalyzed and uncatalyzed reactions is evaluated relative to the 0 min time point of the uncatalyzed reaction. Where possible, exchange data are fit in Sigma Plot using a first-order exponential decay regression.

Results

Figure 2 compares the ability of the Vav2 DH domain and Dbs DH domain to catalyze the exchange of [^{3}H]GDP for GTP on GST–Rac1, GST–Cdc42, and GST–RhoA. As shown previously, the Dbs DH domain did not exhibit exchange activity toward Rac1 when using a fivefold excess of GTPase (Fig. 2A).[17] The Dbs DH domain also failed to catalyze exchange on Rac1 when GEF and GTPases were present in equimolar concentrations (data not shown). In contrast, the Vav2 DH domain stimulated exchange on Rac1, with a half-time of [^{3}H]GDP dissociation ($t_{1/2}$) of approximately 5 min, again using a 5:1 ratio of GTPase:GEF. This exchange rate is comparable to that reported for the bacterial-expressed DH/PH domains of Trio and Tiam-1, Dbl members specific for Rac1.[11,17]

Both the Vav2 DH and the Dbs DH domains were found to be efficient

[25] M. J. Hart, A. Eva, T. Evans, S. A. Aaronson, and R. A. Cerione, *Nature* **354,** 311 (1991).

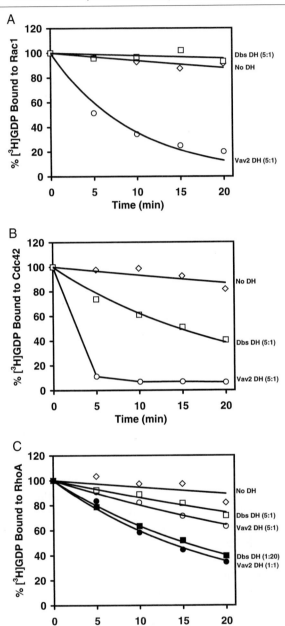

exchange factors toward GST–Cdc42, although a range of activity was observed between them (Fig. 2B). Using a 5 : 1 molar ratio of GTPase : GEF, the Vav2 DH domain essentially completed the exchange of Cdc42 in 5 min, whereas the $t_{1/2}$ for the Dbs DH domain was approximately 15 min. Exchange rates for the Dbs DH domain-catalyzed reaction are at least as rapid as described for the Dbl DH domain expressed in *Spodoptera frugiperda* cells, whereas both the Vav2 and the Dbs DH domains appear to be more efficient than DH/PH domains of the Cdc42-specific Dbl member, Fgd1, in catalyzing nucleotide exchange.[9,26] To our knowledge, the measured activity of the Vav2 DH domain on Cdc42 is the highest reported for any isolated DH domain.

Figure 2C shows a comparison of the Vav2 and Dbs DH domain nucleotide exchange activities on GST–RhoA. Both Dbs and Vav2 show a diminished ability to activate RhoA, as compared to rates when Cdc42 (Dbs and Vav2) or Rac1 (Vav2) was used as substrate. When RhoA is present in 5-fold molar excess, only slight [³H]GDP exchange is observed for both Vav2 and Dbs DH domains. Increasing the concentration of Vav2 to equimolar amounts with RhoA causes an increase in nucleotide exchange by Vav2 such that the $t_{1/2}$ for the reaction was approximately 15 min. In comparison, a 20-fold excess of the Dbs DH domain was required to approach the Vav2 DH domain rate. This same phenomenon was reported for the Trio DH domain-catalyzed exchange of Rac1, where an approximate 40-fold excess of the DH domain was needed to measure a comparable exchange rate ($t_{1/2} \approx 15$ min.).[11] The rate of GST–RhoA exchange by the Vav2 DH domain was approximately 3-fold slower than that reported for Sf9-expressed DH/PH domains of the RhoA-specific Dbl members Lfc and Lsc ($t_{1/2} \approx 5$ min).[7]

Based on the results of Fig. 2, the rank order of the specificity of the DH domains for Rac1, Cdc42, and RhoA can be determined. Dbs is capable of stimulating guanine nucleotide exchange on Cdc42 and RhoA, but not Rac1, where the rank order of activity is Cdc42 > RhoA. Vav2 behaves as a multifunctional exchange factor, proficient at stimulating exchange on Rac1, Cdc42, and RhoA, with the rank order of Cdc42 > Rac1 > RhoA.

---

[26] M. J. Hart, A. Eva, D. Zangrilli, S. A. Aaronson, T. Evans, R. A. Cerione, and Y. Zheng, *J. Biol. Chem.* **269,** 62 (1994).

FIG. 2. Stimulation of [³H]GDP from Rac1, Cdc42, and RhoA by the Vav2 and Dbs DH domains. The time course of [³H]GDP dissociation from 4 μM GST–Rac1 (A), GST–Cdc42 (B), and GST–RhoA (C) was measured in the presence of 0.8 μM (5 : 1 molar ratio of GTPase : GEF) (□) and 80 μM (1 : 20 molar ratio) (■) Dbs DH, 0.8 μM (5 : 1 molar ratio) (○) and 4 μM (1 : 1 molar ratio) (●) expressed Vav2 DH domain, or no DH domain (◇). Data shown are the average of two independent experiments.

These data further show that a range of activities also exist between DH domains, even when having similar substrate preferences.

The contribution of the PH domain to the GEF activity of the Dbs DH domain was investigated next. The catalytic efficiency of the DH domain of Dbs was directly compared to Dbs DH/PH domains by measuring the [^{3}H]GDP exchange of GST–Cdc42. Again using a 5:1 molar ratio of GTPase:GEF, the $t_{1/2}$ for the Dbs DH domain-catalyzed rate was approximately 15 min, whereas the Dbs DH/PH domains completed the exchange reaction within the first measured time point, indicating that the PH domain acts to enhance the catalytic efficiency of the DH domain (Fig. 3). Therefore, whereas the isolated DH domain exhibits GEF activity for Dbs, the DH/PH bidomain shows enhanced activity and comprises the complete catalytic unit. Furthermore, this conclusion may be generalized to other Dbl family members as similar observations were made for Dbl and Trio.[11,25] We have attempted the same comparisons between Vav2 DH and DH/PH domains. Unfortunately the Vav2 DH/PH domain formed aggregates in solution, which hindered these efforts.

Figure 4 shows a comparison of the activity of the Dbs DH and DH/PH domains at various concentrations on GST–RhoA, GST–Rac1, and GST–Cdc42. For RhoA, the DH domain was able to exchange approximately 50% of [^{3}H]GDP after 15 min when using a 20-fold excess of GEF, whereas the DH/PH bidomain exchanged 75% of [^{3}H]GDP using a 5-fold excess of GTPase. The activity observed with the Dbs DH/PH bidomain on RhoA represents over a 100-fold increase in rate enhancement of nucleotide

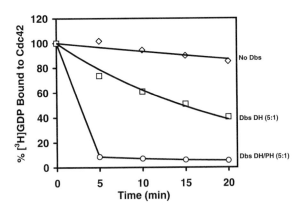

FIG. 3. Time course of Dbs DH and DH/PH domain-stimulated [^{3}H]GDP dissociation from Cdc42. The time course of [^{3}H]GDP dissociation from 4 μM GST–Cdc42 was measured in the presence of 0.8 μM (5:1 molar ratio of GTPase:GEF) Dbs DH domain (□), 0.8 μM (5:1 molar ratio) Dbs DH/PH domain (○), or no Dbs (◇). Data shown are the average of two independent experiments.

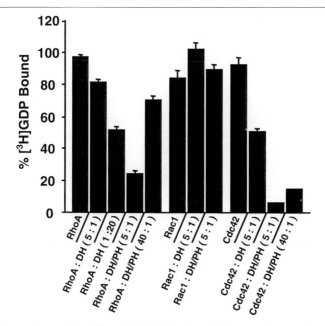

FIG. 4. Comparison of Dbs DH and DH/PH domain abilities to catalyze guanine nucleotide exchange on RhoA, Rac1, and Cdc42. [³H]GDP dissociation from 4 μM GST–RhoA, GST–Rac1, and GST–Cdc42 was measured in the presence of different concentrations of Dbs DH and DH/PH domain protein. Ratios shown are the relative amounts of GTPase:Dbs used in the guanine nucleotide exchange reactions: 0.1 μM Dbs (40:1), 0.8 μM Dbs (5:1), 4 μM Dbs (1:1), or 80 μM Dbs (1:20). Data shown represent the amount of [³H]GDP remaining bound to the GTPase after 15 min. Results represent the average and standard deviation of two independent experiments.

exchange when compared to the DH domain. Also, although greatly reduced, exchange was still apparent using a 40-fold excess of RhoA over the DH/PH bidomain. There was no detectable exchange activity on GST–Rac1 by either the Dbs DH or the DH/PH domains, consistent with published observations on Dbs and the rat Dbs homolog, Ost.[17,27] As shown in Fig. 3, the rate of activation of GST–Cdc42 by the Dbs DH/PH bidomain was measured to be extremely rapid when compared to that of the DH domain alone. Remarkably, the Dbs DH/PH bidomain was able to exchange greater than 80% of [³H]GDP after 15 min using a 40-fold excess of Cdc42. Therefore, the PH domain of Dbs contributes to a uniform rate enhancement of GEF activity toward both RhoA and Cdc42 when compared to the isolated DH domain. Furthermore, although the Dbs DH

[27] Y. Horii, J. F. Beeler, K. Sakaguchi, M. Tachibana, and T. Miki, *EMBO J.* **13,** 4776 (1994).

domain is less active than the DH/PH bidomain, it still maintains GEF activity as well as native specificity toward Rho family members.

Discussion

The Rho family of GTPases is known to regulate multiple cellular processes such as dynamic cytoskeletal reorganization, transcription, membrane trafficking, cell cycle progression, and invasion.[1] Rho GTPases are upregulated by the Dbl family of proteins, Rho-specific GEFs that populate Rho GTPases in their biologically active GTP-bound form *in vivo*. From studies of other Ras superfamily GEFs, the following nucleotide exchange mechanism is proposed for Dbl-related proteins. Presumably, Dbl-related proteins bind to Rho GTPases and disrupt interactions among GTPase, Mg^{2+}, and GDP. Disruption of these interactions, in turn, causes GDP dissociation from the GTPase, resulting in a high affinity Dbl–Rho binary complex. Once the binary complex is formed *in vivo*, GTP, which is present in an approximate 20-fold greater concentration than GDP in the cell, along with Mg^{2+}, destabilizes the binary complex by binding Rho and dissociating Dbl. This reaction results in populating Rho family GTPases in their "active" GTP bound form.[4] Although many Dbl family members are large multidomain proteins, they all share the DH/PH bidomain cassette, in which a DH domain is invariably arranged in tandem with a carboxy-terminal PH domain. The DH domain is considered to possess Rho GEF activity, whereas multiple functions have been ascribed to the PH domain.[3,4] The invariable structural arrangement of the tandem DH and PH domains, however, suggests a functional interdependence between them.

This article described methods for expressing and purifying DH and DH/PH domains in *E. coli* and showed that these proteins have similar activities to other Dbl members described thus far. In particular, we have presented data on the activities of the DH domains from Vav2 and Dbs, as well as Dbs DH/PH tandem domains. Dbs is a Dbl family member that, like Dbl, is specific for Cdc42 and RhoA. However, whereas the Dbs DH domain was moderately active on Cdc42, a vast excess of the exchange factor was needed to detect activity on RhoA. Most intriguing was the discovery that the addition of the PH domain, in the contiguous DH/PH bidomain construct, greatly enhanced the exchange rate of Dbs toward both of these GTPases without altering specificity. The Dbs DH/PH bidomain possessed proficient GEF activity toward Cdc42, as we were able to measure activity when using a 40-fold excess of GTPase in the exchange reaction. *In vivo*, removal of the Dbs PH domain relocalized the protein predominantly to the cytosol from the plasma membrane and completely eliminated its focus-forming activity.[17] Addition of the CAAX motif to the Dbs mutant lacking the PH domain localized the protein to the plasma membrane but only restored transforming activity to approximately 30%. Although the partial restoration of transformation by the Dbs PH domain minus–CAAX mutant could result from mislocalizaton of Dbs at the plasma membrane, *in vitro* exchange data suggest that removal of the PH domain will reduce the ability of Dbs to catalyze nucleotide exchange directly on Cdc42 and RhoA. Taken together, these data support the notion of at least two distinct roles for PH domains within Dbl family members: membrane localization and the enhancement of GEF activity.

Perhaps one of the best studied Dbl family members is Vav, a Dbl member capable of stimulating guanine nucleotide exchange on RhoA, Rac1, and Cdc42. Vav is composed of an array of known protein/protein and protein/lipid interaction domains in addition to DH and PH domains. Vav GEF activity has been proposed to be regulated *in vivo* and *in vitro* by three mechanisms thus far. These mechanisms include the regulation of GEF activity by sequences amino-terminal to the DH domain, tyrosine phosphorylation by the Lck kinase, and binding of PIs to the Vav PH domain.[6,19] Vav2 and Vav are 53% identical in primary sequence and share the same structural domain topology (Fig. 1A). Vav2 apparently shares similar mechanisms of regulation to that of Vav, also being activated by amino-terminal truncation and tyrosine phosphorylation.[28] Because of the high sequence and mechanistic similarities between these family members, it is predicted they may also act as GEFs toward the same subset of Rho GTPases, as seen between other Dbl homologs such as Dbl and Dbs, and Lsc and Lfc. Our *in vitro* exchange data using an isolated DH domain demonstrate that, like Vav, Vav2 can catalyze guanine nucleotide exchange on RhoA, Rac1, and Cdc42, with the order of proficiency being Cdc42 > Rac1 > RhoA. Furthermore, this activity was not dependent on tyrosine phosphorylation, as was demonstrated in full-length Vav.[6] In contrast to results with the isolated DH domain, it has been reported that the amino-terminally truncated and tyrosine-phosphorylated Sf9 cell-expressed Vav2 protein was only active toward RhoG and RhoA-like GTPases *in vitro*.[28] It is possible that regions of Vav2 outside the DH domain are involved in the specific recognition of GTPase substrates, although our results with the Dbs DH and DH/PH domains suggest that the DH domain retains the native GTPase specificity. In support of this, we have shown a requirement for Cdc42, Rac1, and RhoA for Vav2 signaling *in vivo,* consistent with our *in vitro* exchange data.[29] Another explanation may be that although

[28] K. E. Schuebel, N. Movilla, J. L. Rosa, and X. R. Bustelo, *EMBO J.* **17,** 6608 (1998).
[29] K. Abe, K. L. Rossman, B. Liu, K. D. Ritola, D. Chiang, S. L. Campbell, K. Burridge, and C. J. Der, *J. Biol. Chem.* **275,** 10141 (2000).

truncation and phosphorylation may relieve intramolecular regulation on GEF activity, it is only partially alleviated in constructs containing other motifs, and thus partially active. This raises the question as to what further signals may be required *in vivo* for Vav2 to be fully activated? One potential mechanism of regulation could be the enhancement of nucleotide exchange activity by the binding of PIP$_3$ within the PH domain, as is proposed for Vav1.[19]

Data presented herein demonstrated the utility of studying isolated DH and DH/PH domains *in vitro*. For Dbs, comparison of DH and DH/PH domain activities has led to a better understanding of the roles of the PH domain in facilitating guanine nucleotide exchange and in regulating intracellular signaling. The results presented implicate the Dbs DH/PH bidomain as the base catalytic unit in this Dbl-related protein. Studies with the isolated Vav2 DH domain have made it possible to assess substrate specificity that may have not been apparent in the full-length molecule, indicating that regulatory mechanisms may be operating in Vav2 in addition to amino-terminal inhibition and tyrosine phosphorylation. Isolated DH domains can provide useful information on the Dbl member under study, particularly when the DH/PH domain or native protein is not available.

[4] Biochemical Analysis of Regulation of Vav, a Guanine-Nucleotide Exchange Factor for Rho Family of GTPases

By Raymond Mosteller, Jaewon Han, Balaka Das, and Daniel Broek

Introduction

Ras and Rac GTPases are activated by guanine-nucleotide exchange factors (GEFs) related to the *Saccharomyces cerevisiae CDC25* gene product and the Dbl oncoprotein, respectively.[1,2] Several Ras–GEFs have been described, whereas more than 20 proteins structurally related to the catalytic domain of Dbl have been reported. Vertebrate Ras–GEFs are generally bifunctional molecules possessing both a CDC25-related domain and a Dbl homology (DH) domain. Thus, Ras–GEFs have the potential to activate

---

[1] M. S. Boguski and F. McCormick, *Nature* **366,** 643 (1993).

[2] I. P. Whitehead, S. Campbell, K. L. Rossman, and C. J. Der, *Biochim. Biophys. Acta* **1332,** F1 (1997).

Ras and Rac family members. For those Dbl-related molecules where GEF activity has been detected, this activity is specific for one or more of the Rho family GTPases.[2] However, many of the Dbl-related molecules (including Vav), when expressed as recombinant proteins, fail to stimulate the release of GDP from Rho proteins *in vitro*. It is now apparent that the initial failure by us and others to detect a Vav GEF activity for Rho GTPases is due to the lack of essential posttranslational modifications that are necessary to activate an otherwise latent Rho-specific GTPase activity. We suggest that the lack of GEF activity observed for other recombinant forms of the Dbl-related molecules may also be due to the lack of proper posttranslational modifications. Studies of Vav have shown that its GEF activity is under dual control of λ phosphoinositide 3-kinase (PI 3-kinase) and a Src-related tyrosine kinase.[3–5] This article describes the biochemical analysis of Vav and the events contributing to the activation of its Rho family-specific GEF activity.

Overview of Events Regulating GEF Activity of Vav

Vav was initially identified by its ability to induce morphological transformation of NIH 3T3 cells.[6] Subsequently, other oncogenes were found to encode a domain structurally related to one within the Vav protein. Dbl is the first of these oncogene products shown to have a GEF activity that stimulates guanine-nucleotide exchange on a Rho GTPase.[7]

Vav plays a critical role in T-cell activation and its expression is restricted to hematopoietic cell lineages. Vav knockout mice display a severe defect in T-cell function and viability.[8] During T-cell activation, Vav is rapidly phosphorylated on specific tyrosine residues.[9,10] This phosphorylation was the first clue indicating possible events that might activate a latent GEF activity on Vav. We and others have shown that a latent GEF activity of recombinant Vav protein is activated by a tyrosine phosphorylation event mediated by the hematopoietic tyrosine kinase, Lck.[3,4]

[3] P. Crespo, K. E. Schuebel, A. A. Ostrom, J. S. Gutkind, and X. R. Bustelo, *Nature* **385,** 169 (1997).

[4] J. Han, B. Das, W. Wei, L. Van Aelst, R. D. Mosteller, R. Khosravi-Far, J. K. Westwick, C. J. Der, and D. Broek, *Mol. Cell. Biol.* **17,** 1346 (1997).

[5] J. Han, K. Luby-Phelps, B. Das, X. Shu, Y. Xia, R. Mosteller, U. Murali Krishna, J. R. Falck, M. A. White, and D. Broek, *Science* **279,** 558 (1998).

[6] S. Katzav, D. Martin-Zanca, and M. Barbacid, *EMBO J.* **8,** 2283 (1989).

[7] M. J. Hart, A. Eva, T. Evans, S. A. Aaronson, and R. A. Cerione, *Nature* **354,** 311 (1991).

[8] K. D. Fischer, K. Tedford, and J. M. Penninger, *Semin. Immunol.* **10,** 317 (1998).

[9] X. R. Bustelo, J. F. Ledbetter, and M. Barbacid, *Nature* **356,** 68 (1992).

[10] B. Margolis, P. Hu, S. Katzav, W. Li, J. M. Oliver, A. Ullrich, A. Weiss, and J. Schlessinger, *Nature* **356,** 71 (1992).

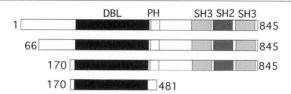

Fig. 1. Structure of full-length Vav and fragments of Vav used in these studies. Full-length Vav encodes a protein of 845 amino acid residues, including a DBL homology (DH) domain, a pleckstrin homology (PH) domain, and three Src homology (SH) domains. The indicated fragments were cloned into the pRESET expression vector (Invitrogen) and expressed in *E. coli* to produce the corresponding His$_6$-tagged derivatives.

Other clues concerning the regulation of Vav come from signaling events reported to lead to Ras-dependent activation of the Rho-related GTPases and from the domain structure of Vav (and other Dbl-related molecules). In vertebrates, Ras proteins have multiple protein targets, including the Raf and PI 3-kinases, which are the best characterized among these downstream targets. Several lines of evidence indicate that the activation of Ras leads to the activation of Rho GTPases.[11] Further, it is known that PI 3-kinase is required for the activation of Rho GTPases.[12,13] Consequently, we hypothesized that either the protein kinase activity or the lipid kinase activity of PI 3-kinase might contribute to the activation of Vav.

Vav has numerous structural domains, including two SH3 domains, an SH2 domain, and a pleckstrin homology (PH) domain (see Fig. 1).[14] All Dbl-related GEFs have a PH domain on the C-terminal side of the DH domain.[2] The invariant juxtaposition of the DH and PH domains in Dbl-related GEFs suggests that there may be an integrated function of these adjacent domains. PH domains have been identified in more than 100 proteins, and many of the PH domains have been demonstrated to bind phospholipid molecules, including substrates and products of PI 3-kinase.[15] Together these observations led us to test whether the substrates or products of PI 3-kinase might influence the GEF activity directly by interaction with Vav or indirectly by controlling phosphorylation of Vav. Our analysis indicates that the substrates and products play both direct and indirect

[11] R. Khosravi-Far, P. A. Solski, G. J. Clark, M. S. Kinch, and C. J. Der, *Mol. Cell. Biol.* **15**, 6443 (1995).
[12] P. T. Hawkins, A. Eguinoa, R.-G. Qiu, D. Stokoe, F. T. Cooke, R. Walters, S. Wennstrom, L. Claesson-Welsh, T. Evans, M. Symons, and L. Stephens, *Curr. Biol.* **5**, 393 (1995).
[13] S. Wennstrom, P. Hawkins, F. Cooke, K. Hara, K. Yonezawa, M. Kasuga, T. Jackson, L. Claesson-Welsh, and L. Stephens, *Curr. Biol.* **4**, 385 (1994).
[14] S. Katzav, *Crit. Rev. Oncogen.* **6**, 87 (1995).
[15] J. M. Kavran, D. E. Klein, A. Lee, M. Falasca, J. F. Isakoff, and E. F. Skolnik, *J. Biol. Chem.* **273**, 30497 (1998).

roles in regulating Vav.[5] The products of PI 3-kinase dramatically enhance the rate at which Vav is phosphorylated by the Src-related Lck tyrosine kinase. The substrate of PI 3-kinase, phosphatidylinositol 4,5-bisphosphate (PI-4,5-P_2), does not enhance the ability of Lck to phosphorylate Vav. However, these substrates appear to play a direct role in regulating Vav GEF activity. The substrate PI-4,5-P_2 completely inhibits the GEF activity of Lck-phosphorylated Vav.

Reagents and Methods

Buffers

Buffer H: 20 mM Tris–HCl, pH 7.9, 5 mM imidazole, 50 mM NaCl
Buffer NB: 20 mM Tris–HCl, pH 7.9, 60 mM imidazole, 0.5 M NaCl, 1% (v/v) Triton X-100, 1 mM phenylmethylsulfonyl fluoride (PMSF), 1 μg/ml leupeptin
Buffer NE: 20 mM Tris–HCl, pH 7.9, 0.5 mM imidazole, 20 mM NaCl
Buffer G: 50 mM Tris–HCl, pH 7.5, 5 mM MgCl$_2$, 20 mM KCl
Buffer K: 10 mM HEPES, pH 7.5, 12 mM MgCl$_2$
Buffer P: 20 mM Tris–HCl, pH 7.5, 50 mM NaCl, 5 mM MgCl$_2$, 0.2% (v/v) Triton X-100
Buffer GSH: 50 mM Tris–HCl, pH 7.5, 5 mM MgCl$_2$, 150 mM NaCl, 1 mM DTT

Purification of Vav Expressed in Escherichia coli

Fragments of the *Vav* cDNA are subcloned into the bacterial expression vector, pRSET-B (Invitrogen, Carlsbad, CA), for the production of hexahistidine (His$_6$)-tagged Vav proteins.[4] Plasmids are transformed into *E. coli* strain BL21 (DE3) or ER2566. Portions of the *Vav* cDNA encoding residues 1–845 (full-length Vav), residues 66–845 (oncogenic Vav), residues 170–845, or residues 170–481 (encompassing the DH domain and part of the PH domain) are expressed as His$_6$-tagged fusion proteins (Fig. 1). Constructs for the expression of Vav proteins beginning at residue 1 or residue 66 fail to express detectable amounts of Vav protein, whereas constructs beginning at residues 170 result in expression of Vav at approximately 0.1–0.5 mg of purified protein per liter of culture.

We examined various culture conditions, including concentration of isopropylthiogalactoside (IPTG), time of induction with IPTG, and temperature during the induction phase. Maximal recovery of soluble Vav proteins is found by the addition of 0.25 mM IPTG to an OD$_{600}$ of 0.2 log phase culture in LB medium containing ampicillin (0.1 mg/ml). The culture is then incubated at 28° for 90 min. The cultures are chilled on ice and centrifuged at 2,500g for 20 min at 4° to pellet the bacteria. The pellet is washed (resuspended and repelleted) with ice-cold buffer H. To the bacterial pellet resuspended in buffer H containing 1% Triton X-100 is added 1 mM (final concentrations) phenylmethylsulfonyl fluoride (PMSF) and 1 μg/ml leupeptin immediately prior to sonication to lyse the bacteria. Bacterial suspensions are sonicated to achieve >90% lysis of the bacteria, being careful not to heat the culture to higher than 6°. For a Branson Sonifier 250 (Branson Sonic Power Company, Danbury, CT) with a 1-cm-diameter probe, bacterial suspensions are sonicated at maximal power for 6 cycles of 20 sec each with a 5- to 10-min incubation in a 5 M NaCl/ice bath between each cycle. The lysed bacterial suspensions are centrifuged at 18,000g for 30 min at 4° to remove cell debris.

The cleared supernatants are incubated with Ni^{2+}-agarose beads (300 μl of a 50% slurry per liter of bacterial culture) at 4° for 5 hr. The Ni^{2+}-agarose beads are washed twice with buffer NB. The agarose beads are then washed twice with buffer H. For experiments using His_6-Vav proteins immobilized on Ni^{2+}-agarose beads, the beads are washed once in buffer G and stored at 0°. The concentration of Vav in the suspension of Ni^{2+}-agarose beads is determined by analysis of an aliquot on Coomassie Brilliant Blue R250-stained sodium dodecyl sulfate–polyacrylamide gel electrophoresis (SDS–PAGE) gels by comparison of the intensity of protein standards. For experiments using soluble Vav proteins, Vav is eluted from the beads by two sequential incubations in buffer NE for 30 min at room temperature. The pooled eluted Vav protein is dialyzed against buffer G and then stored at 0°.

Phosphorylation of Vav by Lck

One microgram of Vav protein bound to Ni^{2+}-agarose beads is washed twice in buffer K. To the beads resuspended in 50 μl of buffer K is added 10 μM ATP (final concentration) and 1–10 units of bovine Lck kinase (Upstate Biotechnologies Inc., Lake Placid, NY). The reaction mixture is then incubated at 37° for 1 hr. The *in vivo* phosphorylation pattern of Vav is complex. Further, coexpression experiments with Vav and Lck suggest that Vav is phosphorylated at multiple sites.[16] However, the major site of Lck-dependent phosphorylation of Vav *in vitro* appears to be tyrosine residue 174 (see Fig. 2).

Influence of Phosphatidylinositides on Phosphorylation of Vav by Lck

For maximal phosphorylation of Vav *in vitro,* water-soluble eight carbon backbone (C_8)-PI-3,4,5-P_3 (20 μM final concentration) is added at the start

---

[16] K. Abe, I. P. Whitehead, J. P. O'Bryan, and C. J. Der, *J. Biol. Chem.* **274**, 30410 (1999).

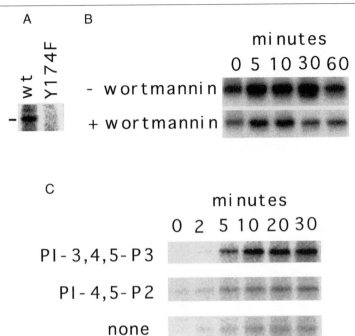

FIG. 2. Lck-dependent phosphorylation of Vav. (A) Wild-type and Y174F mutant His_6-tagged Vav(L) were incubated with baculovirus-produced Lck in the presence of 10 μM [^{32}P]ATP for 30 min and analyzed by SDS–PAGE and autoradiography. The wild-type but not the Y174F mutant Vav protein was phosphorylated under these conditions, indicating that Y174 is the major site of Lck-dependent phosphorylation of Vav. (B) Quiescent NIH 3T3 cells expressing wild-type Vav, labeled with ortho[^{32}P]phosphate, were stimulated with calf serum. Where indicated, 100 nM wortmannin was added 40 min prior to serum addition. Cells were harvested at 0, 5, 10, 30, and 60 min after adding serum. Vav immunoprecipitates prepared from the cell lysates were analyzed by SDS–PAGE and autoradiography. (C) *In vitro* stimulation of Lck-dependent phosphorylation of Vav by the water-soluble phosphoinositide C_8-PI-3,4,5-P_3. Wild-type His_6-tagged Vav(L) was incubated with 10 μM [^{32}P]ATP in the absence of phosphoinositide or in the presence of 50 μM C_8-PI-4,5-P_2 or 50 μM C_8-PI-3,4,5-P_3. At the times indicated, samples were removed and analyzed by SDS–PAGE and autoradiography. Phosphorylation was markedly stimulated by the water-soluble analog product of PI 3-kinase (C_8-PI-3,4,5-P_3) but not the water-soluble analog of the substrate of PI 3-kinase (C_8-PI-4,5-P_2).

of the kinase reaction (see earlier discussion).[5] As seen in Fig. 2, Vav is a poor substrate in the absence of phosphatidylinositides or in the presence of the PI 3-kinase substrate, C_8-PI-4,5-P_2 (50 μM). In contrast, Vav is phosphorylated readily in the presence of the PI 3-kinase product, C_8-PI-3,4,5-P_3. In the presence of 1 unit of Lck, 50 μM C_8-PI-3,4,5-P_3, phosphorylation of Vav (1 μg) is near completion by 45 min and the molar ratio of phosphate to Vav approaches 1:1 (data not shown). A similar degree of

Vav phosphorylation is observed in the presence of 1 unit of Lck and 50 μM C_8-PI-3,4,5-P_3 (data not shown) and in the presence of 100 units of Lck without the addition of C_8-PI-3,4,5-P_3.

The dependence of Vav phosphorylation on PI 3-kinase products can also be assessed in NIH 3T3 cells overexpressing Vav (Fig. 2). NIH 3T3 cells expressing wild-type Vav are grown to 75% confluence in Dulbecco's modified Eagle's medium (DMEM) (GIBCO-BRL, Gaithersburg, MD) containing 15% (v/v) calf serum. Cells are then incubated in DMEM containing 0.5% (v/v) calf serum for 18 hr to induce quiescence. After washing with phosphate-free DMEM, cells are incubated for 6 hr in phosphate-free, serum-free DMEM containing 0.5 mCi/ml ortho[^{32}P]phosphate. Cells are then incubated with or without 100 nM wortmannin for 40 min as indicated. Following the addition of calf serum (15%), cells are harvested at various time point over a range of 0 to 60 min. Cells extracts are then prepared on ice by incubation with lysis buffer (10 mM Tris–HCl, pH 8.0, 150 mM NaCl, 1% Triton X-100, 1 mM sodium vanadate, 100 μM NaF, 1 mM phenylmethylsulfonyl fluoride, 10 μg/ml aprotinin) for 30 min. The anti-Vav antibody conjugated to agarose beads (Santa Cruz Biotechnology, Inc., Santa Cruz, CA) is used for Vav immunoprecipitates. The bead-bound immunocomplexes are washed five times in lysis buffer and resuspended in lysis buffer (200 μl) without Triton X-100. Phosphorylation of Vav can then be determined after SDS–PAGE and autoradiography of the dried gel (Fig. 2).

Prior to using the agarose-bound, phosphorylated Vav in GEF assays or in binding reactions, the beads are washed several times in the appropriate reaction buffer. In addition to placing Vav in the appropriate buffer, the washing steps remove Lck kinase and excess C_8-PI-3,4,5-P_3. Removal of excess C_8-PI-3,4,5-P_3 is essential for GEF reactions monitoring the effects of various phosphatidylinositides (see later). For experiments measuring the extent of Vav phosphorylation by Lck, 4 μl of [^{32}P]ATP (3000 mCi/mmol, 10 mCi/ml, NEN-Dupont, Boston, MA) is added to the kinase reaction, and phosphorylation of Vav is detected after SDS–PAGE and autoradiography of the dried gel.

Binding of Phosphatidylinositides to Vav

The use of water-soluble C_8 derivatives of phosphatidylinositides make possible biochemical analysis of Vav in an aqueous phase.[5] As with their natural counterparts, the C_8-phosphatidylinositides are unstable in aqueous solution. Therefore, the phosphatidylinositides are stored in small aliquots in a lyophilized state at $-70°$. Once resuspended, the phosphatidylinositides are used within 8 hr.

Through its PH domain, Vav binds to water-soluble derivatives of phosphatidylinositides. We found that C_8-PI-4,5-P_2 binds Vav with an apparent K_d of 3–4 μM.[5] Direct and indirect evidence suggests that C_8-PI-3,4,5-P_3 binds with a modestly higher affinity than C_8-PI-4,5-P_2. First, whereas 50 μM C_8-PI-3,4,5-P_3 blocks approximately 80% of the binding of radiolabeled C_8-PI-3,4,5-P_3 (~5 μM) to Vav, 50 μM C_8-PI-4,5-P_2 blocks only 25% of the binding of radiolabeled C_8-PI-3,4,5-P_3. Second, whereas C_8-PI-3,4,5-P_3 stimulates the GEF activity of phosphorylated Vav and C_8-PI-4,5-P_2 potently inhibits GEF activity, we found that Vav GEF activity is activated in the presence of equal concentrations of these phosphatidylinositides.

For binding of water-soluble phosphatidylinositides to Vav, we incubate Ni^{2+}-agarose-bound Vav proteins in buffer P for 1 hr at room temperature in the presence of the phosphatidylinositide. The beads are then washed twice in the same buffer in the absence of the phosphatidylinositide. A similar efficiency of binding phosphoinositides (50 μM) is also observed using buffer K.

Preparation of Small GTPases

The GEF assays and the binding reaction described require soluble GTPases. We utilized glutathione S-transferase (GST)–GTPase fusion proteins for much of our analysis; however, GTPases in the form of nonfusion proteins are likely to work as well in these assays. Two intrinsic properties of the GTPases are critical to GEF assays. First, the stochiometry of GDP binding to the GTPase must be high (a stochiometry of 1 : 1 is best) for the GTPase to serve well in a GEF reaction. The reason for the poor performance in a GEF reaction of a preparation of a GTPase that binds GDP poorly may be due to the fraction of the GTPase that does not bind GDP. GTPases that do not bind GDP, if they retain a structure that binds to the GEF, may bind the GEF irreversibly. This would then prevent the GEF from acting on the GTPases that do bind GDP. The stochiometry of GDP binding varies between each batch of GTPase produced and thus must be determined for each preparation of GTPase. In our experience, some GTPases almost always can be purified to yield a preparation that binds with a stoichiometry of >0.8 molecules of GDP per GTPase molecule (e.g., the H-Ras protein). In our hands, Rho-related GTPases are often a problem with respect to stochiometry of GDP binding; CDC42 is the best among these, RhoA intermediate, and Rac the most problematic. In our experience, a preparation of GTPases with a stoichiometry above 0.7 works well in a GEF reaction. GTPases with stoichiometries between 0.4 and 0.7 often serve as substrates for GEFs, whereas GTPase preparations with a stoichiometry below 0.4 are not useful in GEF reactions.

A second intrinsic property of GTPases critical to GEF reactions is the off rate for bound GDP. A GTPase bound to radiolabeled GDP, when incubated with excess cold GDP or GTP can be observed to release the bound radiolabeled GDP. If the off rate is very fast the GTPase tends not to perform well in a GTPase reaction. This is due in part to the fact that we employ the commonly used filter binding assay (see later) to monitor GEF activity and this assay is cumbersome such that short time points are not possible. For experiments where analysis of GEF activity by monitoring short time points is desired, a method employing fluorescence measurements is an alternative.[17] The intrinsic rates of GDP release vary among different GTPases and even between different preparations of the same GTPase. For, example, Ras routinely has an intrinsic off rate that is rather slow (generally 20% loss of bound GDP after 25 min at 37°). Rho-related GTPases have a higher rate of GDP release such that at 37° as much as 80% of bound GDP is released by 15 min. However, at room temperature the rate of GDP release is significantly slower such that most preparations of the Rho–GTPases release approximately 25% of bound GDP after 25 min. In general, in GEF reactions we use the temperature closest to 37° that results in less that 30% loss of bound GDP after 15 min.

The expression vectors for GST fusions of RhoA, Rac, and CDC42 were provided by Gary Bokoch (The Scripps Institute, La Jolla, CA). These vectors introduced into E. coli strain DH5α express in response to IPTG these GTPases lacking several C-terminal amino acids fused to the C terminus of GST. The culture conditions used to induce expression of these GTPases are the same as that described earlier for the expression of Vav. The method for lysing bacteria expressing GST–GTPases is similar to that described for Vav with the exception of the buffers used. All steps are carried out as close to 0° as possible. For the GST–GTPase expression system, the bacteria are lysed in buffer GSH supplemented with 1% Triton X-100 and $1\mu M$ GDP. PMSF (1 mM) is added immediately prior to sonication. The lysed suspension of bacteria is clarified by centrifugation at 18,000g for 30 min. The supernatants are incubated with glutathione-agarose beads (Sigma) (300 μl of a 50% slurry per 0.5 liters of bacterial culture) for 2 hr with constant rotation of the tube. The agarose beads are washed several times in buffer GSH containing 1% Triton X-100 and 1 μM GDP and twice in buffer GSH containing 1 μM GDP. The GST–GTPases are liberated by two sequential elutions in 1 ml of buffer GSH containing 10 mM reduced glutathione and incubation at 4° for 20 min. The pooled fractions are dialyzed against buffer G. The dialyzed proteins are concentrated using Centricon-30 membranes (Amicon, Danvers, MA) to a concentration of

[17] C. Lenzen, R. H. Cool, and A. Wittinghofer, Methods Enzymol. 255, 95 (1995).

approximately 2 mg/ml. The proteins are stored at $-70°$ in buffer G supplemented with glycerol to 40%.

Preparation of [³H]GDP-Bound GST–GTPase

Fast Exchange Method. The purification of GTPases as described previously generally yields a mixture of the GDP-bound or nucleotide-free states. GEF reactions are generally carried out using GTPases bound to [³H]GDP (10 Ci/mmol; DuPont), whereas the binding assays of GEFs to GTPases are often carried out using nucleotide-free GTPases. Thus, it is necessary to prepare GTPases in each of these states. As magnesium ion is essential for high-affinity binding of nucleotides by GTPases, chelators of magnesium ion can be used to prepare nucleotide-free GTPase.[18] For GTPases in buffers containing 5 mM MgCl$_2$, EDTA is added to a final concentration of 7.5 mM followed by incubation at room temperature for 5 min. Preparation of the nucleotide-free GTPase can also be used as an intermediate in the preparation of GTPases bound to either GDP or GTP. After preparation of the nucleotide-free GTPase, the guanine nucleotide of choice is added to a concentration of 1–3 μM together with the addition of magnesium ion in sufficient quantities to assure a free magnesium ion concentration of approximately 5 mM. The concentration of the excess free nucleotide should be at least 10-fold higher that the concentration of GTPase because the GTPase as purified is likely bound in a near 1 : 1 ratio with GDP. The desired nucleotide-bound state forms spontaneously and rapidly at room temperature in the presence of free magnesium.

Slow Exchange Method. The use of magnesium chelators to prepare nucleotide-free GTPase as an intermediate in preparing the GTPase bound to [³H]GDP is problematic for some GTPases used in GEF reactions. Preparation of Rac–[³H]GDP in this manner can often yield a poor substrate for GEFs. An alternative method for preparing [³H]GDP–GTPase used in GEF reactions is by exploiting the slow spontaneous exchange observed in the presence of magnesium. For example, for Rho GTPases bound to GDP, the off rate of bound nucleotide at $37°$ is such that >75% of bound GDP is released in 30 min. When the concentration of free nucleotide is 1 μM or greater, the nucleotide-free GTPase is rapidly converted to a GTPase bound to the nucleotide in excess. Because the methods described earlier for the purification of GTPases yield predominantly GDP-bound GTPase without free nucleotide, the amount of GDP present in the sample is less than the amount of GTPase. Consequently, a GTPase bound predominantly to [³H]GDP can be prepared by incubating the GTPase in

---

[18] J. Field, D. Broek, T. Kataoka, and M. Wigler, *Mol. Cell. Biol.* **7,** 2128 (1987).

the presence of [³H]GDP (1–3 μM) in a magnesium-containing buffer for 30 min at 37° followed by incubation on ice. For GTPases with faster off rates the time of incubation can be shortened, whereas for GTPases with a slower off rate the incubation times can be increased.

Filter-Binding Assay to Detect [³H]GDP-Bound GTPases

The amount of [³H]GDP bound to a GTPase is determined using a filter-binding assay. The sample containing the radiolabeled GDP-bound GTPase is passed through a nitrocellulose filter (MF-Millipore membrane filters, Millipore, Bedford, MA) using a filtration manifold (Hoeffer Scientific Instruments, San Francisco, CA). The filter is immediately washed three times with 10 ml of ice-cold buffer G. The filters are dried under a heating lamp for 20–30 min, and the amount of radiolabeled material bound is determined by liquid scintillation counting.

In Vitro Binding of Vav to Small GTPases

Small GTPases of the Ras and Rho families are activated by GEFs that function by stabilizing a reaction intermediate composed of the GEF bound tightly to the nucleotide-free GTPase. The subsequent binding of GDP or GTP to the reaction intermediate then allows the nucleotide-bound GTPase to dissociate from the GEF. Some of the GEFs related to the Dbl oncoprotein activate a broad spectrum of the Rho family (Rho, Rac, and CDC42 proteins) while others a subset of these. The high affinity of the GEF for its substrates in their nucleotide-free state makes it possible to ask which GTPases are likely substrates for a GEF.

For example, the DH domain of Sos1 purified from an *E. coli* expression system was used to determine whether it can bind to RhoA, Rac, or CDC42. This binding analysis revealed that the DH domain of Sos1 binds nucleotide-free Rac with high affinity but not the other GTPases (Fig. 3). Comparison

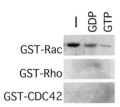

Fig. 3. Binding of mSos1 Dbl domain to Rac. Fifty picomoles of His₆-tagged Sos DH domain [His₆-Sos(DH)] immobilized on Ni²⁺-agarose beads was incubated with 150 pmol of soluble GST–Rac (top), GST–Rho (middle), or GST–Cdc42 (bottom) for 1 hr at 20°. GDP or GTP (1 mM) was included as indicated. The beads were washed, and the GST–GTPases bound to Sos(DH) were detected by SDS–PAGE followed by immunoblotting with anti-GST antibody.

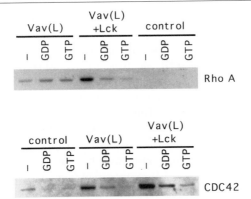

FIG. 4. Binding of Vav(L) and Vav(DH) proteins to the Rho family of GTPases. Fifty picomoles of His₆-tagged Vav(L) or Vav(DH) bound to Ni^{2+}-agarose beads was incubated for 1 hr at room temperature in 300 μl of buffer G (supplemented with 5 mM imidazole and 500 μg/ml BSA) containing 500 pmol of the indicated soluble GST–GTPase (CDC42Hs, Rac1, or RhoA) in the nucleotide-free state, GDP-bound state, or GTP-bound state. Where indicated the Vav protein was phosphorylated with Lck under conditions that achieved saturation of phosphorylation. Unbound GST–GTPase was removed by washing the Ni^{2+}-agarose beads five times in buffer NB, and the amount of bound GST–GTPase was determined by immunoblotting using a GST-specific antibody. Negative ($-$) controls contained Ni^{2+}-agarose beads not complexed to Vav protein.

of Sos1(DH) binding of Rac in its nucleotide-free, GDP-bound, or GTP-bound states revealed a dramatic preference for the nucleotide-free state as expected for a GEF. These results suggest that the Sos1(DH) domain might be a Rac-specific GEF. Biochemical analysis of the GEF activity of Sos1 from mammalian cells revealed activity directed toward Rac but not Rho or CDC42 proteins.[19]

We reported that Vav binds Rho, Rac, and CDC42, suggesting a broad range of activity for Vav (see Fig. 4).[4] This is in agreement with biochemical analysis showing that Vav can stimulate nucleotide exchange on all three of these GTPases. Further, this is in agreement with the ability of Vav to induce morphological changes consistent with the activation of Rho, Rac, and CDC42.[20] Finally, unphosphorylated Vav binds weakly and equally to RhoA in its nucleotide-free, GDP-bound, and GTP-bound states. However, phosphorylation of Vav by Lck, an event that activates its GEF activity for RhoA, yields a molecule that has high affinity for nucleotide-free RhoA without affecting affinity for the nucleotide-bound states. Thus, binding

[19] A. S. Nimnual, B. A. Yatsula, and D. Bar-Sagi, *Science* **279,** 560 (1998).
[20] M. F. Olson, N. G. Pasteris, J. L. Gorski, and A. Hall, *Curr. Biol.* **6,** 1628 (1996).

analysis of GEFs to GTPases may be useful for assessing possible substrates as well as detecting events that activate a latent GEF activity.

For detecting the interaction of a GEF with GTPases, we have used His_6-tagged GEFs bound to Ni^{2+}-agarose beads (20–100 pmol) to bind and precipitate soluble GST fusions of the Rho-related GTPase (100–500 pmol) in a 0.3-ml reaction. The binding reactions (20–37°) are carried out with continual rotation in buffer G. Buffer G supplemented with 1 mM GDP or GTP is used for the analysis of GTPase/GEF interaction in the presence of these nucleotides. Binding reactions are carried out for 1 hr. The Ni^{2+}-agarose beads are washed four times with buffer NB and once with buffer G. GTPases associated with GEFs bound to Ni^{2+}-agarose beads are detected after SDS–PAGE and immunoblotting. For binding assays, the DH domain-containing protein is prepared as described earlier for Vav. Procedures for preparation and purification of GST–Rho–GTPases from *E. coli* expression systems were described previously.

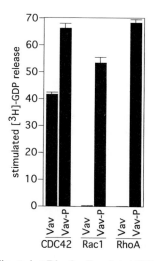

FIG. 5. Vav GEF activity directed at Rho family-related GTPases. Ten picomoles of His_6-tagged Vav(L) or Vav(DH) bound to Ni^{2+}-agarose beads was incubated for 15 min at room temperature in 200 μl of buffer G containing 40 pmol of the indicated soluble [³H]GDP-bound GST–GTPase (CDC42, Rac1, or RhoA) in the presence of 1 mM GTP. Where indicated the Vav protein was phosphorylated with Lck under conditions that achieved saturation of phosphorylation. The amount of [³H]GDP remaining bound to the GST–GTPase protein was determined by the filter-binding assay described in the text. Stimulated GDP release is defined as the percentage of [³H]GDP released in the presence of Vav protein less the percentage of GDP released in control samples not containing Vav or Vav-P.

Detection of Vav GEF Activity

At room temperature, a sample of [³H]GDP-bound GST–Rac protein in the presence of an excess of GTP (0.5 mM final concentration) will, over a period of time, be spontaneously converted to a population of predominantly GTP-bound GST–Rac molecules. GEFs that act on Rac can dramatically enhance the rate of this conversion. The GEF assay using [³H]GDP-bound GST–Rac as a substrate therefore requires a means to assess the amount of [³H]GDP-bound GST–Rac in a sample. A commonly used assay for detecting [³H]GDP-bound GST–Rac exploits the high affinity of proteins (such as [³H]GDP-bound GST–Rac) for nitrocellulose filters and the low affinity of free [³H]GDP for these filters. Thus, in GEF assays the amount of radioactivity captured on the filter decreases over time as [³H]GDP is released from GST–Rac (see the filter-binding assay described earlier).

GEF assays for Vav are typically carried out under the following conditions: 10–40 pmol of [³H]GDP-bound GST–Rac, 1–5 pmol of His$_6$-Vav, and 1 mM GTP in buffer G (50 μl volume per data point). For each data point a 50-μl aliquot is removed at the appropriate time and analyzed by a filter-binding assay (see Fig. 5).

His$_6$-Vav proteins can be added as soluble protein or immobilized on Ni^{2+}-agarose beads. When the latter is used, the tubes are rotated for the duration of the reaction. In addition, when multiple aliquots are taken from a single tube, pipette tips with a >0.5-mm-inner-diameter opening are used to prevent agarose beads from partially clogging the tip. This would result in changing the experimental conditions for subsequent samples taken from the tube.

[5] Activation of Rac1 by Human Tiam1

By GIDEON BOLLAG, ANNE M. CROMPTON, DANIELLE PEVERLY-MITCHELL, GASTON G. M. HABETS, and MARC SYMONS

Introduction

The small GTPase Rac1 becomes activated via catalyzed displacement of bound GDP by a guanine nucleotide exchange factor (GEF). An important GEF for Rac1 was discovered in a screen for genes affecting tumor

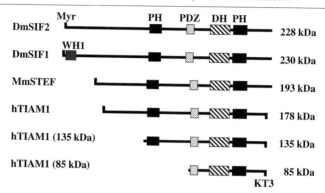

Fig. 1. Schematic representation of the primary structure of full-length human Tiam1 and its homologs *Mus musculus* STEF and *Drosophila melanogaster* SIF1 and SIF2. Also indicated are representations of the truncated 135- and 85-kDa forms of human Tiam1. Domains indicated include pleckstrin homology (PH), PDZ, Dbl homology (DH), and WASP homology 1 (WH1) domains, as described in the text. Also depicted is a putative myristoylation site (Myr) at the amino termini, as well as the appended KT3 epitope tag on recombinant full-length and truncated (135 and 85 kDa) Tiam1 proteins.

invasion and metastasis, and hence named Tiam1.[1] Early experiments with recombinant mouse Tiam1 indicated that partially purified preparations do indeed possess selective GEF activity toward Rac1.[2] Substantial additional evidence now implicates Rac1 activation in regulating the actin cytoskeleton during invasion.[3,4] An important role for Tiam1 is also implied, as Tiam1 is rather ubiquitously expressed in cancer cell lines.[5] Hence, Tiam1 is an attractive target for anticancer therapeutics and development of robust assay methods is of keen interest.

Several Tiam1 homologs with very similar domain organization have been described in recent literature (Fig. 1). All of these proteins share two pleckstrin homology (PH) domains that putatively bind to lipids,[6] a PDZ

[1] G. G. Habets, E. H. Scholtes, D. Zuydgeest, R. A. van der Kammen, J. C. Stam, A. Berns, and J. G. Collard, *Cell* **77,** 537 (1994).

[2] F. Michiels, G. G. Habets, J. C. Stam, R. A. van der Kammen, and J. G. Collard, *Nature* **375,** 338 (1995).

[3] P. J. Keely, J. K. Westwick, I. P. Whitehead, C. J. Der, and L. V. Parise, *Nature* **390,** 632 (1997).

[4] L. M. Shaw, I. Rabinovitz, H. H. Wang, A. Toker, and A. M. Mercurio, *Cell* **91,** 949 (1997).

[5] G. G. Habets, R. W. van der Kammen, J. C. Stam, F. Michiels, and J. G. Collard, *Oncogene* **10,** 1371 (1995).

[6] M. A. Lemmon and K. M. Ferguson, *Curr. Top. Microbiol. Immunol.* **228,** 39 (1998).

domain that typically interacts with protein partners,[7] and a Dbl homology (DH) domain that encodes the GEF active site.[8] In *Drosophila melanogaster*, the Tiam1 homologs are known as the *still life* or *sif* genes, as they regulate fly locomotion and sexual function.[9] Interestingly, SIF1 also has a WASP homology 1 (WH1) domain at its amino terminus; WH1 domain tertiary structure is similar to that of the PH domain, yet WH1 domains typically bind to proline-rich regulators of the actin cytoskeleton.[10] A mammalian homolog to SIF and Tiam1 has been called STEF in mouse[11] and Tiam2 in humans.[12] Throughout, it appears that the primary target of Tiam1 and its homologs is Rac1. Regulation of Tiam1 activity is still poorly understood. Phosphatidylinositol lipids bind to the pleckstrin homology domains of Tiam1[13] and phosphoinositide (PI) 3′-kinase activity is required for Tiam1 function.[14] In addition, phosphorylation of Tiam1 is also predicted to affect its activity.[15]

Because Tiam1 is implicated in invasion, we were interested in exploring the biochemical properties of Tiam1. However, the full-length and various truncated versions of Tiam1 expressed in Sf9 (*Spodoptera frugiperda-9*, fall armyworm ovary) insect cells proved inactive in our *in vitro* assay formats. In exploring the reasons for this inactivity, we happened on the observation that certain lipids or amphiphilic small molecules are capable of stimulating Tiam1 activity *in vitro*.[16] This article describes the characterization of recombinant human Tiam1 expressed in Sf9 cells and assayed in the presence of one of these amphiphilic molecules, ascorbyl stearate (Fig. 2).

[7] Z. Songyang, A. S. Fanning, C. Fu, J. Xu, S. M. Marfatia, A. H. Chishti, A. Crompton, A. C. Chan, J. M. Anderson, and L. C. Cantley, *Science* **275,** 73 (1997).

[8] I. P. Whitehead, S. Campbell, K. L. Rossman, and C. J. Der, *Biochim, Biophys. Acta* **1332,** F1 (1997).

[9] M. Sone, M. Hoshino, E. Suzuki, S. Kuroda, K. Kaibuchi, H. Nakagoshi, K. Saigo, Y. Nabeshima, and C. Hama, *Science* **275,** 543 (1997).

[10] K. E. Prehoda, D. J. Lee, and W. A. Lim, *Cell* **97,** 471 (1999).

[11] M. Hoshino, M. Sone, M. Fukata, S. Kuroda, K. Kaibuchi, Y. Nabeshima and C. Hama, *J. Biol. Chem.* **274,** 17837 (1999).

[12] C. Y. Chiu, S. Leng, K. A. Martin, E. Kim, S. Gorman, and D. M. Duhl, *Genomics* **61,** 66 (1999).

[13] L. E. Rameh, A. K. Arvidsson, K. L. Carraway III, A. D. Couvillon, G. Rathbun, A. Crompton, B. VanRenterghem, M. P. Czech, K. S. Ravichandran, S. J. Burakoff, D. S. Wang, C. S. Chen, and L. C. Cantley, *J. Biol. Chem.* **272,** 22059 (1997).

[14] E. E. Sander, S. van Delft, J. P. ten Klooster, T. Reid, R. A. Van der Kammen, F. Michiels, and J. G. Collard, *J. Cell Biol.* **143,** 1385 (1998).

[15] I. N. Fleming, C. M. Elliott, F. G. Buchanan, C. P. Downes, and J. H. Exton, *J. Biol. Chem.* **274,** 12753 (1999).

[16] A. M. Crompton, L. H. Foley, A. Wood, W. Roscoe, D. Stokoe, F. McCormick, M. Symons and G. Bollag, submitted for publication.

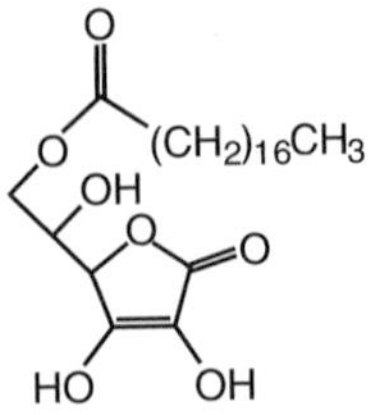

FIG. 2. Chemical structure of the amphiphilic Tiam1 activator, ascorbyl stearate.

Materials

Tiam1 Plasmids for Sf9 Expression

The full-length human Tiam1 cDNA is pieced together from two clones obtained from a human fetal brain library (Stratagene, La Jolla, CA) and from two polymerase chain reaction (PCR) fragments generated from a human hippocampal library (Clontech, Palo Alto, CA) and a human brain library (Clontech). PCR amplification with the antisense 3′-primer GATC-CCGGGTCATGTTTCTGGTTCTGGGATCTCAGTGTTCAGTTTCC-TG is used to add the KT3 epitope tag "PEPET," a stop codon, and a *Sma*I site to the end of the Tiam1 cDNA. Sequencing of this cDNA reveals one coding difference from the published sequence,[5] namely G3005C, resulting in a glutamine at position 844 (where the published sequence indicates a histidine), as well as two PCR-introduced mutations yielding G4739A and G5153A in the final coding sequence. This cDNA is subcloned into the pAcC4 vector[17] for expression in Sf9 insect cells. Because expression of full-length (178 kDa) Tiam1 is exceptionally poor, two truncated versions coding for 135- and 85-kDa forms were created by engineering initiating methionines in front of amino acids 402 and 837, respectively (Fig. 1).

Reagents

Radiolabeled [α-^{32}P]GTP and [α-^{33}P]GTP (both at specific activity of 3000 Ci/mmol) are from Dupont-NEN (Boston, MA). Nonidet P-40 and protease inhibitors (Pefabloc, aprotinin, pepstatin, leupeptin, and E64) are from Roche Molecular Biochemicals (Indianapolis, IN). KT3-sepharose is prepared by coupling anti-KT3 monoclonal antibody to protein G–Sepharose (Pharmacia, Piscataway, NJ) using dimethyl pimelimidate (Pierce, Rockford, IL) as described previously.[18] Ascorbyl stearate is from

[17] B. Rubinfeld, S. Munemitsu, R. Clark, L. Conroy, K. Watt, W. J. Crosier, F. McCormick, and P. Polakis, *Cell* **65**, 1033 (1991).
[18] G. Bollag and F. McCormick, *Methods Enzymol.* **255**, 21 (1995).

TCI America, Inc. (Portland, OR). A 25 mM stock solution of ascorbyl stearate is prepared in 100% ethanol. Twenty-five-fold dilutions of this 25 mM solution are initially made into distilled water to preserve compound solubility; further dilutions are made into the final assay buffer [20 mM HEPES, pH 7.3, 100 mM NaCl, 2 mM MgCl$_2$, 2 mM dithiothreitol (DTT), 0.2 mg/ml bovine serum albumin (BSA)].

Methods

Expression and Purification of Recombinant Proteins

In order to generate recombinant baculoviruses, the Tiam1 expression plasmids are cotransfected with *Autographa californica* nuclear polyhedrosis virus DNA (PharMingen, San Diego, CA) and recombinant viruses selected as detailed previously.[19] A plasmid encoding the human Racl sequence[20] with a 5'-appended coding sequence for the Glu epitope (MEYMPME) is constructed in a pAcC13 vector backbone, and recombinant baculovirus is generated as described.[19] The preparation of Glu affinity-tagged Racl protein from Sf9 cells is performed exactly as described in a previous volume for Glu-tagged Ras proteins.[19] Prior to assay, the Racl · GDP is diluted to 1 μM in assay buffer containing 0.2 μM GDP to ensure that all of the Racl is nucleotide bound.

The 135- and 85-kDa forms of Tiam1 are purified from baculovirus-infected Sf9 cells using KT3-MAb immunoaffinity chromatography.[21] Frozen cell pellets are lysed in 4 volumes of lysis buffer (25 mM Tris, pH 7.8, 1 mM EDTA, 0.5% Nonidet P-40, 1 mM DTT, 0.5 mM Pefabloc, 0.5 μM aprotinin, 2 μM pepstatin, 5 μM leupeptin, 5 μM E64) by stirring for 30 min; after 10 min, glycerol is added to 10% and NaCl is added to 200 mM. The lysate is centrifuged at 20,000g for 45 min at 4°, and the clarified supernatant is passed over a KT3–Sepharose column (1 ml column volume per 4 g of Sf9 cell pellets). The column is washed with wash buffer (25 mM Tris, pH 7.8, 1 mM EDTA, 1 mM DTT, 4 μM leupeptin) using 10 column volumes of wash buffer containing 1% Nonidet P-40, 10 column volumes of wash buffer, and 10 column volumes of wash buffer containing 200 mM NaCl. Purified protein is then eluted with 5 column volumes of elution buffer [25 mM Tris, pH 7.8, 200 mM NaCl, 1 mM EDTA, 1 mM DTT, 4

[19] E. Porfiri, T. Evans, G. Bollag, R. Clark, and J. F. Hancock, *Methods Enzymol.* **255,** 13 (1995).
[20] J. Didsbury, R. F. Weber, G. M. Bokoch, T. Evans, and R. Snyderman, *J. Biol. Chem.* **264,** 16378 (1989).
[21] G. A. Martin, D. Viskochil, G. Bollag, P. C. McCabe, W. J. Crosier, H. Haubruck, L. Conroy, R. Clark, P. O'Connell, R. M. Cawthon, M. A. Innis, and F. McCormick, *Cell* **63,** 843 (1990).

μM leupeptin, 100 $\mu g/ml$ KT3 peptide (TPEPET)]. The protein is dialyzed into storage buffer (25 mM Tris, pH 7.8, 50% glycerol, 100 mM NaCl, 1 mM EDTA, 1 mM DTT, 4 μM leupeptin) and stored at $-80°$. Typical yields are 1 mg per 10 g of Sf9 cell pellets (or 500 ml Sf9 cell culture volume). This protein has been stable for more than 4 years.

Assays of Nucleotide Exchange

Nucleotide exchange onto small GTPases has been described in several different formats in previous volumes: release of [³H]GDP,[22–27] release of [³²P]GDP,[28] binding of [³H]GTP[27], and fluorescent changes on dissociation of the fluorescent nucleotide analog mant-GDP,[22] for example. When using purified proteins, we have consistently found that binding of [α-³²P]GTP or [α-³³P]GTP provides a simple and sensitive format. This assay takes advantage of the observation that all of the small GTPases copurify with bound GDP during conventional or immunopurification.

Measurement of [α-³²P]GTP or [α-³³P]GTP binding to Rac1 in the presence of Tiam1 is achieved by trapping the nucleotide-bound Rac1 on a selection of various filters. Free nucleotide will not bind to these filters in the appropriate buffer. Note that this method relies on preserving the native conformation of Rac1 during the harvesting and wash process. One shortcoming of this procedure is the limitation in possible quenching conditions: generally, we terminate the reaction by diluting in ice-cold filter washing buffer. As described later, use of 96-well filtration plates or filter mats allows for rapid handling of multiple samples in a time-effective manner. This article discusses the use of (1) conventional 2.5-cm cellulose filters in a manifold that accommodates 12 such filters at once, (2) 96-well cellulose filtration plates using a simple vacuum filtration apparatus, and (3) filter mats that can be used with an automated liquid-handling apparatus (harvester). Because of the cost of the harvester and the 96-well scintillation counter, the setup costs for the 96-well assays are higher. However, as mentioned later, because individual filters from the filtration plates can be removed to count in a conventional scintillation counter, the setup cost for such a format is quite similar to that using conventional cellulose filters.

[22] C. Lenzen, R. H. Cool, and A. Wittinghofer, Methods Enzymol. 255, 95 (1995).
[23] J. Downward, Methods Enzymol. 255, 110 (1995).
[24] M. Frech, D. Cussac, P. Chardin, and D. Bar-Sagi, Methods Enzymol. 225, 125 (1995).
[25] A. J. Self and A. Hall, Methods Enzymol. 256, 67 (1995).
[26] Y. Zheng, M. J. Hart, and R. A. Cerione, Methods Enzymol. 256, 77 (1995).
[27] E. Porfiri and J. F. Hancock, Methods Enzymol. 256, 98 (1995).
[28] D. A. Leonard and R. A. Cerione, Methods Enzymol. 256, 98 (1995).

Assay Formats

Assay buffer: 20 mM HEPES, pH 7.3, 100 mM NaCl, 2 mM MgCl$_2$, 2 mM DTT, 0.2 mg/ml BSA

Wash buffer: 20 mM Tris, pH 8, 100 mM NaCl, 5 mM MgCl$_2$ (ice cold)

1. Cellulose Filter Binding Using Single Filter Disks. Reactions are carried out in 20 μl final volume, typically including 5 nM Tiam1, 100 nM Rac1, and 10 μM ascorbyl stearate in assay buffer. Reactions are initiated by adding 0.15 μCi [α-^{32}P]GTP (final concentration 2.5 nM). After 20 min, reactions are stopped by adding 200 μl ice-cold wash buffer; 2.5-cm cellulose membranes (Millipore, Bedford, MA; HA, 0.45 μm) are prewet with wash buffer and assembled into a filtration manifold (Millipore 1225 sampling manifold). Reactions are transferred rapidly to prewet filters and washed four times with 2 ml wash buffer while maintaining a vacuum. After the manifold is disassembled, the filters are removed and dried under a heat lamp. Scintillation cocktail is added to the filters and they are counted for ^{32}P in a conventional scintillation counter.

2. Cellulose Filter Binding Using 96-Well Filter Plates. Reactions are carried out on 96-well polypropylene plates (e.g., from Corning, Corning, NY) in 40 μl final volume, typically including 5 nM Tiam1, 100 nM Rac1, and 10 μM ascorbyl stearate in assay buffer. Reactions are initiated by adding 0.05 μCi [α-^{33}P]GTP (final concentration 1 nM). After 20 min, reactions are stopped by adding 200 μl ice-cold wash buffer; 96-well cellulose filtration plates (Millipore Multiscreen MHAB N45 10) are prewet with wash buffer and aspirated on a vacuum filtration apparatus (e.g., Millipore Multiscreen vacuum manifold). Reactions are transferred rapidly to the filter plates and washed five times with 0.3 ml wash buffer while maintaining a vacuum. The filtration plates are dried under a heat lamp, scintillation cocktail is added, and plates are quantified using a 96-well scintillation counter (e.g., Wallac 1450 Microbeta, PerkinElmer Life Sciences, Turku, Finland). ^{33}P is the preferred label to minimize "cross talk" during scintillation counting. Alternatively, filters can be removed from the filter plates and quantified in a conventional scintillation counter.

3. DEAE Filter Binding Using an Automated Harvester. Reactions are carried out in 50 μl final volume, typically including 5 nM Tiam1, 100 nM Rac1, and 10 μM ascorbyl stearate in assay buffer. Reactions are initiated by adding 0.025 μCi [α-^{33}P]GTP (final concentration 0.2 nM). After 20 min, reactions are harvested onto 96-well DEAE filters (Wallac) using a Skatron Micro96 Harvester (Molecular Devices, Sunnyvale, CA). The harvester is programmed to prewet the filters with wash buffer, load the samples from the 96-well reaction plate, wash continuously for 10 sec, and then dry the filter for 10 sec. Again, the samples are dried under a heat lamp and quantified using a 96-well scintillation counter.

Immunofluorescence

Analysis of cytoskeletal changes by immunofluorescence has been detailed elsewhere in this series.[29] The effects of platelet-derived growth factor on the actin cytoskeleton of porcine aortic endothelial cells have also been detailed.[30] Briefly, cells are grown and treated on glass coverslips, fixed with formaldehyde, and stained with tetramethylrhodamine isothiocyanate-conjugated phalloidin (TRITC-phalloidin).

Characterization of Tiam1 Nucleotide Exchange Activity

We found that full-length Tiam1 is expressed very poorly in Sf9 cells. This poor expression could result from the presence of a proline-, glutamate-, serine-, and threonine-rich (PEST) domain near the amino terminus that often leads to protein instability.[31] We therefore performed most of our experiments using the truncated 85- and 135-kDa forms (Fig. 1). No significant biochemical differences were found between these two proteins; the 135-kDa Tiam1 was used to generate data shown in Fig. 3. The method of filter binding also did not affect the interpretation of the results; data shown in Fig. 3 were generated on DEAE filters using an automated harvester.

Using the assay conditions and format described earlier, we have characterized recombinant human Tiam1 protein (Fig. 3). We detect minimal activity in the absence of a lipid or amphiphilic molecule. We serendipitously discovered that the amphiphilic molecule ascorbyl stearate (Fig. 2) is able to stimulate Tiam1 activity by binding to the C-terminal pleckstrin homology domain.[16] This observation is highlighted by data in Fig. 3: as shown in A–C, ascorbyl stearate increases the Tiam1-dependent binding of GTP to Rac1 dramatically. The reaction is highly dependent on Tiam1 (Fig. 3A), Rac1 (Fig. 3B), and ascorbyl stearate (Fig. 3C). Figure 3D is a representative SDS–PAGE gel indicating the purity of the proteins used in these assays. Under the conditions described previously (5 nM Tiam1, 100 nM Rac1, 10 μM ascorbyl stearate), the reaction is linear up to about 20 min. Note that the graph in Fig. 3C indicates that higher concentrations

[29] A. J. Ridley, *Methods Enzymol.* **256**, 306 (1995).
[30] M. Symons, J. Derry, B. Karlak, S. Jiang, V. Lemahieu, F. McCormick, U. Francke, and A. Abo, *Cell* **84**, 723 (1996).
[31] S. Rogers, R. Wells, and M. Rechsteiner, *Science* **234**, 364 (1986).

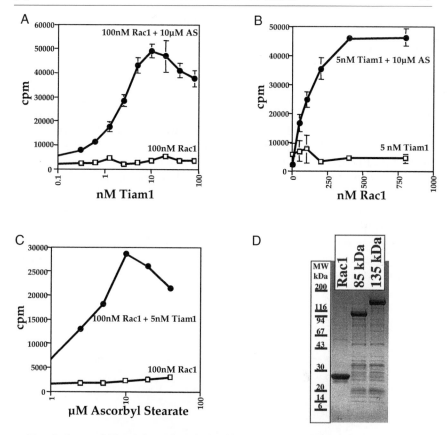

FIG. 3. Assay of Tiam1-dependent nucleotide exchange on Rac1. (A) Dependence on Tiam1 concentration. (B) Dependence on Rac1 concentration. (C) Dependence on ascorbyl stearate (AS) concentration. (D) Purity of recombinant proteins. Two micrograms each of Rac1, 85-kDa hTiam1, and 135-kDa hTiam1 was separated on 4–20% SDS–PAGE gels and stained with Coomassie blue.

of ascorbyl stearate become less effective. We find this consistently and tentatively attribute it to the relatively poor solubility of ascorbyl stearate in this buffer system.

Tiam1, like all Rho family GEFs, contains a pleckstrin homology (PH) domain C-terminally adjacent to the active Dbl homology (DH) domain (Fig. 1). This PH domain is competent to bind lipids such as phosphatidyl-inositides.[13] The requirement for a lipid or amphiphilic reagent for optimal Tiam1 activity indicates a putative role for lipid ligation of the PH domain in directly activating the GEF activity. Indeed, we find that phosphatidylinositides are also able to stimulate Tiam1 activity.[16] We propose that ascorbyl stearate is able to mimic the effects of a physiological Tiam1-activating lipid.

In order to test the effects of ascorbyl stearate on Rac activation in cells, we used porcine aortic endothelial cells. In these cells, the ability of platelet-derived growth factor (PDGF) to induce lamellipodia has been shown to depend on Rac activation.[30] We pretreated these cells with 10 μM ascorbyl stearate and then treated them with 10 nM PDGF. As shown in Fig. 4, the ascorbyl stearate pretreatment resulted in loss of sensitivity to PDGF: PDGF does not induce lamellipodia in the presence of ascorbyl stearate. The Rac activator in PAE cells has not been identified, and ascorbyl stearate possibly induces pleiotropic effects in these cells. It is interesting to note, however, that ascorbyl stearate did not display toxic effects, as it did not significantly affect the growth rate of NIH 3T3 mouse fibroblasts measured over 7 days. Nonetheless, this very preliminary experiment suggests an effect on the Rac pathway that is consistent with ascorbyl stearate antagonizing an endogenous Rac GEF activator.

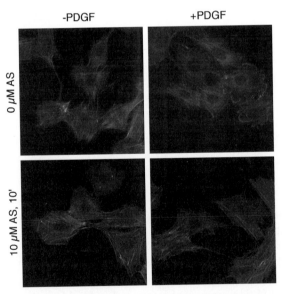

Fɪɢ. 4. Effect of ascorbyl stearate on PDGF-induced lamellipodia. Porcine aortic endothelial cells were treated with or without 10 nM platelet-derived growth factor (PDGF) following pretreatment with vehicle alone (0 μM) or 10 μM ascorbyl stearate (AS). Cells were then fixed and stained with TRITC-phalloidin.

Conclusions

The Tiam1-dependent conversion of Rac1 to its active GTP-bound conformation is likely to be an important step in organizing the actin cytoskeleton of many cells. In addition, Tiam1 may also play an important role in tumor cell invasion, as highlighted by its ability to effect T lymphoma invasion and metastasis.[1] Because the GEF activity of Tiam1 is an intriguing target for anticancer therapeutics, the robust assay methods described here could present useful tools for basic research as well as drug discovery projects. The observation that effective Tiam1 activity is dependent on a lipid or other amphiphilic molecules, such as ascorbyl stearate, also predicts an important role for allosteric regulators in the Rac1 signaling pathway. We propose that the GEF activity of the DH domain of Tiam1 is inhibited by the adjacent PH domain and that ligation of the PH domain results in an active DH conformation.

[6] Activation of Rho GEF Activity by Gα₁₃

By MATTHEW J. HART, WILLIAM ROSCOE, and GIDEON BOLLAG

Introduction

Because the Rho GTPase participates in pathways governing diverse pathologies such as carcinogenesis and cardiovascular dysfunction, an understanding of Rho activity is of keen therapeutic importance. Rho activity is largely determined by guanine nucleotide exchange factors (GEFs) that allow formation of the effective Rho · GTP complex by causing displacement of GDP from the inactive, ambient Rho · GDP complex.[1] A number of GEFs for Rho have been described in the literature, including Lbc,[2] Lfc,[3] Lsc,[4] and PDZ Rho GEF.[5] Among these Rho GEFs, p115 Rho GEF (the human homolog of Lsc) and PDZ Rho GEF display ubiqui-

---

[1] I. P. Whitehead, S. Campbell, K. L. Rossman, and C. J. Der, *Biochim. Biophys. Acta* **1332,** F1 (1997).

[2] D. Toksoz and D. A. Williams, *Oncogene* **9,** 621 (1994).

[3] I. Whitehead, H. Kirk, C. Tognon, G. Trigo-Gonzalez, and R. Kay, *J. Biol. Chem.* **270,** 18388 (1995).

[4] I. P. Whitehead, R. Khosravi-Far, H. Kirk, G. Trigo-Gonzalez, C. J. Der, and R. Kay, *J. Biol. Chem.* **271,** 18643 (1996).

[5] S. Fukuhara, C. Murga, M. Zohar, T. Igishi, and J. S. Gutkind, *J. Biol. Chem.* **274,** 5868 (1999).

tous expression patterns consistent with a role in diverse cellular functions.[5,6]

Close examination of the amino-terminal sequence of p115 Rho GEF (Fig. 1A) reveals a very weak, but significant, homology to the regulator of G-protein signaling (RGS) family of proteins. Although the homology may appear tenuous, most of the hydrophobic residues that form the core of the RGS domain are conserved.[7] RGS domains interact with GTP-bound complexes of the α subunits of heterotrimeric G-proteins and accelerate GTP hydrolysis on cognate α subunits.[8] Accordingly, the RGS domain of p115 Rho GEF was shown to accelerate GTP hydrolysis on $G\alpha_{12}$ and $G\alpha_{13}$ proteins, but not on $G\alpha_s$, $G\alpha_i$, $G\alpha_q$, or $G\alpha_z$.[9]

Heterotrimeric G-proteins are known to activate the Rho GTPase pathway.[10] Therefore, the identification of a $G\alpha_{13}$-responsive RGS domain on p115 Rho GEF implies a functional role for $G\alpha_{13}$ in the activation of Rho. Two important biochemical properties are suggested: (1) $G\alpha_{13}$ should bind directly to p115 Rho GEF and (2) $G\alpha_{13}$ should stimulate the nucleotide exchange activity of p115 Rho GEF. As shown previously,[11] both of these predictions are borne out experimentally. This article describes the reagent development and methodology for measuring these biochemical properties *in vitro*.

Materials

P115 Rho GEF Expression Plasmids

For insect cell expression, p115 Rho GEF cDNAs encoding full-length (FL115) or an amino-terminally truncated version (ΔN115) lacking the RGS domain are subcloned into the pFastBac1 vector (Life Technologies, Inc., Gaithersburg, MD), along with an amino-terminally appended Glu-epitope tag.[6] The resulting constructs encode 115-kDa (FL115) or 95-kDa (ΔN115) proteins (Fig. 1B). For mammalian expression, these same cDNAs are subcloned into a pEXV3 vector with an amino-terminally appended Myc-epitope tag.[6] DNA encoding the p115 Rho GEF RGS domain is

[6] M. J. Hart, S. Sharma, N. elMasry, R.-G. Qiu, P. McCabe, P. Polakis, and G. Bollag, *J. Biol. Chem.* **271,** 25452 (1996).

[7] J. J. Tesmer, D. M. Berman, A. G. Gilman, and S. R. Sprang, *Cell* **89,** 251 (1997).

[8] D. M. Berman and A. G. Gilman, *J. Biol. Chem.* **273,** 1269 (1998).

[9] T. Kozasa, X. Jiang, M. J. Hart, P. M. Sternweis, W. D. Singer, A. G. Gilman, G. Bollag, and P. C. Sternweis, *Science* **280,** 2109 (1998).

[10] D. J. Mackay and A. Hall, *J. Biol. Chem.* **273,** 20685 (1998).

[11] M. J. Hart, X. Jiang, T. Kozasa, W. Roscoe, W. D. Singer, A. G. Gilman, P. C. Sternweis, and G. Bollag, *Science* **280,** 2112 (1998).

subcloned into the pGEX4T-2 vector (Pharmacia, Piscataway, NJ) for expression in *Escherichia coli.*[9]

PDZ Rho GEF Sf9 Expression Plasmids

The full-length PDZ Rho GEF[5] cDNA is pieced together from two clones obtained from a human fetal brain library (Stratagene, La Jolla, CA) and from an EST (Accession AA780424, IMAGE consortium). Sequencing of the full-length clone reveals several differences with the published sequence (Accession AB002378): an insertion of 40 residues (RICEVYSRNPASLLEEQIEGARRRVTQLQLKIQQETGGSV) between Q194 and D195, a deletion of 30 nucleotides resulting in replacement of residues 550–560 (TFHIPLSPVEV) with isoleucine, and two missense differences near the carboxy terminus (S1416G and H1427R). This full-length cDNA (PDZ Rho GEF) is subcloned into the pFastBac1 plasmid with an amino-terminal Glu-epitope tag.

Gα₁₃ Expression Plasmids

Wild-type (Gα₁₃WT) and Q226L (Gα₁₃QL) mutant human Gα₁₃ cDNAs with an appended amino-terminal Glu-epitope tag (MEYMPME) are subcloned into the pFastBac1 vector for insect cell expression. The resulting plasmids encode the epitope tag in-frame with the start methionine such that the first 10 amino acids are MEYMPMEGSM. Subsequent sequencing reveals that the "wild-type" protein actually encodes a L221V mutation; this protein is still active, as the native codon 221 in mouse Gα₁₃ is naturally a valine.

Reagents

Radiolabeled [α-³²P]GTP and α-³³P-GTP (both at specific activity of 3000 Ci/mmol) are from Dupont-NEN (Boston, MA). Protease inhibitors (Pefabloc, aprotinin, and leupeptin), GTPγS, and Nonidet P-40 are from Roche Molecular Biochemicals (Indianapolis, IN). Anti-Glu and anti-Myc (9E10) monoclonal antibodies and ED peptide (N-terminally acetylated EYMPTD, a peptide with high affinity for the anti-Glu antibody) are from BAbCo (Richmond, CA). Glu- and Myc-Sepharoses are prepared by coupling the respective monoclonal antibodies to protein G–Sepharose (Pharmacia, Peapack, NJ) using dimethyl pimelimidate (Pierce, Rockford, IL) as described previously.[12]

---

[12] G. Bollag and F. McCormick, *Methods Enzymol.* **255,** 21 (1995).

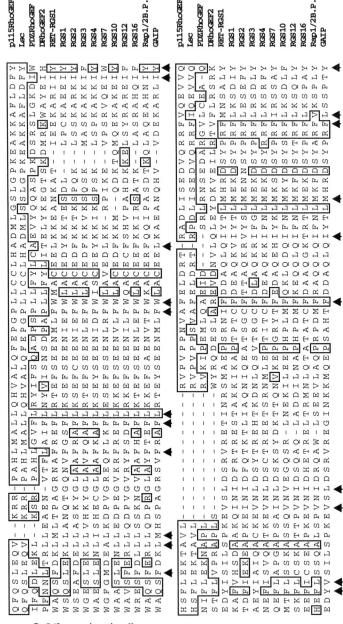

B

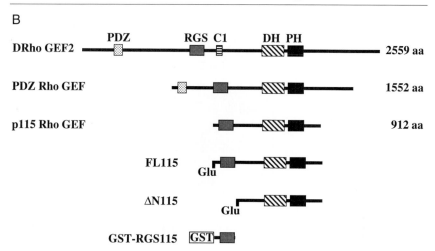

FIG. 1. Identification of an RGS domain on Rho exchange factors. (A) Homology of the p115 Rho GEF RGS domain with RGS domains of conventional RGS proteins. Residues identical to those of p115 Rho GEF are boxed. Arrows indicate residues that putatively form the hydrophobic core of the RGS domain. Sequence accession numbers are as follows: p115 Rho GEF, U64105; Lsc, U58203; PDZ Rho GEF, AB002378; DRho GEF2, AF032870; RET-RGS1, U89254; RGS1, S59049; RGS2, L13391; RGS3, U27655; RGS4, U27768; RGS7, U32439; RGS10, AF045229; RGS12, U92280; RGS16, U70426; Rap1/2 B.P., U85055; and GAIP, P49795. (B) Schematic depiction of the homology domains of DRho GEF2, PDZ Rho GEF p115 Rho GEF, including depictions of the truncated p115 Rho GEF proteins discussed in the text. Regulators of G-protein signaling (RGS), Dbl homology (DH), pleckstrin homology (PH), and PDZ domains are indicated along with the glutathione *S*-transferase (GST) and Glu-epitope tags used for purification. DRho GEF2 also contains a C1 domain homologous to a cysteine-rich domain in protein kinase C.

Methods

Expression and Purification of Recombinant Proteins

Conversion of pFastBac1 plasmids to expressing baculoviruses is performed according to the manufacturer's instructions (Life Technologies, Inc., Gaithersburg, MD).[13] Briefly, the p115 Rho GEF and Gα13 plasmids are first transformed into *E. coli* cells (DH10BAC, Life Technologies, Inc.) harboring a baculovirus shuttle vector (bacmid) and a helper plasmid to allow site-specific transposition. The recombinant bacmid DNA is isolated from the *E. coli* cells and transfected into Sf9 (*Spodoptera frugiperda-9*, fall armyworm ovary) cells for protein expression. For optimal expression of Gα13 proteins, coinfection of Sf9 cells with baculoviruses encoding the

---

[13] V. A. Luckow, S. C. Lee, G. F. Barry, and P. O. Olins, *J. Virol.* **67,** 4566 (1993).

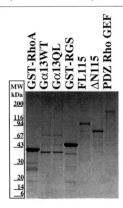

FIG. 2. Purity of recombinant proteins. Two micrograms each of GST–RhoA, $G\alpha_{13}WT$, $G\alpha_{13}QL$, GST–RGS115, FL115, $\Delta N115$, and PDZ Rho GEF was separated on 4–20% SDS–PAGE gels and stained with Coomassie blue.

untagged $G\beta_1$ and $G\gamma_2$ proteins is recommended.[14,15] RhoA fused to gluta-thione S-transferase (GST) is expressed in *E. coli* and purified as described previously.[16] Glu-epitope-tagged RhoA in a pAcC13 vector is converted to recombinant baculovirus and Sf9 cell-expressed protein purified as described in a previous volume for Glu-tagged Ras.[17]

Purification of p115 Rho GEF

Frozen Sf9 cell pellets are lysed in 2.5 volumes buffer 1 [50 mM Tris, pH 8, 1 mM EDTA, 1 mM dithiothreitol (DTT), 1 mM Pefabloc, 10 μM leupeptin, 1 μM aprotinin] containing 1% Nonidet P-40 by sonication for 5 min, using a microtip at 35% output (sonic dismembrator Model 300; Fisher, Pittsburgh, PA). All purification steps are carried out at 4°. The lysate is centrifuged at 20,000g for 20 min, and the clarified supernatant is passed over a Glu-Sepharose column (1 ml column volume per 4 g of Sf9 cell pellets). The column is washed with 20 volumes buffer 1 containing 0.5% Nonidet P-40 and 20 volumes buffer 1 containing 400 mM NaCl before elution in 10 volumes buffer 1 containing 400 mM NaCl and 100 μg/ml ED peptide. Purified proteins are concentrated 10-fold in a Centriprep-30 concentrator (Amicon, Danvers, MA), diluted with 1 volume 100% glycerol

[14] W. D. Singer, R. T. Miller, and P. C. Sternweis, *J. Biol. Chem.* **269,** 19796 (1994).
[15] L. R. Stephens, A. Eguinoa, H. Erdjument-Bromage, M. Lui, F. Cooke, J. Coadwell, A. S. Smrcka, M. Thelen, K. Cadwallader, P. Tempst, and P. T. Hawkins, *Cell* **89,** 105 (1997).
[16] A. J. Self and A. Hall, *Methods Enzymol.* **256,** 3 (1995).
[17] E. Porfiri, T. Evans, G. Bollag, R. Clark, and J. F. Hancock, *Methods Enzymol.* **255,** 13 (1995).

and stored at $-80°$. Typical yields are 1 mg per 2 g of Sf9 cell pellets (or 100 ml Sf9 cell culture volume).

Purification of $G\alpha_{13}$

Frozen Sf9 cell pellets are lysed in 2.5 volumes buffer 1 containing 1% sodium cholate, 0.1 mM GDP, and AMF (5 mM MgCl$_2$, 10 mM NaF, 50 μM AlCl$_3$) by sonication for 5 min as described earlier. Again, all purification steps are carried out at 4°. The lysate is centrifuged at 20,000g for 20 min, and the clarified supernatant is passed over a Glu-Sepharose column (1 ml column volume per 10 g of Sf9 cell pellets). The column is washed with 20 volumes buffer 1 containing 0.5% sodium cholate, 0.1 mM GDP, and AMF, followed by 20 volumes buffer 1 containing 400 mM NaCl and 20 μM GDP. Proteins are then eluted in 10 volumes buffer 1 containing 400 mM NaCl, 20 μM GDP, and 100 μg/ml ED peptide. Purified proteins are concentrated 10-fold in a Centriprep-30 concentrator (Amicon), diluted with 1 volume 100% glycerol, and stored at $-80°$. Typical yields are 1 mg per 10 g of Sf9 cell pellets (or 500 ml Sf9 cell culture volume).

Assay of Direct Interaction of p115 Rho GEF with $G\alpha_{13}$

In order to detect direct interaction of recombinant p115 Rho GEF proteins with endogenous or recombinant $G\alpha_{13}$, COS cell transfections are employed.[18] One microgram each of the appropriate Myc-epitope tagged pEXV3 plasmids is transfected into COS cells by electroporation and then plated in six-well dishes. After 24 hr, cells are lysed by resuspension in 1 ml lysis buffer (20 mM Tris, pH 7.5, 100 mM NaCl, 1 mM EGTA, 1 mM DTT, 0.5% Triton X-100, 30 μg/ml leupeptin, 30 μg/ml aprotinin, 1 mM Pefabloc) in the presence or absence of AMF. Recombinant Myc-epitope-tagged p115 Rho GEF proteins are then immunopurified using Myc-Sepharose resin: 20 μl resin is added to 1 ml lysate in 1.5-ml microcentrifuge tubes; following 1 hr of rotation at 4°, the resin is pelleted by brief centrifugation (4000g, 15 sec 4°), the supernatant is removed, and the resin is washed four times with 1 ml lysis buffer in the presence or absence of AMF as described earlier. Immunopurified protein prepared in the presence of AMF is eluted from the resin by the addition of 20 μl sample buffer [50 mM Tris, pH 8, 0.1% sodium dodecyl sulfate (SDS), 20% glycerol, 5 mM 2-mercaptoethanol)] and heating to 95° for 5 min. Immunopurified proteins prepared in the absence of AMF are then incubated with 0.1 μg of purified $G\alpha_{13}$ for 30 min at 4° in the presence or absence of AMF. The resin is washed and proteins are eluted from the resin as described earlier. Proteins

---

[18] M. J. Hart, M. G. Callow, B. Souza, and P. Polakis, *EMBO J.* **15**, 2997 (1996).

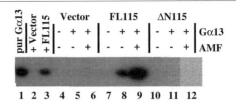

1 2 3   4 5 6   7 8 9 10 11 12

FIG. 3. Direct interaction of p115 Rho GEF with $G\alpha_{13}$. Immunopurification of lysates from COS cells transfected with vector, FL115, or ΔN115 was achieved using antibody to the Myc-epitope. Coimmunopurifications of endogenous $G\alpha_{13}$ (lanes 2 and 3) or added recombinant $G\alpha_{13}$ (lanes 4–12) were performed as described in the text. Recombinant $G\alpha_{13}$ was omitted from lanes 4, 7, and 12 as indicated (-), and activated $G\alpha_{13}$ was prepared using AMF, as indicated. The samples were separated on 8% SDS–PAGE gels and immunoblotted with $G\alpha_{13}$ antiserum. Purified $G\alpha_{13}$ (50 ng, pur $G\alpha_{13}$, lane 1) is included for reference.

separated by 8% SDS–PAGE gel electrophoresis are then immunoblotted with antibody to $G\alpha_{13}$.[14]

The results of such an experiment are shown in Fig. 3. In lane 3 (Fig. 3), FL115 is specifically able to bind to endogenous COS cell $G\alpha_{13}$. In lanes 8 and 9 (Fig. 3), the binding of recombinant $G\alpha_{13}$ to FL115 occurs selectively in the presence of AMF, although some binding in the absence is also observed. Importantly, p115 Rho GEF is unable to bind to $G\alpha_{13}$ if the RGS domain is not present (ΔN-115, lanes 10–12, Fig. 3). All of this data is consistent with the anticipated properties of an RGS domain and its cognate $G\alpha$ subunit: direct binding occurs when the $G\alpha$ subunit is in its activated state (here mimicked by GDP · AlF$_4$). We have observed similar binding of $G\alpha_{13}$ with PDZ Rho GEF, consistent with other reports.[5]

This direct interaction of $G\alpha_{13}$ with Rho GEFs should be transient during endogenous cellular signaling, as the activated GTP-bound $G\alpha_{13}$ would be rapidly inactivated by RGS domain-catalyzed GTP hydrolysis.[9] Consistent with this, the Q226L mutant of $G\alpha_{13}$ induces neoplastic transformation of fibroblasts.[19] Rapid deactivation affords a very sensitive mechanism for controlling the duration of the signal. Clearly, this allows the signaling network to be reset rapidly following stimulation. Therefore, the signal output would dissipate shortly upon removal of the stimulus. In addition, signaling events, such as the coupling of $G\alpha_{13}$ to Rho, that occur at the membrane can persist in the presence of signal; however, the signal would end abruptly, as key components, such as p115 Rho GEF, would return to the cytosol.

[19] T. A. Voyno-Yasenetskaya, A. M. Pace, and H. R. Bourne, *Oncogene* **9,** 2559 (1994).

Assays of Nucleotide Exchange

This section describes experiments in which binding of [α-^{32}P]GTP or [α-^{33}P]GTP to GST–RhoA is measured by filter binding. Related assay protocols are provided elsewhere in this volume.[20] The effects of Gα_{13} are demonstrated more effectively using an assay measuring the dissociation of [α-^{33}P]GDP from Rho in order to avoid complications introduced by Gα_{13} nucleotide binding. Briefly, 200 μM GST–RhoA · GDP is incubated with 0.2 μM [α-^{33}P]GTP in 50 mM HEPES, pH 7.3, 5 mM EDTA, 2 mM MgCl$_2$ at 22°. After 15 min, a 20-fold excess of 50 mM HEPES, pH 7.3, 50 mM NaCl, 2 mM DTT, 0.2 mg/ml BSA, 5 mM MgCl$_2$ is added and GTP hydrolysis is allowed to continue for 60 min at 22°. This Rho · [α-^{33}P]GDP is then incubated with GEF and/or 200 nM Gα_{13} for 20 min at 30° before binding to filters and washing unbound nucleotide through the filters. In these experiments, the final assay buffer consists of 50 mM HEPES, pH 7.3, 50 mM NaCl, 2 mM DTT, 0.2 mg/ml BSA, 100 μM GTPγS, 5 mM MgCl$_2$, 5 mM NaF, 50 μM AlCl$_3$.

Addition of 2 μM Rho · [α-^{33}P]GDP marks the beginning of the assay, and the assay is terminated by adding ice-cold filter wash buffer (20 mM Tris, pH 8, 100 mM NaCl, 5 mM MgCl$_2$) and aspirating through the filter. The addition of 5 mM MgCl$_2$, 5 mM NaF, 50 μM AlCl$_3$ to the assay is included to allow formation of the activated Gα_{13} · GDP · AlF$_4$ complex. The large excess of GTPγS is included to prevent rebinding of α-^{33}P-GDP: the nucleotide-free GST–RhoA will thus bind to GTPγS. The poorly hydrolyzable GTP analog GTPγS is used to prevent GTP hydrolysis and thus allow dissociation of the GEF, which should have a lower affinity for Rho · GTPγS than for Rho · GDP. As described, the assay format can employ single filter disk membranes (Millipore, Bedford, MA; HA, 0.45 μm) with aspiration through a 12-space manifold (Millipore 1225 sampling manifold) or 96-well filter plates (Millipore Multiscreen MHAB N45 10) with aspiration through a vacuum filtration apparatus (e.g., Millipore Multiscreen vacuum manifold).[20]

For experiments evaluating the effect of the p115 Rho GEF RGS domain, a different assay format was employed. Instead of measuring the dissociation of [α-^{33}P]GDP from GST–RhoA, the binding of [α-^{33}P]GTP to Glu-RhoA was determined. In this format, addition of 0.2 nM [α-^{33}P]GTP to 100 nM Glu-RhoA · GDP initiates the reaction in the presence of 20 nM ΔN115 or 40 nM FL115 and a range of GST–RGS115 concentrations. After 25 min at 22°, the reaction is harvested onto GF/A

---

[20] G. Bollag, A. M. Crompton, D. Peverly-Mitchell, G. G. M. Habets, and M. Symons, *Methods Enzymol.* **325** [5] 2000 (this volume).

filtermats (PerkinElmer Life Sciences, Turku, Finland) using a Skatron Micro96 Harvester (Molecular Devices, Sunnyvale, CA). The harvester is programmed to prewet the filters with filter wash buffer, load the samples from the 96-well reaction plate, wash continuously for 10 sec, and then dry the filter for 10 sec. The samples are then dried under a heat lamp and quantified using a 96-well scintillation counter.

$G\alpha_{13}$ Proteins Stimulate p115 Rho GEF Nucleotide Exchange Activity

Using these GEF activity assays, the different versions of p115 Rho GEF and PDZ Rho GEF were compared (Fig. 4A) in the presence or absence of $G\alpha_{13}$. As evident from these data, deletion of the RGS domain from p115 Rho GEF results in increased GEF activity, consistent with a regulatory role for the RGS domain. Accordingly, the addition of $G\alpha_{13}$ stimulates the GEF activity of FL115, but not the ΔN115 isoform (Fig. 4A). The activity of PDZ Rho GEF under these conditions is higher than that of FL115, and the addition of $G\alpha_{13}$ results in a smaller stimulation. Data shown in Fig. 4A involve wild-type $G\alpha_{13}$, and the Q226L form behaves similarly. Presumably, the direct binding of $G\alpha_{13}$ to the RGS domain of p115 liberates the GEF activity.

These data are consistent with an autoinhibitory role for the RGS domain on p115 Rho GEF activity. Indeed, the ΔN115 protein is significantly more transforming than FL115 when tested in a focus formation assay in fibroblasts.[6] As further confirmation of this model, the isolated GST–RGS115 protein was added to exchange reactions in the presence of

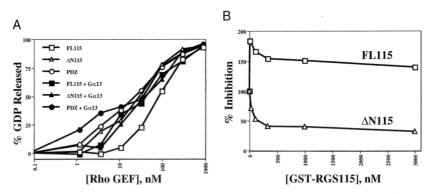

FIG. 4. Stimulation of p115 Rho GEF activity by $G\alpha_{13}$. (A) Comparison of full-length p115 Rho GEF (FL115, squares) with RGS-deleted p115 Rho GEF (ΔN115, triangles) and PDZ Rho GEF (PDZ, circles) in the presence (closed symbols) and absence (open symbols) of 200 nM $G\alpha_{13}$WT. (B) Inhibition of ΔN115 (triangles) but not FL115 (squares) activity by the p115 Rho GEF RGS (GST–RGS115) domain.

either FL115 or ΔN115. As shown in Fig. 4B, GST–RGS115 is able to inhibit the GEF activity of ΔN115 selectively. The stimulation of FL115 activity by low concentrations of GST–RGS115 is seen consistently; this could suggest that intramolecular inhibition by the RGS domain is more effective than the inhibition by exogenously added GST–RGS115. Nonetheless, data and methods summarized here strongly support the signaling model that invokes direct binding of the $G\alpha_{13}$ protein to the RGS domain of p115 Rho GEF and PDZ Rho GEF, resulting in the stimulation of Rho GEF activity.[11]

Conclusions

The Rho GTPase plays a key role in organizing the cytoskeleton of eukaryotic cells.[10] In particular, Rho directs the assembly of actin–myosin stress fibers and focal adhesion complexes that in turn mediate important intra- and intercellular signaling events. Because Rho activity is determined primarily by Rho GEFs, such as those described here, the regulation of Rho GEF activity must play a critical role in these pathways. We have developed robust assays for determining Rho GEF activity and applied these assays to biochemically analyze Rho GEFs.

Binding of extracellular factors (such as lysophosphatidic acid) to heptahelical receptors results in activation of heterotrimeric G-proteins of the α_{13} family. This event triggers a pathway leading to Rho activation. Specifically, the $G\alpha_{13}$ protein binds directly to the RGS domain of p115 Rho GEF, affecting an increase in GEF activity. Because the Rho pathway plays important roles in pathologies such as hypertension and cancer, the $G\alpha_{13}$/Rho GEF interaction makes an attractive therapeutic target.

[7] Purification and Characterization of Guanine Nucleotide Dissociation Stimulator Protein

By Jon P. Hutchinson, Katrin Rittinger, and John F. Eccleston

Introduction

Most guanine nucleotide exchange factors (GEFs) for small GTPases identified so far are large multidomain proteins that possess multiple functionalities. Exchange activity is typically specific to a particular subfamily. For example, Sos and Cdc25, which share a homologous catalytic domain, show exchange activity against Ras subfamily proteins, whereas DH domain proteins are active against the Rho subfamily.[1] However, the guanine nucleotide dissociation stimulator (GDS), which was originally purified from bovine brain and shown to be a positive regulator of Ras subfamily proteins Rap1A and Rap1B,[2] is a relatively small protein containing 558 residues, which has no significant sequence homology to other known GEFs. It does display a weak homology to the *Drosophila* Armadillo protein,[3] which is homologous to the single-domain repeat region of β-catenin. Another distinction between GDS and other exchange factors is that activity has been detected *in vitro* against substrates from different small GTPases subfamilies.

Specificity of GDS

Using filter-binding assays, GDS has been shown to stimulate the exchange of GDP bound to Rap1A, Rap1B, K-Ras, RhoA, and Rac1.[2,4] These early studies were performed with GTPases isolated from eukaryotic sources and indicated that C-terminal lipid modification of the substrate was required for functional interaction with GDS.[5] However, this has been shown not to be an absolute requirement, as GDS activity has been detected against the unprocessed forms of RhoA and, to a lesser extent, Cdc42 and Rac1.[6] More recently, the activity of GDS toward unmodified Rac was found to be dependent on the C-terminal region of the small GTPase, as GDS did not stimulate GDP dissociation from C-terminally truncated Rac, which lacks the last eight residues.[7] C-terminal proteolysis during expression in *Escherichia coli* has been observed with a number of small GTPases[8] and could prevent their functional interaction with GDS. Unmodified full-length GTPases are being used in current studies of the exchange reaction because they are expressed readily in large amounts in *E. coli,* and detergent, which may affect activity, is not required to maintain solubility.[9]

[1] L. A. Quilliam, R. Khosravi-Far, S. Y. Huff, and C. J. Der, *BioEssays* **17**, 395 (1995).
[2] T. Yamamoto, K. Kaibuchi, T. Mizuno, M. Hiroyoshi, H. Shirataki, and Y. Takai, *J. Biol. Chem.* **265**, 16626 (1990).
[3] M. Peifer, S. Berg, and A. B. Reynolds, *Cell* **76**, 789 (1994).
[4] K. Hiraoka, K. Kaibuchi, S. Ando, T. Musha, K. Takaishi, T. Mizuno, M. Asada, L. Ménard, E. Tomhave, J. Didsbury, R. Snyderman, and Y. Takai, *Biochem. Biophys. Res. Commun.* **182**, 921 (1992).
[5] T. Mizuno, K. Kaibuchi, T. Yamamoto, M. Kawamura, T. Sakoda, H. Fujioka, Y. Matsuura, and Y. Takai, *Proc. Natl. Acad. Sci. U.S.A.* **88**, 6442 (1991).
[6] T-H. Chuang, X. Xu, L. A. Quilliam, and G. M. Bokoch, *Biochem. J.* **303**, 761 (1994).
[7] J. P. Hutchinson, unpublished observations (1999).
[8] A. J. Self and A. Hall, *Methods Enzymol.* **256**, 3 (1995).
[9] E. Porfiri and J. F. Hancock, *Methods Enzymol.* **256**, 85 (1995).

Expression and Purification of GDS

GDS was originally isolated from cytosolic extracts of bovine brain.[2] An *E. coli* expression system was constructed from cDNA cloned from a bovine brain cDNA library, and overexpressed GDS was purified by anion-exchange chromatography.[10] More recently, other *E. coli* expression systems for GDS have been described.[6,9] For ease of purification, we expressed GDS as a fusion with glutathione *S*-transferase in *E. coli* using the pGEX-4T1 expression vector.

Construction of Vector

DNA encoding full-length GDS is amplified by polymerase chain reaction (PCR) using the original GDS clone (kindly provided by Professor Y. Takai, Osaka University Medical School, Japan) as template and is cloned into the *Bam*HI/*Xho*I restriction sites of pGEX-4T1 (Amersham Pharmacia).

Growth of Cells

The expression plasmid is transformed into *E. coli* strain BL21. Six liters of LB medium (100 μg ml^{-1} ampicillin) in 500-ml aliquots contained in 2-liter flasks is inoculated with an overnight culture of this transformant (10 ml per flask), and the flasks are shaken vigorously at 37°. Expression is induced at on optical density (OD$_{600}$) of ~0.8 by the addition of 0.5 mM isopropyl-β-D-thiogalactoside. Shaking is continued for 3 hr at 30°. Cells are harvested by centrifugation for 20 min at 4° (3000g).

Purification of Fusion Protein

The cell pellet is resuspended in 100 ml of cold lysis buffer [20 mM Tris–Cl, 300 mM NaCl, 1 mM EDTA, 5 mM dithiothreitol (DTT), pH 7.5] containing protease inhibitors (0.4 mM phenylmethylsulfonyl fluoride, 1 mM benzamidine). Cells are lysed by pulsed sonication (8 min) with ice cooling followed by centrifugation for 1 hr at 4° (125,000g). All subsequent steps are performed at 4°. The supernatant is loaded onto a 20-ml glutathione–Sepharose column (Amersham Pharmacia), preequilibrated with lysis buffer. The column is then washed with lysis buffer containing 500 mM NaCl until baseline absorption is reached. Because the efficiency of cleavage of the column-bound fusion protein with thrombin was found to be poor, the GDS–GST fusion is eluted with lysis buffer containing 10 mM glutathi-

[10] K. Kaibuchi, T. Mizuno, H. Fujioka, T. Yamamoto, K. Kishi, Y. Fukumoto, Y. Hori, and Y. Takai, *Mol. Cell Biol.* **11**, 2873 (1991).

one (pH 7.5) and cleaved in solution as described later. The GDS–GST fusion protein elutes in a volume of ~50 ml.

Cleavage of Fusion Protein

To remove excess glutathione, the fusion protein is dialyzed against 20 mM Tris–Cl, 1 M NaCl, 1 mM DTT (pH 7.5) for 3 hr and then against 20 mM Tris–Cl, 50 mM NaCl, 3 mM CaCl$_2$, 1 mM DTT (pH 7.5) overnight at 4°. The protein is treated with 500 units human thrombin (Calbiochem, La Jolla, CA) at 20° for 2 hr and then passed down a column of glutathione–Sepharose (20 ml) connected in series with a column of p-aminobenzamidine-agarose (2 ml) at 4°. The eluant is concentrated over a YM10 membrane (Amicon, Danvers, MA) in a nitrogen pressure cell to approximately 15 ml, and 3-ml portions are purified further on an S-200 gel-filtration column (60 × 2.6 cm) (Amersham Pharmacia) equilibrated with 20 mM Tris–Cl, 50 mM NaCl, 1mM DTT (pH 7.5). The purified protein is concentrated as described earlier to approximately 50 mg ml^{-1}, snap frozen in aliquots, and stored at −80°. Approximately 100 mg of GDS is obtained from 6 liters of cell culture.

Characterization of GDS

The purity of GDS obtained by this procedure is estimated to be in excess of 95% as judged by Coomassie blue-stained sodium dodecyl sulfate (SDS) polyacrylamide gels. Analytical gel filtration (Superdex 200 HR10/30), performed at 0.4 ml min^{-1} in 20 mM Tris–Cl, 50 mM NaCl, 1 mM EDTA, 1 mM DTT (pH 7.5) with detection of absorption at 280 nm, showed a single component and gave a retention time consistent with a monomeric species. Sedimentation velocity experiments gave $s_{20,w}$ = 4.0 S, and a molecular mass of 59.0 kDa was measured in sedimentation equilibrium experiments. These were performed at 60 μM GDS and three different rotor speeds, using a partial specific volume of 0.748 ml g^{-1} calculated from the amino acid composition. The concentration of GDS is determined by absorbance at 280 nm using an extinction coefficient of 13940 M^{-1} cm^{-1}, which is calculated from the amino acid composition.

Sequencing of GDS

The molecular mass of GDS purified by the method just described, measured by electrospray mass spectrometry, was 42 units less than that calculated from the published sequence.[10] Because the same mass difference was observed with GDS purified from the original clone, it is unlikely that mutations have been introduced during the PCR amplification just described previously. Sequencing of the amplified GDS gene revealed two discrepancies from the published sequence[10]: Cys-335 is present as Ala and Lys-146 as Asn. The molecular mass of GDS with these two substitutions is consistent with mass spectral data. The first of these substitutions has been reported previously in cDNA fragments amplified by PCR from a bovine brain cDNA library.[6] Bovine GDS (Genebank accession number 1707894) shows a high degree of identity with GDS sequences from two other species (95% to human, accession number 1707895; 86% to *Xenopus*, accession number 3702194). Because both of the substitutions identified in the bovine sequence are present in human and *Xenopus* proteins, it is likely that they represent errors in the original published bovine sequence.[10]

Isolation and Characterization of Nucleotide-Free Rho · GDS Complex

In common with other exchange factors, GDS forms a tight complex with small GTPases in the absence of nucleotides. These complexes are valuable in studies of the mechanism of stimulated nucleotide exchange. The nucleotide-free complex Rho · GDS is formed readily by incubating GDS (200 μM) with an excess of Rho · GDP (300 μM) in 20 mM Tris–Cl, 40 mM EDTA, 1 mM DTT (pH 7.5) for 2 hr at 4°. The mixture (~2.5 ml) is fractionated at 4° on an S-200 column (60 × 2.6 cm) equilibrated with 20 mM Tris–Cl, 50 mM NaCl, 1 mM EDTA, 1 mM DTT (pH 7.5) Figure 1 shows the separation of Rho · GDS from Rho and free nucleotide on this

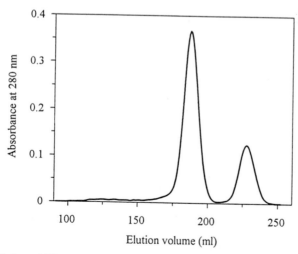

FIG. 1. Isolation of Rho · GDS binary complex. Rho · GDS was separated from Rho and free nucleotide by S-200 gel filtration as described in the text. Elution was monitored by absorbance at 280 nm.

column. Rho · GDS is concentrated to approximately 10 mg ml^{-1} using Centriplus 30 centrifugal concentrators, snap frozen in aliquots, and stored at −80°. Concentrations of Rho · GDS are determined by absorbance at 280 nm using an extinction coefficient of 31,720 M^{-1} cm^{-1}, which is calculated from the amino acid composition. Sedimentation velocity analysis gave $s°_{20,w}$ for Rho · GDS of 4.8 S. Sedimentation equilibrium analysis gave a molecular mass of 76.9 kDa for Rho · GDS, using a partial specific volume of 0.745 ml g^{-1} calculated from the amino acid composition, which confirms the 1:1 stoichiometry of the complex. The complex Rac · GDS can be prepared in a similar fashion from full-length Rac · GDP.

Measurement of GDS Activity

Most work on measuring the activity of guanine nucleotide exchange factors, including GDS, has made use of radioactive guanine nucleotides and filter-binding assays.[2,5,6] The nucleotide exchange activity can be measured by first incorporating the radioactive nucleotide at low magnesium concentration in the presence of EDTA, where it is incorporated rapidly, before increasing the magnesium concentration to above that of the EDTA. Excess nonradioactive nucleotide is then added and the loss of protein-bound radioactive nucleotide is followed by the filter-binding assay. Alternatively, the incorporation of a radioactive nucleotide, often [^{35}S]GTPγS, into the G-protein-containing endogenous GDP is followed in the same way:

$$\text{GTPase} \cdot [^3\text{H}]\text{GDP} + \text{GTP} \Rightarrow \text{GTPase} \cdot \text{GTP} + [^3\text{H}]\text{GDP}$$
$$\text{GTPase} \cdot \text{GDP} + [^{35}\text{S}]\text{GTP}\gamma\text{S} \Rightarrow \text{GTPase} \cdot [^{35}\text{S}]\text{GTP}\gamma\text{S} + \text{GDP}$$

In principle, both methods should give the same rate constants if the displacing nucleotide is in large excess over the GTPase, as in both cases the nucleotide exchange is governed by the dissociation rate constant of GDP from the protein. The effect of catalytic amounts of the exchange factor can then be investigated. In both of the methods described earlier, particularly the second, there will be a large excess of unbound radioactive nucleotide present throughout the displacement reaction, resulting in a high background of radioactivity. Therefore, it is preferable to work with a stoichiometric complex of the GTPase and [^{3}H]GDP and measure the dissociation of the [^{3}H]GDP from this. Under these conditions there is no excess radioactive nucleotide so the background radioactivity is low, and the complex is in well-defined salt conditions.

Preparation of Rho · [³H]GDP

Rho · [^{3}H]GDP is prepared by incubating 0.5 mg Rho · GDP with [^{3}H]GDP (600 kBq, 437 GBq/mmol, Amersham) in 500 μl 20 mM Tris–Cl,

40 mM EDTA, 200 mM ammonium sulfate, 1 mM DTT (pH 7.5) at 20°
for 10 min. After quenching the exchange process by the addition of MgCl$_2$
to a final concentration of 50 mM, the complex is separated from free
nucleotide using a FastDesalt HR10/10 column (Amersham Pharmacia)
equilibrated with 20 mM Tris–Cl, 2 mM MgCl$_2$, 1 mM DTT (pH 7.5)
at 4°, collecting 0.5-ml fractions. Elution is monitored by absorption and
radioactivity, as shown in Fig. 2. The concentration of Rho · [³H]GDP is
determined by absorption at 280 nm using an extinction coefficient of
26,430 M^{-1} cm^{-1}, which is calculated from the amino acid composition
and a stoichiometrically bound equivalent of GDP.

GDS Activity by Filter-Binding Assay

Rho · [³H]GDP (1 μM) is incubated at 30° in assay buffer (20 mM
Tris–Cl, 2 mM MgCl$_2$, 1 mM DTT, pH 7.5). The exchange reaction is
initiated by the addition of GDS and a large excess of GDP (100 μM).
Fifty-microliter aliquots of the reaction mixture are diluted into 1 ml ice-
cold assay buffer and filtered over nitrocellulose membranes (Millipore
Bedford, MA; type HA), which have been soaked previously for at least
10 min in the same buffer. The membrane is washed three times with 2-
ml portions of cold assay buffer and dried thoroughly by placing under
a 300-W lamp. Protein-bound [³H]GDP on the filter is determined by
scintillation counting in 5 ml of 0.5% (w/v) 2-(4-t-butylphenyl)-5-(4-biphe-

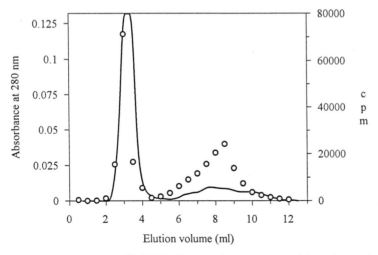

FIG. 2. Preparation of Rho · [³H]GDP. The complex was separated from free nucleotide
by gel filtration on a FastDesalt column as described in the text. Elution was monitored by
absorbance at 280 nm (—) and scintillation counting (○).

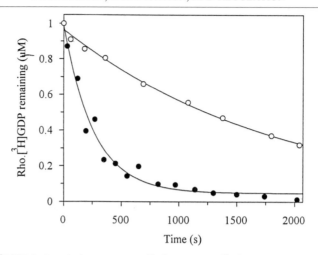

FIG. 3. [^{3}H]GDP dissociation from Rho · [^{3}H]GDP. Rho · [^{3}H]GDP (1 μM) was incubated with GDP (100 μM) in the absence (○) and presence (●) of GDS (200 nM) at 30°. Aliquots were removed and assayed for Rho · [^{3}H]GDP by the filter-binding assay as described in the text. Time courses have been fitted to single exponential functions.

nyl)-1,3,4-oxadiazole (butyl-PBD) in toluene. The measured counts per minute at each time point can be converted to Rho · [^{3}H]GDP concentration using the counts measured in an identical filtration performed before initiation of the exchange process, where Rho · [^{3}H]GDP is known to be 1 μM. Figure 3 shows time courses of [^{3}H]GDP dissociation from Rho · [^{3}H]GDP in the absence and presence of GDS (200 nM) at 30°. Under these reaction conditions, an initial rate of 0.003 μM sec^{-1} can be estimated for the exchange reaction catalyzed by 200 nM GDS.

Use of Fluorescent Nucleotide Analogs to Study Exchange Reaction

The filter-binding assays described earlier allow point-by-point measurements of the nucleotide release process. In some circumstances, it is advantageous to have a continuous assay for nucleotide dissociation, for which ribose-modified fluorescent analogs can be used. The N-methylanthraniloyl derivative of GDP, 2′ (3′)-O-methylanthraniloyl-GDP (mant-GDP), is suitable for such purposes.[11] A disadvantage of this analog is that it exists as an equilibrium mixture of the 2′-O and 3′-O derivatives, which interconvert within minutes at physiological pH and may have different kinetic properties. This can be overcome by the use of deoxy nucleotides. 2′-Deoxy-GDP

[11] D. M. Jameson and J. F. Eccleston, *Methods Enzymol.* **278**, 363 (1997).

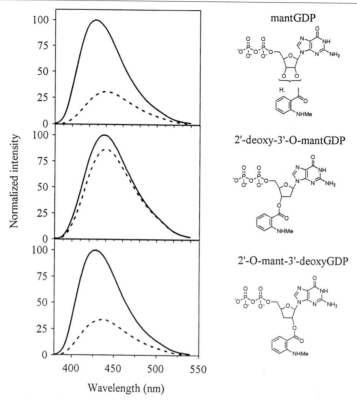

FIG. 4. Fluorescence emission spectra of Rho-bound (—) and free (---) mant nucleotides. After recording the spectrum of the Rho · mant nucleotide complex (2.5 μM), EDTA (10 mM) and GDP (250 μM) were added to displace mant nucleotide. Spectra were recorded on excitation at 366 nm. Excitation bandpass was 4 nm, and emission bandpass was 2 nm.

is available commercially and can be converted easily to 2′-deoxy-3′-O-mant-GDP.[11] 3′-Deoxy-GDP is not available commercially. However, 3′-deoxyguanosine can be converted to 3′-deoxy-GTP in a two-step reaction from which 2′-O-mant-3′-deoxy GTP can be synthesized.[12] Either intrinsic or GAP-activated GTPase activity can be used to convert this to 2′-O-mant-3′-deoxy-GDP. The rationale for the use of mant nucleotides for continuous measurements of nucleotide release is illustrated in Fig. 4, where the fluorescence emission spectra of mant-GDP, 2′-deoxy-3′-O-mant-GDP, and 2′-O-mant-3′-deoxy-GDP both free in solution and bound to RhoA are shown. For mant-GDP and 2′-O-mant-3′-deoxy-GDP there is a large

[12] E. Hamel, *J. Carbohydr. Nucleosides Nucleotides* **4**, 377 (1997).

difference between the fluorescence emission intensities of Rho-bound and free states of the nucleotide (3.5- and 3.0-fold, respectively, at 430 nm). However, 2′-deoxy-3′-O-mant-GDP is less suitable as an analog for following nucleotide dissociation from Rho, as there is only a small difference between the emission intensities of bound and free forms (1.2-fold at 430 nm).

Preparation of Rho · mant-GDP

Rho · mant-GDP is prepared by incubating 0.5 mg Rho · GDP with mant-GDP (50-fold molar excess) in 500 μl 20 mM Tris–Cl, 40 mM EDTA, 200 mM ammonium sulfate, 1 mM DTT (pH 7.5) at 20° for 10 min. After quenching the exchange process by the addition of MgCl$_2$ to a final concentration of 50 mM, the complex is separated from free nucleotide by gel filtration on a PD-10 column (Amersham Pharmacia) equilibrated in 20 mM Tris–Cl, 2 mM MgCl$_2$, 1 mM DTT (pH 7.5) at 4°. Separation on the column can be visualized with a hand-held long wavelength UV lamp (365 nm). The concentration of the GTPase · mant nucleotide complex is determined by absorbance at 354 nm using the extinction coefficient of the mant group of $\varepsilon_{354} = 5700 \ M^{-1} \ cm^{-1}$.[13]

Continuous Assay of mant-GDP Dissociation

A continuous assay of GDS activity is performed in a similar way to the filter-binding assay described previously. Rho · mant-GDP (1 μM) is incubated in the filter-binding assay buffer at 30° in a spectofluorimeter, and the intensity of fluorescence emission from the mant group at 430 nm is followed on excitation of the fluorophore at 366 nm. The exchange reaction is initiated by the addition of GDS and unlabeled GDP (100 μM), and the fluorescence is followed with time. The expected amplitude of the release reaction is known from the difference in intensity between Rho-bound and free mant-GDP (see Fig. 4), from which the concentration of Rho · mant-GDP remaining as the reaction proceeds can be calculated.

A further advantage of the use of fluorescence for monitoring nucleotide release is that measurements can be made on a much shorter time scale than with filter-binding methods if stopped-flow techniques are used. This allows measurements to be made with higher concentrations of GDS. An example is shown in Fig. 5, where Rho · mant-GDP is mixed with excess GDS and GDP. A possible artifact is that the mant fluorophore is susceptible to photobleaching. This effect should be investigated in control experiments on Rho · mant-GDP alone and taken into account in the analysis.

---

[13] T. Hiratsuka, *Biochim. Biophys. Acta* **742**, 496 (1983).

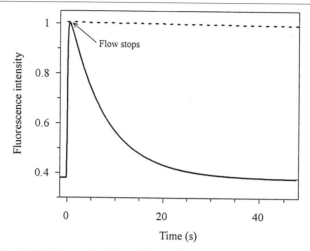

Fig. 5. Mant-GDP dissociation from Rho monitored by fluorescence. Rho · mant-GDP (0.5 μM) was mixed with GDS (10 μM) containing a large excess of GDP (100 μM) in a stopped-flow apparatus at 30°. Concentrations refer to those after mixing. Mant fluorescence intensity was followed on excitation at 366 nm. The decay in mant fluorescence was fitted to an exponential function to give an observed rate constant of 0.13 sec.$^{-1}$ The dashed line shows a control experiment where Rho · mant-GDP (0.5 μM) was mixed with a large excess of GDP (100 μM) in the absence of GDS.

Photobleaching can be reduced by lowering the intensity of the exciting light while retaining an adequate signal-to-noise ratio.

Interpretation of Activity Measurements

Most determinations of the activity of exchange factors are either interpreted as half-times of the release process (which can be converted to observed rate constants by the relationship $k_{obs} = 0.69/t_{1/2}$) or more simply by comparisons of the percentage of nucleotide released under certain conditions at a given time. It is difficult to compare various authors' work because the exchange reactions are often carried out with different concentrations of reagents. This problem arises because the exchange factor-catalyzed nucleotide release process is effectively a two-substrate enzymatic reaction. For a detailed description of the activity of an exchange factor, ideally the K_m and k_{cat} for both substrates (i.e., Rho · GDP/Rho · mant-GDP and GDP) need to be measured. Unless this is done, differences in the observed rate constant of nucleotide dissociation cannot be compared because K_m and k_{cat} effects cannot be separated. However, even detailed steady-state kinetic measurements can give conflicting interpretations as discussed in relation to studies on the eukaryotic protein biosynthetic exchange factor eIf2b.[14] An alternative approach is to make detailed mechanistic studies using presteady-state kinetic measurements,[15–17] but these are beyond the scope of this article.

[14] K. L. Manchester, *Biochem. Biophys. Res. Commun.* **239,** 223 (1997).
[15] C. Klebe, H. Prinz, A. Wittinghofer, and R. S. Goody *Biochemistry* **34,** 12543 (1995).
[16] C. Lenzen, R. H. Cool, H. Prinz, J. Kuhlmann, and A. Wittinghofer, *Biochemistry* **37,** 7420 (1998).
[17] J. P. Hutchinson and J. F. Eccleston, *Biophys. J.* **76,** A 429 (1999).

[8] Purification and Biochemical Activity of *Salmonella* Exchange Factor SopE

By Andrea Friebel and Wolf-Dietrich Hardt

Introduction

Salmonella typhimurium is the bacterial enteropathogen responsible for a large percentage of bacterial food poisoning. It employs a "type III secretion system," a syringe-like apparatus that allows *S. typhimurium* to inject bacterial virulence factors (effectors) directly into the cytosol of cells of the host's gut tissue. These injected effector proteins evoke cellular responses such as apoptosis in macrophages, increased chloride secretion, the production of interleukin 8 (IL-8), or ruffling of the host cell membrane, which eventually leads to bacterial entry into nonphagocytic host cells.[1]

SopE is one of the *S. typhimurium* effector proteins injected into host cells. SopE is a 240 amino acid protein that does not share any recognizable sequence similarity with any known protein of eukaryotic or prokaryotic origin. Transfection of a SopE expression vector or microinjection of purified recombinant SopE protein into Cos cells leads to profuse membrane ruffling and is sufficient to facilitate internalization of a noninvasive *S. typhimurium* mutant.[2] Using a ^{32}P-labeled recombinant version of SopE to screen a λgt11 expression library of HeLa cell cDNA Cdc42 and Rac1 were identified as cellular binding partners for SopE.[2]

Based on data derived from other effector proteins,[3] the N-terminal amino acids of SopE are thought to provide the signals necessary for recog-

[1] J. E. Galán, *Curr. Opin. Microbiol.* **2,** 46 (1999).
[2] W.-D. Hardt, L. M. Chen, K. E. Schuebel, X. R. Bustelo, and J. E. Galán, *Cell* **93,** 815 (1998).
[3] C. J. Hueck, *Microbiol. Mol. Biol. Rev.* **62,** 379 (1998).

TABLE I
CONSTANTS FOR SopE$_{78-240}$-MEDIATED GUANINE NUCLEOTIDE
EXCHANGE ON Cdc42[a]

Multiple turnover kinetics	Binding constants
$K_m = 4.5\ \mu M$	$k_{bind} = 6 \times 10^5\ M^{-1}\ sec^{-1}$
$k_{cat}{}^b = 0.95\ sec^{-1}$	$k_{dissociation}{}^c = 1.4 \times 10^{-4}\ sec^{-1}$
$k_{cat}/K_m = 2 \times 10^5\ M^{-1}\ sec^{-1}$	$K_D = k_{dissociation}/k_{bind} = 0.2\ nM^c$
	$k_{dissociation(20\ \mu M\ GDP)} = 0.6\ sec^{-1}$

[a] Data were taken from Rudolph *et al.*[4]
[b] Under conditions of excess SopE$_{78-240}$, the catalytic rate[4] was
≥2.5 sec^{-1}.
[c] Due to the experimental drift, which may have interfered with
the exact determination of $k_{dissociation}$, these values should be
viewed as a rough estimate.

nition and injection into host cells by the type III secretion system. Deletion analysis has demonstrated that the N-terminal 77 amino acids of SopE are dispensible for the interaction with Rho GTPases and the induction of membrane ruffling or nuclear signaling, whereas deletion of the N-terminal 115 amino acids disrupts SopE function.[2] Detailed biochemical analysis has demonstrated that SopE is a highly efficient guanine nucleotide exchange factor for Cdc42, accelerating exchange rates almost 10^5-fold[4] (Table I). The catalytic parameters for this interaction are strikingly similar to those determined for the interaction of eukaryotic guanine nucleotide exchange factors with their cognate GTP-binding proteins[5,6] (compare Table I). Therefore, recombinant SopE may provide a useful tool in analyzing various aspects of Rho GTPase activation. For this reason, this article provides a detailed description of the preparation and of several biochemical assays to analyze the interaction of SopE with Rho GTPases.

Overproduction and Purification of Recombinant SopE Proteins

In order to analyze the structure and the biochemical function of SopE in more detail, we have established a system for the overproduction and purification of the catalytic domain comprising amino acids 78–240 of the protein.

[4] M. G. Rudolph, C. Weise, S. Mirold, B. Hillenbrand, B. Bader, A. Wittinghofer, and W.-D. Hardt, *J. Biol. Chem.* **274**, 30501 (1999).
[5] C. Lenzen, R. H. Cool, H. Prinz, J. Kuhlmann, and A. Wittinghofer, *Biochemistry* **37**, 7420 (1998).
[6] C. Klebe, H. Prinz, A. Wittinghofer, and R. S. Goody, *Biochemistry* **34**, 12543 (1995).

A 1-ml starter culture of *Escherichia coli* XL1-blue (pSB1188[2]; Fig. 1) in 2× TY (50 μg/ml ampicillin) is used to inoculate 50 ml of 2× TY (50 μg/ml ampicillin), which is shaken vigorously overnight at 37°. One liter of 2× TY without antibiotics is inoculated with 40 ml of the overnight culture and shaken at 37° for 3 hr. When the culture reaches a density of 0.6 A_{600}, the temperature is shifted to 26° and 0.25 mM isopropylthiogalactoside (IPTG) is added for induction. The culture is grown under vigorous aeration for another 3 hr at 26°, and the bacteria are harvested by centrifugation (4°, 6000g, 10 min). The bacterial pellets are snap-frozen in liquid nitrogen and stored below −80°.

Bacteria are thawed on ice and resuspended in 30 ml of buffer a [100 mM NaCl, 1 mM MgCl$_2$, 2 mM dithiothreitol (DTT), 50 mM Tris–HCl, pH 7.6] supplemented with 1 mM phenylmethylsulfonyl fluoride (PMSF) and lysed at 4° in a French pressure cell. The lysate is supplemented with 1 mM ATP, and the cellular debris is removed by two rounds of centrifugation (4°, 10,000g, 15 min) and membrane filtration using a 50-ml syringe and 25-mm filter holders (0.45-μm pore size). To recover the glutathione S-transferase (GST)-fusion protein, the sterile lysate is incubated for 1 hr on a roller mixer at 4° with 700 μl bed volumes of glutathione-Sepharose 4B (Pharmacia, Piscataway, NJ) equilibrated with cold buffer a. The beads are recovered by centrifugation (4°, 80g, 5 min) and washed six times with 10 ml of cold buffer a.

To elute the GST–SopE$_{78-240}$ fusion protein, the beads are treated twice for 1 hr at 4° with 3 ml elution buffer [buffer a supplemented with 10 mM of freshly dissolved glutathione (Sigma, St. Louis, MO)] on a roller mixer for 1 hr at 4°. The eluates are cleared by centrifugation (4°, 100g, 5 min), pooled, passed through a membrane filter (0.2-μm pore size), and dialyzed (8000 molecular weight cutoff) three times against 2 liters of cold buffer b (100 mM NaCl, 1 mM MgCl$_2$, 2 mM DTT, 20 mM Tris–HCl, pH 7.6). Finally, the protein is concentrated by ultrafiltration using a Millipore Ultrafree centrifugation cartridge with a Biomax-5K membrane to yield a final concentration of 1 to 10 mg/ml GST–SopE$_{78-240}$. The protein can be kept at 4° for several weeks. Alternatively, it can be snap-frozen in small aliquots and stored at −80°.

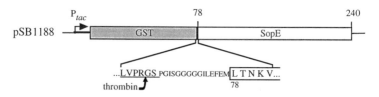

FIG. 1. Map of the recombinant *sopE* gene in pSB1188.

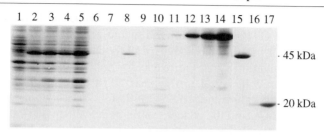

FIG. 2. Control gel for the purification of recombinant SopE$_{78-240}$ and GST–SopE$_{78-240}$ from *E. coli*. The samples were taken at different stages of the purification procedure, analyzed by 10% SDS–PAGE, and stained with Coomassie Brilliant Blue. Lane 1, 50 μl of the culture before addition of IPTG; lane 2, 50 μl of the IPTG-induced culture before centrifugation; lane 3, 2 μl of the resuspended bacteria; lane 4, 2 μl of the cleared lysate before the addition of glutathione beads; lane 5, 6 μl of lysate after GST-fusion proteins had been recovered; lane 6, 10 μl of the first wash; lane 7, 10 μl of the sixth wash; lane 8, 10 μl of the protein eluted with glutathione; lane 9, 10 μl of the protein cleaved off the beads by thrombin; lane 10, molecular weight marker (Benchmark, GIBCO Grand Island, NY); lanes 11–14, 1, 5, 10, and 20 μg bovine serum albumin (BSA); lane 15, 5 μl GST–SopE$_{78-240}$ after concentration by membrane filtration; and lanes 16 and 17, 1 and 5 μl SopE$_{78-240}$ after concentration by membrane filtration.

In order to prepare SopE$_{78-240}$, the GST-fusion protein bound to gluta-thione-Sepharose 4B beads is subjected to thrombin digestion: 700 μl beads are incubated on a roller mixer in 1.5 ml of thrombin buffer (150 mM NaCl, 2.5 mM CaCl$_2$, 5 mM MgCl$_2$, 2mM DTT, 50 mM Tris–HCl, pH 7.6) supplemented with 25 units of thrombin (Pharmacia) at 4° overnight. The supernatant is recovered after centrifugation (4°, 100g, 5 min), and the beads are washed for a second time with thrombin buffer. The supernatants are pooled, treated twice for 30 min with 30 μl benzamidine-Sepharose beads (Pharmacia) to remove the protease, cleared from remaining bead material by membrane filtration (0.2-μm pore size), dialyzed, concentrated to 1–10 mg/ml SopE$_{78-240}$ as judged by the intensity of the Coomassie stain on SDS–polyacrylamide gels (see Fig. 2), and stocked as described earlier for the GST–SopE$_{78-240}$ fusion protein. The yield is generally 2–4 mg SopE$_{78-240}$ per liter of bacterial culture.

Preparation of Rho GTPase Proteins

Cdc42, Cdc42$_{\Delta C}$, and Rac1 are overexpressed and purified as GST-fusion proteins as described.[4,7] The proteins are cleaved off the column by digestion with thrombin protease (Cdc42) or with factor Xa (Rac1). After

[7] A. J. Self and A. Hall, *Methods Enzymol.* **256**, 3 (1995).

proteases have been removed using 20 μl/ml benzamidine-Sepharose beads (Pharmacia), the GTPases are dialyzed in buffer c (50 mM NaCl, 2 mM DTT, 4 mM MgCl$_2$, 10 mM Tris–HCl, pH 7.6), concentrated in a Millipore Ultrafree centrifugation cartridge to yield a final concentration of 1 mg/ml, and snap-frozen in liquid nitrogen as 20-μl aliquots. The quality of the preparation is determined by SDS–PAGE and staining with Coomassie Brilliant Blue, and the concentration of active Rho GTPase is determined by loading with [^{3}H] GDP and analysis in a filter-binding assay[7] ([GTPase] = cpm/μl × 10^{-6} × counter efficiency). Due to the design of the expression vectors, Rac1 carries 7 additional N-terminal amino acids (GIDPGAT) and Cdc42 carries 14 additional N-terminal amino acids (GSRRASVGSKIISA). Cdc42$_{\Delta C}$ (residues 1–178) carries two additional N-terminal residues (GS) and the 13 C-terminal amino acids (PPEPKKSRRCVLL) are absent.

Filter-Binding Assay for SopE-Mediated Guanine Nucleotide Exchange

In the presence of native SopE (or SopE$_{78-240}$), guanine nucleotide exchange rates of [^{3}H]GDP-loaded Cdc42 are enhanced greatly (Table I), whereas guanine exchange rates of [^{3}H]GDP-loaded HaRas remain unaltered. Therefore, SopE-mediated guanine nucleotide exchange can be used to estimate the specific activity of SopE preparations in simple filter-binding assays. Two micrograms of Cdc42 or HaRas is loaded with [^{3}H]GDP by incubation for 10 min at 30° in buffer × (50 mM NaCl, 5 mM MgCl$_2$, 5 mM DTT, 50 mM Tris, pH 7.6) supplemented with 8 μCi [^{3}H]GDP (10 Ci/mmol, 1 mCi/ml, Amersham) and 10 mM EDTA in a total volume of 100 μl. Loading is stopped by cooling on ice and adding 2.1 μl MgCl$_2$ (0.5 M) and 300 μl of ice-cold buffer ×.

To start the guanine nucleotide exchange reactions, 80 μl of the preloaded Cdc42 or HaRas is added to each of four microfuge tubes (prewarmed to 30°) containing 420 μl of buffer × supplemented with 1 mM unlabeled GDP and (1) 10 mM EDTA, (2) 10 μg SopE$_{78-240}$, (3) 10 μg of GST, or (4) 0.5 μg SopE$_{78-240}$. Aliquots of 80 μl are withdrawn at the indicated times (see Fig. 3), added to 800 μl of cold wash buffer (50 mM NaCl, 5 mM MgCl$_2$, 50 mM Tris, pH 7.6), applied to a prewetted nitrocellulose filter (Schleicher & Schuell, Keane, NH BA85, 2.4-cm diameter placed on a Millipore filtration manifold), and washed three times with 3 ml of cold wash buffer. Filters are dried and transferred into 4-ml scintillation tubes, and after the addition of 3 ml of scintillation cocktail (Optiphase super mix, PerkinElmer/Wallac, Gaithersburg, MD), radioactivity bound to the filters is measured in a scintillation counter (1450 Micro beta Trilux,

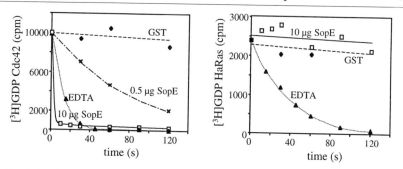

FIG. 3. SopE$_{78-240}$-catalyzed guanine nucleotide exchange of Cdc42. Cdc42 (left) or HaRas (right) was loaded with [^{3}H]GDP, and rates of exchange against a large excess of unlabeled GDP were analyzed in a filter-binding assay. Guanine nucleotide exchange rates were determined in the presence of (1) 10 mM EDTA (▲; dotted line); (2) 10 μg SopE$_{78-240}$ (□, solid line); (3) 10 μg GST (◆, strippled line); or (4) 0.5 μg SopE$_{78-240}$ (x, dotted/strippled line).

Wallac). Data ($t_{1/2}$ = 60 s) obtained with 0.5 μg SopE (Fig. 3) are typical for preparations of SopE$_{78-240}$ purified as described earlier. Less active preparations yield higher values for $t_{1/2}$. Signal changes during the formation of a GST–SopE$_{78-240}$ · Cdc42$_{\Delta C}$ complex in surface plasmon resonance measurements indicate that the preparations may contain a substantial fraction (20–50%) of inactive SopE$_{78-240}$.[4]

Affinity Purification of Rac1 on GST–SopE$_{78-240}$ Column

We wanted to rule out an enzymatic mechanism involving covalent modification of the RhoGTPases. For this purpose, we prepared a Rac1 fraction, 100% of which had been specifically bound to SopE. This was achieved using an affinity purification assay and taking advantage of the high-affinity binding of SopE to Rac1 in the absence of guanine nucleotides and the much weaker binding in the presence of excess GDP.[2] For immobilization, 500 μg of GST–SopE$_{78-240}$ in 2 ml of buffer d (10 mM Tris–HCl, pH 7.6, 150 mM NaCl, 5 mM MgCl$_2$, 1 mM DTT) is circulated for 30 min over a glutathione-Sepharose 4B column (200-μl bed volume). Unbound material is removed by washing for 10 min (0.5 ml/min) with buffer d. Five hundred micrograms of Rac1 (100 μM in buffer d; 0.2 ml/min; Fig. 4b, lane 1) is loaded onto this column. Unbound Rac1 is removed by washing for 10 min (0.5 ml/min; Fig. 4b, lane 2) with buffer d. Finally, bound Rac1 is eluted with 5 ml buffer d (1 mM GDP; 0.5 ml/min; Fig. 4b, lane 3; total yield = 250 μg, i.e., 50% recovery). Unbound GDP is removed by dialysis against 3 × 1 l cold buffer c, and the eluted Rac1 (Rac1$_{el.}$) is concentrated

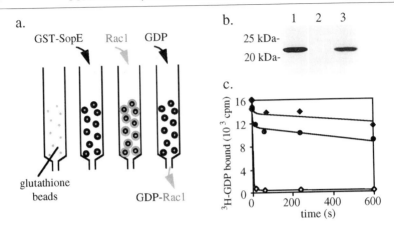

FIG. 4. Affinity purification of Rac1 on a SopE column. (a) Scheme of the affinity purification procedure. (b) 10% SDS–PAGE analysis of the key fractions obtained during the affinity purification procedure. Protein bands were stained using Coomassie blue. Lane 1, 2-μl aliquot of the Rac1 solution before loading onto the GST–SopE$_{78-240}$ column; lane 2, last 100 μl of wash buffer before elution of Rac1 was started; and lane 3, 100-μl aliquot of Rac1$_{el.}$ obtained by elution with 5 ml of buffer d (1 mM GDP). (c) Intrinsic exchange rates of Rac1$_{el.}$ remain unchanged. Rac1$_{el.}$ ($\bullet$) or Rac1 ($\blacklozenge$) was loaded with [^{3}H]GDP, and the rate of [^{3}H]GDP release was measured in buffer $\times$ supplemented with 1 mM GDP. $\Diamond$, [^{3}H]GDP release from Rac1$_{el.}$ in buffer $\times$ supplemented with 1 mM GDP and 1 μM SopE$_{78-240}$. Aliquots were withdrawn as indicated, and the amount of [^{3}H]GDP that had remained bound to Rac1 was determined in a filter-binding assay (© ASBMB, adopted with permission from Rudolph et al.[4]).

in a Millipore Ultrafree centrifugation cartridge and snap-frozen in liquid nitrogen.

Analysis of the intrinsic rate of [^{3}H]GDP release of Rac1$_{el.}$ in a filter-binding assay demonstrates that the rates are indistinguishable from those observed for untreated Rac1[4] (Fig. 4c, $\blacklozenge$ and $\bullet$). When 1 μM SopE$_{78-240}$ is added to the assay, all [^{3}H]GDP is displaced in less than 10 sec, demonstrating that the Rac1$_{el.}$ preparation is still active and that it is capable of interacting efficiently with SopE for a second time (Fig. 4c, $\Diamond$). Therefore, in contrast to the other known bacterial toxins modulating RhoGTPase function,[8] SopE does not act as a modifying enzyme by introducing some type of covalent modification that might increase the intrinsic rate guanine nucleotide exchange. Instead, SopE acts by transient interaction, much like eukaryotic guanine nucleotide exchange factors.

[8] K. Aktories, *Trends Microbiol.* **5**, 282 (1997).

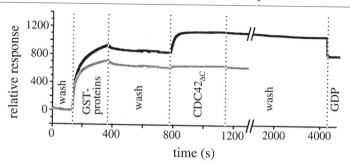

time (s)

Fig. 5. Surface plasmon resonance analysis of the Cdc42$_{\Delta C}$/GST–SopE$_{78-240}$ interaction. α-GST antibodies were covalently attached to the surface of a BIAcore sensor chip. GST–SopE$_{78-240}$ (black line) or GST (gray line) (0.5 μM) was bound to this affinity matrix (sec 150–350). In the presence of 50 nM Cdc42$_{\Delta C}$ (sec 770–1150) the resonance signal increased as indicated. Dissociation of the complex was followed for more than 3200 sec. Then, 20 μM GDP was added to the wash buffer and dissociation of the complex was followed for an additional 500 sec. Changes in plasmon resonance are shown as relative resonance units (© ASBMB, adopted with permission from Rudolph *et al.*[4]).

Surface Plasmon Resonance Assay to Study SopE · Cdc42 Complex

To study the kinetics of formation and dissociation of the complex of SopE with Rho GTPases in more detail, we used the BIAcore system (BIAcore AB, Uppsala, Sweden). This system allows one to study binding kinetics in real time by measuring the change in mass on the surface of a sensor chip. α-GST antibodies are covalently attached to the surface of a BIAcore sensor chip using an amino group coupling procedure[9]; 0.5 μM GST–SopE$_{78-240}$ in buffer e [10 mM HEPES/NaOH, pH 7.4, 150 mM NaCl, 5 mM MgCl$_2$, 0.005% Nonidet P-40 (NP-40)] is bound to this affinity matrix (Fig. 5, black line, sec 150–350; flow rate 5 μl/min; 20°). After washing with buffer e (Fig. 5, sec 351–769), 50 nM Cdc42$_{\Delta C}$ in buffer e is applied to the chip (sec 770–1150; flow rate 5 μl/min, 20°). The increase of the resonance signal is attributable to the formation of the GST–SopE$_{78-240}$ · Cdc42$_{\Delta C}$ complex. Dissociation of the complex in the absence of GDP is very slow (Fig. 5, sec 1151–4200). Due to experimental drift, it is impossible to determine the exact value of the dissociation rate constant, and $k_{\text{dissociation}}$ of 1.4×10^{-4} sec^{-1} (Table I) should only be viewed as a rough estimate. In the presence of 20 μM GDP, the complex dissociates more than 1000 times faster (Fig. 5, after sec 4201, Table I). This effect is well known for eukaryotic guanine nucleotide exchange factors.[5] At the end of the assay, all noncovalently bound material is removed from the matrix by washing

[9] D. J. O'Shannessy, M. Brigham-Burke, and K. Peck, *Anal. Biochem.* **205**, 132 (1992).

with 10 mM glycine hydrochloride, pH 2.0, and 0.05% SDS, leaving the immobilized immunoglobulin at full activity.

To eliminate the contribution of nonspecific binding, equivalent control experiments using immobilized GST instead of GST–SopE$_{78-240}$ (Fig. 5, gray line) are used to determine specific signal changes. Specific binding and dissociation curves are obtained by subtracting GST-controls from the GST–SopE$_{78-240}$ curves to take into account unspecific binding of Cdc42 to GST. Parameters are fitted using a single exponential with BIAevaluation software (version 2.1).

The observed rates for formation of the GST–SopE$_{78-240}$ complex with Cdc42$_{\Delta C}$ are dependent on the concentration of Cdc42$_{\Delta C}$ applied (data not shown). This has been used to determine the association rate constant for the formation of the GST–SopE$_{78-240}$ · Cdc42$_{\Delta C}$ complex[4] (Table I).

Specificity of SopE

Results from multiple and single turnover kinetic measurements, as well as analysis of the kinetics of formation and dissociation of the complex between SopE and Cdc42, have demonstrated that SopE is a highly efficient guanine nucleotide exchange factor.[4] The kinetic and thermodynamic parameters of SopE-mediated guanine nucleotide exchange on Cdc42 are strikingly similar to the parameters that have been determined for the catalytic domains of guanine nucleotide exchange factors of eukaryotic origin.[4-6] Results from filter-binding assays indicate that Rac1 is recognized by SopE with similar efficiency. Under conditions of excess enzyme (1 μM SopE$_{78-240}$, 0.2 μM [^{3}H]GDP-Rac1), the guanine exchange was completed within 10 sec.[4] It has also been reported that SopE is able to accelerate guanine nucleotide exchange rates of Rac2, RhoA, and RhoG, whereas GTPases of other Ras subfamilies were not affected by SopE.[2] However, the catalytic performance was not analyzed in any of these cases. In addition, several Rho GTPases, such as RhoA, are known for their strong tendency to adopt inactive conformations. Further studies are in progress to analyze the determinants required for efficient guanine nucleotide exchange by SopE.

Further Assay Systems to Analyze SopE Function

A variety of assay systems have been established to analyze SopE function and the potential role of the *Salmonella* protein in pathogenesis. These assays are only mentioned briefly and the reader is referred to the original publications. The biological function of SopE has first been analyzed using *Salmonella* strains carrying a mutation in the *sopE* gene. These bacterial strains had a reduced capacity to induce rearrangements of the host cell actin cytoskeleton and were slightly less invasive than wild-type bacteria.[10,11] Microinjection of purified GST–SopE protein and transfection with SopE expression vectors were sufficient to induce profuse membrane ruffling and to activate JNK signaling cascades in Cosl cells.[2,12] Further biophysical methods for the detailed analysis of the interaction of SopE with Rho GTPases have also been established, including fluorescence measurements using fluorescent guanine nucleotide analogs to follow SopE-mediated guanine nucleotide exchange on Cdc42. This method has been described in great detail in another publication.[4]

Acknowledgments

We thank M. Rudolph and B. Bader for help with the biophysical assays and J. E. Galán and J. Heesemann for scientific support. We thank M. Hensel, K. Ruckdeschel, and M. Aepfelbacher for critical review of the manuscript. Financial support from the DFG and the BMBF is acknowledged.

[10] M. W. Wood, R. Rosqvist, P. B. Mullan, M. H. Edwards, and E. E. Galyov, *Mol. Microbiol.* **22,** 327 (1996).
[11] W.-D. Hardt, H. Urlaub, and J. E. Galán, *Proc. Natl. Acad. Sci. USA* **95,** 2574 (1998).
[12] L. M. Chen, S. Bagrodia, R. A. Cerione, and J. E. Galán, *J. Exp. Med.* **189,** 1479 (1999).

[9] Stimulation of Rho GDI Release by ERM Proteins

By Akiko Mammoto, Kazuo Takahashi, Takuya Sasaki, and Yoshimi Takai

Introduction

The Rho small GTPase family is implicated in various cell functions, such as cell shape change, cell motility,and cytokinesis, through reorganization of the actin cytoskeleton.[1,2] The Rho family consists of the Rho subfamily (RhoA, RhoB, RhoC), the Rac subfamily (Racl, Rac2, Rac3), the Cdc42 subfamily (Cdc42Hs, G25K) and RhoD, RhoE, and RhoG. Rho GDP dissociation inhibitor (GDI) is a general regulator of these Rho family members that forms a complex with their GDP-bound inactive form and inhibits their activation.[1] In addition, Rho GDI induces dissociation of the

[1] Y. Takai, T. Sasaki, K. Tanaka, and H. Nakanishi, *Trends Biochem. Sci.* **20,** 272 (1995).
[2] A. Hall, *Science* **279,** 509 (1998).

GDP-bound form of Rho family members from membranes and inhibits their association with membranes,[3-5] suggesting that Rho GDI regulates the cycling of the Rho proteins between the membranes and the cytosol. The GDP-bound form of Rho family members is activated by Rho guanine nucleotide exchange factors (GEFs), such as Db1, Rom1/2, and Frabin, all of which share a common sequence motif, the Dbl homology domain.[6-8] The GDP-bound form complexed with Rho GDI is hardly activated by these Rho GEFs in a cell-free system,[1,4,7-9] suggesting the presence of another factor that initiates the activation of Rho family members by reducing Rho GDI activity.

We have found that the Rho family members regulate the ezrin/radixin/moesin (ERM)-CD44 system,[10,11] which has been implicated in reorganization of the actin cytoskeleton.[12-15] The ERM family consists of three closely related proteins: ezrin, radixin, and moesin.[12-15]

Ezrin was first identified as a constituent of microvilli,[16] radixin as a barbed, end-capping actin-modulating protein from isolated junctional fractions,[17] and moesin as a heparin-binding protein.[18] ERM molecules consist of three domains: a globular domain in the N-terminal half, followed by the extended helical domain and the charged C-terminal domain.[12] Functionally, the N-terminal domain interacts with the plasma membrane and the C-terminal domain interacts with actin filaments.[12-15] ERM are translocated to the plasma membrane, probably through interaction with the

[3] M. Isomura, A. Kikuchi, N. Ohga, and Y. Takai, *Oncogene* **6,** 119 (1991).
[4] D. Leonard, M. J. Hart, J. V. Platko, E. Alessandra, W. Henzel, T. Evans, and R. A. Cerione, *J. Biol. Chem.* **267,** 22860 (1992).
[5] T. Sasaki, M. Kato, and Y. Takai, *J. Biol. Chem.* **268,** 23959 (1993).
[6] R. A. Cerione and Y. Zheng, *Curr. Opin. Cell Biol.* **8,** 216 (1996).
[7] K. Ozaki, K. Tanaka, H. Imamura, T. Hihara, T. Kameyama, H. Nonaka, H. Hirano, Y. Matsuura, and Y. Takai, *EMBO J.* **15,** 2196 (1996).
[8] M. Umikawa, H. Obaishi, H. Nakanishi, K. Satoh-Horikawa, K. Takahashi, I. Hotta, Y. Matsuura, and Y. Takai, *J. Biol. Chem.* **274,** 25197 (1999).
[9] H. Yaku, T. Sasaki, and Y. Takai, *Biochem. Biophys. Res. Commun.* **198,** 811 (1994).
[10] K. Takaishi, T. Sasaki, T. Kameyama, Sa. Tsukita, Sh. Tsukita, and Y. Takai, *Oncogene* **11,** 39 (1995).
[11] M. Hirao, N. Sato, T. Kondo, S. Yonemura, M. Monden, T. Sasaki, Y. Takai, Sh. Tsukita, and Sa. Tsukita, *J. Cell Biol.* **135,** 37 (1996).
[12] M. Arpin, M. Algrain, and D. Louvard, *Curr. Opin. Cell Biol.* **6,** 136 (1994).
[13] Sa. Tsukita, S. Yonemura, and Sh. Tsukita, *Curr. Opin. Cell Biol.* **9,** 70 (1997).
[14] Sa. Tsukita, S. Yonemura, and Sh. Tsukita, *Trends Biochem. Sci.* **22,** 53 (1997).
[15] Sa. Tsukita and S. Yonemura, *J. Biol. Chem.* **274,** 34507 (1999).
[16] A. Bretscer, *J. Cell Biol.* **97,** 425 (1983).
[17] Sa. Tsukita, Y. Hieda, and Sh. Tsukita, *J. Cell Biol.* **108,** 2369 (1989).
[18] W. Lankes, A. Giesmacher, J. Grunwald, R. Schwartz-Albiez, and R. Keller, *Biochem. J.* **251,** 831 (1988).

cytoplasmic domain of integral plasma membrane proteins, such as CD44, providing the actin filament association sites.[13-15] When cells are treated with agonists, the GDP-bound form of RhoA, which is in the cytosol in a complex with Rho GDI, is activated to the GTP-bound form and translocated to the same areas of the plasma membrane as those to which ERM are translocated.[10] We have found that Rho GDI and CD44 are coimmunoprecipitated with moesin from baby hamster kidney cell lysates and that Rho subfamily members stimulate the interaction of ERM with the plasma membrane.[11] Moreover, we have shown that Rho GDI interacts directly with ERM, initiating the activation of Rho subfamily members by reducing Rho GDI activity.[19] These results suggest that ERM, as well as Rho GDI and Rho GEFs, are involved in the activation of Rho family members, which then regulate reorganization of the actin cytoskeleton through the ERM system.

This article first describes the procedures for the purification of recombinant full-length and truncated mutants of radixin, the isolation of the RhoA–Rho GDI complex, and the cell-free assay for radixin activity toward Rho GDI. It then describes the procedures for the analysis of radixin-induced activation of Rho subfamily members in intact cells.

There are at least three isoforms of Rho GDI: Rho GDIα, -β, and -γ. Rho GDIα, originally isolated from the bovine brain,[20] is expressed ubiquitously,[21] whereas Rho GDIβ, also called Ly/D4-GDI, is expressed abundantly in hematopoietic cells.[22,23] Rho GDIγ is expressed predominantly in brain, lung, kidney, testis, and pancreas.[24,25] This article describes Rho GDIα; Rho GDIα is simply written here as Rho GDI.

Materials

Dithiothreitol (DTT), Nonidet P-40, and Triton X-100 are from Nacalai Tesque (Kyoto, Japan). EDTA and 3-[(3-cholamidopropyl) dimethylam-

[19] K. Takahashi, T. Sasaki, A. Mammoto, K. Takaishi, T. Kameyama, Sa. Tsukita, Sh. Tsukita, and Y. Takai, *J. Biol. Chem.* **272,** 23371 (1997).
[20] T. Ueda, A. Kikuchi, N. Ohga, J. Yamamoto, and Y. Takai, *J. Biol. Chem.* **265,** 9373 (1990).
[21] Y. Fukumoto, K. Kaibuchi, Y. Hori, H. Fujioka, S. Araki, T. Ueda, A. Kikuchi, and Y. Takai, *Oncogene* **5,** 1321 (1990).
[22] J.-M. Lelias, C. N. Adra, G. M. Wulf, J.-C. Guillemot, M. Khagad, D. Caput, and B. Lim, *Proc. Natl. Acad. Sci. U.S.A.* **90,** 1479 (1993).
[23] P. Scherle, T. Behrens, and L. M. Staudt, *Proc. Natl. Acad. Sci. U.S.A.* **90,** 7568 (1993).
[24] G. Zalcman, V. Closson, J. Camonis, N. Honoré, M.-F. Rousseau-Merck, A. Tavitian, and B. Olofsson, *J. Biol. Chem.* **271,** 30366 (1996).
[25] C. N. Adra, D. Manor, J. L. Ko, S. Zhu, T. Horiuchi, L. V. Aelst, R. A. Cerione, and B. Lim, *Proc. Natl. Acad. Sci. U.S.A.* **94,** 4279 (1997).

monio]-1-propanesulfonic acid (CHAPS) are from Dojindo Laboratories (Kumamoto, Japan). Human thrombin (T3010) is from Sigma (St. Louis, MO). Reduced glutathione, (p-amidinophenyl)methanesulfonyl fluoride (APMSF), isopropyl-β-D-thiogalactopyranoside (IPTG), and formaldehyde are from Wako Pure Chemicals (Osaka, Japan). Guanosine 5'-(3-O-thio)triphosphate (GTPγS) is from Boehringer Mannheim (Indianapolis, IN). A glutathione S-transferase (GST) expression vector, pGEX-2T, glutathione-Sepharose 4B, protein A-Sepharose CL-4B, and [^{3}H]GDP (518 GBq/mmol) are from Amersham-Pharmacia Biotech Inc. (Milwaukee, WI). [^{35}S]GTPγS (40.7 TBq/mmol) is from Du Pont-New England Nuclear (Boston, MA). BA-85 nitrocellulose filters (pore size, 0.45 μm) are from Schleicher & Schuell (Dassel, Germany). All other chemicals are of reagent grade.

Lipid-modified RhoA is purified from the membrane fraction of *Spodoptera frugiperda* ovary cells (Sf9 cells), which are infected with baculovirus carrying the cDNA of human RhoA.[26] The purified sample is dissolved in a buffer containing 20 mM Tris–HCl, pH 7.5, 5 mM MgCl$_2$, 1 mM EDTA, 1 mM DTT, and 0.6% CHAPS. Bovine Rho GDI is expressed and purified as a GST-fusion protein from *Escherichia coli,* and the GST carrier is cleaved off by digestion with thrombin as described.[27]

Plasmids for expression of a series of truncated mouse radixin fusion proteins are constructed as follows. The cDNA fragment, containing Nr-fragment (amino acids 1–280), Nr1-fragment (amino acids 1–318), Cr-fragment (amino acids 281–584), or full-length radixin (amino acids 1–584), with the *Bam*HI sites upstream of the initiator methionine codon and downstream of the termination codon is synthesized by polymerase chain reaction (PCR). These fragments are digested with *Bam*HI and inserted into the *Bam*HI-cut pGEX-2T. *Escherichia coli* strain DH5α is transformed with these plasmids, and the resulting strains are used as sources for GST-Nr-, Nr1-, Cr-fragment, and full-length radixin.

Mammalian expression plasmids (pSRα neo-myc and pEFBOS-HA) are generated to express fusion proteins with the N-terminal myc and HA epitopes, respectively. To generate pSRαneo-myc-Rho GDI and pEFBOS-HA-Nr-fragment, their cDNA constructs are made by the PCR using specific oligonucleotide primers and inserted into pSRαneo-myc or pEFBOS-HA.

---

[26] T. Mizuno, K. Kaibuchi, T. Yamamoto, M. Kawamura, T. Sakoda, H. Fujioka, Y. Matsuura, and Y. Takai, *Proc. Natl. Acad. Sci. U.S.A.* **88,** 6422 (1991).
[27] K. Tanaka, T. Sasaki, and Y. Takai, *Methods Enzymol.* **256,** 41 (1995).

Methods

*Purification of GST-Nr-, Nrl-, Cr-Fragment, and
Full-Length Radixin*

Buffers used in the purification of GST-tagged full-length and truncated mutants of radixin are as follows:

Buffer A: 25 mM Tris–HCl, pH 7.5, 0.5 mM EDTA, 1 mM DTT, 10% sucrose, and 10 μM APMSF

Buffer B: 25 mM Tris–HCl, pH 7.5, 0.5 mM EDTA, and 1 mM DTT

Buffer C: 50 mM Tris–HCl, pH 8.0, and 5 mM reduced glutathione

Recombinant radixin is purified by the following steps: (1) cultivation of *E. coli* and induction of GST–radixin, (2) preparation of crude supernatant, (3) affinity purification of GST–radixin, and (4) cleavage of GST–radixin with thrombin.

Cultivation of E. coli and Induction of GST–Radixin

DH5α *E. coli* transformed with pGEX-2T-radixin are cultured at 37° in 500 ml of LB medium containing 50 μg of ampicillin/ml to an OD_{595} of 0.4. After addition of IPTG at a final concentration of 0.1 mM, cells are cultured further at 30° for 4 hr. All procedures after this step should be performed at 0°–4°. Cells are harvested, suspended in 20 ml of phosphate-buffered saline (PBS), and washed with 20 ml of PBS. The cell pellet can be frozen at −80°.

Preparation of Crude Supernatant

The cell pellet is thawed quickly at 37° and suspended in 20 ml of ice-cold buffer A, and the cell suspension is sonicated at a setting of 60 by the Ultrasonic Processor (Taitec, Tokyo, Japan) on ice for 30 sec four times at 30-sec intervals. The homogenate is centrifuged at 100,000g at 4° for 1 hr. The supernatant is used for the affinity purification.

Affinity Purification of GST–Radixin

Glutathione-Sepharose 4B beads are packed into a 5-ml disposable syringe (bed volume, 1 ml). The beads are washed with 15 ml of buffer B and equilibrated with 15 ml of buffer B. Twenty ml of the crude supernatant prepared as described earlier is applied to the column, and the pass fraction is reapplied to the column. After the column is washed with 20 ml of buffer B, GST–radixin is eluted with 5 ml of buffer C.

Cleavage of GST–Radixin with Thrombin

The eluate containing GST-radixin (5 ml) is dialyzed against 500 ml of PBS or buffer B three times. This dialyzed sample in PBS is used for microinjection. To 5 ml of the dialyzed sample in buffer B, 5 μl of thrombin (5 units/μl) is added. The reaction mixture is incubated at 25° for 30 min and is then kept on ice. Another glutathione-Sepharose 4B column (bed volume, 1 ml) is prepared and equilibrated with 15 ml of buffer B, and the sample containing cleaved GST–radixin is applied to the column. The pass fraction in which GST is removed is reapplied to the same column. The pass fraction is stored at −80° until use for the cell-free assay.

Isolation of RhoA–Rho GDI Complex

RhoA (100 pmol) is incubated with or without 1 μM [^{3}H]GDP in a reaction mixture (1.2 ml) containing 20 mM Tris–HCl, pH 7.5, 5 mM MgCl$_2$, 10 mM EDTA, 1 mM DTT, and 0.25% CHAPS for 20 min at 30°. After incubation, 50 μl of 375 mM MgCl$_2$ is added to give a final concentration of 20 mM to prevent the dissociation of [^{3}H]GDP from RhoA, and the mixture is cooled immediately on ice. The samples of three same preparations are combined and concentrated to 100 μl with a Centricon-10 concentrator (Amicon, Denvers, MA). One-third of the concentrate containing [^{3}H]GDP–RhoA or GDP–RhoA (100 pmol) is then incubated with Rho GDI (100 pmol) in a reaction mixture (50 μl) containing 20 mM Tris–HCl, pH 7.5, 13 mM MgCl$_2$, 7 mM EDTA, 1 mM DTT, and 0.17% CHAPS for 30 min at 4°. The sample is then subjected to gel filtration using a Superdex 75 PC 3.2/30 column (Amersham Pharmacia Biotech Inc.) equilibrated with 20 mM Tris–HCl, pH 7.5 containing 5 mM MgCl$_2$, 1 mM EDTA, 1 mM DTT, and 0.1% CHAPS. [^{3}H]GDP–RhoA or GDP–RhoA complexed with Rho GDI is detected by protein staining.

Properties of Radixin

Assay for Activity of Radixin to Regulate GDP/GTP Exchange Reaction of RhoA Complexed with Rho GDI

The activity of radixin to regulate the GDP/GTP exchange reaction of RhoA complexed with Rho GDI is assayed by measuring either the dissociation of [^{3}H]GDP from [^{3}H]GDP–RhoA or the binding of [^{35}S] GTPγS to GDP–RhoA.

DISSOCIATION ASSAY. The [^{3}H]GDP–RhoA–Rho GDI complex (50 nM) is incubated for 20 min at 30° with various concentrations of Nr-fragment, Cr-fragment, or full-length radixin in a reaction mixture (50 μl) containing 100 μM GTP, 30 mM Tris–HCl, pH 7.5, 10 mM EDTA, 1 mM DTT, 5 mM MgCl$_2$, 0.12% CHAPS, and 0.2 mg/ml bovine serum albumin. The reaction is stopped by adding 2 ml of an ice-cold solution containing 20 mM Tris–HCl, pH 7.5, 25 mM MgCl$_2$, and 100 mM NaCl to the reaction mixture, followed by rapid filtration on BA-85 nitrocellulose filters and washing with 2 ml of the same solution four times. Radioactivity trapped on the filters is measured by liquid scintillation counting.

BINDING ASSAY. The GDP–RhoA–Rho GDI complex (50 nM) is incubated with various concentrations of Nr-fragment, Cr-fragment, or full-length radixin in a 50-μl reaction mixture containing 20 mM Tris–HCl, pH 7.5, 5 mM MgCl$_2$, 10 mM EDTA, 1 mM DTT, and 1 μM [^{35}S]GTPγS. The reaction is started by adding [^{35}S]GTPγS, and incubation is performed for 20 min at 30°. The reaction is stopped, and the radioactivity trapped on the filters is counted as described earlier.

In these two types of experiments, Nr-fragment reduces the Rho GDI activity in a dose-dependent manner, whereas neither Cr-fragment nor full-length radixin affects Rho GDI activity. The same inhibitory effect of Nr-fragment is observed when Racl or Cdc42 is used as a substrate for Rho GDI. The amino acid sequence of the N-terminal fragment is highly conserved within ERM (~85% identical for any pair). Consistently, the N-terminal fragments of ezrin (amino acids 1–280) and moesin (amino acids 1–280) show the same inhibitory effects on Rho GDI activity toward RhoA, Racl, and Cdc42.

Assay for Nr-Fragment-Induced Activation of Rho Subfamily Members in Intact Cells

To examine whether ERM regulate Rho GDI activity in intact cells, immunoprecipitation and immunofluorescence microscopy are done.

IMMUNOPRECIPITATION OF EXPRESSED PROTEINS. Transient expression of myc-Rho GDI with or without HA-Nr-fragment is carried out using pSRαneo-myc-Rho GDI and pEFBOS-HA or pEFBOS-HA-Nr-fragment in COS-7 cells. The cells are plated at a density of 5 × 10^5 cells/60-mm dish and are incubated for 18 hr. The cells are then cotransfected with 2 μg of pSRαneo-myc-Rho GDI and 2 μg of pEFBOS-HA or pEFBOS-HA-Nr-fragment using the DEAE-dextran method. Immunoprecipitation is performed at 48 hr after the transfection. The cells are washed with PBS twice, lysed in a lysis buffer (containing 20 mM Tris–HCl, pH 7.5, 5 mM MgCl$_2$, 1 mM EDTA, 1 mM DTT, and 10 μM APMSF), and sonicated. The cell lysate is centrifuged at 100,000g for 1 hr to prepare the cytosol fraction. Myc-Rho GDI is precipitated with 3 μg of an anti-Myc monoclonal

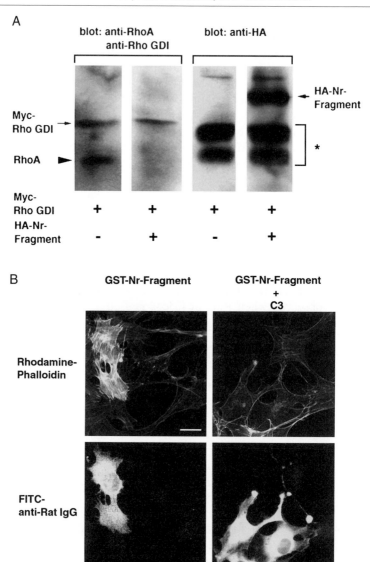

Fig. 1. Nr-fragment-induced activation of RhoA in intact cells. (A) Nr-fragment-induced dissociation of RhoA from Rho GDI and subsequent formation of Rho GDI-Nr-fragment complex in intact COS-7 cells. Myc-Rho GDI was expressed with or without HA-Nr-fragment in COS-7 cells and immunoprecipitated with an anti-Myc antibody. Immunoprecipitated Myc-Rho GDI, RhoA, and HA-Nr-fragment were detected with anti-Rho GDI polyclonal, anti-RhoA polyclonal, and anti-HA monoclonal antibodies, respectively. The results shown are representative of three independent experiments. The light chain of the anti-Myc antibody is shown by an asterisk. (B) Nr-fragment-induced formation of stress fibers in intact Swiss

antibody bound to 20 μl of protein A-Sepharose CL-4B, followed by centrifugation and extensive washing with the lysis buffer in the presence of 1% Nonidet P-40. Comparable amounts of the pellets are subjected to SDS–polyacrylamide gel electrophoresis, and the separated proteins are transferred electrophoretically to a nitrocellulose membrane sheet. The sheet is processed using the ECL detection kit (enhanced chemiluminescence, Amersham-Pharmacia Biotech Inc.) to detect Myc-Rho GDI, RhoA, and HA-Nr-fragment with the anti-Rho GDI polyclonal, anti-RhoA polyclonal, and anti-HA monoclonal antibodies as primary antibodies, respectively. In these experiments, endogenous RhoA is coimmunoprecipitated with Myc-Rho GDI from the lysate of the COS-7 cells transiently expressing Myc-Rho GDI alone (Fig. 1A). However, from the lysate of cells transiently expressing Myc-Rho GDI and HA-Nr-fragment, endogenous RhoA is not coimmunoprecipitated with Myc-Rho GDI, but HA-Nr-fragment is coimmunoprecipitated with Myc-Rho GDI, suggesting that the RhoA complexed with Rho GDI is replaced by HA-Nr-fragment to form the Rho GDI-HA-Nr-fragment complex in intact cells.

IMMUNOFLUORESCENCE MICROSCOPY. Swiss 3T3 cells are plated at a density of 1×10^5 cells/35-mm grid dish in 1 ml of Dulbecco's modified Eagle's medium (DMEM, ICN Inc., Costa Mesa, CA) containing 10% fetal bovine serum (GIBCO-BRL, Gaithersburg, MD) and are incubated for 3 days. The medium is then changed to a serum-free medium, and the cells are further incubated for 24 hr. After incubation, microinjection is performed. All proteins used are concentrated to 5 mg/ml with a Centricon-10 concentrator (Amicon, Inc.). Each sample to be tested is comicroinjected with 2.5 mg/ml rat immunoglobulin G (IgG) into cells as described.[19] GST-Nr-fragment is microinjected at 4 mg/ml, and its intracellular concentration is ~10 μM. C3 is microinjected at 40 μg/ml, and its intracellular concentration is ~0.23 μM. The cells are fixed at 30 min after the microinjection with 3.7% (v/v) formaldehyde in PBS for 10 min. The fixed cells are permeabilized with 0.2% (w/v) Triton X-100 in PBS for 10 min. After being soaked in 10% fetal bovine serum/PBS for 1 hr, the cells are treated for 1 hr with fluorescein isothiocyanate-conjugated goat anti-rat IgG (Chemicon International, Inc., Temecula, CA) and rhodamine-labeled phalloidin in 10% fetal bovine serum/PBS for detection of microinjected cells and actin

---

3T3 cells. GST-Nr-fragment or GST-Nr-fragment plus C3 was comicroinjected with rat IgG into Swiss 3T3 cells. The cells were stained with fluorescein isothiocyanate (FITC)-conjugated goat anti-rat IgG or rhodamine-phalloidin and analyzed by confocal microscopy. The results shown are representative of three independent experiments. Bar: 25 μm. Reproduced from Takahashi et al.[19]

filaments, respectively. After being washed with PBS three times, the cells are examined using an LSM 410 confocal laser-scanning microscope (Carl Zeiss, Oberkochen, Germany). In these experiments, serum-starved Swiss 3T3 cells have very few stress fibers, but microinjection of GST-Nr-fragment into these cells induces the formation of prominent stress fibers (Fig. 1B), but not the formation of lamellipodia and filopodia or membrane ruffling. This response is inhibited by comicroinjection with C3, which is known to ADP-ribosylate the Rho subfamily members and to inhibit their functions. Microinjection of GST does not show any effect. These results suggest that ERM indeed initiate the activation of Rho subfamily members through Rho GDI to induce the formation of stress fibers in intact cells.

Comments

At the free Mg^{2+} concentration of 0.5 μM used in the cell-free assay described here, the rate of GDP dissociation from RhoA is so fast that Rho GDI activity and the inhibitory effect of Nr-fragment on the Rho GDI activity can be detected easily. At the high Mg^{2+} concentration (mM order), Rho GEFs, such as Dbl and Rom1/2, stimulate the GDP/GTP exchange reaction of GDP–RhoA free of Rho GDI, but not that of GDP–RhoA complexed with Rho GDI. Nr-fragment also reduces this Rho GDI activity to inhibit the Rho GEF-stimulated GDP/GTP exchange reaction.

It has been reported that the C-terminal fragment of ezrin interacts with the N-terminal fragment longer than amino acids 1–296[28] and that the C-terminal fragment of radixin has been shown to interact with the N-terminal fragment containing amino acids 1–318.[29] Consistently, Cr-fragment reduces the activity of Nrl-fragment (amino acids 1–318) to inhibit Rho GDI activity, but not that of Nr-fragment (amino acids 1–280), although both Nrl- and Nr-fragment show the inhibitory effect on Rho GDI activity with a similar efficacy. It is now accepted that this intramolecular interaction between the N-terminal and the C-terminal regions of ERM suppresses their activity in intact cells. Some activation signal may release this suppression of ERM activity, resulting in their interactions with the plasma membrane and Rho GDI and with actin filaments at their N- and C-terminal regions, respectively, although the activation mechanism of ERM remains to be clarified.

Although Nr-fragment inhibits Rho GDI activity toward RhoA, Rac1, and Cdc42 in a cell-free assay, Nr-fragment initiates the activation of Rho, but not Rac or Cdc42 in intact cells. These results suggest that there is a

[28] R. Gary and A. Bretcher, *Mol. Biol. Cell* **6,** 1061 (1995).
[29] M. Magendantz, M. D. Henry, A. Lander, and F. Solomon, *J. Biol. Chem.* **270,** 25324 (1995).

mechanism by which ERM induce the selective activation of each Rho GDI substrate in intact cells. This mechanism is currently unknown, but one possible candidate involved in this mechanism is a Rho GEF specific for each Rho family member, such as Lbc, Tiam-1, and FGD1.

[10] Expression and Activity of Human Prenylcysteine-Directed Carboxyl Methyltransferase

By Edwin Choy and Mark Philips

Introduction

Prenylcysteine-directed carboxyl methyltransferase (pccMT) is the third of three enzymes that sequentially modify the C terminus of CAAX proteins such as Ras and the Rho family of Ras-related GTPases. Nascent CAAX proteins are localized in the cytosol where they encounter one of several prenyltransferases[1] that modify the CAAX cysteine with a farnesyl (C_{15}) or geranylgeranyl (C_{20}) lipid. Prenylated CAAX proteins are directed to the endomembrane where they encounter a prenyl-CAAX-specific protease, Rce1,[2,3] that cleaves the AAX amino acids. The newly C-terminal prenylcysteine residue then becomes a substrate for pccMT, also localized in the endomembrane,[4] that methylesterifies the α-carboxyl group of the prenylcysteine. The net effect of this sequence of posttranslational modifications is to create a hydrophobic domain at the C terminus of otherwise hydrophilic proteins and to thereby engage pathways for specific targeting of CAAX proteins to cellular membranes.[5] The precise nature of these pathways and how specificity for one membrane over another is accomplished remain unresolved.

Although prenylcysteine carboxyl methylation contributes significantly to the hydrophobicity of farnesylated proteins, this is not the case with geranylgeranylated proteins,[6] suggesting a more specific role for this revers-

[1] P. J. Casey and M. C. Seabra, *J. Biol. Chem.* **271,** 5289 (1996).
[2] V. L. Boyartchuk, M. N. Ashby, and J. Rine, *Science* **275,** 1796 (1997).
[3] W. K. Schmidt, A. Tam, K. Fujimura-Kamada, and S. Michaelis, *Proc. Natl. Acad. Sci. U.S.A.* **95,** 11175 (1998).
[4] Q. Dai, E. Choy, V. Chiu, J. Romano, S. Slivka, S. Steitz, S. Michaelis, and M. R. Philips, *J. Biol. Chem.* **273,** 15030 (1998).
[5] E. Choy, V. K. Chiu, J. Silletti, M. Feoktistov, T. Morimoto, D. Michaelson, I. E. Ivanov, and M. R. Philips, *Cell* **98,** 69 (1999).
[6] J. R. Silvius and F. l'Heureux, *Biochemistry* **33,** 3014 (1994).

ible modification in protein–protein interactions, perhaps analogous to protein phosphorylation. Elucidation of the precise biological role of prenylcysteine carboxyl methylation has been confounded by the observation that whereas yeast require ras homologs for growth, strains of yeast that lack pcCMT (*Saccharomyces cerevisiae* Ste14) are not defective in growth. Nevertheless, pcCMT has been highly conserved through evolution, suggesting an important, if not essential, biological role, and recent observations on Ras trafficking in mammalian cells suggest that pcCMT is required for the proper membrane targeting of Ras.[5]

Although pcCMT activities have been described in mammalian tissues and cells for a decade, until recently the only pcCMT characterized at the molecular level was Ste14p of *S. cerevisiae*. We have now cloned and expressed human pcCMT from an HL-60 cell cDNA library.[4] This article describes assays of pcCMT utilizing membranes derived from either myeloid cells or tissue culture cells that overexpress human pcCMT.

In Vitro Methyltransferase Assay

Human pcCMT is a polytopic integral membrane protein. The enzymatic activity of pcCMT is sensitive to detergent and therefore the protein cannot be extracted from membranes in active form. Consequently, preparations of membranes from cells relatively rich in endogenous pcCMT or enriched by the overexpression of recombinant pcCMT serve as the source of enzyme for methylation assays.

Membranes derived from brain tissue and myeloid cells have been found to possess high levels of endogenous pcCMT activity.[7,8] Neutrophils and HL-60 cells are convenient sources of membranes rich in pcCMT. Human neutrophils isolated from the blood of normal volunteers, when available, offer several advantages over cultured cells as a source of pcCMT-rich membranes. These include rapid isolation of large numbers of cells ($1–3 \times 10^8$ per 150 ml of whole blood) and reduced tissue culture costs. The disadvantages of using freshly isolated human neutrophils include the requirement for a human subjects protocol (expired blood bank donations are not suitable for the isolation of neutrophils because of their very short *ex vivo* shelf life), the precautions needed when working with human blood products, and the requirement for a relatively expensive nitrogen bomb (see later) for cell disruption to avoid lysis of protease-laden granules. If tissue culture is to be used, the human promyelocytic leukemia cell line

---

[7] M. H. Pillinger, C. Volker, J. B. Stock, G. Weissmann, and M. R. Philips, *J. Biol. Chem.* **269**, 1486 (1994).

[8] C. Volker, R. A. Miller, W. R. McCleary, A. Rao, M. Poenie, J. M. Backer, and J. B. Stock, *J. Biol. Chem.* **266**, 21515 (1991).

TABLE I
ACTIVITY OF ECTOPICALLY EXPRESSED HUMAN pcCMT

Source of membranes	Specific activity[a] (pmol/mg·min)	Fold increase
Neutrophils	3.8	(3)
COS-1 untransfected	1.1	1
COS-1 + vector	1.1	1
COS-1 + human pcCMT	49.6	45

[a] Assayed by AFC methylation as described in text.

HL-60 offers the advantage of relatively rapid growth to high density in suspension cultures, but like neutrophils, HL-60 cells require a nitrogen bomb for high-quality membrane preparations.

Although COS-1 cells express relatively low levels of endogenous pcCMT, they are the cell of choice for the preparation of membranes expressing recombinant pcCMT because of their ease of transfection and the high levels of expression that can be achieved with vectors that contain robust promoters and a simian virus 40 (SV40) origin of replication (e.g., pcDNA3.1). Although it is more difficult to harvest COS-1 cells in numbers approaching human neutrophils or even HL-60 cells, overexpression of recombinant pcCMT can yield membranes with a specific activity 10- to 20-fold that obtained from myeloid cells (Table I). Homogenization by physical shearing (e.g., Dounce) is adequate for preparing membranes from COS-1 cells because they lack the protease-rich granules of myeloid cells.

Human Neutrophils

Isolation of human neutrophils from whole blood of healthy volunteers has been described in detail elsewhere.[9] Although isolated neutrophils have a bench life measured in hours, pcCMT activity is well preserved for up to 1 year in membrane preparations stored at $-80°$. One hundred and fifty milliliters of blood from a single human donor can yield $1–2 \times 10^8$ cells and $100–300$ μg of membranes.

HL-60 Cells

HL-60 cells (obtained from ATCC, Rockville, MD) are maintained in Dulbecco's modified Eagle's medium (DMEM) supplemented with 10% fetal calf serum (FCS; Life Technologies) at $37°$ in 5% CO_2. Cells are split $1:20$ twice a week. Because HL-60 are nonadherent cells that settle on the

[9] M. R. Philips and M. H. Pillinger, *Methods Enzymol.* **256**, 49 (1995).

bottom of culture flasks, maximal yields of cells can be obtained with rotating or spinner flasks, although incubating T75 flasks on rockers achieves similar results. Under these conditions, HL-60 cells will grow, doubling approximately once every 24 hr, to a density of 5×10^4 cells/ml of medium. Thus, 500 ml of medium can yield up to 2.5×10^7 cells.

Preparation of Myeloid Cell Membranes by Nitrogen Cavitation

Because neutrophils and HL-60 cells contain numerous granules rich in potent proteases, nitrogen cavitation is the optimal method of cell disruption.[10] This method yields vesiculated light membranes (plasma membrane and endomembrane) without disrupting the granules.

Neutrophils or HL-60 cells to be fractionated are washed twice with cell buffer (150 mM NaCl, 5 mM KOH, 1.3 mM CaCl$_2$, 1.2 mM MgCl$_2$, 10 mM HEPES, pH 7.4) and are then resuspended in 18 ml of ice-cold relaxation buffer [100 mM KCl, 3 mM NaCl, 3.5 mM MgCl$_2$, 1 mM ATP, 2 mM phenylmethylsulfonyl fluoride (PMSF), 10 μg/ml leupeptin, 10 μg/ml chymostatin, 10 μg/ml pepstatin A, 27 μg/ml aprotinin, 10 mM HEPES, pH 7.3]. As described previously,[9] cell suspensions are stirred at 350 psi in a nitrogen bomb (Parr Instruments, Moline, IL) for 20 min at 4° and then cavitated by dropwise collection into a tube containing a volume of 20 mM EGTA calculated to yield a final concentration of 1.25 mM. The cavitate is centrifuged at 3000 rpm to remove nuclei and unbroken cells. The postnuclear supernatant is layered on enough ice-cold 40% sucrose ($\eta_D = 1.389$ at 20°, $\rho = 1.150$ g/cm^3) in 20 mM Tris–HCl, pH 7.4, containing 27 μg/ml aprotinin to fill a 12-ml Ultra-Clear (Beckman) polycarbonate ultracentrifuge tube. The tube is spun in an SW-41 rotor (Beckman) for 2 hr at 4° at 35,000 rpm. Cytosol is collected from the layer above the sucrose. This cytosol, dialyzed against 20 mM Tris–HCl, pH 7.4, containing 3μg/ml aprotinin and then concentrated by Centricon (Millipore, Bedford, MA), can be used as a source for p21s (see later). Light membranes (plasma membrane and endomembranes) are collected as an opalescent band just below the cytosol/sucrose interface. The granular pellet is discarded. The membrane fraction is suspended in several volumes of ice-cold 20 mM Tris–HCl, pH 7.4, containing 27 μg/ml aprotinin, freeze/thawed (ethanol in dry ice) in the same buffer five to seven times to release cytosol contaminating the reclosed membrane vesicles, pelleted at 4° at 150,000g, resuspended in 100–200 μl of the same buffer, and stored in aliquots at −80°. The membrane protein is assayed with the Bicinchoninic Acid (BCA) assay (Pierce, Rockford, IL) using 0.1% sodium dodecyl sulfate (SDS) in 150 mM NaCl as the diluent and bovine serum albumin (BSA) as the standard. Approxi-

[10] M. R. Philips, S. B. Abramson, S. L. Kolasinski, K. A. Haines, G. Weissmann, and M. G. Rosenfeld, *J. Biol. Chem.* **266**, 1289 (1991).

mately 150 μg of membrane proteins (enough for 30–50 methylation reactions) can be obtained from 10^8 cells.

Ectopic Expression of Human pcCMT in COS-1 Cells

COS-1 cells (obtained from ATCC) are grown in 5% CO_2 at 37° in DMEM containing 10% FCS and antibiotics. The day before transfection, cells must be split and seeded at 10^6 per 10-cm plate in 6 ml of DMEM containing antibiotics and 10% FCS. The cells should be grown to 30% confluence on the day of transfection, which will yield 5×10^6 cells per 10-cm plate at the time of harvesting.

The pcCMT cDNA cloned into a mammalian expression vector, e.g., pcDNA3.1,[4] can be transfected into COS-1 cells with high efficiency using SuperFect reagent (Qiagen, Valencia, CA). Although lipid-based transfection methods such as LipofectAMINE (Life Technologies, Rockville, MD) give transfection and expression efficiencies similar to SuperFect, the lipid can perturb the membrane and is best avoided when membrane preparations are desired. Dilute 10 μg of vector DNA in 300 μl of DMEM without serum and antibiotics and mix gently. Add 40 μl of SuperFect reagent to the DNA solution and mix by pipetting five times. Incubate the sample at room temperature for 10–15 min. While incubating, wash the cells gently with phosphate-buffered saline (PBS) prewarmed to 37°. Add 3 ml of DMEM (containing serum and antibiotics) to the transfection mixture. Mix by pipetting and immediately transfer onto the cells in a 10-cm dish. Incubate cells with the transfection mixture for 3 hr at 37° in 5% CO_2. Remove the medium with the transfection complex and wash three times with prewarmed PBS. Add fresh DMEM containing serum and antibiotics and place in incubator. Seventy-two hours after the initiation of transfection, cells should be 95% confluent. These cells can be harvested by scraping for membrane preparation.

Preparation of COS-1 Membranes by Homogenization

Scrape cells into 1 ml of ice-cold homogenizing buffer [HB: 10 mM Tris–HCl, pH 7.4, 10 mM KCl, 1 mM dithiothreitol (DTT), 2 mM PMSF, 27 μg/ml aprotinin, 10 μg/ml antipain, 10 μg/ml pepstatin, 10 μg/ml chymostatin]. Keeping cells cold on ice or working in a cold room at 4°, homogenize cells in a tight-fitting Dounce homogenizer with 30 strokes. Immediately transfer the homogenate to a clean tube containing 1/10 volume of 2.5 M sucrose to make the homogenate isotonic.

Centrifuge homogenate at 600g for 5 min at 4°. Collect and save the first postnuclear supernatant (PNS). Resuspend the pellet in 0.5 ml of HB, rehomogenize, adjust the tonicity with 1/10 volume 2.5 M sucrose, and

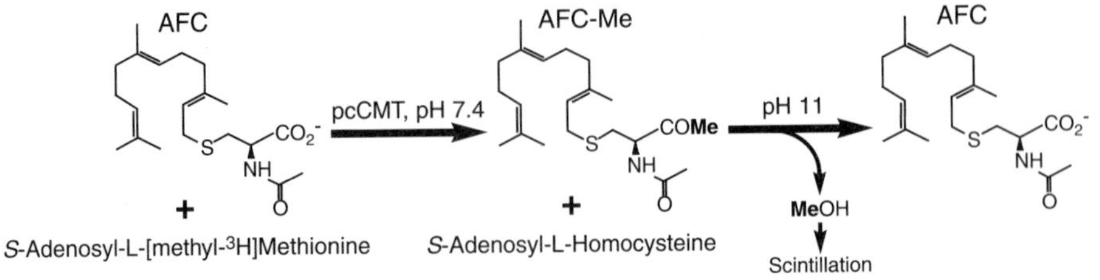

FIG. 1. Carboxyl methylation of N-acetyl-S-*trans,trans*-farnesylcysteine (AFC) by pcCMT. The two reactions shown form the basis of a rapid, quantitative, and reproducible assay for pcCMT activity. In the first reaction the methyl acceptor, AFC, and the methyl donor, S-adenosyl-L-[*methyl*-^{3}H]methionine ([^{3}H]AdoMet), are incubated with membranes containing pcCMT at pH 7.4, resulting in the formation of N-acetyl-S-*trans,trans*-farnesylcysteine methyl ester (AFC-Me) that can be extracted from unreacted [^{3}H]AdoMet in n-heptane. In the second reaction the pH is raised to 11, resulting in hydrolysis of AFC-Me and liberation of volatile [^{3}H]methanol that can be quantitated by partition into scintillation fluid. The radiolabeled methyl group is shown in boldface type.

centrifuge at 600g for 5 min at 4°. Collect the second PNS and combine it with the first.

Centrifuge the combined PNS at 8000g for 10 min at 4° to remove mitochondria. Collect the postmitochondrial supernatant (PMS), transfer it to clean tubes, and centrifuge at 150,000g for 90 min at 4°. Resuspend the pellet (crude membranes) in 100–500 μl 25 mM Tris–HCl, pH 7.4, containing 27 μg/ml aprotinin. Measure membrane protein concentration with the BCA assay (Pierce) using 0.1% SDS in 150 mM NaCl as diluent. Ten 10-cm plates of confluent COS-1 cells yield up to 3 mg of membrane protein, enough for up to 3000 methylation assays. Store membrane suspensions in aliquots at −80°.

In Vitro AFC Methylation Assay

Although a pcCMT assay has been described that utilizes as a methyl acceptor synthetic peptides that terminate in a farnesylcysteine and thereby mimic the C-terminal domains of partially processed CAAX proteins,[11] this assay requires custom synthesis and, unless the peptide is epitope tagged for affinity separation, relatively expensive S-adenosyl-L-[*methyl*-^{14}C]methionine as the methyl donor. A convenient alternative is a rapid, reproducible, and quantitative assay for pcCMT activity that utilizes small farnesylcysteine analogs such as N-acetyl-S-*trans,trans*-farnesyl-L-cysteine (AFC) as a substrate (methyl acceptor) (Fig. 1) that can be separated easily

[11] R. C. Stephenson and S. Clarke, *J. Biol. Chem.* **265**, 16248 (1990).

from the unreacted methyl donor.[12] This assay has been described in detail previously.[9] AFC from BioMol is prepared as a 100 mM stock solution in dimethyl sulfoxide (DMSO) and stored at $-20°$. Prepare a 1 mM working solution by diluting with distilled H_2O. Combine in a 1.5-ml Eppendorf tube: (a) membrane or fraction to be tested for pcCMT activity (e.g., 1 μg of COS-1 membranes) or a control without membrane; (b) 5 μl 1 mM AFC; (c) 3 μl adenosyl-L-S-[^{3}H]methionine ([^{3}H]AdoMet, 60 Ci/mmol, 0.55 mCi/ml, NEN Life Science Products); (d) 12.5 μl 4× TE buffer (4×: 200 mM Tris–HCl, pH 8.0, 4 mM NaEDTA); and (e) distilled H_2O to 50 μl. Incubate the reaction for 30 min at $37°$, tapping occasionally to mix. Terminate the reaction by adding an equal volume (50 μl) of 20% TCA and vortex for 10 sec.

To the terminated reaction add 400 μl of n-heptane and vortex for 10 sec. Centrifuge the reaction mixture at 16,000 rpm for 2 min. The pellet will contain proteins, the bottom aqueous layer will contain unreacted [^{3}H]AdoMet, and the top organic layer will contain [^{3}H]AFC-methyl ester. The organic n-heptane fraction should contain few radiolabeled species other than [^{3}H]AFC-methyl ester and therefore, in principle, could be counted directly without alkaline hydrolysis and quantitation of volatilized [^{3}H]methanol. However, the possibility of methylation of small molecules in the membrane preparation that might partition into n-heptane warrants the following added step, which makes the assay specific for carboxyl methylation.

Remove 300 μl of the organic (n-heptane) layer from the top of the tube, being careful not to allow mixing or entry of the pipette tip into the aqueous phase. Transfer the organic layer to a fresh 1.5-ml Eppendorf tube and dry by rotary evaporation. Add 100 μl 1 N NaOH, which will promote alkaline hydrolysis of the [^{3}H]AFC-methyl ester to produce volatile [^{3}H]methanol. Promptly transfer the uncapped Eppendorf tube into a scintillation vial containing approximately 7 ml of scintillation fluid (Ecoscint). The cap of the open Eppendorf tube can be used to keep the tube upright in the scintillation vial. Do not allow mixing of the Eppendorf contents with the scintillation fluid. Quickly close the scintillation vial and set aside for 24–72 hr in order to allow the volatilized [^{3}H]methanol to partition into the organic scintillation fluid. Preliminary readings can be obtained after 24 hr but $\geq$48 hr should be allowed for definitive results. While incubating, do not disturb the tubes in order to assure that the scintillation fluid does not splash into the Eppendorf tubes. Determine counts per minute (cpm) in the tritium channel of a scintillation counter.

The assay just described typically yields cpm levels 1000-fold higher

---

[12] C. Volker, R. A. Miller, and J. B. Stock, *Methods* **1**, 283 (1990).

than background (no enzyme control) and is therefore very sensitive. However, the concentration of [³H]AdoMet is far below its K_m and therefore the reaction rate is below maximal. Accordingly, for kinetic studies of pcCMT, the final concentration of [³H]AdoMet should be increased with unlabeled AdoMet (Sigma) to at least 15 μM. The cpm detected can be converted into specific activity (pmol AFC methylated/mg membrane protein·min) with the following formula:

Specific activity =

$$\frac{cpm[^3H]methanol/(cpm[^3H]Adomet/pmol\ AdoMet)}{(fraction\ heptane\ recovered = 0.75)(efficiency\ of\ methanol\ partition = 0.8)(mg\ membrane)(min\ reaction)}$$

Carboxyl Methylation of p21s

Endogenous CAAX proteins such as Rho GTPases (p21s) derived from cells can be used as substrates in an *in vitro* pcCMT assay instead of small molecule methyl acceptors such as AFC. Although such an assay is less quantitative than AFC methylation, the p21s methylation assay has the advantage of allowing direct visualization of methylated substrates in a polyacrylamide gel (Fig. 2A). Bacterially expressed recombinant CAAX proteins cannot serve directly as substrates unless a system is available for stoichiometric *in vitro* prenylation and AAX proteolysis. Endogenous Ras proteins cannot be utilized because they require sufficient detergent for extraction that renders pcCMT inactive.[7] We have found that cytosol derived from neutrophils, as described earlier, contains several Rho family GTPases that are prenylated but remain soluble by forming a complex with GDI and can serve as substrates for pcCMT.[4] Using such substrates, the reaction is accelerated considerably in the presence of a nonhydrolyzable GTP analog.[13] Unfractionated cytosol can be used as a source of p21s. Alternatively, partially purified Rho GTPase/GDI complexes derived from neutrophil cytosol can be utilized in this system (Fig. 2B).

Combine 1.5 μCi of [³H]AdoMet with a pcCMT-containing membrane (e.g., 1 μg of pcCMT-transfected COS-1 cell membranes) in 5 μl of 4× TE buffer containing 400 μM GTPγS and bring the total volume to 20 μl with distilled H$_2$O. Add the p21s-containing fraction (e.g., 100 μg of dialyzed, concentrated neutrophil cytosol) to the reaction mixture and incubate at 37° for 30 min. In some reactions, 100 μM AFC can be included in the methylation mixture as a competitive inhibitor to increase specificity (a

[13] M. R. Philips, M. H. Pillinger, R. Staud, C. Volker, M. G. Rosenfeld, G. Weissmann, and J. B. Stock, *Science* **259**, 977 (1993).

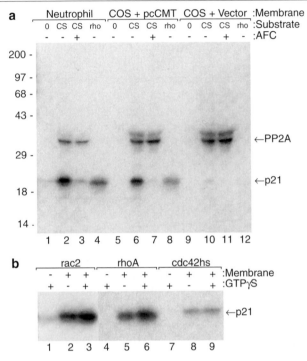

Fig. 2. Carboxyl methylation of Ras-related GTPases by COS-1 cell membranes expressing ectopic human pcCMT. (a) Neutrophil cytosol (CS) containing a mixture of Rho GTPases complexed with RhoGDI (lanes 2, 3, 6, 7, 10, and 11) or RhoA–GDI complexes partially purified from CS (lanes 4, 8, and 12) were incubated with the methyl donor S-adenosyl-L-[methyl-³H]methionine and light membranes derived from nitrogen cavitation of human neutrophils (lanes 1–4) or COS-1 cells transiently transfected with either human pcCMT (lanes 5–8) or vector alone (lanes 9–12). In some reactions (lanes 3, 7, and 11) a competitive pcCMT inhibitor, AFC, was included. Reaction products were analyzed by SDS–PAGE and fluorography. The position of carboxyl methylated p21s is indicated as is the position of protein phosphatase 2A (PP2A) that is methylated on a C-terminal leucine by a cytosolic enzyme unrelated to pcCMT. (b) Analysis as in (a) showing GTP sensitivity (±10 μM GTPγS) of carboxyl methylation of partially purified human neutrophil rac2, rhoA, and cdc42hs by COS-1 cell membranes expressing human pcCMT.

cytosolic protein carboxyl methyltransferase that methylates the C-terminal leucine in PP2a is AFC insensitive[4]). Performing the reaction in the presence or absence of GTPγS also increases specificity, as methylation of p21s is GTP sensitive. The reaction is terminated by adding 20 μl SDS sample buffer. The reaction products are analyzed by SDS–polyacrylamide gel electrophoresis and fluorography. The pH of the running buffer of the standard Laemmli gel system (pH 8.9) is sufficient to hydrolyze all aspartyl carboxyl methyl esters, leaving only protein carboxyl methyl esters on the α-carboxyl group of C-terminal residues such as prenylcysteine. Carboxyl methylation of radiolabeled proteins can be verified and quantitated by excision of the bands of interest from gels dried onto filter paper (1-mm strips can be excised with a razor). The excised bands are placed in an Eppendorf tube and rehydrated with 100 μl 1N NaOH to hydrolyze α-carboxyl methyl esters, and the volatilized [^{3}H]methanol that is produced is quantitated by partition into organic scintillation fluid as described earlier for the dried n-heptane residue containing [^{3}H]AFC-methyl ester.

Methylation of CAAX Proteins in Intact Cells

Assays of prenylcysteine carboxyl methylation on proteins in intact cells offer the advantage of analyzing the reaction under physiologic conditions in resting, stimulated, and genetically manipulated cells and, when combined with immunoprecipitation, add substrate specificity to the system. The disadvantages in relation to the in vitro assay include a relatively weak signal and the high cost of L-[methyl-^{3}H]methionine, required as a precursor of the methyl donor [^{3}H]AdoMet in metabolic labeling, as AdoMet cannot enter cells. Although we have detected prenylcysteine carboxyl methylation of endogenous p21s using this method,[13] we have found that overexpressing by transfection of CAAX substrates, particularly those with epitope tags that facilitate immunoprecipitation, gives quantitative and reproducible results (Fig. 3). The use of EGFP as the tag used for the immunoprecipitation of CAAX proteins affords the additional advantage of allowing one to directly visualize by epifluorescence the transfection efficiency and the level of expression. Epitope tags used more commonly for immunoprecipitation (e.g., HA, FLAG, and Myc) can also be used. In the case of Ras where an avid immunoprecipitating monoclonal antibody is available (Y13-259), an epitope tag is unnecessary. However, overexpression of Ras is still required for the intact cell methylation assay as endogenous levels are too low.

Construction of pEGFP CAAX Constructs

Using the cDNA of choice (wild-type or mutant Ras proteins, Ras-related GTPases, other CAAX proteins, or C-terminal fragments thereof), the coding sequence is PCR amplified with 5' EcoRI and 3' ApaI linkers designed into the primers (Table II). The amplification product is subcloned into the eukaryotic expression vector pEGFP-C3 (Clontech). The CAAX cysteine and/or palmitoylated cysteines can be mutated to serine, when applicable, by the design of the 3' PCR primer.

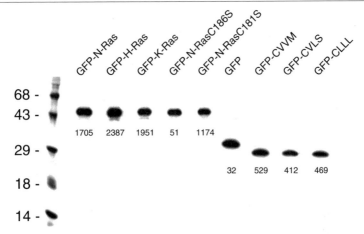

Fig. 3. Carboxyl methylation of ectopically expressed epitope-tagged CAAX constructs in metabolically labeled COS-1 cells. COS-1 cells were transfected with GFP alone or GFP extended at the C terminus with *ras* gene products or fragments thereof, as indicated. Twenty-four hours after transfection the cells were incubated with L-[*methyl-*^{3}H]methionine to label intracellular pools of AdoMet. The cells were lysed, and GFP-tagged proteins were immunoprecipitated and analyzed by SDS–PAGE and fluorography. The vast majority of the radioactivity that generates the bands on the flourogram is due to L-[*methyl-*^{3}H]methionine incorporated into the primary sequence of the proteins. To analyze carboxyl methylation the bands were excised from the gel and subjected to alkaline hydrolysis and quantitation of volatile [^{3}H]methanol, and the results (cpm) are given as numbers below each band.

DNA inserts of 33 bp or less used to produce GFP-CAAX constructs are generated by synthesizing complementary positive and negative oligonucleotides (with appropriate restriction site overhangs) that have a 5′ *Bgl*II, a 3′ stop codon, and a 3′ *Apa*I site in their sequence (Table III). These single-stranded oligonucleotides are purified by agarose gel electrophoresis. Each complementary strand (20 n*M*) is mixed in 1.5-ml Eppendorf tubes, and 20 μl of 1× ligation buffer (Life Technologies) is added and heated to 100° for 5 min. The tubes are placed on ice, and 1 U of T4 ligase (Life Technologies) and 1 μg of pEGFP-C3 vector linearized by double digestion with *Eco*RI and *Apa*I are added to the tube and incubated at 14° for 12 hr. The complementary strands anneal to form linear dsDNA and are subsequently ligated into the *Eco*RI and *Apa*I sites of pEGFP-C3.

Transfection and Metabolic Labeling

COS-1 cells are seeded into six-well plates at a density of 1.5 × 10^5 cells per well. After 24 hr, the cells are transfected with the various EGFP-CAAX fusion constructs using SuperFect (Qiagen) as described previously

TABLE II

PCR PRIMERS FOR AMPLIFICATION FROM Ras cDNAs OF INSERTS FOR pEGFP-C3 TO PRODUCE EGFP-TAGGED Ras CONSTRUCTS

Primer	Sequence
Forward Nras	5'-CTT CGA ATT CTG ACT GAG TAC AAA CTG GTG-3'
Reverse Nras	5'-TTA GGG CCC TTA CAT CAC ACA TGG CAA-3'
Forward Nras C186S	5'-TTA GGG CCC TTA CAT CAC AGA TGG CAA-3'
Forward Nras C181S	5'-GGA GGG CCC TTA CAT CAC ACA TGG CAA TCC CAT TGA ACC-3'
Forward Kras	5'-CTT CGA ATT CTG ACT GAA TAT AAA CTT GTG-3'
Reverse Kras	5'-GCG GGG CCC TTA CAT AAT TAC ACA CTT TGT-3'
Forward Hras	5'-CTT CGA ATT CTG ACC GAA TAC AAG CTT GTT-3'
Reverse Hras	5'-GCA GGG CCC TCA GGA GAG CAC ACA CTT GCA-3'

TABLE III
COMPLEMENTARY OLIGONUCLEOTIDES FOR CONSTRUCTION OF SHORT
DOUBLE-STRANDED DNA INSERTS[a]

Primer	Sequence
Forward CVVM	5'-GAT CTC TGT GTG GTG ATG TAA GGG CC-3'
Reverse CVVM	5'-C TTA CAT CAC CAC ACA GA-3'
Forward CVLS	5'-GAT CTC TGC GTT CTG TCT TAA GGG CC-3'
Reverse CVLS	5'-C TTA ACA CAG AAC GCA GA-3'
Forward CVIM	5'-GAT CTC TGT GTA ATT ATG TAA GGG CC-3'
Reverse CVIM	5'-C TTA CAT AAT TAC ACA GA-3'

[a] For generation of GFP extended at the C terminus with a CAAX motif using the 5'-*Bgl*II and 3'-*Apa*I sites of pEGFP-C3.

for expression of pcCMT except the volumes can be reduced to give a 0.5 ml final volume of transfection mixture. LipofectAMINE (Life Technologies) used according to the manufacturer's instructions gives results similar to those using SuperFect. Control cells should be transfected with EGFP alone to establish a background. Twenty-four hours after transfection, COS-1 cells are ready for metabolic labeling. Expression of EGFP-CAAX fusion constructs can be verified by direct visualization in an epifluorescence microscope with filters optimized for green fluorescent protein (GFP) or fluorescein isothiocyanate (FITC). Cells are washed twice with PBS (prewarmed to 37°) and methionine starved by incubating in 1 ml methionine-free DMEM containing 10% dialyzed FCS at 37° in 5% CO_2 for 30 min. Next, 200 μCi L-[*methyl-^{3}*H]methionine (NEN Life Science Products) is added directly to the medium and the cells are incubated for 30 min. The cells are then washed three times in cold PBS and can be lysed directly in the six-well plates with 1 ml RIPA buffer (20 mM Tris–HCl, pH 7.5, 150 mM NaCl, 1% NP-40, 0.1% SDS, 0.1% Na-deoxycholate (DOC), 0.5 mM EDTA, 10 μg/ml leupeptin, 1 mM PMSF, 27 μg/ml aprotinin, 1 mM DTT). The samples are now ready for immunoprecipitation but can be stored for several days at $-20°$.

Immunoprecipitation, SDS–PAGE, and Analysis of Carboxyl Methylation

Immediately prior to immunoprecipitation the RIPA lysates should be centrifuged at 15,000 rpm for 15 min to remove detergent-insoluble material. One microliter of rabbit polyclonal anti-GFP antiserum (Molecular Probes, Eugene, OR) is sufficient to immunoprecipitate GFP-tagged proteins from 1 ml of RIPA lysate. The antiserum is added and the mixture is rotated at 4° for 1 hr. Next, 20 μl of a 50% slurry (in PBS) of protein A-conjugated agarose beads (Life Technologies) is added to the lysate/ antiserum mixture and mixed at 4° for 1 hr. The agarose beads are then washed five times with PBS, pelleted by centrifugation at 15,000 rpm for 5 min, and eluted by boiling in 20 μl of SDS sample buffer. Eluates are loaded onto 10 or 14% Tris–glycine gels (e.g., Novex) and subjected to electrophoresis. Immunoprecipitated proteins labeled by L-[*methyl-* ^{3}H]methionine are identified by fluorography. Because the L-[*methyl-* ^{3}H]methionine that is not converted to [^{3}H]AdoMet will metabolically label all nascent proteins, most GFP-tagged CAAX proteins will be visible by fluorography whether they are carboxyl methylated or not. Therefore, the carboxyl methylation of labeled proteins must be determined by alkaline hydrolysis of bands excised from the dried gel as described earlier for *in vitro* methylation of p21s.

Densitometric analysis of the fluorogram gives an accurate determination of protein expression (D). The relative methylation of different EGFP-CAAX constructs (M) can be determined by normalizing cpm to protein expression. Accordingly, $M = (\mathrm{cpm}/D) \times [(6 + N)/6]$, where N is the number of methionine residues in the protein or peptide fused to the EGFP that contains six methionines.

[11] Purification and Evaluation of Large Clostridial Cytotoxins That Inhibit Small GTPases of Rho and Ras Subfamilies

By MICHAEL MOOS and CHRISTOPH VON EICHEL-STREIBER

Small GTP-binding proteins are key players in the regulation of signal-transducing networks of eukaryotic cells.[1] Their regulatory role renders GTPases ideal targets for bacterial toxins aiming to either kill or influence the behavior of the cells. In recent years a growing number of toxins modifying mainly GTPases of the Rho subfamily have been identified, e.g., CNF of *Escherichia coli,* DNT of *Bordetella dermonecroticum,* ExoS of *Pseudomonas aeruginosa,* C3 of *Clostridium botulinum,* and the group of large clostridial cytotoxins. Of these, CNF and DNT activate GTPase

---

[1] D. T. Denhardt, *Biochem. J.* **318,** 729 (1996).

TABLE I
SUBSTRATES AND COSUBSTRATES OF LCTs

| | Substrate | | | | | | | Co substrate | |
| | Rho subfamily | | | Ras subfamily | | | | | |
Toxin	Rho	Rac	Cdc-42	Ras	R-Ras	Rap	Ral	UDP-Glc	UDP-GlcNAc
TcdA-10463	+	+	+			+		+	
TcdB-10463	+	+	+					+	
TcdB-1470	+				+	+	+	+	
TcdB-8864	+				+	+	+	+	
TcsL-82	+			+	+	+	+	+	
TcsH-82	+	+	+					+	
Tcnα-19402	+	+	+			+			+

Rho,[2,3] whereas all other toxins inactivate different members of the Rho and Ras subfamilies of small GTP-binding proteins.[4]

This article deals with the purification and enzymatic activities of large clostridial cytotoxins (LCTs). LCTs are glycosyltransferases that inactivate GTPases of the Rho and Ras subfamilies by covalently coupling a sugar moiety (mostly glucose) to the conserved threonine residue in region switch 1 of the GTPases (T35 in Ras).[5,6] This glycosylation functionally inactivates the GTPases, leads to the collapse of the actin cytoskeleton, and ultimately induces apoptosis of the cells.[7-9] In contrast to exoenzyme C3, which acts solely on Rho, all LCTs inactivate a distinct subset of GTPases (Table I). Rac is the only substrate common to all LCTs. Whether this has structural reasons[10] or whether Rac is the best target to induce apoptosis is currently not known.[11]

[2] Y. Horiguchi, N. Inoue, M. Masuda, T. Kashimoto, J. Katahira, N. Sugimoto, and M. Matsuda, *Proc. Natl. Acad. Sci. U.S.A* **94**, 11623 (1997).
[3] G. Schmidt, P. Sehr, M. Wilm, J. Selzer, M. Mann, and K. Aktories, *Nature* **387**, 725 (1997).
[4] K. Aktories, *Trends Microbiol.* **5**, 282 (1997).
[5] I. Just, J. Selzer, M. Wilm, C. von Eichel Streiber, M. Mann, and K. Aktories, *Nature* **375**, 500 (1995).
[6] C. von Eichel Streiber, P. Boquet, M. Sauerborn, and M. Thelestam, *Trends Microbiol.* **4**, 375 (1996).
[7] Y. R. Mahida, S. Makh, S. Hyde, T. Gray, and S. P. Borriello, *Gut* **38**, 337 (1996).
[8] G. M. Calderon, J. Torres-Lopez, T. J. Lin, B. Chavez, M. Hernandez, O. Munoz, A. D. Befus, and J. A. Enciso, *Infect. Immun.* **66**, 2755 (1998).
[9] C. Fiorentini, A. Fabbri, L. Falzano, A. Fattorossi, P. Matarrese, R. Rivabene, and G. Donelli, *Infect. Immun.* **66**, 2660 (1998).
[10] S. Mueller, C. von Eichel-Streiber, and M. Moos, *Eur. J. Biochem.* **266**, 1073 (1999).
[11] T. Joneson and D. Bar-Sagi, *Mol. Cell. Biol.* **19**, 5892 (1999).

With one exception, all cell lines tested to date are susceptible to LCTs.[12] The toxins are taken up by eukaryotic cells through receptor-mediated endocytosis. The cellular receptors for the LCTs are assumed to be ubiquitously expressed glycolipids[13] or glycoproteins[14]; their molecular nature, however, remains to be defined. From the acidic endosomes the LCTs escape to the cytosol where they are enzymatically active.[15] Uptake via endosomes is, however, no prerequisite for the enzymatic activity of LCTs, as the toxins can also be microinjected and still retain their cytotoxic potential.[5]

On the basis of an earlier method for the isolation of LCTs,[16] this article presents an improved method for the purification of toxins TcdA-10463 and TcdB-10463 and describes the assays necessary to evaluate the cytotoxic potential of LCTs.

Materials

All media used for culturing of bacteria are from Difco Laboratories (Detroit, MI), and material for culturing of eukaryotic cells is from GIBCO-BRL (Gaithersburg, MD). All other material is from standard sources (Sigma-Aldrich, Steinheim, Germany; Roth, Karlsruhe, Germany) except if specially noted.

Purification of TcdA-10463 and TcdB-10463

Because of their unusual codon usage (60% AT content) and their size (250–308 kDa), LCTs are not expressed efficiently in *E. coli*. Therefore we purifiy the toxins from the supernatant of the respective *Clostridium difficile, C. sordellii,* or *C. novyi* strains. Purification of LCTs involves four essential steps: (i) growth of the anaerobic bacteria, (ii) fractionated precipitation of proteins from the supernatant, (iii) purification via a sucrose density gradient, and (iv) purification via ion-exchange chromatography. This article concentrates on purification of toxins TcdA-10463 and TcdB-10463; however, other LCTs are purified accordingly.

[12] M. Flores-Diaz, A. Alape-Giron, B. Persson, P. Pollesello, M. Moos, C. von Eichel-Streiber, M. Thelestam, and I. Florin, *J. Biol. Chem.* **272**, 23784 (1997).

[13] S. Teneberg, I. Lonnroth, J. F. Torres Lopez, U. Galili, M. O. Halvarsson, J. Angstrom, and K. A. Karlsson, *Glycobiology* **6**, 599 (1996).

[14] C. Pothoulakis, R. J. Gilbert, C. Cladaras, I. Castagliuolo, G. Semenza, Y. Hitti, J. S. Montcrief, J. Linevsky, C. P. Kelly, S. Nikulasson, H. P. Desai, T. D. Wilkins, and J. T. LaMont, *J. Clin. Invest.* **98**, 641 (1996).

[15] I. Florin and M. Thelestam, *Microb. Pathog.* **1**, 373 (1986).

[16] C. von Eichel-Streiber, U. Harperath, D. Bosse, and U. Hadding, *Microb. Pathog.* **2**, 307 (1987).

Growth of C. difficile VPI10463

One colony of freshly grown *C. difficile* is inoculated in 5 ml of 2× brain–heart infusion medium (BHI medium) and grown overnight at 37° in an anaerobic chamber. The next morning, 800 μl of this culture is trans-fered into a dialysis bag (molecular weight cutoff 12,000–14,000; Medicell, London, UK) containing 200 ml of sterile NaCl (0.9%, w/v). The dialysis bag is then placed in 1.3 liter of 2× BHI medium, and bacteria are incubated at 37° in an anaerobic chamber. Thus bacterial growth is restricted to the inside of the dialysis bag. After 3 days the content of the dialysis bag is centrifuged (25 min, 9000*g*, 4°) to separate bacteria and supernatant.

There are two main reasons for the inoculation of *C. difficile* in the dialysis bag: (a) due to their large size, all toxins produced will be retained inside the dialysis bag, thus raising the concentration of TcdA-10463 and TcdB-10463 in the supernatant. (b) The extracellular signals that induce toxin production are not fully understood. However, toxin production in *C. difficile* seems to be coupled to spore formation[17] and might be triggered by stress (unpublished data, 1997). This is in agreement with the fact that the toxin yield is increased if *C. difficile* cultures are allowed to reach the stationary phase as compared to continuous flow cultures (unpublished data, 1996). The increased toxin yield seems to be triggered by the space restrictions within the dialysis bag that add to the stress of the bacteria.

It is important to note that only *C. difficile* can be grown within a dialysis bag, as *C. sordellii* and *C. novyi* produce large amounts of gas during growth, which destroys the dialysis bag. For their cultivation the just-de-scribed system is reversed in a way that the bacteria are grown in 1–1.5 liter NaCl (0.9%, w/v) into which a dialysis bag containing 3.2× BHI medium is placed. At the end of the growth period (3 days, 37°, anaerobic chamber) the NaCl solution is collected and its volume is reduced by passage through a capillary dialyzer (Hemoflow F60S, Fresenius, Bad Homburg, Germany).

Fractionate Precipitation of Supernatant

An initial separation of the toxins from contaminating supernatant pro-teins is achieved by fractionate precipitation with $(NH_4)_2SO_4$. Therefore, $(NH_4)_2SO_4$ is added to the supernatant to a final content of 45% (w/v). The mixture is stirred at least for 3 hr at 4° and is subsequently centrifuged (30 min, 9000*g*, 4°) to collect the precipitated proteins. Optionally, a second precipitation with 65% (w/v) $(NH_4)_2SO_4$ can be performed. The precipitated proteins are resuspended in 8 ml of 50 m*M* Tris–HCl, pH 7.5. The 45%

[17] T. Hundsberger, V. Braun, M. Weidmann, P. Leukel, M. Sauerborn, and C. von Eichel-Streiber, *Eur. J. Biochem.* **244**, 735 (1997).

fraction should be enriched for TcdA-10463 and the 65% fraction for TcdB-10463.

Purification by Passage through a Sucrose Density Gradient

A further purification of the toxins from *C. difficile* is accomplished by passing the precipitated proteins through a sucrose density gradient where they are separated according to their specific density. A five-step sucrose gradient ranging from 10 to 45% sucrose in 50 mM Tris–HCl, pH 7.5, is freshly prepared inside an ultracentrifuge tube. Four milliliters of the 10% sucrose solution is placed at the bottom of the tube; the remaining solutions (4 ml each) are underlayed in increasing order of their sucrose content. Finally, about 4 ml of the resuspended *C. difficile* proteins is loaded on top, the sucrose gradient is then centrifuged for 3.5 hr at 100,000g and 4°. Subsequently, the gradient is separated into 4-ml fractions. Five to 10 μl of each fraction is run on a SDS–PAGE or, alternatively, a cytotoxicity assay (see later) may be performed to identify toxin-containing fractions.

Purification by Ion-Exchange Chromatography

For the final purification the toxin-containing fractions are passed through an ion-exchange column (Mono Q PC 1.6/5, Amersham-Pharmacia, Freiburg, Germany). Prior to loading, the fractions are diluted 1:2 with 50 mM Tris–HCl, pH 7.5, and then the column is washed with 50 mM Tris–HCl, pH 7.5, until extinction reaches basal values. A gradient ranging from 50 to 700 mM NaCl is run (ΔNaCl 5 mM/ml), and the eluate is collected in fractions of about 2 ml. These fractions are subsequently analyzed for the presence of toxin and contaminating proteins by SDS–PAGE (Fig. 1) and Western blot. Toxin TcdA-10463 is usually found in fractions 16 to 24 (150–270 mM NaCl) and TcdB-10463 in fractions 50 to 60 (530–650 mM NaCl, Fig. 1).

The toxin is stable in the presence of 20% glycerin at $-20°$ for several years. However, repeated cycles of freezing and thawing should be avoided. At 4° the toxins are stable for 2–4 weeks.

This article can offer only a brief introduction into the purification of TcdA-10463 and TcdB-10463. The purification of all other LCTs is similar; however, conditions have to be adjusted slightly. LCTs are currently unavailable, commercially; however, TGC biomics (http://www.tgc-biomics.com) is planning to offer some LCTs and corresponding antibodies in the future.

Even though this method allows purification of TcdA-10463 and TcdB-10463 from most other proteins of the culture supernatant, the purified toxins still contain traces of contaminating proteins (Fig. 1). A 39-kDa

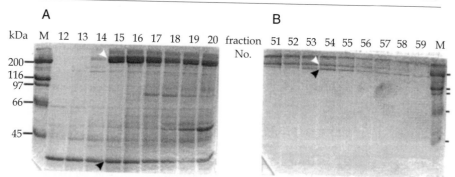

FIG. 1. SDS–PAGE of purified TcdA-10463 (A) and TcdB-10463 (B). The toxins were purified as described, and 10 μl of each fraction was separated by 10% SDS–PAGE. Fraction numbers from the ion-exchange chromatography are indicated above the gel. White arrowheads designate the toxins, and black arrowheads indicate the typical contaminants: 39-kDa protein for TcdA-10463 (A) and 150-kDa protein for TcdB-10463 (B).

protein is a characteristic contaminant of purified TcdA-10463, whereas a protein of 150 kDa is commonly copurified with TcdB-10463. As the protein content does not strictly correlate with the enzymatic activity of the toxins, we usually determine enzymatic and cytotoxic units of the purified toxins in order to measure the toxicity of the LCTs. These assays test the toxic potential of the LCTs *in vitro* and on eukaryotic cells.

Assaying Enzymatic Activities of LCTs on GTPases

LCTs specifically glycosylate GTPases in the cytosol of eukaryotic cells. The glycosyltransferase reaction can also be performed in an *in vitro* assay. Therefore the toxins are incubated in the presence of the appropriate GTPases and radioactively labeled cosubstrate. The glycosylation reaction transfers the radioactive sugar onto the GTPases, which are subsequently analyzed. This radioactive labeling of small GTPases has been used successfully to identify the catalytic domain of the LCTs[18] and for the analysis of structures involved in the interaction between GTPases and LCTs.[10]

Thirty micromolar UDP-[14C]glucose (specific activity 0.25 μCi) is used to perform the *in vitro* glucosylation reaction with TcdB-10463. Usually, UDP-[14C]glucose is supplied in an ethanol/water mix that might inhibit the enzymatic reaction, e.g., ICN Biomedicals (Eschwege, Germany) provides the radioactively labeled sugar in a volume of 0.5 ml (ethanol/water

---

[18] A. Wagenknecht-Wiesner, M. Weidmann, V. Braun, P. Leukel, M. Moos, and C. von Eichel-Streiber, *FEMS Microbiol. Lett.* **152**, 109 (1997).

2:8, specific activity 50 μCi). To adjust the concentration of the labeled sugar, we dilute the UDP-[^{14}C]glucose with 1.5 ml of an ethanol/water mix (2:8). This mixture can be stored at $-20°$. For each assay, 10 μl UDP-[^{14}C]glucose is vacuum dried until all liquid has evaporated in order to remove the ethanol and reduce the total volume. Control experiments demonstrated that no radioactivity is lost during the procedure (unpublished data, 1996). To the dried UDP-[^{14}C]glucose, a master mix containing the following components is added: 500 ng of the respective GTPase, 3 μl GDP (50 mM), 3 μl 10× MPBS (1.5 M NaCl, 160 mM Na$_2$HPO$_4$, 40 mM NaH$_2$PO$_4$), 1.5 μl bovine serum albumin (BSA) (2 mg/ml), 0.3 μl dithiothreitol (DTT) (10 mM), 3 μl KCl (100 mM), 1.2 μl MgCl$_2$ (100 mM), and 1 μl MnCl$_2$ (30 mM). The final volume is adjusted to 30 μl, and the glycosylation reaction is started by the addition of the appropriate amount of TcdB-10463 (see later) and incubated for 1 hr at $37°$. To stop the reaction, the samples are boiled (5 min, $95°$) in the presence of 15 μl 2× Laemmli buffer (4% SDS, 12.5% glycerin, 375 mM Tris–HCl, pH 6.8, 5% 2-mercaptoethanol, bromochlorophenol blue). The proteins are separated by 12.5% SDS–PAGE, and the gel is fixed and stained by standard methods and then incubated for 20 min in Amplify (Amersham-Pharmacia, Freiburg, Germany) to improve signal strength. The gel is mounted on a 3MM (Whatman, Clifton, NJ) paper support and dried. The glucosylation reaction is analyzed either by autoradiography or using a phosphoimaging system.

The typical result of such an experiment is shown in Fig. 2. Figure

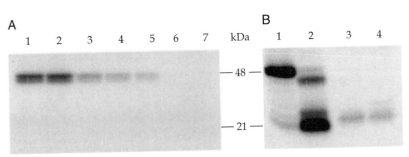

Fig. 2. *In vitro* glucosylation reaction by TcdB-10463. (A) Racl (500 ng) was used as substrate to demonstrate the effects of different amounts of TcdB-10463 on the glucosylation reaction. TcdB-10463 was used either at a protein concentration of 30 μg/ml (lane 1) or diluted 1:5 (lane 2), 1:25 (lane 3), 1:50 (lane 4), 1:100 (lane 5), 1:1000 (lane 6), and 1:10000 (lane 7) in 1× MPBS. (B) Cdc42-GST (lane 1) or Cdc42 (lane 2) was glucosylated by TcdB-10463. Alternatively, the cell lysate from DON cells was used as substrate for TcdB-10463 (lane 3) or TcsL-1522 (lane 4). Additional bands in lane 2 result from incomplete cleavage of the GST–Cdc42 fusion protein. (For TcdB-10463: 1 enzymatic unit equals 2 = 20 × 10^3 TCD.)

2A demonstrates the effects of different amounts of TcdB-10463 on the glucosylation of GTPase Rac. Here the signal strength is increasing with the amount of TcdB-10463 added until there is sufficient toxin in the assay to glucosylate all GTPases within 1 hr. Hence the characteristics of a saturation curve are observed: a further increase of the amount of TcdB-10463 does not increase signal strength (lanes 1 and 2, Fig. 2A). Thus we define one enzymatic unit of TcdB-10463 as the amount of toxin necessary to glucosylate 500 ng Rac-GST completely within 1 hr at 37°. This definition can be applied accordingly to all other LCTs as all the toxins glycosylate GTPase Rac. Generally we use 1–10 U LCT in standard glycosylation reactions.

Considering the glycosylation assay, several other important aspects should be kept in mind. (a) glycosylation of GST-fusion proteins: Most recombinant GTPases are expressed as glutathione S-transferase (GST)-fusion proteins.[19] Figure 2B compares the efficiency of glycosylation of the GST-Cdc42 fusion protein to cleaved Cdc42 (lanes 1 and 2). Usually, fusion with GST does not affect glycosylation of the GTPases. Some evidence may, however, indicate that certain GTPases are glucosylated only in their cleaved forms[20,21] (unpublished results). (b) Usage of eukaryotic cell lysate: Apart from recombinant GTPases, the eukaryotic cell lysate can also be used as substrate in the glycosylation reaction (Fig. 2B, lanes 3 and 4). The lysate is prepared by detaching 10–15×10^6 cells from their substrate, washing them in PBS, and resuspending them in 300 μl of 1$\times$ MPBS (150 mM NaCl, 16 mM Na$_2$HPO$_4$, 4 mM NaH$_2$PO$_4$). The cells are lysed by sonication, cell debris is removed by centrifugation (5 min, 10,000g, 4°), and the supernatant is aliquoted and stored at -20°. Fifteen microliters is commonly used in the glycosylation reaction. Using a eukaryotic cell lysate as substrate for glycosylation by TcdB-10463 and TcsL-1522 gives different band patterns corresponding to different inactivated GTPases (Fig. 2B, lanes 3 and 4, Table I). This assay has also been used in a modified way to demonstrate the differential glycosylation of GTPases by different LCTs.[22] Cells were incubated with a specific toxin (e.g., TcdA-10463) until all cells were intoxicated, the cells were subsequently lysed, and the lysate was used as substrate in the radioactive glycosylation assay with a different toxin (e.g., TcdB-10463).

[19] A. J. Self and A. Hall, *Methods Enzymol.* **256**, 3 (1995).
[20] F. Hofmann, G. Rex, K. Aktories, and I. Just, *Biochem. Biophys. Res. Commun.* **227,** 77 (1996).
[21] M. R. Popoff, E. Chaves-Olarte, E. Lemichez, C. von Eichel-Streiber, M. Thelestam, P. Chardin, D. Cussac, B. Antonny, P. Chavrier, G. Flatau, M. Giry, J. de Gunzburg, and P. Boquet, *J. Biol. Chem.* **271,** 10217 (1996).
[22] E. Chaves-Olarte, M. Weidmann, C. Eichel-Streiber, and M. Thelestam, *J. Clin. Invest.* **100,** 1734 (1997).

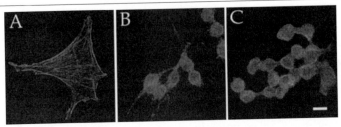

FIG. 3. Cytopathic effects of TcdB-10463 and TcsL-1522 on NIH 3T3 cells. The actin cytoskeleton of NIH 3T3 cells was stained with fluorescein isothiocyanate (FITC)-phalloidin using standard methods. Cells were either left untreated (A) or treated with 0.1 TCD (5-50 × 10⁻⁴ enzymatic units) TcdB-10463 (B) or 1 TCD TcsL-1522 (C) for 24 hr. Bar: 2 μm.

Assaying Cytotoxic Activities of LCTs on Cells

LCTs have been used in a number of studies to evaluate the role of small GTPases in eukaryotic cells. Mostly their influence on the actin cytoskeleton was assessed, e.g., the role of the Rho subfamily of GTPases in actin depolymerization during axonal growth.[23] Furthermore, TcdA-10463 was used to show that a balance of signaling by GTPases Rho, Rac, and Cdc42 regulates the cellular cytoskeleton.[24] Additional studies dealt with downstream effects of the GTPases, e.g., in experiments using a selection of different LCTs the role of Rho and Ral on activation of PLD and PLC was demonstrated.[25–27] In general, these experiments were conducted by simply adding the toxins to the growth medium of the cells and measuring the effects after glycosylation of the GTPases.

Cytopathic Effects of LCTs

The cytotoxic effects of LCTs on eukaryotic cells are characterized by the breakdown of the actin cytoskeleton. However, depending on the substrates of the toxins, different cytopathic effects (CPE) can be observed.[6] To date, two different CPE types have been identified: (a) D-type CPE induced by toxins TcdA-10463, TcdB-10463, and Tcnα-19402 (Fig. 3B) where the cells show a characteristic astrocyte-like shape, a small cell body

[23] F. Bradke and C. G. Dotti, Science 283, 1931 (1999).
[24] J. P. Moorman, D. Luu, J. Wickham, D. A. Bobak, and C. S. Hahn, Oncogene 18, 47 (1999).
[25] M. Schmidt, C. Bienek, U. Rumenapp, C. Zhang, G. Lummen, K. H. Jakobs, I. Just, K. Aktories, M. Moos, and C. von Eichel-Streiber, Naunyn Schmiedebergs Arch Pharmacol. 354, 87 (1996).
[26] U. Rumenapp, M. Schmidt, S. Olesch, S. Ott, C. von Eichel-Streiber, and K. H. Jakobs, Biochem. J. 334, 625 (1998).
[27] M. Schmidt, M. Voss, M. Thiel, B. Bauer, A. Grannass, E. Tapp, R. H. Cool, J. de Gunzburg, C. von Eichel-Streiber, and K. H. Jakobs, J. Biol. Chem. 273, 7413 (1998).

with long protrusions, and where the cells remain firmly attached to the substrate, and (b) S-type CPE induced by toxins TcsL-1522, TcdB-1470, and TcdB-8864 (Fig. 3C) where the cells are completely rounded without any protrusions. Characteristically, intoxicated cells are no longer attached to the substrate (about 90% of the TcsL-1522-treated cells are in solution), a feature that has to be remembered whenever a constant number of cells is a prerequisite for an assay.

Cytotoxicity Assay

The toxic potential of LCTs on eukaryotic cells generally differs between different cell lines. Their cytotoxicity is dependent on a number of variables: (a) the type of cells used, (b) the density that the cell are seeded with, (c) the period of time that the cells are incubated with the toxins, (d) the way that the assay is read, and (e) the amount of toxin used. Whereas variables (a)–(d) can be freely chosen and yet kept constant, this article has shown that preparations of TcdB-10463 or other LCTs contain contaminating proteins. Thus the high protein content of a specific toxin preparation does not necessarily reflect an equally high cytotoxic potential of the preparation. To rule out any possible influence of contaminants, we use a standard assay, where all variables except the toxin concentration are kept constant, to determine the toxic potential of a specific toxin preparation.

Five thousand Chinese hamster ovary (CHO) cells per well are seeded onto a 96-well plate in a volume of 100 μl and incubated for 16 hr under standard growth conditions [37°, 5% (v/v) CO_2, Dulbecco's modified Eagle's medium (DMEM) nutrient mix F-10 supplemented with 2 mM L-glutamine, 50 units/ml penicillin/streptomycin, and 5% fetal calf serum]. The toxins, diluted in growth medium, are then added to the cells to give a final dilution of 10^{-4} to 10^{-9} for TcdB-10463 and 10^{-1} to 10^{-6} for TcsL-1522. The cells are incubated with the toxin for 24 hr under standard growth conditions and are subsequently analyzed under the microscope for the percentage of cells rounded. To quantify, we photograph a representative area of each well and determine the ratio of rounded versus elongated cells by counting. Based on these data dose/response, graphs of the toxins can be drawn as depicted in Fig. 4. We define 1 cytotoxic unit (TCD, tissue culture dose) as the amount of toxin necessary to induce 100% rounding of CHO cells within 24 hr. Thus 1 TCD equals a dilution of 10^{-7} (Fig. 4A) and 10^{-1} (Fig. 4B) for TcdB-10463 and TcsL-1522, respectively. However, determination of these values has to be repeated for each individual toxin preparation.

If cells other than CHO cells are used, it is essential to determine the cytotoxicity of the LCTs on these cells. Therefore the cells are intoxicated with different toxin concentrations and a dose/response graph is drawn.

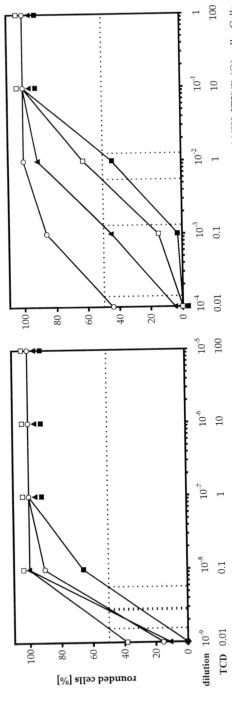

FIG. 4. Toxicity of TcdB-10463 (A) and TcsL-1522 (B) on CHO cells (■), NIH 3T3 (□), NIH 3T3$^{v\text{-}ras}$ (▲), and NIH 3T3$^{v\text{-}src}$ (○) cells. Cells were treated as described in the text, and the experiments were repeated at least three times. Undiluted, TcdB-10463 was at a concentration of 1 mg/ml and TcsL-1522 at a concentration of 80 μg/ml. Dashed lines indicate 50% rounded cells (horizontal line) or the TD$_{50}$ values of the respective cell lines (vertical lines).

In our experience, TD_{50} values (the amount of toxin necessary to round 50% of the cells within 24 hr) are a reliable measure to compare the cyto-toxic potential of LCTs on different cell lines. TD_{50} values can be deter-mined easily from the dose/response graph. Figure 4 compares the cytotoxicity of TcdB-10463 and TcsL-1522 on NIH 3T3 cells either untrans-formed or transformed with v-Ras or v-Src. The cytotoxic potential of TcdB-10463 is almost identical on all three cell lines ($TD_{50}^{NIH\ 3T3}$ = 0.02 TCD; $TD_{50}^{NIH\ 3T3v\text{-}src}$ = 0.03 TCD; $TD_{50}^{NIH\ 3T3v\text{-}ras}$ = 0.03 TCD; Fig. 4A). In contrast, significant differences can be observed after intoxication with TcsL-1522. Interestingly, NIH 3T3 cells have a lower sensitivity toward TcsL-1522 than the two transformed cell lines ($TD_{50}^{NIH\ 3T3}$ = 0.6 TCD; $TD_{50}^{NIH\ 3T3v\text{-}src}$ = 0.01 TCD; $TD_{50}^{NIH\ 3T3v\text{-}ras}$ = 0.1 TCD; Fig. 4B). These results indicate that transformation with v-Ras does not protect cells against the Ras-inactivating toxin TcsL-1522. In addition, the transformed cells are even more susceptible to TcsL-1522 than NIH 3T3 cells. We hypothesize that this effect is based on an intrinsic imbalance in the signal-transducing networks of the transformed cell lines, rendering them more sensitive to the actions of TcsL-1522.

In conclusion, our data demonstrate that the cytotoxic potential of LCTs is dependent on the specific cell line used and that the optimal toxin concentration has to be determined for each cell line individually.

[12] Rho GTPase-Activating Toxins: Cytotoxic Necrotizing Factors and Dermonecrotic Toxin

By GUDULA SCHMIDT and KLAUS AKTORIES

Introduction

Rho proteins are the preferred eukaryotic substrates of various bacterial protein toxins and exoenzymes that inactivate GTPases by glucosylation (e.g., large clostridial cytotoxins) and ADP-ribosylation (e.g., C3-like trans-ferases), respectively.[1,2] It has been shown that Rho proteins are not only inhibited but also activated by bacterial protein toxins,[3–5] Rho-activating

[1] K. Aktories, *Trends Microbiol.* **5,** 282 (1997).
[2] G. Schmidt and K. Aktories, *Naturwissenschaften* **85,** 253 (1998).
[3] G. Schmidt, P. Sehr, M. Wilm, J. Selzer, M. Mann, and K. Aktories, *Nature* **387,** 725 (1997).
[4] G. Flatau, E. Lemichez, M. Gauthier, P. Chardin, S. Paris, C. Fiorentini, and P. Boquet, *Nature* **387,** 729 (1997).

toxins are the *Escherichia coli* cytotoxic necrotizing factors (CNF1 and CNF2) and the dermonecrotic toxin (DNT) from *Bordetella* species.[5,6]

CNF1 and CNF2 are ~115-kDa proteins (1014 amino acids) with ~90% identity in their amino acid sequences.[7] The agents induce necrosis after dermal injection, an effect that led to the designation of the toxins,[8] and they are lethal in animals after systemic application (LD_{50} ~20 ng in mice). CNF1 and CNF2 are cytotoxic for a wide variety of cells such as Hep-2 human epidermoid carcinoma cells, HeLa cells, NIH 3T3 or Swiss 3T3 fibroblasts.[8] In contrast, hematopoietic cells appear to be affected much less by the toxins. The most typical effects caused by CNF in cultured cells are the increase in the formation of actin stress fibers, the cell enlargement with the formation of giant cells, and multinucleation.[7-9]

The dermonecrotizing toxin (DNT) is an ~160-kDa protein (1451 amino acids) produced by *Bordetella bronchiseptica, B. pertussis,* and *B. parapertussis.*[10,11] The toxin is considered to be a virulence factor for turbinate atrophy in porcine atrophic rhinitis.[12] In cell culture, DNT induces stress fiber formation and multinucleation similar to CNF.

The underlying mechanism of the actions of CNF and DNT has been elucidated.[3-5] Both toxins cause deamidation of glutamine-63 of Rho, thereby inhibiting the intrinsic and GAP-stimulated GTP hydrolysis, resulting in a constitutively activated form of GTPases. Not only RhoA but also other members of the Rho family, e.g., Rac and Cdc42, are modified by both toxins *in vitro* and in intact cells.[5,6] This same molecular mechanism induced by both toxins is reflected in the significant amino acid sequence homology of the catalytically active C-terminal domains of the toxins.[10,13] In addition to the deamidating activity, both toxins possess transglutaminase

[5] Y. Horiguchi, N. Inoue, M. Masuda, T. Kashimoto, J. Katahira, N. Sugimoto, and M. Matsuda, *Proc. Natl. Acad. Sci. U.S.A.* **94**, 11623 (1997).

[6] M. Lerm, J. Selzer, A. Hoffmeyer, U. R. Rapp, K. Aktories, and G. Schmidt, *Infect. Immun.* **67**, 496 (1998).

[7] E. Oswald, M. Sugai, A. Labigne, H. C. Wu, C. Fiorentini, P. Boquet, and A. D. O'Brien, *Proc. Natl. Acad. Sci. U.S.A.* **91**, 3814 (1994).

[8] A. Caprioli, V. Falbo, L. G. Roda, F. M. Ruggeri, and C. Zona, *Infect. Immun.* **39**, 1300 (1983).

[9] Y. Horiguchi, T. Senda, N. Sugimoto, J. Katahira, and M. Matsuda, *J. Cell Sci.* **108**, 3243 (1995).

[10] K.E. Walker and A.A. Weiss, *Infect. Immun.* **62**, 3817 (1994).

[11] G. D. Pullinger, T. E. Adams, P. B. Mullan, T. I. Garrod, and A. J. Lax, *Infect. Immun.* **64**, 4163 (1996).

[12] R. M. Roop II, H. P. Veit, R. J. Sinsky, S. P. Veit, E. L. Hewlett, and E. T. Kornegay, *Infect. Immun.* **55**, 217 (1987).

[13] E. Lemichez, G. Flatau, M. Bruzzone, P. Boquet, and M. Gauthier, *Mol. Microbiol.* **24**, 1061 (1997).

activity, thereby catalyzing the exchange of the γ-carboxamide group of glutamine for a primary amine. With respect to transglutaminase activity, DNT appears to be more active than CNF.[14,15] Interestingly, the toxins show no homology to known transglutaminases but share catalytically essential cysteine and histidine residues with them.[14]

This article describes the purification of the holotoxin CNF1 and the catalytic domains of CNF1 and DNT and reports on methods for determination of the biological and biochemical activities of the toxins.

Cloning and Purification of CNF1

The CNF1 gene with flanking *Bam*HI and *Eco*RI sites is generated by Polymerase chain reaction (PCR) from the vector PISS 391[16] and cloned in frame into the expression vector pGEX2TGL + 2. The glutathione *S*-transferase (GST) fusion protein was expressed in BL21 cells and isolated from the lysates by affinity chromatography with glutathione-Sepharose (Pharmacia, Piscataway, NJ) as described later for the active fragment. The holotoxin has to be eluted as a GST fusion protein with an excess of glutathione (10 mM glutathione, 50 mM Tris–HCl, pH 7.5) for 10 min at room temperature.

Activity of Toxin

The activity of the holotoxin can be analyzed by the incubation of mammalian cells with 300–500 ng ml^{-1} purified GST–CNF1 in culture medium. After 3 hr of toxin treatment, subconfluent HeLa cells or NIH 3T3 fibroblasts begin to flatten and, after 24 hr, most of the cells are polynucleated (Fig. 1). Stress fibers can be stained with rhodamine-conjugated phalloidin after 6 to 24 hr of treatment. Therefore, CNF1-treated and control cells (seeded on glass coverslips) are washed three times with phosphate-buffered saline (PBS) and fixed with 4% formaldehyde, 0.1% Tween 20 in PBS for 10 min at room temperature. After intensive washing the cells are incubated with rhodamine-conjugated phalloidin (1 unit per coverslip) at room temperature for 1 hr in a humified atmosphere and washed again. Thereafter the cells are analyzed by fluorescence microscopy [as bleaching preservative KAISER'S glycerol gelatin (Merck, Darmstadt) was used] (Fig. 1). Hematopoietic cells such as T cells or RBL cells (rat basophilic leukemia cells) in our hands do not appear to be affected by the toxins.

[14] G. Schmidt, J. Selzer, M. Lerm, and K. Aktories, *J. Biol. Chem.* **273,** 13669 (1998).
[15] G. Schmidt, U.-M. Goehring, J. Schirmer, M. Lerm, and K. Aktories, *J. Biol. Chem.* **274,** 31875 (1999).
[16] V. Falbo, T. Pace, L. Picci, E. Pizzi, and A. Caprioli, *Infect. Immun.* **61,** 4909 (1993).

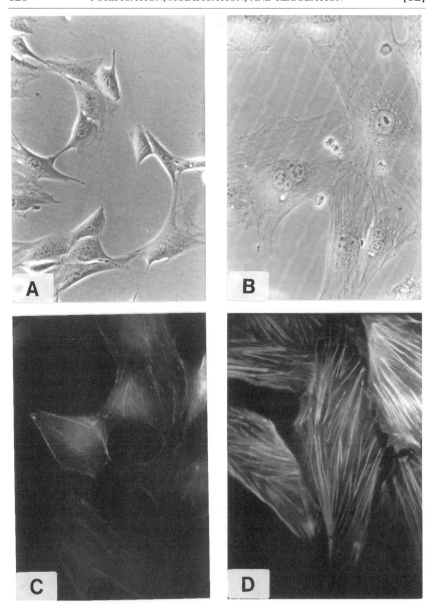

Fig. 1. Induction of multinucleation and stress fiber formation in NIH 3T3 fibroblasts after treatment with CNF1: NIH 3T3 fibroblasts were treated without (A, C) and with CNF1 (400 ng ml^{-1}; B, D) for 24 hr. Cells were then analyzed by phase-contrast microscopy (A, B; 40×) or after staining with rhodamine-phalloidine by fluorescence microscopy (C, D; 100×).

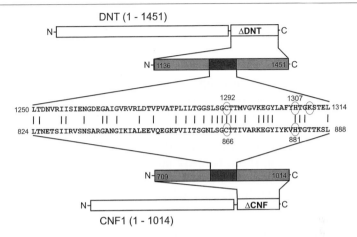

FIG. 2. Catalytically active fragments of DNT (ΔDNT) and CNF1 (ΔCNF1): DNT and CNF1 consist of 1451 and 1014 amino acids, respectively. The active fragments (ΔDNT and ΔCNF1) covering amino acids 1136 through 1451 and 709 through 1014 are located at the C termini of the toxins. They show significant sequence identity (45%) in the aligned region (dark bars). The catalytic essential amino acids are marked.

Cloning and Purification of Active CNF1 (aa709–1014) and DNT (aa1136–1451) Fragments

For construction of the active CNF1 fragment (ΔCNF1; Fig. 2) consisting of amino acids 709 through 1014 of the toxin, the pGEX-CNF1 vector[3] is digested by *Bam*HI and *Sna*BI and purified from agarose gel (Jet sorb, Genomed). Sticky ends are filled up by Klenow enzyme in the presence of 0.25 mM dNTP each for 30 min at 30°. The vector is religated and transformed into BL21 cells by heat shock at 42°.

DNA coding for the active DNT fragment (ΔDNT) consisting of amino acid residues 1136 through 1451 can be amplified from the plasmid DNT 103[10] by PCR introducing *Bam*HI and *Eco*RI sites. The PCR product is then purified from agarose gel (Jet sorb, Genomed) and amplified in the pCR II Vector (Invitrogen) by means of TA cloning. From this vector the DNT fragment can be cut with *Bam*HI and *Eco*RI, purified and ligated into the digested pGEX vector. Expression of the GST fusion proteins in BL21 cells growing at 37° in LB medium is induced by adding 0.2 mM isopropyl-β-D-thiogalactopyranoside (final concentration) at OD 0.5. Three hours after induction, cells are collected and lysed by sonication in lysis buffer (20 mM Tris–HCl, pH 7.4, 10 mM NaCl, 5 mM MgCl$_2$, and 1% Triton X-100) and purified by affinity chromatography with glutathione-Sepharose (Pharmacia). Loaded beads are washed two times in washing buffer A (20 mM Tris–HCl, pH 7.4, 10 mM NaCl, and 5 mM MgCl$_2$) and washing buffer B (150 mM NaCl, 50 mM Tris–HCl, pH 7.5) at 4°. ΔCNF1 is then eluted from the beads by thrombin digestion (200 μg/ml thrombin, 150 mM NaCl, 50 mM triethanolamine hydrochloride, pH 7.5, and 2.5 mM CaCl$_2$) for 45 min at room temperature. Thrombin must be removed by incubation with benzamidine-Sepharose beads for 5 min at room temperature and subsequent centrifugation. In the case of DNT the toxin fragment is degraded by thrombin digestion. Thus it has to be eluted as a GST fusion protein with an excess of glutathione (10 mM glutathione, 50 mM Tris–HCl, pH 7.5) for 10 min at room temperature. It has to be mentioned that in contrast to ΔCNF1 the DNT fragment has to be stored at 4° and loses activity during storage. It can be used *in vitro* assays for about 1 week.

Identification of Deamidated or Transglutaminated GTPases

Changes in Migration of Rho GTPases in SDS–PAGE

Because deamidation and transglutamination induce changes in the migration of GTPases in SDS–PAGE, the modification of RhoA can be analyzed by gel electrophoresis. It was recognized quite early that treatment of culture cells with CNF results in an apparent increase of the molecular mass of Rho in SDS–PAGE.[7] This change in migration behavior is the consequence of deamidation of RhoA at glutamine-63 and, therefore, is also observed with the recombinant mutant Q63E RhoA. The reason for the relatively large change in the apparent molecular mass (about 1 kDa upward shift) is not clear, as deamidation increases the molecular mass by only 1 Da. In contrast to the upward shift induced by deamidation, transglutamination of RhoA with ethylenediamine as cosubstrate causes a downward shift of the GTPase in SDS–PAGE. However, this downward shift is not observed with all cosubstrates. For example, in the presence of dansylcadaverine, no shift is observed. Notably, deamidation or transglutamination of Rac and Cdc42 does not cause significant changes in their migration behavior.

Small GTPases are incubated with GST–ΔCNF1 or GST–ΔDNT in the presence or absence of ethylenediamine (50 mM) in transglutamination buffer [20 mM Tris–HCl, pH 7.5, 5 mM MgCl$_2$, 8 mM CaCl$_2$, 1 mM dithiothreitol (DTT), and 1 mM EDTA] for 30 min at 37°. As controls, Rho proteins are incubated without the toxins but in the presence of ethylenediamine or buffer. The molar ratio of toxin: RhoA, 1 to 1000, is sufficient for full modification. SDS sample buffer is added and the probes are incubated for 2 min at 95° before running on a 12.5% SDS gel (Fig. 3). Best results are obtained with SDS from Pierce.

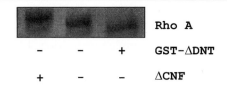

FIG. 3. Effects of ΔCNF1 and GST–ΔDNT on migration of RhoA in SDS–PAGE. RhoA (10 μM) was incubated with GST–ΔDNT (0.1 μM) or ΔCNF1 (0.1 μM) in the presence of 2 mM ethylenediamine (ED) for 15 min at 37°. Thereafter, the proteins were analyzed by SDS–PAGE. The downward shift indicates transglutamination, whereas the upward shift indicates deamidation. Rho proteins incubated in the presence of ethylenediamine, but without the toxins, show no change of migration.

Labeling with the Fluorescent Cosubstrate Dansylcadaverine

A fast method for screening recombinant proteins as toxin substrates is the direct labeling of the acceptor glutamine with typical transglutaminase cosubstrates, for example, with the fluorescent amine derivative dansylcadaverine.

Therefore, proteins are incubated with the toxin (CNF1, DNT) in reaction buffer containing dansylcadaverine (saturated solution), 20 mM Tris–HCl, pH 8.0, 5 mM $MgCl_2$, 8 mM $CaCl_2$, 1 mM DTT, and 1 mM EDTA for up to 60 min at 37°. As a control, the proteins are incubated without toxins but in the presence of cosubstrate.

The samples are applied to SDS–PAGE, and labeling is analyzed under UV light before staining and drying the gel. Due to the low sensitivity of the method and the preference of CNF1 to catalyze deamidation (DNT acts preferably as a transglutaminase), up to 5 μg protein per lane is needed and the pictures must be integrated about 20 to 50 times (Eagle Eye II, Stratagene, Heidelberg, Germany) when analyzing CNF activity.

Mass Spectrometry

Mass spectrometric analysis of proteolytic peptides of Rho GTPases allows the detection of the 1-Da shift caused by CNF1-induced deamidation of Gln-63 of RhoA (Gln-61 of Rac and Cdc42). Similarly, the mass shifts of 43 and 71 Da after treatment of the GTPase with the toxins in the presence of putrescine and ethylenediamine, respectively, are detected by mass analysis. For this purpose, Rho proteins are modified by the toxins as described previously. Rho GTPases modified are analyzed by SDS–PAGE (minigel system, e.g., from Bio-Rad, Hercules, CA). The proteins are stained, and the Rho protein band is excised. The excised gel plugs of RhoA are destained for 1 hr at 50° in 40% acetonitrile/60% hydrogen carbonate (50 mM, pH 7.8) in order to remove Coomassie Blue, gel buffer,

SDS, and salts. The plug is then dried in a vacuum centrifuge for 15 min. Thereafter, 30 μl digestion buffer with trypsin is added and digestion is carried out for 12 hr at 37°. For sample preparation, 4-hydroxy-α-cyanocinnamic acid (HCCA; Aldrich, Milwaukee, WI) is recrystallized from hot methanol and stored in the dark. A saturated matrix solution of HCCA in a 1 : 1 solution of acetonitrile/aqueous 0.1% trifluoroacetic acid is prepared. Two microliters of the proteolytic peptide mixture is mixed with 2 μl saturated matrix containing marker peptides [5 pmol human ACTH(18–39) clip (MW 2466; Sigma, St. Louis, MO) and 5 pmol human angiotensin II (MW 1047, Sigma), respectively] for internal calibration. Using the dried-drop method of matrix crystallization, 1 μl of the sample matrix solution is placed on the MALDI stainless steel target and is allowed to air-dry for several minutes at room temperature, resulting in a thin layer of fine granular matrix crystals. MALDI/TOF-MS can be performed on a Bruker Biflex mass spectrometer equipped with a nitrogen laser to desorb and ionize the samples. Mass spectra are recorded in the reflector positive mode in combination with delayed extraction. External calibration is used routinely, and internal calibration with two points that bracketed the mass range of interest is additionally performed to consolidate peptide masses further. The computer program MS-digest (Peter Baker and Karl Clauser, UCSF Mass Spectrometry Facility) can be used for computer-assisted comparison of tryptic peptide mapping data with the expected set of peptides. As an example, the RhoA peptide (aa52–68) of untreated RhoA, deamidated RhoA, and transglutaminated RhoA (ethylenediamine as cosubstrate) is shown in Fig. 4.

Measurement of Ammonia

A more sensitive method for screening recombinant proteins as toxin substrates is the measurement of ammonia, which is produced during deamidation as well as during transglutamination (Fig. 5). It has to be mentioned that proteins purified by ammonium sulfate precipitation must be dialyzed extensively to remove ammonia. For measurement of ammonia, a coupled enzymatic reaction is performed based on the Ammonia Test Combination for Food Analysis (Boehringer Mannheim). In the presence of glutamate dehydrogenase (GlDH) and nicotinamide-adenine dinucleotide (NADH), ammonia reacts with 2-oxoglutarate to L-glutamine. NADH is oxidized to NAD$^+$. The amount of oxidized NADH is stoichiometric to the amount of ammonia in the test tube. For the reaction, NADH is diluted to a concentration of 50 μM with triethanolamine buffer containing 2-oxoglutarate, 20 mM Tris–HCl, pH 7.5, 10 mM MgCl$_2$, 1 mM DTT, 1 mM EDTA, 100 μM GDP, and stabilizers. Ten units of GlDH and RhoA (final concentration

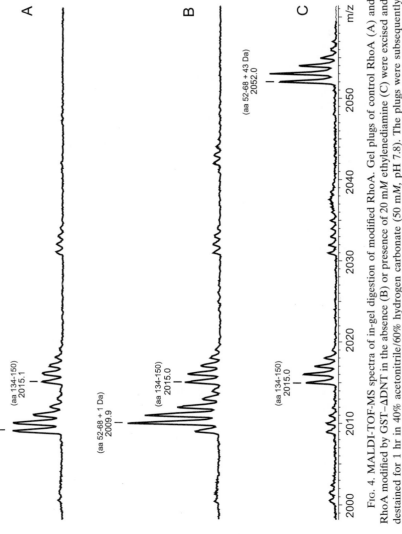

FIG. 4. MALDI-TOF-MS spectra of in-gel digestion of modified RhoA. Gel plugs of control RhoA (A) and RhoA modified by GST–ΔDNT in the absence (B) or presence of 20 mM ethylenediamine (C) were excised and destained for 1 hr in 40% acetonitrile/60% hydrogen carbonate (50 mM, pH 7.8). The plugs were subsequently dried in a vacuum centrifuge for 15 min. Thereafter, trypsin digestion was carried out for 12 hr at 37°. (A) The RhoA peptide Gln[52]-Arg[68] (2009 Da) is shown. (B) Deamidation of Gln-63 of RhoA by GST–ΔDNT results in a mass shift of the peptide of 1 Da. (C) Transglutamination of Gln-63 of RhoA by GST–ΔDNT in the presence of ethylenediamine results in a mass shift of the peptide of 43 Da. From Schmidt et al.,[15] with permission.

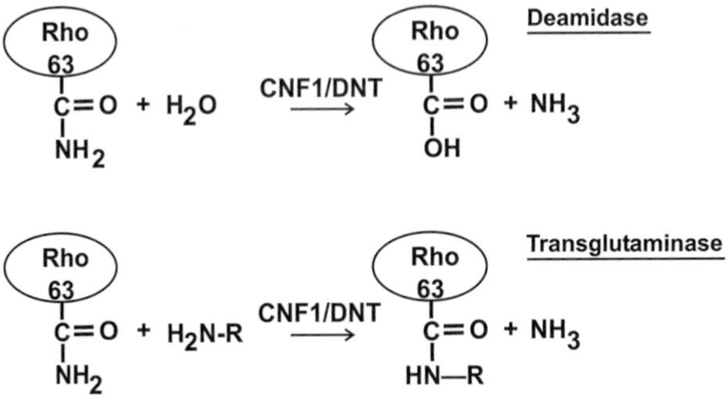

FIG. 5. Molecular mechanism of Rho activation by CNF1 and DNT. Gln-63 of Rho proteins is deamidated by the toxins with water as the cosubstrate. In the presence of primary amines serving as cosubstrates, the GTPases are transglutaminated. With CNF1, up to 25% of the proteins can be transglutaminated (with DNT more than 90%) dependent on the cosubstrate used. During deamidation and transglutamination ammonia is produced.

10 μM) is added. After addition of the toxin, the decrease in NADH fluorescence can be monitored in a Perkin-Elmer (Norwalk, CT) LS-50B luminescence spectrometer. The emission is measured at 460 nm with excitation at 340 nm (Fig. 6).

The method just described allows screening for toxin substrates but is not optimal for quantitative analysis of the ammonia release due to the low k_{cat} of the reporter enzyme GlDH. Thus, an assay to measure ammonia independently of the velocity of the reporter enzyme was performed as follows: Rho proteins (200 μM) are incubated with the toxins in reaction buffer containing 20 mM Tris–HCl, pH 8.0, 5 mM MgCl$_2$, 8 mM CaCl$_2$, 1 mM DTT, and 1 mM EDTA at 37°. The reaction is stopped after different time intervals by incubation for 1 min at 95°. Denatured protein is removed by centrifugation (1 min, 14,000g) and the ammonia produced is measured in the supernatant.

Inhibition of Toxic Activity

It was reported that CNFs and DNT contain a catalytically important cysteine residue.[14,15] Exchange or modification of this residue leads to an inactive protein toxin. Cysteine residues can be modified by alkylating reagents such as N-ethylmaleimide or iodoacetamide. The method is as follows.

GST–ΔDNT is incubated with N-ethylmaleimide (NEM) is 50 mM Tris–HCl, pH 7.5, for 30 min at room temperature. Surplus NEM is then inactivated by adding dithiothreitol in a molar ratio of 10:1 (DTT:NEM) for 10 min. For modification of RhoA, the GTPase is incubated with NEM-

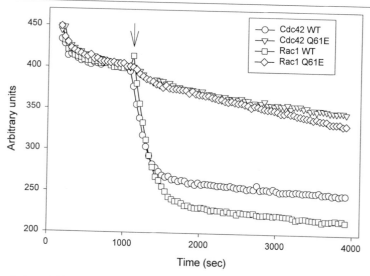

FIG. 6. Ammonia release. Release of ammonia from Rho proteins induced by ΔCNF1 can be measured in a coupled enzymatic reaction with GlDH as the reporter enzyme. Therefore, Rac 1 or Cdc42 and as controls the corresponding Q61E mutants (each 10 μM) were incubated in a buffer containing NADH, α-ketoglutarate, and GlDH. After equilibration at 37°, ΔCNF1 (1 μM, final concentration) was added (arrow) to the samples. Release of ammonia is measured by the decrease in NADH.

treated or untreated toxin in the presence of 50 mM ethylenediamine in transglutamination buffer (20 mM Tris–HCl, pH 7.5, 5 mM MgCl$_2$, 8 mM CaCl$_2$, 1 mM DTT, and 1 mM EDTA) for 30 min at 37°.

Influence of Deamidation or Transglutamination of Rho Proteins on Their GTPase Activity

It has been reported that deamidation and transglutamination of glutamine-63 of Rho (glutamine-61 of Rac and Cdc42) lead to the blockade of intrinsic and GAP-stimulated GTPase activity.[15] Thus, toxin-induced modification of small GTPases can be analyzed by measuring their GTPase activity. Therefore, toxin-treated and control Rho proteins are loaded with [γ-^{32}P]GTP for 5 min at 37° in loading buffer (50 mM Tris–HCl, pH 7.5, 10 mM EDTA, and 2 mM DTT). MgCl$_2$ (12 mM, final concentration) and unlabeled GTP (2 mM, final concentration) are added. For GAP stimulation, p50RhoGAP is added to RhoA (2 μM, final concentration) and the mixture is incubated at 37° or 16° for 4 min (or specific time intervals). GTPase activity is analyzed by a filter-binding assay.

Conclusions

CNF and DNT cause deamidation and/or transglutamination of Rho GTPases. Studies suggest that DNT possesses a higher transglutaminase activity than CNF and may act in intact cells as a transglutaminase rather than as a deamidase. However, so far the precise cellular cofactor of the transglutamination of Rho GTPases is not known. Rho-activating toxins may be of potential importance as pharmacological tools to study Rho GTPases in intact cells. For this purpose, CNF appears to be more suitable because its cellular action is well defined and CNF is much more stable than DNT. The disadvantage of CNF (and of DNT) as a tool is the broad substrate specificity of the toxin, which modifies all members of the Rho GTPase family. It was shown that the protein substrate recognition by CNF1 is defined by a small amino acid sequence covering the switch II region of the Rho GTPases.[17] Introduction of this peptide into Ras resulted in a chimeric protein susceptible for CNF1, which could be activated by deamidation. Whether this approach can be used to construct other small GTPases sensitive for activation by CNF1 remains to be studied.

---

[17] M. Lerm, G. Schmidt, U.-M. Goehring, J. Schirmer, and K. Aktories, *J. Biol. Chem.* **274,** 28999 (1999).

Section II

Purification and Activity of GTPase Targets

[13] Measurement of Gtpase · Effector Affinities

By Danny Manor

Introduction

Targets and modulators of GTPase signaling are usually identified using biochemical methodologies such as affinity precipitations, expression cloning, and genetic interactions screens. Although these methods are invaluable for the initial identification of effectors, they are inherently qualitative. The emerging picture of the complex signaling cascades in which GTPases participate evokes numerous protein–protein interactions with multiple branching points and overlapping binding interfaces. Thus, the selectivity of interactions between any two proteins is often determined by subtle differences in binding affinities, raising the need for precise quantitative characterization of each binding event. Additionally, the availability of modern structural and molecular genetics tools enables us to pinpoint with atomic detail the location and role of each peptidic component in the binding interface. Consequently, a highly sensitive readout of the binding reaction is necessary to confirm or refute a role for a particular residue suggested by other methods.

The purpose of this article is to describe the basic methodologies and protocols used to quantitatively measure the strength of protein–protein interactions that take place in GTPase-mediated signaling cascades. We will focus on methodologies that employ fluorescent nucleotides, as these reagents are easy to produce and were shown to report accurately on interaction of the Cdc42, Ras, and Rac proteins. An alternative strategy involves attachment of fluorescent reporter groups to the GTPase by covalent modification of the polypeptide itself. The advantage of such experimental system is mainly the optimization of the fluorophore group to match the specific experimental needs (e.g., resonance energy transfer experiments[1]); however, protein modifications usually involve lengthy characterization to ascertain that the modified protein is completely functional, and hence will not be discussed here. While the methods described here are based solely on fluorescence approaches, the reader should be advised that other methods that do not involve chemical modifications do exist (such

---

[1] T. Nomanbhoy, D. A. Leonard, D. Manor, and R. A. Cerione, *Biochemistry* **35,** 4602 (1996).

as ones based on scintillation proximity technology[2]) and yield reliable estimates for relative affinities.

Utilization of N-Methylanthraniloyl (Mant) Nucleotides as Fluorescent Probes

General Considerations

Most interactions of small GTPases are determined by the nature of the nucleotide to which they are tightly bound (GDP or GTP). Therefore, all affinity measurements should include quantification of the binding interaction to the GTPase while in the GTP-bound state, as well as while it is in the GDP-bound state. In this regard, a nonhydrolyzable analog of GTP must be employed when measuring affinities of the activated form of the GTPase in order to eliminate the various side reactions. It should be noted that although basal GTP hydrolysis rates are slow relative to the time frame of most experiments, this activity should not be neglected. This is especially critical in cases where an interacting protein stimulates (e.g., GAPs) or inhibits (e.g., Rho-GDI and some targets) GTP hydrolysis. In such cases, use of Mant-GTP can complicate interpretation in two ways. First, GTP hydrolysis was shown to affect Mant fluorescence intensity[3–6] in ways that may be additive or subtractive to the signal that accompanies complex formation. Second, if GTP hydrolysis occurs during the measurement, a mixed population of GTPase · GTP and GTPase · GDP species, each with a drastically different affinity toward the effector, will be present in experiments, rendering accurate analysis virtually impossible. Thus, the use of a nonhydrolyzable analog of GTP, such as GTPγS, GppNHp, or GppCHp, is required. For technical considerations (cost and commercial availability) as well as the similar affinity of GTPases to GTP and to GppNHp, we chose to utilize the latter GTP analog in all affinity measurements.

[2] G. Thompson, D. Owen, P. A. Chalk, and P. N. Lowe, *Biochemistry* **37**, 7885 (1998).
[3] S. E. Neal, J. F. Eccleston, A. Hall, and M. R. Webb, *J. Biol. Chem.* **263**, 19718 (1988).
[4] K. J. M. Moore, M. R. Webb, and J. F. Eccleston, *Biochemistry* **32**, 7451 (1993).
[5] H. Rensland, A. Lautwein, A. Wittinghofer, and R. S. Goody, *Biochemistry* **30**, 11181 (1991).
[6] M. R. Ahmadian, U. Hoffmann, R. S. Goody, and A. Wittinghofer, *Biochemistry* **36**, 4535 (1997).

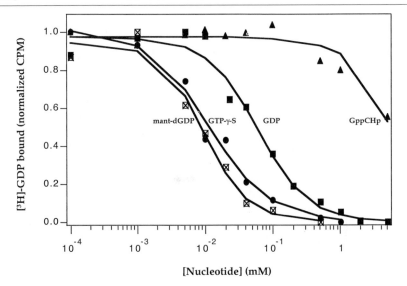

[Nucleotide] (mM)

Fig. 1. Affinity of different guanine nucleotides toward Cdc42Hs. One micromolar Cdc42Hs · GDP was incubated with 1 μM [3H]GDP in buffer B supplemented with 25 mM EDTA and various concentrations of the indicated nucleotides. Following a 30-min incubation at room temperature, MgCl$_2$ was added to 35 mM, and samples were filtered through nitrocellulose filters (BA-85, Schleicher & Schuell, Keene, NH). Protein-bound radioactivity was determined by scintillation counting as described elsewhere.[11]

Suitability of Mant Nucleotides as Fluorescent Reporter Groups

Despite the extensive use of Mant nucleotides in Ras,[4,5] Cdc42,[7,8] and Racl[9] when using a derivatized nucleotide as a probe, one must confirm that the chemical modification does not interfere with the proper "fit" of the nucleotide to the binding pocket of the GTPase. This can be directly verified in a competition assay in which the relative efficacy of different nucleotides to displace [3H]GDP from the GTPase is examined. Thus, purified GTPase (Cdc42Hs in Fig. 1) is incubated with an equimolar concentration of [3H]GDP in the presence of increasing concentrations of unlabeled, chemically modified nucleotides. The experiment is done in the presence of 25 mM EDTA to facilitate equilibration between bound and

[7] D. A. Leonard, T. Evans, M. Hart, R. A. Cerione, and D. Manor, *Biochemistry* **33**, 12323 (1994).
[8] D. A. Leonard, R. Sartoskar, J. W. Wu, R. A. Cerione, and D. Manor, *Biochemistry* **36**, 1173 (1997).
[9] J. Freeman, A. Abo, and D. J. Lambeth, *J. Biol. Chem.* **271**, 19794 (1996).

free nucleotides.[10] After equilibration is achieved, the bound nucleotide is "fixed" in the binding pocket by the addition of 35 mM MgCl$_2$, and protein-bound radioactivity is measured using nitrocellulose filtration and scintillation counting.[11] In such competition experiments, nucleotides with higher affinity toward the GTPase are able to displace the bound, radioactive nucleotide at lower concentrations, thus yielding relative estimates for their affinity toward the protein. Data shown in Fig. 1 demonstrate that the affinity of Mant-dGDP to Cdc42 is at least as tight as that of GTP, thus reaffirming that modification of the ribose ring of GDP with the (bulky) anthraniloyl group does not result in any significant perturbation to the nucleotide-binding site. This result was further verified in the case of Ras, where the three-dimensional structure of Ras bound to Mant-GppNHp was determined, showing that the fluorescent probe is extending out of the protein surface, with no significant alteration to the binding mode of other regions of the nucleotide.[12]

Synthesis of Mant Nucleotides

The following protocol for the synthesis of Mant nucleotides is after the procedure outlined by Hiratsuka,[13] as modified by John et al.[14]

Reagents

Nucleotide (dGDP or GppNHp; Sigma Chemical Co., St. Louis, MO)
N-Methylisatoic anhydride (NIMA; Molecular Probes, Eugene, OR)
Triethylamine (fresh bottle; Aldrich-Sigma Chemical Co., St. Louis, MO)
CO$_2$ tank with aerating stone
DEAE-Sepharose fast flow (Amersham-Pharmacia, Piscataway, NJ)

Buffer Preparation. As Mant nucleotides are eluted from ion-exchange columns at high salt concentrations, a volatile buffer system is used. One molar triethylammonium bicarbonate (TEAB) is prepared by purging 0.5 liter distilled water with CO$_2$ for 1 hr. Then 140 ml of triethylamine is added, the solution is further purged with CO$_2$ for an additional 2 hr, and distilled water is added to a final volume of 1 liter. The solution should be between pH 7.5 and 8.0.

---

[10] A. Hall and A. J. Self, *J. Biol. Chem.* **261,** 10963 (1986).
[11] M. J. Hart, A. Eva, T. Evans, S. A. Aaronson, and R. A. Cerione, *Nature* **354,** 311 (1991).
[12] A. J. Scheidig, S. M. Franken, J. E. Corrie, G. P. Reid, A. Wittinghofer, E. F. Pai, and R. S. Goody, *J. Mol. Biol.* **253,** 132 (1995).
[13] T. Hiratsuka, *Biochim. Biophys. Acta* **742,** 496 (1983).
[14] J. John, R. Sohmen, J. Feuerstein, R. Linke, A. Wittinghofer, and R. S. Goody, *Biochemistry* **29,** 6058 (1990).

Synthesis. A 25 mM solution of dGDP or GppNHp in 1.7 ml of water is prepared and mixed with 44 mg of NIMA in a 5-ml scintillation vial with a small stirring bar ("flea") over a stirring plate, and the pH is measured with a pH microelectrode. As the reaction progresses, the pH of the mixture drifts toward acidic values and aliquots of 1 M NaOH are added to maintain a pH value of 9.6. The solution gradually clarifies, and the pH stabilizes. Reaction is complete when the pH remains at a value of 9.6 for 30 min. The solution is neutralized by adding dilute hydrochloric acid to a pH value of 7.4 and can be frozen at $-80°$ until the next step (referred to as a "crude product" later).

Chromatography. Forty milliliters DEAE-Sepharose is washed with cold distilled water in a 2.5 × 40-cm column and then equilibrated with 0.2 M TEAB. The "crude" Mant reaction product is loaded on the column and washed with 0.2 M TEAB until the absorbance at 260 nm plateaus at zero. The column is then developed with a linear salt gradient (0.2 M → 1.0 M TEAB, 350 ml each), and 5-ml fractions are collected. The Mant nucleotide is the last, broadest peak to elute during this gradient. Eluted peaks are then analyzed by determination of absorbance at 254 and 360 nm. Pure Mant nucleotides have a value of 2.6–2.8 for the $A_{254}:A_{360}$ ratio.

The Mant nucleotide peak is pooled, shell-frozen in liquid nitrogen, and lyophilized to remove excess salt. The lyophilizate is then washed with methanol, shell-frozen, and relyophilized two more times before final dissolution in 2 ml distilled water. The concentration of the nucleotide is measured by absorbance ($\varepsilon_{350} = 5700\ M^{-1}\ cm^{-1}$), and the final product is stored at $-80°$ (typical concentration ca. 2–4 mM).

Formation of GTPase · Mant Nucleotide Complexes

Expression and purification Cdc42Hs have been described in detail previously[8] and will not be discussed here. We found that the same expression/purification strategy (expression in a pET vector, metal chelation chromatography followed by proteolytic cleavage of the affinity tag and anion-exchange separation) consistently produces high-yield, high-purity (>95% by SDS–PAGE) preparations for other GTPases such as Rac1, RhoA, and H-Ras. Proteins are stored in buffer A (20 mM HEPES, 5 mM MgCl$_2$, 2 mM sodium azide, pH 8.0) supplemented with 40% glycerol at $-20°$ until use. Interacting proteins (targets and activity modulators) are expressed and purified using a variety of methodologies that are highly protein specific and have to be determined empirically.

The purified GTPase is transferred to buffer B [20 mM Tris, 1 mM dithiothreitol (DTT), 1 mM sodium azide, 1 mM EDTA, pH 8.0] by either dialysis or ultrafiltration (Centricon-10; Amicon, Danvers, MA), and EDTA is added to 25 mM. A 40-fold molar excess of the desired Mant nucleotide is then added, and the nucleotide exchange is allowed to proceed for 2 hr at room temperature. For nucleotides with a very low affinity (e.g., GppChp), the alkaline phosphatase-facilitated exchange method of John *et al.*[14] is much preferable, as it ensures complete exchange. After 2 hr, MgCl$_2$ is added (from a 1 M stock) to a final concentration of 35 mM, and the excess Mant nucleotide is removed by filtration through a Sephadex G-25 (Superfine) column (0.75 × 40 cm) equilibrated with buffer A. Protein:Mant stoichiometry is determined from absorbance measurements using an extinction coefficient of (ε_{350}) of 5700 M^{-1} cm^{-1} for the Mant moiety[13] and protein determination assay (Bradford Protein Reagent; Bio-Rad, Hercules, CA). The complex is concentrated by ultrafiltration to ca. 10 mg protein/ml and stored in buffer A supplemented with 40% (v/v) glycerol at $-20°$.

GTPase · Effector Complex Formation Assays

The fluorescent properties of Mant nucleotides can be utilized to assay protein–protein interactions in one of two ways, which will be discussed separately here: fluorescence anisotropy and environmental sensitivity. While traditionally either one or the other of these methods is utilized as readouts for complex formation, it is worthwhile to consider the two assays as complementary, as each of them is subject to a distinct set of limitations.[15]

Environmental Sensitivity Assays

Mant nucleotides comprise convenient, spectrally distinct (λ_{exc} = 360 nm; λ_{em} = 440 nm) fluorescent reporters of the nucleotide-binding site in GTPases. At this region of the visible spectrum, very little interference from other protein moieties exists, and the anthranoilyl moiety appears to be well suited for monitoring subtle changes in the vicinity of the nucleotide-binding pocket. This sensitivity is exemplified by the changes in Mant-dGDP excitation and emission spectra upon moving from the aqueous milieu of a buffer into the binding pocket of a small GTPase such as Cdc42Hs.

As shown in Fig. 2, transition from the polar environment of the buffer into the nucleotide-binding pocket of the small GTPase Cdc42Hs is accompanied by a large (ca. 200%) change in the quantum yield for fluorescence. This property can be utilized for assaying the nucleotide exchange reaction, the key activation step of all GTPases, which is catalyzed by upstream

[15] D. A. Leonard, R. Lin, R. A. Cerione, and D. Manor, *J. Biol. Chem.* **273,** 16210 (1998).

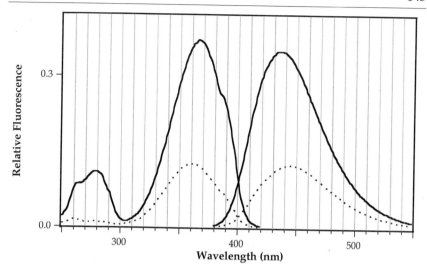

FIG. 2. Fluorescence spectra of Mant-dGDP in solution and when bound to Cdc42Hs. Mant-dGDP (0.15 μM) was incubated with 0.9 μM Cdc42Hs in buffer C, and excitation spectra (emission 440 mm, left-hand curves) and emission spectra (excitation 366 nm, right-hand curves) were recorded before and after addition of 15 mM EDTA (lower and upper spectra, respectively). Spectral resolution: 2 nm.

guanine nucleotide exchange factors.[7,16] Once in the binding pocket of the GTPase, the Mant nucleotide can also report further perturbations and changes that are induced in the binding pocket upon complexation with other proteins (e.g., targets and activity modulators).

Protein–Protein Interactions

  Reagents

   Buffer C (4× stock; 80 mM HEPES, 20 mM MgCl$_2$, 600 mM NaCl, pH 7.4, filtered through a 0.2-μm syringe filter)
   GTPase · Mant nucleotide (prepared as described earlier, in buffer B + glycerol, about 10 mg/ml)
   Distilled water (filtered through a 0.2-μm syringe filter)
   Spectrofluorimeter with a thermostatted and stirred cuvette chamber
   Small-volume pipettor (1–10 ml, preferably positive displacement type, such as Drummond Scientific, Broomall, PA)
   Procedure. Two hundred microliters of buffer C is added to a quartz fluorescence cuvette in which 600 μl water is stirred inside the fluorimeter

[16] R. A. Cerione and Y. Zheng, *Curr. Opin. Cell Biol.* **8**, 205 (1996).

cuvette chamber. GTPase complexed with the appropriate Mant nucleotide (Mant-dGDP or Mant-GppNHp) is added to a final concentration of 100 nM, and the emission from the Mant moiety is monitored with time (λ_{exc} = 360 nm; λ_{em} = 440 nm; spectral resolution = 4 nm; time resolution = 5 sec; dwell time = 1 sec). After the signal is stabilized, aliquots of the purified effector are added sequentially. Titration is done in such a way as to cover a wide concentration range, starting from well below the estimated K_d and ending with complete saturation of the signal change, with at least 10 data points in between. For correct data analysis it is imperative that complete saturation occurs, i.e., that more the one data point is collected after the end point is reached. A limitation that is inherent in fluorescence-based assays is the limited sensitivity (relative to radioisotope-based methods). A good signal-to-noise ratio in Mant fluorescence experiments requires utilization of >50 nM GTPase. This is of some concern, as many biological interactions occur within the single nanomolar affinity range. In such cases, it is impossible to span the correct concentration range during the titration, and one should be aware that only upper limit values are obtained for the apparent dissociation constants.

A typical environmental sensitivity experiment is shown in Fig. 3, in which Cdc42Hs complexed with Mant-GppNHp is mixed with an interacting peptide from the target mPAK3 (termed PBD).[8,17] In Fig. 3A, the time course of the experiment is shown, where the emission at 440 nm from Cdc42Ha · Mant-GppNHp (excitation at 360 nm) is monitored following sequential additions of the PBD (denoted by black arrows). Following saturation, the reversibility of the reaction is ascertained by the addition of excess (nonfluorescent) Cdc42Hs · GTP, which complexes the PBD and restores the fluorescence to its initial level. Quantitative treatment of such data is shown in Fig. 3B, where fluorescence data are plotted against the final concentration of added PBD and are fitted to a bimolecular reaction model (see later). Dissociation constants extracted from the fits are shown for Cdc42Hs bound to either Mant-dGDP or Mant-GppNHp and show a 100-fold higher affinity of PBD toward the activated form of Cdc42Hs.

Anisotropy-Based Assays

The ability of a fluorescent molecule to alter the geometric polarization of the excitation light is a function of a number of physical parameters, including the rotational volume of the fluorophore.[18] Thus, fluorescence

[17] S. Bagrodia, S. Taylor, C. Creasy, J. Chernoff, and R. Cerione, *J. Biol. Chem.* **270**, 22731 (1995).
[18] J. R. Lackowicz, "Principles of Fluorescence Spectroscopy." Plenum Press, New York, 1983.

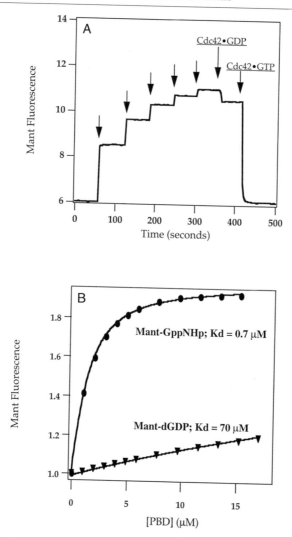

FIG. 3. Fluorescence assay for the interaction of Cdc42Hs · Mant nucleotides with the P21-binding domain (PBD) of mPAK3. (A) One micromolar Cdc42Hs · Mant-GppNHp was added to the fluorescence cuvette and Mant fluorescence (emission 440 nm; excitation 360 nm) was monitored. At the times indicated by the arrows, PBD was added to final concentrations of 1.0, 1.95, 2.98, 3.90, and 4.99 μM. To assess the reversibility and selectivity of the PBD $\rightleftharpoons$ Cdc42Hs · Mant-GppNHp interaction, wild-type (GDP-bound) Cdc42Hs and the Cdc42Hs (Q61L) mutant (GTP-bound) were added at the indicated times to final concentration of 50 μM. (B) Fluorescence titration of Cdc42Hs · Mant nucleotides with PBD. One micromolar of either Cdc42Hs · Mant-GppNHp (circles) or Cdc42Hs · Mant-dGDP (triangles) was titrated with the indicated amounts of PBD while monitoring Mant fluorescence. Solid lines represent fits of the data to a bimolecular association model as described in the text.

anisotropy of Mant nucleotides bound to GTPases provide for a sensitive assay for the binding of GTPases by their effectors, in cases where a significant change in molecular weight accompanies complex formation (i.e, when the molecular weight of the interacting protein is similar or larger then that of the GTPase). Essentially, the experiment is carried out in an similar manner to that described previously for environmental sensitivity assays, and the only difference is in the mechanical setup of the spectrofluorimeter. Due to the loss of optical throughput by the polarizing filters, GTPase · Mant nucleotide complexes have to be present at a somewhat higher concentration in the cuvette (0.5–1.0 μM), and optical slits are maintained fully open (16 nm spectral resolution).

Instrumentation. Although T-format (dual-channel) instruments are easier to use in anisotropy measurements and, generally, their manufacturer's driving software directly calculates the fluorescence anisotropy value (r), L-format (single-channel) instruments can be used with similar accuracy and sensitivity. Detailed operational protocols for the two systems are described in the instruments' operating manual and discussed in detail elsewhere.[18]

Procedure. The general experimental protocol is identical to that described earlier for environmental sensitivity assays. Following each effector addition, 5–10 separate readings of the anisotropy value are taken and averaged.

Data Analysis. Fluorescence titration data are fitted to a simple bimolecular association model. The association between the GTPase (C) and an effector (E)

$$C + E \rightleftharpoons C \cdot E \qquad (1)$$

can be described by a simple bimolecular interaction equation,

$$[CE] = \frac{[E]_T[C]}{[C] + K_d} \qquad (2)$$

where CE is the concentration of the formed complex at equilibrium, C is the concentration of free GTPase in solution, E_T is the total effector concentration, and K_d is the equilibrium dissociation constant. Because it cannot be assumed that only a small, negligible fraction of the total GTPase is bound (i.e., that $C \cong C_T$) and because the concentration of free GTPase is not measured easily, it is necessary to fit the data to the quadratic form of the solution. Thus, fluorescence titration data are fitted to the following expression:

$$F = \left(\frac{-K_d + E_T + C_T) + \sqrt{(K_d + E_T\,C_T)^2 - 4C_T E_T}}{-2C_T} (F_f - F_0) \right) + F_0 \qquad (3)$$

where F is the fluorescence intensity, E_T is total effector concentration at any point in the titration, and F_0 and F_f are the fluorescence intensities at the starting point and end point of the titration, respectively. When analyzing anisotropy data, anisotropy values (r, r_0, r_f) replace the fluorescence intensity terms (F, F_0, F_f) in Eq. (3). Fitting of data is done using a commercially available data analysis software such as Igor for Macintosh (WaveMetrics Inc., Eugene, OR).

[14] Purification and in Vitro Activity of Rho-Associated Kinase

By Mutsuki Amano, Yuko Fukata, Hiroaki Shimokawa, and Kozo Kaibuchi

Introduction

Rho-associated kinase (Rho kinase) was identified as a GTP · Rho-binding protein by affinity column chromatography of bovine brain membrane extract fraction.[1] Rho kinase was also identified as ROKα[2] and as ROCK2.[3] ROKβ[4]/ROCK1[5] is an isoform of Rho kinase/ROKα/ROCK2. The kinase domain of Rho kinase at the NH2-terminal end has 72% sequence homology with that of myotonic dystrophy kinase.[1] Rho kinase has a putative coiled-coil domain in its middle portion and a PH-like domain at its COOH-terminal end. GTP-bound Rho interacts with the COOH-terminal portion of the coiled-coil domain and activates the phosphotransferase activity of Rho kinase.[1]

Rho kinase is involved in the formation of stress fibers and focal adhesions,[4,6,7] the regulation of smooth muscle contraction,[8–10] neurite retrac-

[1] T. Matsui, M. Amano, T. Yamamoto, K. Chihara, M. Nakafuku, M. Ito, T. Nakano, K. Okawa, A. Iwamatsu, and K. Kaibuchi, EMBO J. 15, 2208 (1996).

[2] T. Leung, E. Manser, L. Tan, and L. Lim, J. Biol. Chem. 270, 29051 (1995).

[3] O. Nakagawa, K. Fujisawa, T. Ishizaki, Y. Saito, K. Nakao, and S. Narumiya, FEBS Lett. 392, 189 (1996).

[4] T. Leung, X. Q. Chen, E. Manser, and L. Lim, Mol. Cell. Biol. 16, 5313 (1996).

[5] T. Ishizaki, M. Maekawa, K. Fujisawa, K. Okawa, A. Iwamatsu, A. Fujita, N. Watanabe, Y. Saito, A. Kakizuka, N. Morii, and S. Narumiya, EMBO J. 15, 1885 (1996).

[6] M. Amano, K. Chihara, K. Kimura, Y. Fukata, N. Nakamura, Y. Matsuura, and K. Kaibuchi, Science 275, 1308 (1997).

[7] T. Ishizaki, M. Naito, K. Fujisawa, M. Maekawa, N. Watanabe, Y. Saito, and S. Narumiya, FEBS Lett. 404, 118 (1997).

tion,[11-13] and cytokinesis.[14] To understand the mode of action of Rho kinase, the probable substrates of Rho kinase have been identified by *in vitro* assay, and some of them, such as the myosin light chain (MLC) and myosin binding subunit (MBS) of myosin phosphatase, have been demonstrated to be physiological substrates of Rho kinase.

This article describes the procedures of purification and *in vitro* kinase assay of Rho kinase.

Materials

Glutathione-Sepharose and Mono Q 5/5 column are from Amersham Pharmacia Biotech (UK). GTPγS is from Roche Diagnostics (Germany). 3-[(3-Cholamidopropyl)dimethylammonio]propanesulfonic acid (CHAPS) is from Dojin (Japan). Leupeptin and (*p*-amidinophenyl)methanesulfonyl fluoride are from Wako Chemical Co. Ltd. (Japan). [γ-^{32}P]ATP is from Amersham Pharmacia Biotech. rsk kinase S6 substrate peptide (RRRLSSLRA) and PKC substrate peptide (RFARKGSLRQKNV-HEVK) are synthesized. Other materials and chemicals are obtained from commercial sources. Human RhoA cDNA is cloned into the *Bam*HI site of the pGEX-2T vector (Amersham Pharmacia Biotech) and glutathione *S*-transferase (GST)-RhoA is produced in *Escherichia coli*. Bovine Rho kinase cDNA corresponding to 6–553 amino acids, human myotonic dystrophy kinase-related Cdc42-binding kinase β (MRCKβ) cDNA corresponding to 1–550 amino acids, and human protein kinase N (PKN) corresponding to 581–942 amino acids are cloned into the *Bam*HI site of pAcYM1-GST, and recombinant baculoviruses are produced.[15]

[8] M. Amano, M. Ito, K. Kimura, Y. Fukata, K. Chihara, T. Nakano, Y. Matsuura, and K. Kaibuchi, *J. Biol. Chem.* **271**, 20246 (1996).
[9] K. Kimura, M. Ito, M. Amano, K. Chihara, Y. Fukata, M. Nakafuku, B. Yamamori, J. Feng, T. Nakano, K. Okawa, A. Iwamatsu, and K. Kaibuchi, *Science* **273**, 245 (1996).
[10] Y. Kureishi, S. Kobayashi, M. Amano, K. Kimura, H. Kanaide, T. Nakano, K. Kaibuchi, and M. Ito, *J. Biol. Chem.* **272**, 12257 (1997).
[11] M. Amano, K. Chihara, N. Nakamura, Y. Fukata, T. Yano, M. Shibata, M. Ikebe, and K. Kaibuchi, *Genes Cells* **3**, 177 (1998).
[12] H. Katoh, J. Aoki, A. Ichikawa, and M. Negishi, *J. Biol. Chem.* **273**, 2489 (1998).
[13] M. Hirose, T. Ishizaki, N. Watanabe, M. Uehata, O. Kranenburg, W. H. Moolenaar, F. Matsumura, M. Maekawa, H. Bito, and S. Narumiya, *J. Cell Biol.* **141**, 1625 (1998).
[14] Y. Yasui, M. Amano, K. Nagata, N. Inagaki, H. Nakamura, H. Saya, K. Kaibuchi, and M. Inagaki, *J. Cell Biol.* **143**, 1249 (1998).
[15] Y. Matsuura, R. D. Possee, H. A. Overton, and D. H. Bishop, *J. Gen. Virol.* **68**, 1233 (1987).

Methods

Loading of GTPγS onto GST–RhoA

The loading reaction is carried out in reaction mixture [20 mM Tris–HCl, pH 7.5, 10 mM EDTA, 1 mM dithiothreitol (DTT), 5 mM MgCl$_2$, 0.3% CHAPS, 30 μM GTPγS containing 10 μM GST–RhoA] for 20 min at 30°. At the same time, the binding efficiency of GST–RhoA with GTPγS is determined by the use of [^{35}S]GTPγS on a small scale.

Under these conditions, nearly 1 mol of GTPγS is bound with 1 mol of GST–RhoA. The reaction is stopped by the addition of MgCl$_2$ to give a final concentration of 15 mM.

Preparation of Bovine Brain Membrane Extract Fraction

Fresh bovine brain gray matter, 200 g, is cut into small pieces with a pair of scissors and suspended in 600 ml of homogenization buffer [25 mM Tris–HCl, pH 7.5, 1 mM DTT, 5 mM EGTA, 10 mM MgCl$_2$, 10 μM (p-amidinophenyl)methanesulfonyl fluoride, 1 μg/ml leupeptin, 10% sucrose]. The suspension is homogenized with a Potter–Elvehjem Teflon–glass homogenizer and filtered through four layers of gauze. The homogenate is centrifuged at 20,000g for 30 min at 4°. The precipitate is suspended in 400 ml of homogenizing buffer to prepare the crude membrane fraction. Proteins in this fraction are extracted by addition of an equal volume of homogenization buffer containing 4 M NaCl. After shaking for 1 hr at 4°, the membrane fraction is centrifuged at 20,000g for 1 hr at 4°. The supernatant is dialyzed against buffer A (20 mM Tris–HCl, pH 7.5, 1 mM EDTA, 1 mM 2-mercaptoethanol, 5 mM MgCl$_2$) three times. Solid ammonium sulfate is then added to a final concentration of 40% saturation. The 0–40% precipitate is dissolved into 20 ml of buffer B (20 mM Tris–HCl, pH 7.5, 1 mM EDTA, 1 mM DTT, 5 mM MgCl$_2$), dialyzed against buffer B three times, and used as the membrane extract.

Purification of Bovine Rho Kinase

Twenty-four nanomoles of GTPγS · GST–RhoA is incubated with 1 ml of glutathione-Sepharose for 1 hr at 4° with rotating, and the beads are packed onto disposable column (Pierce, Rockford, IL). The beads are washed with 10 ml of buffer B, 10 ml of buffer B containing 200 mM NaCl, and 10 ml of buffer B containing 1% CHAPS and are equilibrated with 10 ml of buffer B. The membrane extract (20 ml; about 300 mg of protein) is passed through a 1-ml glutathione-Sepharose column. The pass-through fraction is loaded onto the glutathione-Sepharose column containing

GTPγS · GST–RhoA. After the column is washed with 10 ml of buffer B and subsequent 10 ml of buffer B containing 200 mM NaCl, proteins are eluted by the addition of 10 ml of buffer B containing 1% CHAPS, and 1-ml fractions each are collected. Rho kinase appears in fractions 1–10 (Fig. 1). About 20 μg of Rho kinase is obtained in these fractions. Fractions 3–10 are used as partial purified Rho kinase. Rho kinase is also eluted in the fractions of buffer B containing 200 mM NaCl, but they include PKN and p190Rho–GAP. Fractions 2–10 of buffer B containing 1% CHAPS are mixed and diluted with an equal volume of buffer B and are subjected to a Mono Q 5/5 column equilibrated with buffer B. After washing with 10 ml of buffer B, proteins are eluted with a linear gradient of NaCl (0–0.5 M) in 15 ml of buffer B, and 0.5-ml fractions each are collected. Rho kinase appears as a single peak in fractions 10–12 (used as purified Rho kinase).

Purification of Baculovirus-Expressed GST–Rho Kinase-Catalytic Domain (CAT)

Three flasks (175 cm²) of *Spodoptera frugiperda* ovary (Sf9) cells are infected with baculovirus carrying GST–Rho kinase–CAT and are incubated at 27° for 3 days.[6,15] Three days after the infection, most of the cells are detached from the flask. The cells are collected and washed with phosphate-buffered saline. The cells are lysed with 5 ml of buffer C (50 mM Tris–HCl, pH 7.5, 2 mM EGTA, 1 mM DTT) containing 10 μM (*p*-amidinophenyl)methanesulfonyl fluoride, 1 μg/ml leupeptin, and 10% sucrose by sonication and are centrifuged at 100,000g for 1 hr at 4°. The supernatant is applied on a 0.4-ml glutathione-Sepharose column, and the

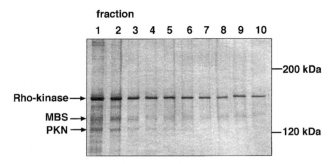

FIG. 1. Elution profile of Rho kinase on a glutathione-Sepharose column containing GTPγS · GST–RhoA. Twenty microliters of each fraction eluted with buffer B containing 1% CHAPS is applied per lane and visualized by silver staining. Rho kinase appears in fractions 1–10. PKN appears in fractions 1 and 2. In many cases, a small amount of MBS also appears in fractions 1–10.

pass fraction is reapplied onto the column. After washing with 4 ml of buffer C, bound proteins are eluted with 2 ml of buffer C containing 10 mM glutathione (used as recombinant constitutively activated Rho kinase). About 300 μg of GST–Rho kinase–CAT is obtained. If necessary, the buffer is changed by dialysis or by concentrators. GST–MRCKβ–CAT or GST–PKN–CAT is also prepared by the same method.[16]

Kinase Assay

The kinase reaction is carried out in 50 μl of kinase buffer [50 mM Tris–HCl, pH 7.5, 1 mM EDTA, 1 mM EGTA, 1 mM DTT, 5 mM MgCl$_2$ (0.1% CHAPS in case of using partially purified Rho kinase)] containing 10–100 μM [γ-^{32}P]ATP (0.2–20 GBq/mmol), substrate, and purified Rho kinase or GST–Rho kinase–CAT. The reaction is started by the addition of ATP. After incubation at 30°, the reaction mixture is boiled in SDS sample buffer and resolved by sodium dodecyl sulfate–polyacrylamide gel electrophoresis (SDS–PAGE). The radiolabeled bands are visualized by autoradiography or by an imaging analyzer (Fuji, Japan). To examine the phosphorylation of peptide (such as rsk kinase S6 substrate peptide; RRRLSSLRA), the reaction is carried out at a concentration of 40 μM peptide. After incubation at 30°, the reaction mixture is spotted onto a Whatman (Clifton, NJ) p81 paper and the paper is washed with 75 mM phosphoric acid three times. The incorporation of ^{32}P into the peptide is assessed by scintillation counting.

Inhibition of Activity of Rho Kinase by Drugs

The activity of Rho kinase is inhibited by various drugs, including staurosporine, Y-27632, and HA-1077, or by the COOH-terminal portion of Rho kinase *in vitro*.[16–18] Inhibition of the activity of GST–Rho kinase–CAT by protein kinase inhibitor Y-27632 is shown (Fig. 2). The kinase reaction is carried out as described earlier in 50 μl of kinase buffer containing 100 μM ATP, 40 μM rsk kinase S6 substrate peptide, and 0.5 pmol of GST–Rho kinase–CAT or 100 μM ATP, 1 μM of MLC, and 0.2 pmol of GST–Rho kinase–CAT with indicated concentrations of Y-27632. After incubation for 10 min at 30°, the reaction is stopped and the incorporation of ^{32}P into

[16] M. Amano, K. Chihara, N. Nakamura, T. Kaneko, Y. Matsuura, and K. Kaibuchi, *J. Biol. Chem.* **274,** 32418 (1999).
[17] M. Uehata, T. Ishizaki, H. Satoh, T. Ono, T. Kawahara, T. Morishita, H. Tamakawa, K. Yamagami, J. Inui, M. Maekawa, and S. Narumiya, *Nature* **389,** 990 (1997).
[18] H. Shimokawa, M. Seto, N. Katsumata, M. Amano, T. Kozai, T. Yamawaki, K. Kuwata, T. Kandabashi, K. Egashira, I. Ikegaki, T. Asano, K. Kaibuchi, and A. Takeshita, *Cardiovasc. Res.* **43,** 1029 (1999).

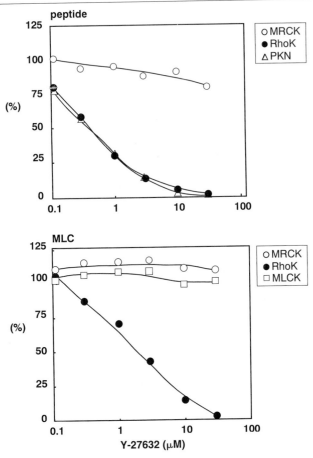

FIG. 2. Inhibition of the activity of GST–Rho kinase–CAT by Y-27632. The activities of recombinant GST–Rho kinase–CAT (closed circle), GST–MRCKβ–CAT (open circle), GST–PKN–CAT (open triangle), and native MLC kinase (MLCK) (open square) are assayed in the presence of the indicated concentrations of Y-27632. The S6 peptide or PKC peptide is used as the substrate for Rho kinase and MRCKβ or for PKN, respectively (top). MLC is used as the substrate for Rho kinase, MRCKβ, or MLCK (bottom).

the substrates is assessed. Under these conditions, about 250 or 5 pmol of phosphate is incorporated into 2 nmol of peptide or 50 pmol of MLC, respectively, without inhibitor. Rho kinase activity is inhibited by Y-27632, and apparent IC_{50} values of Y-27632 are about 0.5 μM for peptide and about 2.5 μM for MLC, respectively. For comparison, a kinase reaction using GST–MRCKβ–CAT, GST–PKN–CAT, or MLC kinase is performed.

Comments

This article described the purification method and *in vitro* assay of Rho kinase. The purified Rho kinase or GST–Rho kinase–CAT is split into small aliquots and stored at −80°. Repeated freezing and thawing decreases both the activity and the Rho-dependent activation of Rho kinase and completely loses the activity of GST–Rho kinase–CAT. Both partially purified Rho kinase and purified Rho kinase are activated by GTPγS-loaded GST–RhoA produced in *Escherichia coli* and more efficiently by that produced in the baculovirus system. It is also noted that the activity of Rho kinase is affected by some lipids, including arachidonic acid, besides Rho.[19]

Acknowledgments

We thank Dr. M. Ito (Mie University School of Medicine, Japan) for providing MLC and MLC kinase and Yoshitomi Pharmaceutical Industries Ltd. (Japan) for providing Y-27632.

[19] J. Feng, M. Ito, Y. Kureishi, K. Ichikawa, M. Amano, N. Isaka, K. Okawa, A. Iwamatsu, K. Kaibuchi, D. J. Hartshorne, and T. Nakano, *J. Biol. Chem.* **274,** 3744 (1999).

[15] Purification and *in Vitro* Activities of p21-Activated Kinases

By Charles C. King, Abina M. Reilly, and Ulla G. Knaus

Introduction

PAKs (p21-activated kinases) were first identified as targets of the Rho GTPases Rac and Cdc42 and have subsequently been implicated in various cellular processes regulated by Rac and Cdc42.[1,2] A comparison of PAK isoforms reveals highly conserved functional domains, mainly the N-terminal regulatory domain, including the p21-binding (PBD) domain, an autoinhibitory domain, several proline-rich motifs, and the C-terminal catalytic serine/threonine kinase domain. Activation of PAKs requires a release mechanism to counteract kinase autoinhibition by intramolecular interac-

[1] S. Bagrodia and R. A. Cerione, *Trends Cell Biol.* **9,** 350 (1999).
[2] U. G. Knaus and G. M. Bokoch, *Int. J. Biochem. Cell Biol.* **30,** 857 (1998).

tions,[3-6] which mask the catalytic domain. Generally, the open and active PAK kinase configuration is obtained by binding of active Rho GTPases (Rac, Cdc42)[7] or lipids[8] to the regulatory domain. Other modes of PAK kinase regulation exist, involving proteolytic cleavage releasing the catalytic domain[9-11] and interaction with G-protein $\beta\gamma$ subunits,[12,13] adapter proteins containing src homology domains,[14-18] or phosphatases,[19] to name a few.

This article describes methods to generate recombinant, constitutively active PAKs for substrate phosphorylation studies and wild-type PAKs for kinase activation studies. Regulation of PAK activity using dominant active mutants or *in vitro* GTP-labeled forms of Rac and Cdc42, which bind to the PBD domain [including the minimally required CRIB motif[20,21] ISxP(x)$_{2-4}$ FxHxxHVG], will be illustrated. We will also introduce a second mechanism, stimulation of PAK kinase activity with lipid compounds, which leads to PAK autophosphorylation and substrate phosphorylation with similar kinetics as GTPase stimulation.[8] Identification of PAKs and detection of the PAK activation state in cells can be achieved by in-gel kinase assays with

[3] Z.-S. Zhao, E. Manser, X.-Q. Chen, C. Chong, T. Leung, and L. Lim, *Mol. Cell. Biol.* **18,** 2153 (1998).

[4] J. A. Frost, A. Khokhlatchev, S. Stippec, M. White, and M. H. Cobb, *J. Biol. Chem.* **273,** 28191 (1998).

[5] H. Tu and M. Wigler, *Mol. Cell. Biol.* **19,** 602 (1999).

[6] F. T. Zenke, C. C. King, B. P. Bohl, and G. M. Bokoch, *J. Biol. Chem.* **274,** 32565 (1999).

[7] E. Manser, T. Leung, H. Salihuddin, Z.-S. Zhao, and L. Lim, *Nature* **367,** 40 (1994).

[8] G. M. Bokoch, A. M. Reilly, R. H. Daniels, C. C. King, A. Olivera, S. Spiegel, and U. G. Knaus, *J. Biol. Chem.* **273,** 8137 (1998).

[9] T. Rudel and G. M. Bokoch, *Science* **276,** 1571 (1997).

[10] N. Lee, H. MacDonald, C. Reinhard, R. Halenbeck, A. Roulston, T. Shi, and L. T. Williams, *Proc. Natl. Acad. Sci. USA* **94,** 13642 (1997).

[11] B. N. Walter, Z. Huang, R. Jakobi, P. T. Tauzon, E. S. Alnemri, G. Litwack, and J. A. Traugh, *J. Biol. Chem.* **273,** 28733 (1998).

[12] T. Leeuw, C. Wu, J. D. Schrag, M. Whiteway, D. Y. Thomas, and E. Leberer, *Nature* **391,** 191 (1998).

[13] J. Wang, J. A. Frost, M. H. Cobb, and E. M. Ross, *J. Biol. Chem.* **274,** 31641 (1999).

[14] M. L. Galisteo, J. Chernoff, Y. C. Su, E. Y. Skolnik, and J. Schlessinger, *J. Biol. Chem.* **271,** 20997 (1996).

[15] G. M. Bokoch, Y. Wang, B. P. Bohl, M. A. Sells, L. A. Quilliam, and U. G. Knaus, *J. Biol. Chem.* **271,** 25746 (1996).

[16] W. Lu, S. Katz, R. Gupta, and B. J. Mayer, *Curr. Biol.* **7,** 85 (1997).

[17] R. H. Daniels, P. S. Hall, and G. M. Bokoch, *EMBO J.* **17,** 754 (1998).

[18] W. Lu and B. J. Mayer, *Oncogene* **18,** 797 (1999).

[19] R. S. Westphal, R. Lane Coffee, A. Marotta, S. L. Pelech, and B. E. Wadzinski, *J. Biol. Chem.* **274,** 687 (1999).

[20] P. D. Burbelo, D. Drechsel, and A. Hall, *J. Biol. Chem.* **270,** 29071 (1995).

[21] G. Thompson, D. Owen, P. A. Chalk, and P. N. Lowe, *Biochemistry* **37,** 7885 (1998).

specific substrates,[22] by ^{32}P labeling and tryptic phosphopeptide mapping or MALDI-MS analysis, both of the latter methods will be described in this article.

Expression and Purification of Recombinant PAKs

Purification protocols for endogenous PAKs from rat brain extracts or neutrophil cytosol have been published in detail[7,23,24] and will not be commented any further herein. The generation of recombinant wild-type and constitutively active PAK proteins is the method of choice to either establish relevant upstream, PAK-regulatory stimuli or to screen cell extracts or expression libraries for PAK substrates. Expression and purification of recombinant PAK from *Escherichia coli* or baculovirus-infected *Spodoptera frugiperda* (Sf9) cells is achieved readily using N-terminal epitope tags such as glutathione *S*-transferase (GST) or six consecutive histidine residues. These tags allow one-step purification on affinity resins containing immobilized glutathione (for GST tags) or metal affinity resins [for hexa-histidine (His$_6$) tags], which interact specifically with the fusion protein. By optimizing the buffers used for loading and elution of the beads, nonspecific binding or retention of contaminating host proteins is minimized and 90–100% pure recombinant PAK protein is prepared. We encountered difficulties to subsequently cleave the purified PAK fusion protein with human thrombin (GST) or rTEV protease (His) to remove the epitope tag. Treatment with these enzymes caused proteolytic breakdown of PAK. It is therefore preferable to use the intact fusion protein. Expression of wild-type PAK in *E. coli* leads independently of the size or location of the epitope tag to constitutive kinase activity. We assume that an activating phosphorylation event takes place during the expression. Variations in the *E. coli* strain, the growth temperature, or induction conditions will not affect constitutive PAK activity. Certain mutations in the catalytic domain of the kinase abolish *E. coli*-induced PAK kinase stimulation and allow isolation of a wild-type protein with low basal kinase activity.[3] The attenuated PAK kinase (PAK1 L404S) undergoes autophosphorylation identical to wild-type PAK upon Rac/Cdc42 activation *in vitro*. Recombinant PAK, expressed and purified from Sf9 cells, is partially active and can be activated by GTP-Rac or GTP-Cdc42 two- to threefold.

To study upstream regulators and activation mechanisms of PAK, one

[22] C. C. King, L. A. Sanders, and G. M. Bokoch, *Methods Enzymol.* **325** [29] (2000) (this volume).

[23] G. A. Martin, G. Bollag, F. McCormick, and A. Abo, *EMBO J.* **14**, 1970 (1995).

[24] E. Manser, T. Leung, and L. Lim, *Methods Enzymol.* **256**, 215 (1995).

has to obtain wild-type PAK with low levels of basal kinase activity. We routinely express recombinant N-terminal epitope-tagged PAKs (Myc, hemagglutinin, His), subcloned into various expression vectors via lipid-mediated transient transfection in mammalian cells. Optimal choice of the cell line will combine high-expression efficiency with almost undetectable PAK kinase activity under unstimulated conditions. The epitope tag facilitates the detection of PAK expression by immunoblotting without using PAK-specific antibodies and allows immunoprecipitation of the kinase. Immobilized PAK on beads can be directly subjected to *in vitro* activation followed by kinase assays. We advise against epitope tags placed after the last C-terminal residue of PAK, which in our experience disturbed GTPase-mediated kinase activation. All plasmids encoding wild-type PAK should be propagated at 30° in *E. coli* to avoid lethal effects, enabling selection of PAK point mutants with altered activity.[25] We currently use two *E. coli* strains, SURE or Solopak Gold (Stratagene, La Jolla, CA), which exhibit acceptable growth rates at this temperature.

Methods

Recombinant GST-PAK is expressed from pGEX vectors (Amersham-Pharmacia Biotech, Piscataway, NJ) in *E. coli* strain BL21, induced with 0.2 mM isopropyl-β-D-thiogalactopyranoside (IPTG) for 3–5 hr at 30°, and purified on glutathione beads according to the manufacturer's protocol (Amersham-Pharmacia) except for the use of PAK lysis buffer I [50 mM HEPES, pH 7.5, 5 mM $MgCl_2$, 150 mM NaCl, 20% (v/v) glycerol, 5 mM 2-mercaptoethanol, 1 mM phenylmethylsulfonyl fluoride, 1 μg/ml of aprotinin and pepstatin]. Recombinant histidine-PAK is expressed from the pQE-30 vector (Qiagen, Valencia, CA) in *E. coli* strain M15 or from pPROEX-HTb (Life Technologies, Rockville, MD) in *E. coli* strain BL21 or SURE, induced as described earlier, and purified on either Ni^{2+}-NTA beads (Qiagen) or Co^{2+}-containing TALON beads (Clontech Lab., Palo Alto, CA). Adjust the PAK lysate (PAK lysis buffer I) to a final concentration of 10 mM imidazole and 0.5% Nonidet P-40 (NP-40) before incubating 1 hr with the metal affinity beads at 4°. After several washes with the same buffer, elute stepwise with increasing amounts of imidazole. The majority of pure His-PAK will elute at 100 mM imidazole. Dialyze the pure PAK protein into 50 mM HEPES, pH 7.5, 5 mM $MgCl_2$, 100 mM NaCl, 5 mM 2-mercaptoethanol, 20% glycerol at 4°, snap freeze in liquid nitrogen, and store in aliquots at −70°. Recombinant histidine-PAK expressed from pAcHLT-A,B,C in Sf9 cells using the BaculoGold system (BD PharMingen,

[25] E. Manser, H. Y. Huang, T. H. Loo, X. Q. Chen, J. M. Dong, T. Leung, and L. Lim, *Mol. Cell. Biol.* **17**, 1129 (1997).

San Diego, CA) is essentially purified as described for *E. coli*-derived His-PAK.

Recombinant myc-PAK cDNA, subcloned into mammalian expression vectors (pCMV, pRK5, pJ3H), is transiently transfected via lipid-mediated methods (Lipofectamine Plus, Life Technologies) according to the manufacturer's protocol. Use 1–2 μg of DNA for each 35-mm dish or 5–7 μg for a 100-mm dish and transfect HeLa or Cos cell lines at 70–85% confluency. After expression for 24–40 hr, wash cells twice in phosphate-buffered saline at room temperature, and scrape into ice-cold PAK lysis buffer II (25 mM Tris–HCl, pH 7.5, 1 mM EDTA, 0.1 mM EGTA, 5 mM MgCl$_2$, 1 mM dithiothreitol, 150 mM NaCl, 10% glycerol, 2% Nonidet P-40, 50 IU/ml aprotinin, 1 mM phenylmethylsulfonyl fluoride, 2 μg/ml leupeptin). Solubilize the cells for 10 min on ice, vortex, and spin for 5 min at 14,000 rpm at 4°. Collect the supernatant, quantitate the protein content of the lysate, and freeze in aliquots.

Assess the expression and purity of the purified recombinant PAK proteins with gel electrophoresis followed by Coomassie blue staining or Western blotting with anti-tag or anti-PAK antibodies. PAK autophosphorylation and substrate phosphorylation will serve as an indicator for PAK kinase activity. Subject purified or immunoprecipitated, unstimulated PAK to *in vitro* kinase reaction as described later.

Activation of PAKs by GTPases

Certain GTPases of the Rho family, including Rac 1-3, Cdc42, and TC10, bind in their active GTP-bound form to the PBD domain of PAK in its inactive conformation, thereby reversing a negative regulatory, intramolecular interaction and inducing autophosphorylation and kinase activity.[1] To elucidate potential regulatory mechanism exerting this effect on PAK, the use of recombinant wild-type PAK in unstimulated form is necessary. The interaction of proteins with PAK can be determined directly with purified compounds *in vitro* or by coexpression experiments in cells. The binding of proteins at different activation or processing states (GDP/GTP form of GTPases, nonphosphorylated versus phosphorylated form of kinases, etc., lipid modifications) is detectable by overlay techniques, by immunoprecipitations, or by immobilizing one of the proteins on beads. In most of the cases when PAK kinase stimulation occurs, direct binding between the activating protein and PAK has been observed. To assess PAK kinase activation, by GTPases for example, the GTPase is incubated in its active (GTP) form with the wild-type kinase, which is immobilized on beads. The reaction mixture is subjected to a kinase reaction with or without inclusion of a substrate, followed by gel electrophoresis and autoradiography. The use of constitutively active GTPase mutants allows one to perform

1 2 3

65 kDa

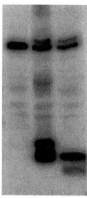

22 kDa

FIG. 1. *In vivo* activation of PAK1 by constitutively active Rac1 and Cdc42. Wild-type PAK1 was coexpressed for 24 hr with pRK5 vector (1), Myc-Rac1V12 (2), and Myc-Cdc42L61 (3) in HeLa cells. Stimulation of PAK1 results in slower migrating, hyperphosphorylated bands (65 kDa), which are visualized by immunoblotting with anti-Myc antibodies. Expression levels of two differently processed forms of Rac1 and Cdc42 are shown at about 22 kDa in the same blot. Note: Rac1 contains an additional tag, altering its size and mobility.

coexpression experiments with wild-type PAK in adherent cells to detect *"in vivo"* PAK stimulation. Transient transfection of active Rac1V12 or Cdc42V12 with PAKwt causes activation of the kinase.[26] The cell lysates can be analyzed for kinase activity using *in vitro* kinase assays (with or without substrate) and/or in-gel kinase assays with PAK immunoprecipitates or, as demonstrated in Fig. 1, by Western blotting. In Fig. 1, the GTPase and PAK are resolved easily by electrophoresis, but both contain a Myc tag allowing for simultaneous detection by immunoblotting. Activation of PAK correlates with the appearance of several hyperphosphorylated bands migrating slower than wild-type PAK.[26] If radiolabeled Protein A is used for detection, the ratio of lower to upper bands can be quantitated and can be adjusted to the varying expression level of different GTPases.

Methods

For *in vitro* kinase activation studies, we use HeLa or COS cell lysates overexpressing PAKwt. Immunoprecipitate PAK (50 μl Cos lysate/IP) with anti-Myc (9E10) antibody (2 μl/IP) on protein G beads (50 μl 1 : 1 slurry/ IP, Amersham-Pharmacia) at 4° and incubate equal aliquots of washed beads with 0.5–1 μg of freshly prepared, GTPγS-labeled GTPase for 2 min

[26] U. G. Knaus, Y. Wang, A. M. Reilly, D. Warnock, and J. H. Jackson, *J. Biol. Chem.* **273,** 21512 (1998).

on ice. The preparation of purified GTPases, the determination of their activity, and *in vitro* labeling of Rho GTPases with guanine nucleotides have been described elsewhere in detail.[27-29] Perform kinase assays by incubating the PAK–GTPase complex on beads with 20 μM ATP and 5 μCi of $[\gamma$-^{32}P]ATP in kinase buffer (50 mM HEPES, pH 7.5, 10 mM MgCl$_2$, 2 mM MnCl$_2$, 0.2 mM dithiothreitol). To measure substrate phosphorylation, add 1–2 μg of pure substrate (myelin basic protein, histone 4) to the kinase reaction. Incubate for 20–30 min at 30° with gentle shaking and stop by adding Laemmli sample buffer (for analysis on SDS–PAGE and autoradiography) or by adding ice-cold 2× PAK lysis buffer II (for quantitative analysis of substrate phosphorylation on phosphocellulose filters, Pierce, Rockford, IL).

To assess PAK activation by coexpression with active GTPases in cell lines, transfect tagged, constitutive active Rac or Cdc42 mutants (i.e., RacV12, RacL61) with tagged wild-type PAK for 24 hr. Adjust the DNA ratio between both plasmids in preliminary transfection experiments to ensure equal expression.[26] Subject cell lysates in modified PAK lysis buffer II (including 5 μM okadaic acid and 1 μM microcysteine LR) to SDS–PAGE and Western blotting to detect PAK mobility shifts and kinase activity (use ^{125}I-labeled protein A for detection) or determine kinase activity of immunoprecipitated PAK with in-gel kinase assays.

Activation of PAKs by Lipids

Certain biologically active lipids can stimulate PAK kinase activity *in vitro* and in adherent cells.[8] Similar results with slightly different lipid specificity have been shown for the PAK-related *Acanthamoeba* myosin I heavy chain kinase (MIHCK).[30,31] C$_2$-ceramides, which were inactive in the first two studies, are reported to either activate PAK in lipid-treated T cells[32] or to inhibit chemoattractant-stimulated PAK activation in neutrophils.[33] *In vitro* activation of wild-type PAK or PAK mutants revealed that the binding site of lipids on PAK is in close proximity or overlapping with the GTPase-binding site.[8] Detection of hyperphosphorylated PAK bands generated by

[27] A. J. Self and A. Hall, *Methods Enzymol.* **256**, 3, (1995).
[28] A. J. Self and A. Hall, *Methods Enzymol.* **256**, 67 (1995).
[29] E. Manser, T. Leung, and L. Lim, *Methods Mol. Biol.* **84**, 295 (1998).
[30] H. Brzeska, R. Young, U. Knaus, and E. D. Korn, *Proc. Natl. Acad. Sci. U.S.A.* **96**, 394 (1999).
[31] S.-F. Lee, A. Mahasneh, M. de la Roche, and G. P. Cote, *J. Biol. Chem.* **273**, 27911 (1998).
[32] S. Kaga, S. Ragg, K. A. Rogers, and A. Ochi, *J. Immunol.* **160**, 4182 (1998).
[33] J. P. Lian, R. Huang, D. Robinson, and J. A. Badwey, *J. Immunol.* **161**, 4375 (1998).

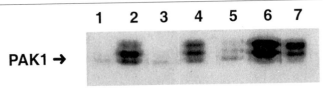

FIG. 2. Stimulation of PAK1 autophosphorylation by active Cdc42 and lipids *in vitro*. Wild-type PAK1 immunoprecipitated from Cos cell lysates was incubated with buffer (1), GTPγS-labeled Cdc42 (2), 10 μM asialoganglioside G_{M1} (3), 10 μM monosialoganglioside G_{M1} (4), 10 μM monosialoganglioside G_{M3} (5), 10 μM disialoganglioside G_{D1} (6), or 10 μM disialoganglioside G_{D3} (7) and subjected to kinase assays. PAK1 autophosphorylation was visualized on 7% SDS–PAGE by autoradiography. Other lipids were tested elsewhere.[8]

in vitro activation of immunoprecipitated PAKwt with gangliosides is shown in Fig. 2.

Methods

Prepare Cos cell lysates overexpressing wild-type PAK in PAK lysis buffer II and generate PAK immunoprecipitates on protein-G beads. Lipid stocks are made in the appropriate solvent such as $CHCl_3$, ethanol, or dimethyl sulfoxide (DMSO). Prior to use, dry lipids under N_2 and resuspend by sonication in a bath sonicator in 50 mM HEPES, pH 8.0, 1 mM EDTA or in Tris–HCl, pH 7.5, until a translucent suspension is obtained. Generate mixed lipid vesicles by mixing the lipid stocks in molar ratios varying from 1:10 to 1:3 prior to sonication in buffer. Incubate lipids or mixed lipid vesicles with PAK immunoprecipitates in kinase buffer, adding 1–2 μg of substrate. Incubation conditions and analysis of PAK are as described for *in vitro* GTPase-mediated activation of PAK.

Analysis of PAK Activation and Autophosphorylation by Two-Dimensional Peptide Mapping

Activated PAK is autophosphorylated on seven or eight sites, which have been mapped in PAK1 to S21, S57, S144, S149, S199, S204, and T423[25] and in PAK2 to S19, S20, S55, S141, S165, S192, S197, and T402.[34] Sites S141, S165, and T402 in PAK2 need the presence of active Rac or Cdc42 for phosphorylation, whereas autophosphorylation of the other sites takes place with MgATP alone.[34] Phosphorylation of the threonine residue in the PAK kinase domain (T423 in PAK1, T402 in PAK2) is critical for the catalytic function of the protein kinase.[6,25,34,35] Determination of

[34] A. Gatti, Z. Huang, P. T. Tuazon, and J. A. Traugh, *J. Biol. Chem.* **274,** 8022 (1999).
[35] J.-S. Yu, W.-J. Chen, M.-H. Ni, W.-H. Chan, and S.-D. Yang, *Biochem. J.* **334,** 121 (1998).

(auto)phosphorylation sites was achieved by the tryptic digest of ^{32}P-labeled purified or immunoprecipitated PAKs in their inactive or active form, followed by either two-dimensional (2D) phosphopeptide mapping on TLC plates[25,35,36] or 2D gel electrophoresis.[34,37] These methods can be easily used to compare phosphorylation sites of various stimuli on immunoprecipitated wild-type PAK. Furthermore, they provide the means to analyze endogenous PAK, activated by unidentified stimuli, ortho[^{32}P]phosphate-labeled cells, or to identify PAK-related kinases with similar catalytic domains. In both cases, commercially available anti-PAK antibodies directed to the N or C terminus can be used for immunoprecipitation. A 2D phosphopeptide map of PAK1 stimulated with Rac1-GTPγS is shown in Fig. 3. This procedure was performed according to the procedure of Boyle et al.[38] with minor modifications.

Methods

Separate phosphorylated PAK bands on 6.5% polyacrylamide gels, transfer to nitrocellulose, and detect by autoradiography. Excise each band separately from the nitrocellulose, soak in PVP-360 for 30 min at 37°, wash four times with H_2O, and resuspend in 50 mM NH_4HCO_3, pH 8.3, containing 1 μg trypsin for 14 hr at 37°. Remove the supernatant containing the tryptic peptides and lyophilize. Resuspend the lyophilized samples in 2 μl of pH 1.9 electrophoresis buffer [2.2% (v/v) formic acid, 8% (v/v) acetic acid], spot onto 100-μm-coated cellulose plates (EM Science, Gibbstown, NJ) in 0.5-μl aliquots, and electrophorese for 40 min at 1300 V in a Multiphor II horizontal electrophoresis unit (Amersham-Pharmacia) in pH 1.9 electrophoresis buffer. Air dry plates and chromatograph in buffer containing 62.5% (v/v) isobutyric acid, 1.9% (v/v) *n*-butanol, 4.8% (v/v) pyridine, and 2.9% (v/v) glacial acetic acid. ^{32}P-labeled tryptic peptides can be detected by autoradiography on Kodak (Rochester, NY) X-AR film. To analyze phosphopeptide sequences, scrape separated, radioactive spots from the cellulose plate, elute the peptides with 0.1% trifluoroacetic acid, and subject to high-performance liquid chromatography (reversed-phase C_{18} column). The column is equilibrated in 2% acetonitrile containing 0.1% trifluoroacetic acid, and peptides are eluted at a flow rate of 1 ml/min with a linear gradient of 2–50% acetonitrile containing 0.1% trifluoroacetic acid developed over 100 min. Monitor the elution of peptides at A_{219} and by Cerenkov counting of the fractions. Determine the sequence of radioactive peaks representing the phosphopeptide by either automated sequencing or manual Edman degradation.

[36] G. E. Benner, P. B. Dennis, and R. A. Masaracchia, *J. Biol. Chem.* **270**, 21121 (1995).
[37] A. Gatti and J. A. Traugh, *Anal. Biochem.* **266**, 198 (1999).
[38] W. J. Boyle, P. van der Geer, and T. Hunter, *Methods Enzymol.* **201**, 110 (1991).

A

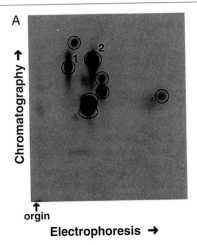

↑
orgin
Electrophoresis →

B

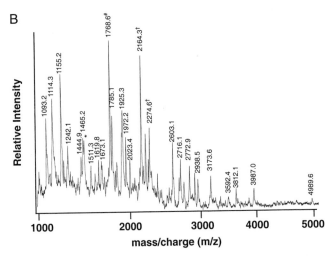

FIG. 3. Analysis of PAK1 by two-dimensional peptide mapping and MALDI-TOF mass spectroscopy. (A) A typical two-dimensional phosphopeptide map of PAK1 stimulated with GTPγS-loaded Rac1. Major phosphorylation sites are circled. Sites corresponding to threonine 423 peptides are numbered 1 and 2. (B) Unstimulated PAK1 was analyzed in a 1.14-m MALDI-TOF mass spectrophotometer with dynamic extraction (Dynamo ThermoBioAnalysis, BioMolecular Instruments, Sante Fe, NM). Peptides were analyzed by the ProFound database at Rockefeller University (http://www.prowl.rockefeller/Prowl/pepmap.html). External calibration was based on a mixture of angiotensin, renin tetradecapeptide, and insulin. Internal calibration was based on autolytic peaks of trypsin (labeled with a dagger). Peaks containing oxidized methionines are labeled with asterisks. The peak labeled with a pound sign is a tryptic fragment containing the carboxy-terminal portion of the Myc tag and the amino-terminal portion of PAK1.

Analysis of PAK Activation and Autophosphorylation by
MALDI-TOF MS

The technique of matrix-assisted laser desorption/ionization mass spectroscopy (MALDI MS) is becoming a mainstream technique for the analysis of phosphorylation sites on proteins. This method is similar to analysis of PAK by 2D peptide mapping, but does not require the use of radioisotopes. Additionally, one can use less than 1 μg of protein to obtain information about the phosphorylation state of multiple amino acids on PAK in one step.[39] This section details a simple method for the analysis of PAK phosphorylation sites by MALDI-TOF MS after an *in vitro* kinase assay based on the procedure of Zhang *et al.*[40] Although this technique can be adapted for use on endogenous PAK within cells, this technique is based on the use of transiently transfected Myc-PAK activated by Cdc42-GTPγS *in vitro*. Figure 3B shows a representative spectrograph of Myc-PAK1 after trypsin digestion.

Methods

Fifty microliters of Cos lysate expressing PAKwt provides sufficient protein for one MALDI-MS analysis sample. Prepare two identical samples for each condition (i.e., PAKwt, PAKwt+Cdc42-GTPγS) by anti-Myc immunoprecipitation followed by treatment with or without GTPγS-labeled Cdc42 as described previously. Load each sample onto a 7% SDS polyacrylamide gel, separating the samples by a prestained molecular weight marker (Life Technologies). The duplicates are used to generate two identical parts of the gel. Run the gel at 25 mA through the stacking gel and 40 mA through the separating gel. Cut the gel in half and transfer one-half to nitrocellulose for anti-Myc immunoblotting. The other half should soak in 10% acetic acid/50% methanol. Compare the position of PAK on the immunoblot with the position of PAK on the gel, based on the position of the prestained molecular weight markers. Excise the band of interest and place into a sterile Eppendorf tube. Soak the gel piece in water for 30 min, changing the liquid every 10 min. Wash the gel piece three times with 500 μl NH$_4$HCO$_3$ (25 mM, pH 7.8–8.4) and grind medium fine with a Kontes dispensable pestle. Cover the gel piece with 25 mM NH$_4$HCO$_3$ (about 150 μl) and add 3 μl of trypsin (0.1 μg/μl, dissolved in 1 mM HCl, sequencing grade, Roche Molecular Biochemicals, Indianapolis, IN). Vortex for 2 min at room temperature and incubate from 1 hr to overnight at 37°.

[39] J. Szczepanowska, X. Zhang, C. J. Herring, J. Qin, E. D. Korn, and H. Brzeska, *Proc. Natl. Acad. Sci. U.S.A.* **94**, 8503 (1997).
[40] X. Zhang, C. J. Herring, P. R. Ramano, J. Szczepanowska, H. Brzeska, A. G. Hinnebusch, and J. Quinn, *Anal. Chem.* **70**, 2050 (1998).

To extract the peptides, add 100 μl of 100% acetonitrile to each tube, vortex for 2 min, centrifuge at 14,000 rpm for 1 min at room temperature, and remove the supernatant into a fresh Eppendorf tube with a gel-loading pipette tip. Repeat the procedure with 100-μl aliquots of 100% acetonitrile three times. Centrifuge the collected supernatants again to pellet any remaining gel pieces. Remove all but the last 25 μl of supernatant into a fresh tube and purge with nitrogen gas for 30 min at room temperature. Quick-freeze the supernatant in liquid nitrogen and lyophilize. Resuspend the sample in 10 μl of 0.1% trifluoroacetic acid and extract the peptides into 3 μl of 50% acetonitrile according to the manufacturer's protocol (Millipore, Bedford, MA). Resuspend the sample in 3 μl of 50% acetonitrile. Prepare a saturated solution of the matrix (2,5-dihydroxybenzoic acid) in 50% acetonitrile. Add 1.5 μl of the matrix to the sample, mix thoroughly, spot onto a Dynamo MALDI-MS plate, and analyze according to published procedures.[41]

Comments

Four PAK isoforms have been identified to date. Whereas PAK4 lacks a regulatory N terminus and is constitutively active,[42] the other three forms are autoinhibited and display very similar activation characteristics when stimulated by GTPases or lipids. This is expected due to their overall structural homology, including an identical CRIB domain and highly homologous adjacent regions. PAK 1–3 sequences differ particularly in a region located between the p21-binding domain and the catalytic domain, enabling recognition and cleavage of PAK2 by CPP32 (caspase 3)[9–11] but not of the two other PAK isoforms. This cleavage causes the release of a constitutively active PAK2 catalytic domain during apoptosis. Differences in *in vitro* activation of PAK 1-3 have not been reported and it is likely that the methods described herein are applicable for novel PAK isoforms.

Acknowledgments

We acknowledge Dr. Gary Bokoch and Yan Wang for collaborative studies and excellent technical assistance. This work was supported by the NIH (U.G.K.) and a fellowship of the American Arthritis Foundation (C.C.K.).

[41] W. T. Moore, *Methods Enzymol.* **289**, 520 (1997).
[42] A. Abo, J. Qu, M. S. Cammarano, C. Dan, A. Fritsch, V. Baud, B. Belisle, and A. Minden, *EMBO J.* **17**, 6527 (1998).

[16] Stimulation of Phospholipase C-β_2 by Rho GTPases

By DARIA ILLENBERGER, ILONA STEPHAN, PETER GIERSCHIK,
and FRIEDER SCHWALD

Introduction

By binding to their cell surface receptors, many extracellular signaling molecules, including growth factors, hormones, and chemoattractants, elicit intracellular responses by activating inositol phospholipid-specific phospholipase C (PLC). PLC activation results in the hydrolysis of phosphatidylinositol 4,5-bisphosphate (PI-4,5-P$_2$) to form D-*myo*-inositol 1,4,5-trisphosphate (IP$_3$) and *sn*-1,2-diacylglycerol (DG). Both act as important intracellular second messengers. IP$_3$ increases release of Ca^{2+} from intracellular stores and DG activates protein kinase C (PKC). Furthermore, regulation of the level of PI-4,5-P$_2$ in the plasma membrane represents an important intracellular signal, modulating the activity of several signal-transducing proteins such as phospholipase D and actin regulatory proteins. Mammalian PLC isozymes can be divided into three major families: β, γ, and δ. PLC-β and PLC-γ enzymes are activated by G-protein-coupled receptors (GPCRs) and protein-tyrosine kinase-linked receptors, respectively. Regulation of PLC-δ by cell surface receptors has not been reported so far. Stimulation of PLC-β, of which four isozymes (β_1–β_4) are known, is mediated by α subunit members of the G$_q$ subfamily and, except for PLC-β_4, by G-protein $\beta\gamma$ dimers.[1,2]

We have shown that cytosolic preparations of myeloid differentiated HL-60 cells and of bovine neutrophils contain a soluble PLC, which is activated by the poorly hydrolyzable GTP analog GTPγS.[3,4] The protein mediating GTPγS-dependent PLC stimulation in neutrophils was separated from endogenous PLC and functionally reconstituted with recombinant PLC-β_2.[4] Purification of this PLC-stimulating protein revealed that it consisted of two components of apparent molecular masses of 23 and 26 kDa, respectively. The purified heterodimer stimulated recombinant PLC-β_2, but not PLC-β_1 and PLC-δ_1. The molecular masses of the constituents of the dimer, together with the observation that it stimulated a deletion mutant of PLC-β_2, PLC-$\beta_2\Delta$, lacking a C-terminal region necessary for stimulation

[1] S. G. Rhee and Y. S. Bae, *J. Biol. Chem.* **272**, 15045 (1997).
[2] M. Katan, *Biochim. Biophys. Acta* **1436**, 5 (1997).
[3] M. Camps, C. Hou, K. H. Jakobs, and P. Gierschik, *Biochem. J.* **271**, 743 (1990).
[4] D. Illenberger, F. Schwald, and P. Gierschik, *Eur. J. Biochem.* **246**, 71 (1997).

by G-protein α_q subunits, indicated that a small GTPase associated with a regulatory protein mediates the stimulation of PLC-β_2.[4] Results of the amino acid sequence analysis of proteolytic fragments of both components of the purified heterodimer[5] led to the conclusion that the soluble PLC-β_2-stimulating GTP-binding protein of bovine neutrophils is a heterodimer of a Rho GTPase and the Rho-specific guanine nucleotide dissociation inhibitor LyGDI (also known as D4-GDI or Rho-GDIβ).[6,7] Using recombinant Rho GTPases and LyGDI, we demonstrated that GTPγS-activated Cdc42Hs-LyGDI, but not RhoA-LyGDI, stimulates the activity of PLC-β_2.[5] PLC-β_2 stimulation was also observed for GTPγS-activated Rac1,[5] Rac2,[8] and G25K (unpublished results) and did not require the presence of LyGDI. However, PLC-β_2 stimulation clearly required both C-terminal processing of Cdc42Hs and Rac1 and integrity of their effector domain in the switch I region.[5] The C-terminal processing of PLC-β_2-stimulating Rho GTPases involves geranylgeranylation, proteolysis, and carboxymethylation of the C-terminal cysteine.[9] Which of these modifications is crucial for PLC-β_2 stimulation is currently unknown. Our results identified PLC-β_2, which is abundant in neutrophils and human promyelocytic HL-60 cells (10, unpublished results), as a novel effector of the Rho GTPases Cdc42Hs, G25K, Rac1, and Rac2. Neutrophils are activated by chemoattractants, which bind to their receptors generating multiple intracellular second messengers and signaling molecules through the activation of phospholipases C, A$_2$, and D, protein kinases, lipid kinases, and small GTP-binding proteins. In particular, the Rho family of GTPases plays essential roles in modulating neutrophil functions.[11-13] They regulate superoxide production, adhesion, secretion, and cell migration. PLC-β_2 activation by Cdc42Hs, G25K, Rac1, and Rac2 could be involved in processes requiring changes in Ca^{2+} concentration or PI-4,5-P$_2$ levels. In addition, PLC stimulation has been implicated in the chemoattractant-mediated activation of mitogen-activated protein kinases (MAPKs),[14] which are known to be regulated by Cdc42Hs/G25K and Rac1/Rac2.[7]

[5] D. Illenberger, F. Schwald, D. Pimmer, W. Binder, G. Maier, A. Dietrich, and P. Gierschik, *EMBO J.* **21**, 6241 (1998).

[6] P. Scherle, T. Behrens, and L. M. Staudt, *Proc. Natl. Acad. Sci. U.S.A.* **90**, 7568 (1993).

[7] L. Van Aelst and D. D'Souza-Schorey, *Genes Dev.* **11**, 2295 (1997).

[8] K. Scheffzek, I. Stephan, O. N. Jensen, D. Illenberger, and P. Gierschik, *Nature Struct. Biol.* **7**, 122 (2000).

[9] J. A. Glomset and C. C. Farnsworth, *Annu. Rev. Cell. Biol.* **10**, 181 (1994)

[10] A. V. Smrcka and P. C. Sternweis, *J. Biol. Chem.* **268**, 9667 (1993).

[11] G. M. Bokoch, *Blood* **86**, 1649 (1995).

[12] V. Benard, G. M. Bokoch, and B. A. Diebold, *Trends Pharmacol. Sci.* **20**, 365 (1999).

[13] F. Sánchez-Madrid and M. A. del Pozo, *EMBO J.* **18**, 501 (1999).

[14] M. J. Rane, S. L. Carrithers, J. M. Arthur, J. B. Klein, and K. R. Mcleish, *J. Immunol* **159**, 5070 (1997).

This article describes the purification of recombinant PLC-$\beta_2\Delta$, the purification of native and recombinant heterodimers of LyGDI and Rho GTPases, and the *in vitro* assays used to measure GTPγS binding to and stimulation of PLC-β_2 by Rho GTPases. Because wild-type PLC-β_2 is stimulated to a lower maximal extent (approximately 2.5-fold) by Cdc42Hs or Rac1 than PLC-$\beta_2\Delta$ (approximately 5-fold),[5] the purification of PLC-$\beta_2\Delta$ is described here in detail. The basis of this difference is not clear at present. We have previously produced recombinant heterodimers of LyGDI and Rho GTPases by extracting C-terminally modified Rho GTPases from membranes of insect cells with the glutathione S-transferase (GST) fusion protein of LyGDI.[5] We now routinely produce LyGDI–Rho heterodimers by coinfection of insect cells with recombinant baculoviruses encoding LyGDI and Rho GTPases, respectively, and by subsequent purification of the dimers from the soluble fraction. The advantages of this procedure are a higher yield of recombinant heterodimers and the avoidance of the GST fusion to LyGDI.

Expression of Recombinant PLC-$\beta_2\Delta$ in Insect Cells

The cDNA of the deletion mutant PLC-$\beta_2\Delta$ (PLC-β_2[F819-E1166])S2A is obtained by the polymerase chain reaction (PCR) overlap–extension method using human PLC-β_2 cDNA as template.[15] The resulting construct still contains the C-terminal recognition site (Q1167-L1881) for the PLC-β_2 antibody.[16] The 5'-noncoding region of the cDNA is removed by PCR amplification of the coding region using the oligonucleotide primers 5'-TTTGAATTCCACCATGGCTCTGCTCAACCCTGTC-3' (upstream, sense) and 5'-TTTGAATTCAGAGGCGGCTCTCCTGGGC-3' (downstream, antisense). The upstream primer introduces an *Eco*RI and a *Nco*I restriction enzyme site into the cDNA. The latter modification causes the serine-to-alanine replacement in position 2 of the protein. The PLC-$\beta_2\Delta$ cDNA is cloned into the *Eco*RI site of the baculovirus transfer vector pVL1393 (Invitrogen, Carlsbad, CA). Recombinant baculoviruses are generated by cotransfecting the recombinant plasmid DNA with linear wild-type Baculogold viral DNA (Pharmingen, San Diego, CA) into *Spodoptera frugiperda* (Sf9) cells. The recombinant baculovirus is amplified through two cycles of infection of 6×10^6 *Trichoplusia ni* 5B1-4 (High five cells, Invitrogen) cells. For production of recombinant PLC-$\beta_2\Delta$, *T. ni* cells are grown in suspension culture in Insect-XPRESS medium (BioWhittaker, Walkersville, MD) supplemented with 0.2% (v/v) Pluronic F-68 (GIBCO-

[15] P. Schnabel, M. Camps, A. Carozzi, P. J. Parker, and P. Gierschik, *Eur. J. Biochem.* **217**, 1109 (1993).

[16] M. Camps, A. Carozzi, P. Schnabel, A. Scheer, P. J. Parker, and P. Gierschik, *Nature* **360**, 684 (1992).

BRL, Gaithersburg, MD) and 2.5 μg/ml amphotericin B (Fungizone, GIBCO-BRL) in a 1800-ml Fernbach culture flask. Cells (1 × 10^9) are incubated with recombinant baculovirus at a multiplicity of infection (MOI) of 10 in 300 ml of medium at 80 rpm on a rotary shaker with an amplitude of 25 mm. Three days after infection, cells are pelleted and resuspended in 10 ml of buffer A [20 mM Tris–HCl, pH 7.5, 2 mM EDTA, 2 μg/ml soybean trypsin inhibitor (STI), 3 mM benzamidine, 1 μM pepstatin, 1 μM leupeptin, 0.1 mM phenylmethylsulfonyl fluoride (PMSF), 1 μg/ml aprotinin]. Cells are homogenized using a 7-ml glass/glass homogenizer. The lysate is centrifuged at 100,000g for 1 hr at 4° and the supernatant is passed through 0.45-μm pore-size nitrocellulose filters.

Purification of Recombinant PLC-$\beta_2\Delta$

The filtered supernatant (10 ml, 200 mg of protein) is applied to a column (1.6 × 8 cm) of heparin Sepharose CL-6B (Pharmacia, Piscataway, NJ) that was equilibrated with buffer B [20 mM Tris–HCl, pH 8.0, 1 mM EDTA, 1 mM dithiothreitol (DTT), 0.2 mM PMSF, 3 mM benzamidine, 10 μM leupeptin, 2 μM pepstatin, 1 μg/ml aprotinin, 16 μg/ml L-1-chloro-3-(4-tosylamido)-4-phenyl-2-butanone (TPCK), 16 μg/ml L-1-chloro-3-(4-tosylamido)-7-amino-2-heptanone hydrochloride (TLCK), 2 μg/ml STI]. The flow rate is 1 ml/min. After application of the sample, the column is washed with 20 ml of buffer B and eluted with a linear gradient (60 ml) of NaCl (0–900 mM) in buffer B, followed by 10 ml of buffer B containing 900 mM NaCl. Fractions of 1 ml are collected and analyzed by SDS–PAGE and immunoblotting using antibodies reactive against PLC-β_2.[16] Fractions containing maximal levels of PLC-$\beta_2\Delta$ are pooled (15 ml, 40 mg protein), concentrated approximately 10-fold by pressure filtration in a stirred cell equipped with an Amicon (Danvers, MA) PM10 membrane, diluted 4-fold with buffer B, and applied to a Mono Q HR 5/5 column (Pharmacia), which was equilibrated with buffer B containing 100 mM NaCl. The flow rate is 0.5 ml/min. After application of the sample, the column is washed with 5 ml of buffer B containing 100 mM NaCl and eluted with a linear gradient (10 ml) of NaCl (100–900 mM) in buffer B, followed by 5 ml of buffer B containing 900 mM NaCl. Fractions of 500 μl are collected into tubes containing 50 μl of 10-fold concentrated protease inhibitors as listed earlier. Fractions containing PLC-$\beta_2\Delta$ (1 ml, 10 mg of protein) are diluted 4-fold with buffer C (20 mM Tris–HCl, pH 7.5, 100 μM EDTA, 1 mM DTT, 0.1 mM PMSF, 3 mM benzamidine, 16 μg/ml TPCK, 16 μg/ml TLCK). The sample (2 ml, 5 mg of protein) is applied to a column (0.5 × 0.5 cm) of hydroxylapatite (Calbiochem, La Jolla, CA, HPLC grade) connected to a SMART micropurification chromatography system (Pharmacia), which was equilibrated with buffer C. The flow rate is 100 μl/min. After application of the sample, the resin is washed with 0.5 ml of buffer C and eluted with a linear gradient (5 ml) of K_2HPO_4/H_3PO_4, pH 7.5 (0–500 mM), in buffer C. PLC-$\beta_2\Delta$ elutes in a single peak at approximately 110 mM K_2HPO_4/H_3PO_4. The purity of the preparation is assessed by SDS–PAGE and protein staining. Typically, at least 95% purity is achieved. We noticed a degradation of purified PLC-$\beta_2\Delta$ on storage and/or freezing and thawing. Attempts to resolve these fragments from full-length PLC-$\beta_2\Delta$ by several chromatographic procedures were unsuccessful. Aliquots of the purified PLC-$\beta_2\Delta$ are snap-frozen in liquid nitrogen and stored at $-80°$.

Purification of Native and Recombinant LyGDI–Rho Heterodimers

Native heterodimers are purified from soluble fractions of neutrophils, which are prepared from bovine blood, homogenized, and fractionated into soluble and particulate constituents.[17] Recombinant heterodimers are produced by coinfection of Sf9 cells with recombinant baculoviruses encoding human LyGDI and human Rho GTPases, respectively. All baculoviruses were produced using Baculogold DNA and pVL 1393 transfer vectors containing the respective coding sequence without noncoding sequences ligated into the *Bam*HI/*Eco*RI site.[5] The quantity of baculoviruses used for coinfection is adjusted by analyzing expression of the recombinant proteins by immunoblotting to avoid an excess of free monomeric LyGDI (see later). Polyclonal antibodies against LyGDI and Rho GTPases are from Santa Cruz Biotechnology (Santa Cruz, CA). Cells are harvested 3 days after infection, pelleted by centrifugation at 250g, and resuspended in buffer D (20 mM Tris–HCl, pH 7.5, 1 mM EDTA, 1 mM MgCl$_2$, 1 mM DTT, 1 $\mu$$M$ GDP, 0.1 mM PMSF) supplemented with 3 mM benzamidine, 1 $\mu$$M$ pepstatin A, 1 $\mu$$M$ leupeptin, 2 μg/ml STI, and 1 μg/ml aprotinin. Homogenization of cells is performed with a 7-ml glass/glass homogenizer on ice, and separation into a soluble and particulate fraction is carried out by centrifugation of the lysate at 100,000g for 1 hr at 4°.

Soluble fractions of bovine neutrophils or baculovirus coinfected insect cells are concentrated by pressure filtration in a stirred cell equipped with an Amicon PM10 membrane. The resulting concentrate (3.5 ml, 100–200 mg protein) is passed through 0.45-μm pore-size nitrocellulose filters and subjected to gel-permeation chromatography on a HiLoad 26/60 Superdex 75 prep grade column (Pharmacia), equilibrated and run in buffer D supple-

---

[17] P. Gierschik, D. Sidiropoulos, A. M. Spiegel, and K. H. Jakobs, *Eur. J. Biochem.* **165,** 185 (1987).

mented with 100 mM NaCl. The flow rate is 1 ml/min. Fractions of 5 ml
are collected and analyzed for GTPγS binding and their ability to confer
GTPγS-dependent stimulation to PLC-$\beta_2\Delta$ (see later and Fig. 1). Gel per-
meation is chosen as the first purification step in order to separate endogenous PLCs from the heterodimers (cf. ref. 4). As expression of recombinant
Rho GTPases and LyGDI in insect cells not only results in the formation
of soluble heterodimers, but also of soluble unmodified monomeric Rho
GTPases, gel permeation also allows the separation of these molecules.
Figure 1 shows that PLCβ_2 stimulation is observed only when both modified
Rac2 and LyGDI are present (fractions 10 and 11), but not in the presence
of monomeric, soluble, and, hence, unmodified Rac2 (fractions 16–19).
This result confirms the requirement of C-terminal modifications of Rho
GTPases for PLCβ_2 stimulation. In additional experiments, we observed

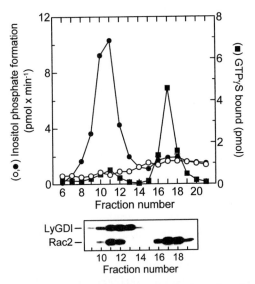

FIG. 1. Partial purification of PLC-$\beta_2\Delta$-stimulating heterodimers by gel-permeation chromatography. *Spodoptera frugiperda* (Sf9) cells were coinfected with baculoviruses encoding Rac2
and LyGDI, respectively. Soluble proteins (3 ml, 100 mg protein) were subjected to chromatography on a Superdex 75 pg. Aliquots of the indicated fractions (10 μl) were supplemented
with soluble proteins of Sf9 insect cells infected with baculovirus encoding PLC-$\beta_2\Delta$ (0.4 μg
protein/sample). Incubations were performed for 1 hr at 25° at 150 nM free Ca^{2+} in the
absence (open circles) or presence (closed circles) of 100 μM GTPγS with phospholipid
vesicles containing PI-4,5-P$_2$. Samples were analyzed for inositol phosphates as described in
the text. Aliquots (10 μl) of fractions were also analyzed for [^{35}S]GTPγS binding (closed
squares), and immunoblotting was performed using antibodies reactive against Rac2 or
LyGDI (bottom).

that recombinant LyGDI expressed in the absence of Rho GTPases elutes from the gel-filtration column in the same fractions as heterodimeric LyGDI–Rho (fractions 10–13 in Fig. 1), i.e., with an apparent molecular mass of approximately 48 kDa.[4] This chromatographic behavior may result from the less compact structure of monomeric LyGDI or from homodimer formation. Elution of monomeric or homodimeric LyGDI in fractions 10–13 peaking in fractions 11 and 12 may cause the reduced [^{35}S]GTPγS binding to Rac2-LyGDI and the reduced PLC-$\beta_2\Delta$ stimulation measured for fraction 12, which contains high amounts of Rac2. In addition, LyGDI clearly inhibits [^{35}S]GTPγS binding to Rac2, resulting in lower amounts of bound [^{35}S]GTPγS to the dimer compared with monomeric Rac2 (cf. fractions 9–13 and fractions 16–18). Immunoblotting of Rac2 (Fig. 1, bottom) also shows that fraction 18 contains monomeric Rac2, which apparently binds less [^{35}S]GTPγS under the conditions used here than fraction 17.

Fractions containing maximal levels of PLC-$\beta_2\Delta$-stimulating activity are pooled (10 ml, approximately 5 mg of protein), diluted fourfold with buffer D, and applied to a 1-ml Mono Q HR 5/5 column, which was equilibrated with buffer D. The flow rate is 0.5 ml/min. After application of the sample, the column is washed with 10 ml of buffer D, and proteins are eluted with a linear gradient (10 ml) of NaCl (0–500 mM) in buffer D, followed by 5 ml buffer D containing 500 mM NaCl. Fractions of 500 μl are collected. PLC-$\beta_2\Delta$-stimulating activity elutes as a broad peak at approximately 100 mM NaCl. Figure 2 shows that this broad peak contains only one peak of Rac2 but two peaks of LyGDI, indicating the presence of two LyGDI species, of which only the Rac2–LyGDI dimer (approximately 50 μg protein) stimulates PLCβ_2. We have noted that the second peak (e.g., fraction 23 in Fig. 2), which most likely contains monomeric (or homodimeric) LyGDI, strongly inhibits both [^{35}S]GTPγS binding to monomeric Rac2 and PLCβ_2 stimulation by heterodimeric LyGDI–Rho. The purity of the heterodimers in fractions 20–22 is assessed by SDS–PAGE and protein staining. Typically, the only proteins detected are the 22-kDa Rac2 and the 26-kDa LyGDI proteins on staining of the gel with Coomassie blue (Fig. 2, right).

Peak fractions containing the native heterodimer from bovine neutrophils are not homogeneous at this stage[4] and are further purified and concentrated by hydroxylapatite chromatography. Fractions containing maximal activity (approximately 1 mg protein in 1 ml) are diluted twofold with buffer D and applied to a column (0.5 × 6 cm) of hydroxylapatite (Calbiochem, HPLC grade), which was equilibrated with buffer D. The flow rate is 0.5 ml/min. The resin is washed with 8 ml of buffer D and eluted with a linear gradient (10 ml) of K_2HPO_4/H_3PO_4, pH 7.5 (0–250 mM), in buffer D. Fractions of 500 μl are collected. A single peak containing

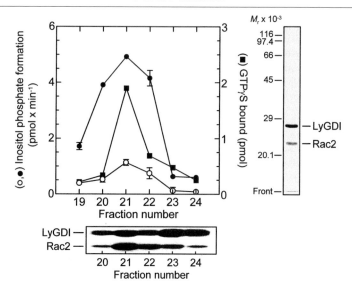

FIG. 2. Purification of the Rac2-LyGDI heterodimer. PLC-$\beta_2\Delta$-stimulating fractions obtained by gel permeation were purified by anion-exchange chromatography on Mono Q as described in the text. Aliquots (10 μl) of the indicated fractions were either reconstituted with soluble proteins from insect cells expressing PLC-$\beta_2\Delta$ (0.4 μg protein/sample) and then incubated in the absence (open circles) or presence (closed circles) of 100 μM GTPγS with phospholipid vesicles containing PI-4,5-P$_2$ or analyzed for [^{35}S]GTPγS binding (closed squares) (top left). Aliquots of fractions (10 μl) were subjected to SDS–PAGE, and immunoblotting was performed using antibodies reactive against Rac2 or LyGDI (bottom). A sample containing purified Rac2–LyGDI was subjected to SDS–PAGE and staining with Coomassie blue (top right).

PLC-$\beta_2\Delta$ stimulating activity (approximately 300 μg of protein) elutes at approximately 40 mM K$_2$HPO$_4$/H$_3$PO$_4$. All samples are frozen in liquid nitrogen and stored at $-80°$.

[^{35}S]GTPγS Binding

[^{35}S]GTPγS binding assays are performed at 30° in an incubation mixture (40 μl) containing 10 μl of Superdex 75 or Mono Q fractions, 25 mM HEPES–NaOH, pH 8.0, 1 mM EDTA, 1 mM DTT, 20 mM MgCl$_2$, 100 mM NaCl, 0.1% (v/v) GENAPOL C-100 (Calbiochem), and 100 nM [^{35}S]GTPγS (296 GBq/mmol). The incubation is terminated after 6 hr by the addition of 2 ml of ice-cold buffer containing 50 mM Tris–HCl, pH 8.0, 100 mM NaCl, and 5 mM MgCl$_2$ and rapid filtration through nitrocellulose filters with a pore size of 0.45 μM (Advanced Microdevices, Ambala Cantt., India) followed by four washes with 2 ml each of ice-cold buffer. Nonspecific binding is defined as the binding not competed for by 10 μM unlabeled GTPγS. Under these conditions, GTPγS binding reaches a plateau after 5–6 hr of incubation.

Phospholipase C Assay

The PLC-$\beta_2\Delta$-stimulating activity of both native and recombinant heterodimers is determined by an *in vitro* assay performed in a volume of 60 μl of 50 mM HEPES–NaOH, pH 7.2, 3 mM EGTA, 70 mM KCl, 2 mM DTT, and 150 nM free Ca^{2+} at 25°. Ten microliters of the soluble heterodimers is supplemented with 10 μl of either the soluble fraction of PLC-$\beta_2\Delta$– baculovirus-infected insect cells or purified PLC-$\beta_2\Delta$. The reactions are initiated by adding 20 μl of phospholipid micelles. To prepare the micelles, a mixture of phosphatidylethanolamine, PI-4,5-P$_2$, and [^{3}H]PI-4,5-P$_2$ in chloroform is evaporated to dryness under a stream of N$_2$ in a 14 × 100-mm Pyrex glass tube at room temperature. The lipids are resuspended by continuous vortex mixing for 30 min in buffer (20 μl/assay). After vortexing, the lipid suspension is sonicated for 15 min in a bath-type sonicator.[18] The final concentrations in 60 μl volume are 33.4 μM PI-4,5-P$_2$ (185 GBq/mmol) and 536 μM phosphatidylethanolamine. After incubating the mixture for 1 hr, the reactions are terminated by the addition of 350 μl ice-cold chloroform/methanol/concentrated HCl (100/100/0.6) followed by the addition of 100 μl of 1 M HCl, 5 mM EGTA. Phase separation is accelerated by centrifugation (15,000g, 30 sec) in a tabletop centrifuge. The formation of water-soluble inositol phosphates is measured by supplementing a 200-μl aliquot of the aqueous (upper) phase with 4 ml of scintillation fluid followed by scintillation counting.[19] Under these conditions, not only PLC-β_2 stimulation by LyGDI–Rho dimers but also the stimulation by monomeric processed Rho GTPases can be measured. Monomeric processed Rho GTPases stimulate PLC-$\beta_2\Delta$ to a similar extent as LyGDI–Rho dimers (approximately 5- to 10-fold). We have noticed that the PLC assay used previously for the measurement of PLC-β_2 stimulation by LyGDI–Rho dimers[4,19] is not suitable for studying posttranslationally modified monomeric Rho GTPases due to the presence of detergent.

---

[18] P. Gierschik and M. Camps, *Methods Enzymol.* **238,** 181 (1994).
[19] A. J. Self and A. Hall, *Methods Enzymol.* **256,** 67 (1995).

Comments

(i) Purified LyGDI–Rho heterodimers stimulate both crude and highly purified PLC-$\beta_2\Delta$, suggesting that this stimulation is mediated by direct protein–protein interaction. (ii) Protein concentrations specified in the method section are estimated according to Bradford. We noticed that these values are 10- to 20-fold higher than those determined by [^{35}S]GTPγS binding. Protein concentrations can also be determined by comparing the samples with bovine serum albumin standards after electrophoresis on SDS–polyacrylamide [12.5% (w/v)] gels and staining with Coomassie Brilliant Blue R (Sigma). The concentration of recombinant heterodimers determined by this method is 5- to 10-fold higher than that determined by [^{35}S]GTPγS binding. A similar discrepancy has been reported by others[18] and is also observed when monomeric Rho GTPases, solubilized from membranes of baculovirus-infected Sf9 cells, are assayed. In this case, the concentrations determined by [^{35}S]GTPγS binding according to the method specified earlier are the same as those determined in the presence of 10 mM EDTA as described by Self and Hall (1995).[19] (iii) We also use [^{35}S]GTPγS binding to estimate the concentration of Rho GTPases activated by GTPγS under conditions of the PLC assay. These measurements yield the same concentrations for the LyGDI–Rac2 dimer as those obtained with the [^{35}S]GTPγS-binding method described previously. However, monomeric Rho GTPases show a different behavior. While similar results are obtained for Cdc42Hs, concentrations of Rac1 and Rac2 are 2.5- and 3-fold, respectively, higher than those determined by the [^{35}S]GTPγS-binding method described earlier. This difference is conceivably due to the presence of PI-4,5-P$_2$, which may differentially affect the dissociation of GDP from monomeric Rho GTPases.[20] (iv) Activation of LyGDI–Rho heterodimers by GTPγS under conditions of the PLC assay used here absolutely requires the presence of PI-4,5-P$_2$. PI-4,5-P$_2$, as well as many other biologically active lipids, has been shown to reduce the inhibitory action of RhoGDIs.[21] Nevertheless, our results show that excess LyGDI inhibits the activation of Rho GTPases even in the presence of PI-4,5-P$_2$. In addition, we noticed that PLC-β_2 stimulation by recombinant heterodimers is independent of whether LyGDI or RhoGDI is present in the dimer. Thus, the approximately 10-fold higher GDI activity of RhoGDI seems not to be exhibited under the conditions of the PLC assay. (v) In our hands, recombinant PLCβ_2 affects neither guanine nucleotide exchange on Rho GTPases nor GTP-hydrolyzing activity of Rho GTPases. Importantly, these results are

[20] Y. Zheng, J. A. Glaven, W. J. Wu, and R. A. Cerione, *J. Biol. Chem.* **271**, 23815 (1996).
[21] T. -H. Chuang, B. P. Bohl, and G. M. Bokoch, *J. Biol. Chem.* **268**, 26206 (1993).

observed independent of whether wild-type PLC-β_2 or PLC-$\beta_2\Delta$ is used. (vi)
The purification method described here not only proved to be satisfactory in
studying PLCβ_2 stimulation by recombinant LyGDI–Rho heterodimers,
but also enabled us to crystallize the LyGDI–Rac2 dimer and to analyze
its structure by X-ray crystallography.[8]

[17] Regulation of Phospholipase D1 Activity
by Rho GTPases

By MICHAEL A. FROHMAN, YASUNORI KANAHO, YUE ZHANG,
and ANDREW J. MORRIS

Introduction

Mammalian phospholipase D (PLD) is a membrane-associated enzyme
activated by a wide variety of agonists that signal through G-protein-coupled or tyrosine kinase receptors.[1,2] PLD is best known for its ability to
hydrolyze phosphatidylcholine (PC), the most abundant cellular phospholipid, to yield choline and phosphatidic acid (PA). PLD can also catalyze
a transphosphatidylation reaction in the presence of primary alcohols leading to the formation of phosphatidyl alcohol (Ptd alcohol) at the expense
of PA. This reaction is key for the classic *in vivo* PLD assay,[3] as Ptd alcohol
is relatively inert and stable in cells and can be formed only through the
action of PLD, whereas PA can be made via at least four cellular pathways
and is metabolized rapidly after its formation. The efficiency with which
PLD carries out the transphosphatidylation reaction also suggests that PLD
may have an endogenous role to generate other less well-understood lipids,
such as bis-PA,[4,5] by the fusion of PC and diacylglycerol [the endogenous
stimulator of protein kinase C (PKC) and as well the product of dephos-
phorylating PA].

Cellular roles for PA are under intense investigation and include action
as (1) a structural lipid that mediates physical changes in membrane curva-

[1] M. A. Frohman, J. Engebrecht, and A. J. Morris, *Chem. Phys. Lipids* **98**, 127 (1999).
[2] J. H. Exton, *Biochim. Biophys. Acta* **1439**, 121 (1999).
[3] M. J. O. Wakelam, M. Hodgkin, and A. Martin, *in* "Signal Transduction Protocols"
(D. A. Kendall and S. J. Hill, eds.), p. 271. Humana Press, Totowa, NJ, 1995.
[4] W. J. van Blitterswijk and H. Hilkmann, *EMBO J.* **12**, 2655 (1993).
[5] T. C. Sung, R. Roper, Y. Zhang, S. A. Rudge, R. Temel, S. M. Hammond, A. J. Morris,
B. Moss, J. Engebrecht, and M. A. Frohman, *EMBO J.* **16**, 4519 (1997).

ture at the neck of vesicles,[6] (2) a binding moiety that recruits coat proteins to promote vesicle formation[7] or mediates Raf1-kinase membrane localization,[8] (3) a signaling lipid that activates NADPH oxidase and phosphatidylinositol-4-phosphate 5-kinase [the enzyme that generates phosphatidylinositol 4,5-bisphosphate (PIP_2)],[9] and (4) an intermediate signaling lipid that can be converted to diacylglycerol or lyso-PA (LPA, a mitogen).

The regulation of mammalian PLD is complex and not fully understood at present. The two isoforms, PLD1 and PLD2, are seemingly activated differently, although the details and degree of difference are not yet resolved.[10] In brief, purified recombinant PLD1 exhibits a very low basal activity when examined using the most popular *in vitro* PLD assay (developed by Brown and colleagues[11]) and is stimulated approximately 500-fold in the presence of a combination of ARF family proteins, Rho family proteins (including Rac1 and Cdc-42), and classic forms of PKC.[12] Each of the activators can stimulate PLD1 independently (from 10- to 50-fold), although the effect when they are combined is synergistic and is probably important in the context of *in vivo* signaling. Each activator is thought to mediate activation directly at a distinct site of interaction on PLD1, although these sites have yet to be defined at a satisfactory level of resolution. PLD2 exhibits a high level of basal activity in the *in vitro* assay and is not dramatically stimulated further in the presence of any of the PLD1 activators.

Stimulation of PLD1 or endogenous PLD by Rho can also be demonstrated *in vivo*. Interference with Rho signaling blocks PLD activation in many settings, whereas overexpression of dominant active alleles of Rho, or overexpression of wild-type Rho coupled with agonist stimulation, leads to increased PLD (PLD1) activation.[13,14] Because RhoA physically interacts

[6] A. Schmidt, M. Wolde, C. Thiele, W. Fest, H. Kratzin, A. V. Podtelejnikov, W. Witke, W. B. Huttner, and H. D. Soling, *Nature* **401**, 133 (1999).

[7] N. T. Ktistakis, H. A. Brown, M. G. Waters, P. C. Sternweis, and M. G. Roth, *J. Cell Biol.* **134**, 295 (1996).

[8] M. A. Rizzo, K. Shome, C. Vasudevan, D. B. Stolz, T.-C. Sung, M. A. Frohman, S. C. Watkins, and G. Romero, *J. Biol. Chem.* **274**, 1131 (1999).

[9] A. Honda, M. Nogami, T. Yokozeki, M. Yamazaki, H. Nakamura, H. Watanabe, K. Kawamoto, K. Nakayama, A. J. Morris, M. A. Frohman, and Y. Kanaho, *Cell* **99**, 521 (1999).

[10] M. A. Frohman, T.-C. Sung, and A. J. Morris, *Biochim. Biophys. Acta* **1439**, 175 (1999).

[11] H. A. Brown, S. Gutowski, C. R. Moomaw, C. Slaughter, and P. C. Sternweis, *Cell* **75**, 1137 (1993).

[12] S. M. Hammond, J. M. Jenco, S. Nakashima, K. Cadwallader, S. Cook, Y. Nozawa, M. A. Frohman, and A. J. Morris, *J. Biol. Chem.* **272**, 3860 (1997).

[13] S.-K. Park, J. J. Provost, C. D. Bae, W.-T. Ho, and J. H. Exton, *J. Biol. Chem.* **272**, 29263 (1997).

[14] Y. Zhang, Y. A. Altshuller, S. A. Hammond, F. Hayes, A. J. Morris, and M. A. Frohman, *EMBO J.* **18**, 6339 (1999).

with PLD1,[5,15] it has been widely assumed that the stimulation of PLD by RhoA *in vivo* is mediated by direct interaction. However, studies have proposed that RhoA may also stimulate PLD indirectly through one or more of the many other RhoA-mediated downstream effector signaling pathways.[16]

This article describes common assays used to measure interaction with or stimulation of PLD1 by RhoA. Preparation of RhoA itself is described in less detail, as it is presented in a previous volume (Volume 256) of this series.

Protocols

Preparation of PLD1 for Use in in Vitro PLD Assay

Three different basic methods with some further variations have been developed to overexpress PLD1 in mammalian, yeast, and Sf9 (*Spodoptera frugiperda* ovary) insect cells.

Mammalian Cells. Use of a wide variety of expression plasmids and cell lines has been reported. Our laboratories use the cytomegalovirus (CMV) promoter-driven plasmid pCGN, which appends an N-terminal HA epitope to the PLD1 protein generated (available on request, MAF), and the simian virus 40 (SV40) promoter-driven plasmid pTB701-FL (IBI), which appends an N-terminal FLAG-epitope (available on request, YK). It should be noted that peptide tags can be added to the N terminus of mammalian PLD or inserted into the central "loop" region without known detrimental effects, but not to the C terminus, which results in a complete loss of enzymatic activity.[17,18]

Our laboratory uses COS-7 cells and HEK-293 cells to overexpress mammalian PLD1 and PLD2. We have determined optimal conditions for transfection of these cells with our expression plasmids using a particular commercially available preparation of cationic lipid (Lipofectamine PLUS, from BRL, Gaithersburg, MD) as detailed later. Investigators interested in modifying the following protocols for use with different vectors, cell lines, or transfection methods should keep in mind that reoptimization may be required to achieve acceptable results.

[15] M. Yamazaki, Y. Zhang, H. Watanabe, T. Yokozeki, S. Ohno, K. Kaibuchi, H. Shibata, H. Mukai, Y. Ono, M. A. Frohman, and Y. Kanaho, *J. Biol. Chem.* **274,** 6035 (1999).

[16] M. Schmidt, M. Voß, P. A. Oude Weernink, J. Wetzel, M. Amano, K. Kaibuchi, and K. H. Jakobs, *J. Biol. Chem.* **274,** 14648 (1999).

[17] T.-C. Sung, Y. M. Altshuller, A. J. Morris, and M. A. Frohman, *J. Biol. Chem.* **274,** 494 (1999).

[18] T.-C. Sung, Y. Zhang, A. J. Morris, and M. A. Frohman, *J. Biol. Chem.* **274,** 3659 (1999).

1. On day 1 (morning), pass cells (COS-7, HEK-293) using standard protocols and plate to a density of ~10–20% confluence in Dulbecco's modified Eagle's medium (DMEM) containing Pen-Strep (penicillin–streptomycin), glutamine, and 10% fetal calf serum (FCS) in 35-mm tissue culture (TC) dishes. In some cases, more reproducible assay data are achieved using polylysine-coated dishes instead of regular TC dishes.

2. On day 2 (the cells should be ~70% confluent by the afternoon), transfect overnight using Lipofectamine PLUS in 1 ml of Opti-mem media with 1 μg plasmid per dish according to the supplier's recommendations. The next morning, replace transfection media with 1 ml of media containing 10% FCS and continue the culture for another 24 hr.

3. On day 4, wash the cells with cold phosphate-buffered saline (PBS) and place the TC dishes on ice. Add 60 μl of homogenization buffer and detach the cells using a disposable rubber policeman. Recover the cells using a Pipetman and transfer to an Eppendorf tube on ice; keep cold (on ice) at all times but do not freeze. Sonicate the cells using a micro tip on a low setting (1–2) for 10 sec. Keep the tube in ice water at all times to avoid generation of excess heat, which will denature the proteins.

4. Each PLD assay sample requires 10 μl of the lysate; a 5- to 10-μl aliquot should be added to 2× SDS–PAGE sample loading buffer containing 8 M urea, denatured at room temperature, and reserved for Western analysis. The lysates retain full activity for several days if kept cold (but not frozen).

Homogenization buffer: PBS containing protease inhibitor cocktail (Roche)

Yeast Cells: Induction and Protein Extraction from Schizosaccharomyces pombe Yeast (for pREP plasmids). An expression system for rat PLD1 in *S. pombe* using the pREP vector and analysis using a modified *in vivo* assay technique was developed by Katayama and colleagues.[19] This section describes an extension of their system to generate PLD1 protein for use in the standard *in vitro* assay system. Basic information about expression of pREP plasmids in *S. pombe* and recipes for media and plates can be obtained at http://pingu.salk.edu/users/forsburg/plasmids.html. pREP-human PLD1 and mouse PLD2 are available on request from MAF.

[19] K. Katayama, T. Kodaki, Y. Nagamachi, and S. Yamashita, *Biochem. J.* **329,** 647 (1998).

1. Inoculate 5 ml EMM medium (+thiamin) with a colony from a plate freshly streaked (not more than 2 weeks old) with *S. pombe* transformed stably with pREP-PLD1. Grow overnight at 32°.

2. The next morning, wash the cells three times with EMM medium (−thiamin). Resuspend in a small volume. Inoculate 50 ml of EMM medium (−thiamine) with the overnight culture. Grow 24 hr at 32°.

3. Harvest the cells by centrifuging at 2500–3000 rpm for 5 min at room temperature in a clinical centrifuge. Resuspend in 20 ml of buffer A and incubate for 15 min at room temperature.

4. Centrifuge as before. Resuspend the cells in a microfuge tube in 1 ml of buffer B containing Zymolase (25 μg/ml) and incubate for 30 min at 30°.

5. Centrifuge the cells at 1000g for 5 min at room temperature. Resuspend in 1 ml buffer B. Repeat spin.

6. Resuspend the cells in 200 μl buffer B (containing 50 μg/ml leupeptin and 1 mM EDTA).

7. Sonicate on ice at medium setting using a microprobe two times for 10 sec each time.

8. Centrifuge at 500–1000g for 5 min. Collect supernatant. Keep at 4°.

Notes and Reagents

Buffer A: 200 mM Tris, pH 7.0, 20 mM dithiothreitol (DTT), 1 mM phenylmethylsulfonyl fluoride (PMSF)

Buffer B: 200 mM Tris, pH 7.0, 10 mM DTT, 1mM PMSF, 500 mM KCl

Leupeptin: use 1 μl of a 10-mg/ml solution (store stock at −20°)

Zymolase: Store stock solution (5 mg/ml) at −80°

Add Zymolase, leupeptin, DTT, and PMSF to buffers just before use. The exact length of time required for the Zymolase incubation may have to be determined empirically. PLD activity remains detectable for at least 1 week if the lysate is kept at 4°. Suggested starting point: use 10 μl of lysate for PLD assay or Western blot analysis.

Sf9 Cells: Expression and Purification of PLD1 Using a Baculoviral Vector. High-titer stocks of recombinant baculoviruses for expression of PLD1 have been described previously[12] and are available on request (AJM).

1. To express PLD1, monolayers of exponentially growing Sf9 cells (3 × 10^7 cells/225-cm^2 flask, cultured at 27° in complete Grace's medium supplemented with lactalbumin, yeastolate, and 10% fetal bovine serum containing antibiotic and antimycotic agents, generally two flasks of cells are used for each purification) are infected with recombinant baculoviruses at a multiplicity of 10 for 1 hr with gentle rocking.

2. The virus-containing medium is then removed and replaced with fresh supplemented Grace's medium. The infected cells are grown for 48 hr, the medium is removed, and the cells are washed once with ice-cold PBS.

3. The cells are then lysed on ice by the addition of 5 ml/225-cm^2 flask of ice-cold lysis buffer containing 150 mM NaCl, 50 mM Tris, pH 8.0, 1% Nonidet P-40 (NP-40), 1 mM EGTA, 0.1 mM benzamidine, 0.1 mM PMSF, 10 μg/ml pepstatin A, and 10 μg/ml leupeptin. After 30 min on ice, the cells are scraped up and the suspension is centrifuged at 50,000g for 30 min at 4°.

4. The supernatant obtained (10 ml) is mixed with 0.5 ml of immunoaffinity resin (either anti-PLD1 resin or antitag resin if expressing an epitope-tagged PLD1) and kept at 4° with constant agitation for 1 hr. The resin is sedimented by gentle centrifugation and unbound protein is removed. The resin is washed three times with 25 volumes of lysis buffer. After the final wash, the resin is resuspended in 5 ml of lysis buffer and placed in a 10-ml Bio-Rad (Hercules, CA) disposable chromatography column.

5. The resin is washed with 10 ml of 10 mM phosphate buffer, pH 6.8, containing 1% β-D-octyglucoside (β-DOG). Bound protein is eluted with 100 mM glycine, pH 3.0, containing 1% β-DOG as 3 × 0.5-ml fractions (or, optionally, using the PLD1 or epitope peptide against which the antiserum was made to elute the PLD1). The eluant is collected on ice into tubes containing 0.075 ml of 1 M phosphate buffer, pH8.0. ARF-stimulated PLD activity in the fractions is determined to identify the appropriate tubes to pool.

Note: Purified PLD1 exhibits very limited stability in our hands. Activity decreases noticeably each day and little is left after 3–4 days. There is a report that purified His-tagged rat PLD1 can be stabilized using glycerol,[13] but it should be noted that this can present problems for subsequent analysis.

Preloading Rho (or ARF) Proteins in Vitro with GTPγS

Due to the nature of Rho as a GTPase, preparations of it eventually consist of GDP-loaded protein regardless of the source or manner in which it was generated. The ability of Rho to stimulate PLD1 is absolutely dependent on it being in the GTP conformation, so it is essential to exchange the GDP for GTP prior to carrying out the *in vitro* PLD assay. The nonhydrolyzable GTP analog, GTPγS, is used to lock Rho into the activated conformation subsequent to the exchange reaction. The following procedure represents a modification of previously published protocols.[20] The modification (suggested by P. Chardin, Inserm, Nice) consists of adding the GTPγS during the first incubation step, when the EDTA-stimulated exchange rate is high, rather than in the second step, after the Mg^{2+} has restored the stability of the nucleotide–protein interaction. The efficacy of activation is not known, as it is not simple to determine the fraction of GTP-bound Rho, as opposed to ARF, which can be measured in a filter assay.[20]

An additional important note is that unlike the interaction of Rho with many of its other downstream effectors, stimulation of PLD1 by Rho is absolutely dependent on the Rho being geranylgeranylated, suggesting that the Rho may have to be membrane associated. Curiously, the physical interaction between Rho and PLD1 does not require this posttranslational modification,[15] nor does Rho stimulation of PLD1 require that the PLD1 be membrane associated,[18] raising the alternate possibility that the geranylgeranylation or some alteration it makes in the presentation of Rho (perhaps in the context of associated lipids) may be directly involved in activating PLD1.

Regardless of the underlying mechanism, the requirement for geranylgeranylation means that bacterially generated Rho cannot be used to stimulate PLD1 activity unless it is geranylgeranylated in vitro using a transferase,[21] which is not generally the method of choice (as opposed to making Rho in Sf9 cells using baculovirus expression) because the transferase is not available commercially.

Combine 100 μl ARF or Rho (at approximately 1–2 mg/ml), 1 μl 500 mM EDTA (final concentration 5 mM), and 0.13 μl 100 mM GTPγS (final concentration 0.13 mM). Incubate on ice for 20 minutes. Add 1 μl 1 M $MgCl_2$ (final concentration 10 mM). Incubate at 37° for 5 min, divide into aliquots, and store at −80°.

In Vitro PLD Assay

The assay (adapted from Brown et al.[11]) consists of adding PLD and activators to a reaction mixture of lipid vesicles containing tritiated PC (where the label is on the choline moiety) and buffer for a fixed period of time. At the end of the assay, the unhydrolyzed labeled lipid is pelleted using bovine serum albumin (BSA) to adsorb the lipid and trichloroacetic acid (TCA) to precipitate it. The labeled choline released by PLD remains soluble in this setting, and scintillation determination of the supernatant accordingly indicates the amount of PLD activity that was present during the assay period.

[20] J. O. Liang and S. Kornfeld, *J. Biol. Chem.* **272,** 4141 (1997).
[21] C. D. Bae, D. S. Min, I. N. Fleming, and J. H. Exton, *J. Biol. Chem.* **273,** 11596 (1998).

The assay mixture consists 50 μl lipids (tritiated PC plus premade lipid mixture), 20 μl 5× assay buffer, 10 μl PLD source, 1 μl RhoA (1–2 μg/ml, GTPγS preloaded), and 19 μl H$_2$O. Total reaction: 100 μl. Assemble on ice. Incubate for 30 min at 37°. Transfer to ice. Add 100 μl BSA (10 mg/ml) and 200 μl 10% TCA (in that order) to stop the reaction. Mix. Centrifuge for 5 min at maximum speed, transfer supernatant into scintillation vials, and count. *Note:* Always set up a negative control (i.e., a sample lacking a PLD source) and, in addition, determine the total radioactivity.

Lipid Preparation. Wash 2.33 μl [^{3}H]PC ([^{3}H]dipalmitoylphosphatidylcholine [choline-methyl-^{3}H], NEN Life Science Products, Boston, MA) by adding 100 μl methanol, 100 μl chloroform, and 90 μl water, vortexing, and centrifuging. Add the lower phase (organic) to 100 μl of premade lipid mixture in chloroform in a 15-ml polypropylene tube. Dry completely under nitrogen. Add 2.5 ml 10 mM HEPES, pH 7.4. Sonicate at high power for 30 sec with a microprobe sonicator.

Premade Lipid Mixture. PE : PIP$_2$: PC = 16 : 1.4 : 1; 100 μM total lipid in each assay sample. Dilute the combined stock lipids with chloroform so that 2 μl is required for each assay sample. The addition of PIP$_2$ to the mixture can cause it to become cloudy because of the limited solubility of PIP$_2$ in the methanol present in the PE and PC stocks. This does not cause problems for the assay.

5× Assay Buffer. 250 mM HEPES, pH 7.5, 15 mM EGTA, 400 mM KCl, 5 mM DTT, 15 mM MgCl$_2$, 10 mM CaCl$_2$

HPLC Purification of PtdIns4P and PtdIns(4,5)P$_2$

PI4,5P$_2$ is required for the *in vitro* PLD assay. This lipid is expensive and commercial preparations are of highly variable quality. Here, we describe a method to prepare it at a fraction of the commercial cost. In brief, a brain phospholipid fraction enriched in acidic lipids (PS, PI, and the polyphosphoinositides) is purified by methanol precipitation and acid extraction, the lipids are converted to their ammonium salt forms, and the material is fractionated by anion-exchange high-performance liquid chromatography (HPLC). Fractions containing these lipids are pooled, the lipids are extracted under acid conditions, reconverted to their ammonium salt forms, and their mass is determined by phosphate analysis. This procedure produces 100 mg each of PtdIns4P and PtdIns4,5P$_2$ from 2 g of starting material.

Sample Preparation. Two grams of bovine brain lipid extract (Type I, Folch Fraction I, Sigma, St. Louis, MO) is divided into two 50-ml polypropylene tubes and dissolved (each) in 12 ml CHCl$_3$. Twenty-two milliliters of MeOH is added to each tube, which are placed on ice for 5 min. The phospholipid precipitate that forms is centrifuged (1000g for 10 min) and the pellets are dissolved in 15 ml CHCl$_3$, 15 ml methanol, and 5 ml 1 M HCl to give a single phase. A further 8.5 ml of 1 M HCl is added to each tube and the sample is mixed vigorously and then centrifuged to give two phases. The lower phases are recovered, pooled together in a new 50-ml polypropylene tube, and concentrated ammonium hydroxide (20% in methanol) is added dropwise to bring the pH to 7 (assessed by spotting on pH paper). The sample is evaporated to dryness (either under a stream of nitrogen or in a centrifugal evaporator concentrator).

HPLC. The lipids are separated by anion-exchange HPLC using a preparative Econosil NH$_2$ column (250 × 10 mm) with a guard column of the same material (Alltech Associates). The column is protected by a preparative scale in-line filter, and a sacrificial 250 × 5-mm column of silica is placed between the pumps and the injector. The basic solvent system used is 20:9:1 (v/v/v) CHCl$_3$:methanol:H$_2$O (solvent A). Solvent B is solvent A containing 1 M ammonium acetate. Anhydrous preparations of CHCl$_3$ and methanol are strongly acidic and should not be used. ACS grade reagents are suitable. The solvents are mixed and filtered through a 0.22-μm nylon filter before use. The column is eluted at 2 ml/min with a 160-ml linear gradient of 0–100% solvent B followed by 20 ml of solvent B before reequilibration with solvent B. Our standard protocol is shown in the tabulation.

Time (min)	Elution volume (ml)	Flow rate (ml/min)	% solvent B
0	0	2	0
10	20	2	0
90	180	2	100
100	200	2	100
102	204	2	0
120	240	2	0

The dried lipids are resuspended in 10 ml of solvent A 20:8:1. Any insoluble material is removed by centrifugation. If desired, approximately 100,000 dpm each of [^{3}H]PtdIns4P and [^{3}H]PtdIns4,5P$_2$ (American Radiolabeled Chemicals, St. Louis, MO) are added to the sample as a tracer. The sample is applied to the column as 4 × 2.5-ml injections (using a 5-ml loading loop) at 5-min intervals. After the final injection, the elution protocol described earlier is begun. The eluant is collected as 2-ml (1 min) fractions. With a new column and fresh solvents, PI4P elutes at approximately 75 min and PI(4,5)P$_2$ at 93 min. Phosphoinositides are identified by phosphate determination (see later) or, if a radioactive tracer has been added, by removing 10 μl of each fraction for liquid scintillation counting. With a new column, 90% of each lipid will be contained in five fractions.

Product Extraction. Prepare synthetic upper and lower phases by mixing CHCl$_3$, methanol, and 0.1 *M* HCl in the ratio 1 : 1 : 0.9 in a separating funnel. Fractions containing the purified lipids are pooled together in 50-ml polypropylene tubes. For each milliliter of eluant, add 0.412 ml of 3 *M* HCl and 0.276 ml of CHCl$_3$. Mix thoroughly and centrifuge to separate the phases. Remove the lower phase to a fresh tube. There may be some white crystals of ammonium chloride at the interface. Do not worry if some of this material is inadvertently taken with the lower phase, it can be removed by washing the lower phase with the synthetic upper phase. Back extract the remaining upper phase with 0.4 volumes of synthetic lower phase, mix, centrifuge, and combine the lower phase with the first lower phase obtained. Wash the combined lower phases with an equal volume of synthetic upper phase, mix, centrifuge, and remove the lower phase to a fresh tube. Neutralize by adding ammonium hydroxide in methanol (as described earlier) and evaporate to dryness. The lipids are resuspended in solvent A for storage at $-20°$. The solubilities of these lipids are such that they can be stored as 10 m*M* solutions. If the lipids come out of solution during storage, warm the solutions to room temperature and mix before using.

Mass Determination. Ten- to 50-μl samples of purified phosphoinositides are wet digested in perchloric acid and phosphate determined by a standard colormetric procedure. The molecular weights of PI4P (diammonium salt) and PI(4,5)P$_2$ (triammonium salt) are 1001 and 1098, respectively. PtdIns4P contains 2 mol of phosphate/mole lipid, whereas PtdIns4,5P$_2$ contains 3 mol phosphate/mole lipid.

In Vivo PLD Assay

This assay is adapted from Wakelam *et al.*[3]

1. On day 1 (morning), pass cells (COS-7, HEK-293) using standard protocols and plate to a density of ~10–20% confluence in DMEM containing Pen-Strep, glutamine, 10% FCS in 35-mm TC dishes. In some cases, more reproducible assay data are achieved using polylysine-coated plates instead of regular TC dishes.

2. On day 2 (morning), transfect using Lipofectamine PLUS in 1 ml Opti-mem media with 1 μg plasmid per dish according to the supplier's recommendations. After 4 hr, replace the transfection media with 1 ml of media containing 10% FCS (containing label, see next step) and continue the culture for another 24 hr.

3. At the start of the 24-hr culture in media with FCS, label the cells using [^{3}H]palmitate. The [^{3}H]palmitate is from American Radiola-

beled Chemicals (St. Louis, MO; ART 129-palmitic acid [9,10^3-H(N)]). Dry 50 μl of 10 mCi/ml [^{3}H]palmitate in a vacuum Speed-Vac. Resuspend the [^{3}H]palmitate resuspended in 100 μl of sterile TE (5 mCi/ml) and sonicate for 30 sec on high power in a water sonicator prior to use. Use 0.4 μl resuspended [^{3}H]palmitate per milliliter of media.

4. At the end of the 24 hr, replace the labeling media with 2 ml of warm, fresh Opti-mem media for 1 to 2 hr, following which the media is replaced with fresh Opti-mem media containing 0.3% butanol and optionally various combinations of stimulators, such as receptor agonists or the PKC activator PMA (100 nM final concentration).

5. At an appropriate stopping point (typically 15–30 min), replace the stimulating media with 300 μl ice-cold methanol and place the dishes on ice. Scrape the cells off using a rubber policeman and rinse and rescrape the dishes with a second 300 μl of cold methanol, which is combined with the first methanol extract.

6. Mix the extraction with 0.5 ml chloroform and vortex well. After sitting at room temperature for 15 min, add 0.4 ml of water. Vortex the samples again and centrifuge for 5 minutes. Dry the organic (bottom) phase by Speed-Vac vacuum centrifugation and resuspend in 2 × 25 μl of chloroform:methanol (19:1) containing 50 μg of authentic (unlabeled) PtdBut (dipalmitate; Avanti Polar Lipids, Alabaster, AL) for spotting onto TLC plates (Whatman LK5DF). (Resuspend in 25 μl; spot on plate; rinse tube with second 25 μl, and spot again in same place.) During summer (in high humidity) it may help to bake the plate (1 hr) before spotting samples to remove any moisture that could cause the TLC run to smear.

7. Load samples 2 cm from the bottom of the plate. The plate is developed in the upper phase of a mixture of ethyl acetate:2,2,4-trimethylpentane (isooctane):acetic acid:water (110:50:20:100; PtdButR_f of 00.4). Mix the solutions in a separating funnel (in the hood) and discard the lower phase. The TLC tank is used unlined. This takes about 1.5 hr. After the TLC run is completed, move the plate (without its rack) to an iodine tank to stain for about 15–30 min. Use a soft pencil to mark the position of the PtdBut. Dry the plate in the hood until all the yellowish staining is gone; scrape the plate in the marked region and count the powder in scintillation fluid (5 ml).

In Vivo Association Assay to Measure Interaction of RhoA^{Val14} and PLD1

In addition to stimulation of PLD1 activity by RhoA, it is also useful to be able to examine their physical interaction. The protocols described here detail how to demonstrate the interaction in COS-7 cells and yeast using the 362 amino acid C-terminal PLD1 fragment (D4 fragment) in which the RhoA-binding site is located.[5,15]

1. Culture COS-7 cells until semiconfluent (70–80% confluent) in DMEM containing 5% FCS in a 15-cm dish.
2. Rinse with 10 ml of PBS. Remove the PBS and add 5 ml of trypsin/EDTA solution to the dish. Incubate at 37° until the cells detach from the dish; add 10 ml of DMEM containing 5% FCS. Collect the cells and centrifuge at 1500 rpm for 2 min. Discard the supernatant and suspend cells in 0.48 ml of PBS.
3. Transfer the cell suspension to a cuvette (0.4 cm cuvette, Bio-Rad), add plasmid DNAs [10 μl of 1 μg/μl pTB701-FL-D4 fragment (Flag-epitope-tagged PLD1 D4 fragment, available on request from YK) and 1 μg/μl pEF-BOS-HA-RhoA^{Val14} (HA epitope-tagged dominant active RhoA, available on request from YK)], and mix well.
4. Electroporate using 220 V and 975 μF. Transfer the cell suspension to a 15-cm dish and add 15 ml of DMEM containing 5% FCS. Culture the cells for 48 hr at 37°.
5. Rinse thoroughly with PBS and then add 10 ml of ice-cold PBS. Scrape cells using a rubber policeman and transfer the cell suspension to a 50-ml Falcon tube. Add 5 ml of ice-cold PBS to the dish, scrape again, and add to the tube. Centrifuge at 1500 rpm for 2 min at 4° and remove the supernatant.
6. Resuspend the cells in 4 volumes of PBS (against 1 volume of the cell pellet) and transfer the suspension to a 1.5-ml Eppendorf tube (the cell pellet should be at least 50 μl in volume).
7. Add an equal volume of 2× lysis buffer (40 mM Tris–HCl, pH 7.5, 2 mM EDTA, 2 mM EGTA, 10 mM MgCl$_2$, 0.4 mM PMSF, 40 μg/ml leupeptin, 20,000 IU/ml aprotinin, 0.75% NP-40) and incubate with rotation for 30 min at 4°.
8. Centrifuge at 10,000g for 10 min at 4° and transfer the supernatant to a new 1.5-ml Eppendorf tube. Add 30 μl of 50% (w/v) slurry of anti-FLAG M2 antibody affinity resin (Sigma) and incubate with rotation for 2 hr at 4°. Wash the pellet three times with 1 ml of 1× lysis buffer.
9. After the final wash, resuspend the resin in 2× sample buffer (or 1× sample buffer containing 8 M urea for full-length PLD1). (If the resin was suspended in 2× sample buffer, then boil the sample for 5 min and centrifuge at 10,000 rpm for 2 min before using for SDS–PAGE. If the resin was suspended in 1× sample buffer containing 8 M urea, then incubate at 37° for 30 min and centrifuge

at 10,000 rpm for 2 min before using for SDS–PAGE, but do not boil.)

10. Carry out SDS–PAGE separation of the proteins using standard protocols and visualize the Rho or PLD proteins using anti-HA and anti-Flag monoclonal antibodies (Sigma) following the supplier's recommendations.

Two-Hybrid Interaction of RhoA with PLD1

Standard protocols for the two-hybrid assay have been described elsewhere and will not be presented here. Instead, assuming the reader has a basic understanding of the system, we will discuss the important factors specific to the RhoA:PLD1 interaction. All published vectors are available on request (YK or MAF).

As described in two published reports from our laboratories,[5,15] wild-type Rho does not interact with PLD1 in yeast, presumably because it is predominantly in the GDP conformation. In contrast, dominant active Rho[Val14] does. In addition, full-length PLD1 does not collaborate with dominant active Rho[Val14] to activate transcription of the yeast two-hybrid reporter genes, either because full-length PLD1 is expressed at low levels that are too low or because it is too large or membrane associated to enter the nucleus efficiently. In contrast, a C-terminal 362 amino acid fragment interacts quite strongly (blue color is easily detectable within 10–20 min using the standard β-Gal filter assay), and a larger C-terminal fragment (amino acids 325–1074) also activates the reported genes, although not as strongly.

Geranylgeranylation of Rho is not required for a successful interaction, although it does appear to increase the reporter gene activation. Whether this signifies a stronger interaction or some alternative trivial explanation, such as higher levels of expression of Rho, is not known. In our laboratories, we use the LexA two-hybrid assay system. PLD1 and Rho have been tested only as fusions in which the LexA DNA-binding protein or the VP16 activator domain were placed at the N terminus of PLD1 and Rho. Rho and PLD1 can be placed on either two-hybrid vector, confirming that the interaction is not dependent on an altered confirmation induced by the specific fusion partner.

Acknowledgments

We thank Pierre Chardin and laboratory members for helpful discussion. This work was supported by NIH grants to MAF (GM54813) and AJM (GM50388) and by a research grant from the Ministry of Education, Science, Sports and Culture, Japan, to YK.

[18] *In Vitro* Interaction of Phosphoinositide-4-phosphate 5-Kinases with Rac

By KIMBERLEY TOLIAS and CHRISTOPHER L. CARPENTER

Introduction

Rho family Gtpases regulate a variety of cellular processes, including actin cytoskeletal organization, gene transcription, and membrane trafficking.[1] To better understand the molecular mechanisms that underlie Rho Gtpase function, it is essential to identify their downstream targets. Due to an intense research effort in the past few years, a number of candidate effector proteins have now been identified.[2] This article discusses the characterization of one of these targets; phosphatidylinositol-4-phosphate 5-kinases (PIP5Ks).

There are two classes of phosphatidylinositol phosphate kinases (PIPKs) that synthesize phosphatidylinositol-4,5-bisphosphate (PtdIns-4,5-P_2) by two different routes. Type I PIPKs catalyze the phosphorylation of phosphatidylinositol 4-phosphate (PtdIns-4-P) at the D-5 position of the inositol ring and are termed PIP5Ks. Type II PIPKs synthesize PtdIns-4,5-P_2 by phosphorylating PtdIns-5-P rather than PtdIns-4-P and have therefore been renamed phosphatidylinositol-5-phosphate 4-kinases (PIP4Ks).[3–7] To date, three distinct mammalian PIP5K isoforms (α, β, and γ) and three distinct PIP4K isoforms (α, β, and γ) have been cloned.[3,8–13] Although similar to each other, the PIPKs share surprisingly little homology with other lipid or protein kinases. They are homologous, however, to two yeast genes,

[1] A. Hall, *Science* **279**, 509 (1998).

[2] L. Van Aelst and C. D'Souza-Schorey, *Genes Dev.* **11**, 2295 (1997).

[3] J. C. Loijens and R. A. Anderson, *J. Biol. Chem.* **271**, 32937 (1996).

[4] C. E. Bazenet, A. R. Ruano, J. L. Brockman, and R. A. Anderson, *J. Biol. Chem.* **265**, 18012 (1990).

[5] L. E. Ling, J. T. Schulz, and L. C. Cantley, *J. Biol. Chem.* **264**, 5080 (1989).

[6] G. H. Jenkins, P. L. Fisette, and R. A. Anderson, *J. Biol. Chem.* **269**, 11547 (1994).

[7] L. E. Rameh, K. T. Tolias, B. Duckworth, and L. C. Cantley, *Nature* **390**, 192 (1997).

[8] I. V. Boronenkov and R. A. Anderson, *J. Biol. Chem.* **270**, 2881 (1995).

[9] N. Divecha, O. Truong, J. J. Hsuan, K. A. Hinchliffe, and R. F. Irvine, *Biochem. J.* **309**, 715 (1995).

[10] H. Ishihara *et al., J. Biol. Chem.* **271**, 23611 (1996).

[11] A. M. Castellino, G. J. Parker, I. V. Boronenkov, R. A. Anderson, and M. V. Chao, *J. Biol. Chem.* **272**, 5861 (1997).

[12] H. Ishihara *et al., J. Biol. Chem.* **273**, 8741 (1998).

[13] T. Itoh, T. Ijuin, and T. Takenawa, *J. Biol. Chem.* **273**, 20292 (1998).

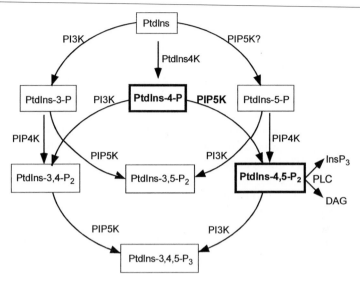

FIG. 1. Synthetic pathways of all known phosphoinositides in mammalian cells. The enzyme thought to be primarily responsible for the synthesis of each phosphoinositide is indicated. The pathway in which PIP5Ks catalyze the conversion of PtdIns-4-P to PtdIns-4,5-P_2 is highlighted. PI3K, phosphoinositide 3-kinase; PtdIns4K, phosphatidylinositol 4-kinase.

MSS4 and FAB1,[14,15] a *Drosophila* gene, skittles,[16] and a FAB1-like gene in *Caenorhabditis elegans*.[3] The crystallization of one of these family members, PIP4Kβ, revealed that three residues thought to be essential for ATP binding and catalysis in the PIPKs are conserved with protein kinases.[17] Based on *in vitro* studies with purified enzymes, mammalian PIP5Ks have been discovered to be relatively promiscuous with respect to substrate utilization.[18,19] For instance, although their preferred substrate is PtdIns-4-P, PIP5Ks can catalyze the phosphorylation PtdIns-3,4-P_2 to PtdIns-3,4,5-P_3 and PtdIns-3-P to produce PtdIns-3,5-P_2, PtdIns-3,4-P_2, and PtdIns-3,4,5-P_3 (Fig. 1).

PtdIns-4,5-P_2 regulates a number of proteins that control actin polymerization and structure. PtdIns-4,5-P_2 stimulates actin polymerization by binding to capping proteins, such as gelsolin, to dissociate them from the barbed

[14] S. Yoshida, Y. Ohya, A. Nakano, and Y. Anraku, *Mol. Gen. Genet.* **242**, 631 (1994).
[15] A. Yamamoto *et al.*, *Mol. Biol. Cell* **6**, 525 (1995).
[16] B. A. Hassan *et al.*, *Genetics* **150**, 1527 (1998).
[17] V. D. Rao, S. Misra, I. V. Boronenkov, R. A. Anderson, and J. H. Hurley, *Cell* **94**, 829 (1998).
[18] X. Zhang *et al.*, *J. Biol. Chem.* **272**, 17756 (1997).
[19] K. F. Tolias *et al.*, *J. Biol. Chem.* **273**, 18040 (1998).

ends of actin filaments and binding to profilin to release G-actin.[20] PtdIns-4,5-P$_2$ binding to vinculin and ezrin, radixin, and moesin (ERM) proteins allows these proteins to link actin filaments to the plasma membrane.[20] PtdIns-4,5-P$_2$ binding to α-actinin stimulates its ability to cross-link actin filaments. As regulators of PtdIns-4,5-P$_2$ levels, PIPKs may also control actin dynamics.

Because Rho GTPases also regulate actin cytoskeletal organization, several laboratories have investigated whether PIPKs might be downstream effectors. PIP5K α and β were found to interact directly with the C terminus of Rac1 in a GTP-independent manner.[21-23] In addition, Rac was shown to stimulate PIP5K activity in permeabilized platelets and neutrophil extracts.[23-25] Rho was also found to associate with an unidentified PIP5K in a GTP-independent manner[26] and to stimulate PIP5K activity in a fibroblast lysate.[27] The ability of Rho family GTPases to bind to and regulate PIP5Ks indicates that PIP5Ks play a role in Rho GTPase signaling pathways. PIP5Kα has been demonstrated to promote actin assembly downstream of Rac and the thrombin receptor in permeabilized platelets.[23]

This article describes an *in vitro*-binding assay that has been used for examining the interaction between Rac and the PIP5K. It also outlines a protocol for assaying PIP5K activity and analyzing the resulting lipid products.

Protein Purification

In the *in vitro*-binding assay outlined here, glutathione S-transferase (GST) fusion proteins of Rac, Rho, and Cdc42 bound to glutathione–agarose beads are used to capture PIP5Ks derived from various sources. The method for purifying GST–Rho GTPase fusion proteins from *Escherichia coli* has been described previously.[28] Because PIP5Ks are ubiquitously expressed enzymes,[3,10,12] they can be obtained from many cell lines and tissues. For a generic source of PIP5Ks, we have found that homogenized

[20] P. A. Janmey, *Annu. Rev. Physiol.* **56,** 169 (1994).

[21] K. F. Tolias, L. C. Cantley, and C. L. Carpenter, *J. Biol. Chem.* **270,** 17656 (1995).

[22] K. F. Tolias, A. D. Couvillon, L. C. Cantley, and C. L. Carpenter, *Mol. Cell. Biol.* **18,** 762 (1998).

[23] K. F. Tolias, J. H. Hartwig, H. Ishihara, Y. Shibasaki, L. C. Cantley, and C. C. Carpenter, *Curr. Biol.* **10,** 153 (2000).

[24] J. H. Hartwig *et al., Cell* **82,** 643 (1995).

[25] S. H. Zigmond, M. Joyce, J. Borleis, G. M. Bokoch, and P. N. Devreotes, *J. Cell Biol.* **138,** 363 (1997).

[26] X.-D. Ren *et al., Mol. Biol. Cell* **7,** 435 (1996).

[27] L. D. Chong, A. Traynor-Kaplan, G. M. Bokoch, and M. A. Schwartz, *Cell* **79,** 507 (1994).

[28] A. J. Self and A. Hall, *Methods Enzymol.* (1995).

rat brain works well in our assay. A rat brain is chopped into small pieces with a razor blade and is then homogenized with a Dounce homogenizer for approximately 2 min in 30–50 ml of ice-cold lysis buffer A [50 mM Tris, pH 7.5, 50 mM NaCl, 5 mM MgCl$_2$, 1% Nonidet P-40 (NP-40), 10% glycerol, 1 mM dithiothreitol (DTT), and 4 μg/ml each of leupeptin, pepstatin, and 4-(2-aminoethyl)benzenesulfonyl fluoride (AEBSF)]. The homogenate is cleared by centrifugation at 4° for 15 min at 15,000 rpm.

To analyze the interaction of Rac with specific PIP5Ks, we used murine PIP5K α, β, and γ isoforms expressed in either *E. coli* or in transiently transfected mammalian cells.[23] GST fusion proteins of PIP5Ks are produced by inoculating Luria broth (LB) containing 50 μg/ml ampicillin with *E. coli* transformed with various PIP5K isoforms, cloned into the pGEX-4T2 vector. After shaking overnight at 37°, the cultures are diluted 1 : 10 into fresh LB/ampicillin and grown for an additional hour at 37°. To induce protein expression, isopropyl-β-D-thiogalactopyranoside (IPTG) is added to a final concentration of 0.1 mM, and the cells are incubated for 5–6 hr at 25°. We have found that incubation at 25°, rather than 37°, following protein induction increases protein expression and stability. Cells are collected by centrifugation at 5000 rpm for 10 min and then subjected to one freeze-thaw cycle at −80°. The cells can be kept at −80° for several days before thawing and lysis. Once thawed, the pellets are resuspended in protein purification buffer (50 mM Tris, pH 7.5, 150 mM NaCl, 1 mM EDTA, 1 mM DTT, and 4 μg/ml of leupeptin, pepstatin, and AEBSF) and then sonicated on ice (3 × 20 sec). After the addition of NP-40 to a final concentration of 1%, the cells are centrifuged for 10 min at 15,000 rpm, and then the supernatants are incubated with glutathione–agarose beads (1 ml of a 50% slurry/500 ml culture) for 1 hr at 4°. Beads are washed three times with ice-cold protein purification buffer plus 1% (NP-40) and twice with TNE (50 mM Tris, pH 7.5, 50 mM NaCl, and 1 mM EDTA).

If necessary, the proteins are cleaved from GST by incubating the beads in thrombin digestion buffer [50 mM Tris, pH 8.0, 150 mM NaCl, 2.5 mM CaCl$_2$, 5 mM MgCl$_2$, 1 mM (DTT)] with thrombin (2.5 units/mg protein) overnight at 4° with gentle rocking. The resulting supernatant containing cleaved proteins is incubated with *p*-aminobenzamidine agarose beads to remove thrombin and is then concentrated using a Centricon (Amicon, Danvers, MA) filter. The concentrated proteins are washed several times by diluting with TNE plus 1 mM DTT and reconcentrating to approximately 300 μl (0.2 mg/ml). Protein concentrations are determined by comparison to bovine serum albumin (BSA) standards on SDS–PAGE gels stained with Coomassie blue. Proteins are stored at −80° in storage buffer [50 mM Tris, pH 7.5, 150 mM NaCl, 5 mM MgCl$_2$, 1 mM DTT, and 50% (v/v) glycerol].

To prepare specific PIP5Ks expressed in mammalian cells, 293T cells are transiently transfected with plasmid DNA encoding HA-tagged PIP5Ks (cloned into the pEBB mammalian expression vector). Transfections can be performed using Lipofectamine (Life Technologies, Inc., Grand Island, NY), Superfect (Qiagen, Valencia, CA), or calcium phosphate and a total of 4–10 μg DNA per 10-cm plate or 2–4 μg DNA per six-well plate. After 18 hr, transfected cells are rinsed with phosphate-buffered saline (PBS) and scrape harvested in lysis buffer A. We have found that allowing transfections to continue longer than 24 hr results in reduced protein expression. After incubating the cells in lysis buffer for 5 min at 4°, the lysate is clarified by centrifugation at 15,000 rpm for 10 min.

In Vitro-Binding Assay

To examine the interaction between Rho family GTPases and PIP5Ks, GST and GST–Rho GTPase fusion proteins (10 μg each) bound to glutathione–agarose beads are incubated at 4° with 0.7 ml rat brain homogenate (3 mg/ml), 0.7 ml PIP5K-transfected cell lysate (2 mg/ml), or 10 μl recombinant cleaved PIP5K (0.2 mg/ml) for 1 hr with constant rocking. The binding assay is done in the presence of 50 mM rather than 150 mM NaCl because we have found that the interaction is sensitive to ionic strength. This suggests that the interaction is primarily ionic, a possibility that is supported by the fact that the C terminus of Rac, which mediates the association with the PIP5Ks, is positively charged.[22] Because the interaction between Rac (and Rho) and PIP5Ks is GTP independent, preloading Rho GTPases with guanine nucleotides is optional. Following incubation, the bead are centrifuged briefly at 5000 rpm and washed twice with ice-cold lysis buffer A and twice with TNM (50 mM Tris, pH 7.5, 50 mM NaCl, and 5 mM MgCl$_2$). TNM is used to reduce the effective concentration of NP-40, as PIP5K activity is inhibited by high concentrations of detergent. PIP5K bound to the beads is detected by immunoblotting or by assaying for associated PIP5K activity as described later. Western blotting is performed by subjecting the proteins associated with the GSH beads to SDS–PAGE, transferring the proteins to nitrocellulose, and then probing the blots with the anti-HA antibody 12CAG or anti-PIP5K antibodies.[23] Blots are developed using ECL (enhanced chemiluminescence) reagents (NEN, Boston, MA).

Phosphatidylinositol-Phosphate Kinase Assay

The presence of PIP5K bound to GST–Rac beads can be detected by assaying the beads for PIP5K activity. In addition, lipid kinase assays can be performed directly on recombinant PIP5Ks expressed in either bacteria or mammalian cells. PIP5Ks (in ≤ 30 μl initial volume) are assayed in 1.5-ml Eppendorf tubes at 25° in a 50-μl final reaction volume containing 50 mM Tris, pH 7.5, 30 mM NaCl, 10 mM MgCl$_2$, μM PtdIns-4-P, 133 μM phosphoserine (PS), and 50 μM [γ-^{32}P]ATP (1–10 μCi/assay). The lipid substrates are prepared as 5× stocks by drying the lipids (stored in chloroform at −80°) under nitrogen, resuspending them in 10 mM HEPES, 0.1 mM EDTA, and then sonicating the resuspended lipids in a cup horn sonicator for 5 min. Sonicated lipids are added to the PIPKs. The reaction is started by adding a 5× stock of [γ-^{32}]ATP/MgCl$_2$ and briefly mixing the contents of the tube. The reactions are stopped after 5–10 min by adding 80 μl of 1 N HCl, followed by 160 μl chloroform/methanol (1:1) to extract the lipids. The reaction is mixed with a vortex and then centrifuged briefly to separate the phases. Phosphorylated lipids, obtained from the lower chloroform phase, are transferred to a new tube and then separated by thin-layer chromatography (TLC) or by high-performance liquid chromatography (HPLC) as described later.

Theoretically, PIP5Ks can be distinguished from PIP4Ks in kinase assays by using either PtdIns-4-P or PtdIns-5-P as a substrate. Unfortunately, commercially available PtdIns-4-P preparations contain PtdIns-5-P[7] and purified PtdIns-5-P or PtdIns-4-P is not readily available. One can however, rely on other enzymatic differences to discriminate between PIP5Ks and PIP4Ks. For instance, PIP5Ks and PIP4Ks can be distinguished based on the ability of phosphatidic acid (PA) to specifically stimulate PIP5K activity.[6,29] PA stimulation can be assessed by assaying the PIPKs in the presence of 0.1% Triton X-100 and 80 μM of PA.

As mentioned previously, PIP5Ks can phosphorylate other phosphoinositides in $vitro$ in addition to PtdIns-4-P, such as PtdIns, PtdIns-3-P, and PtdIns-3,4,-P$_2$.[18,19] To examine the ability of the PIP5Ks to phosphorylate these substrates, PtdIns-4-P should be replaced in the kinase assay with PtdIns, PtdIns-3-P, or PtdIns-3,4-P$_2$. When PtdIns is used, no carrier lipid (PS) is necessary, but PS is necessary when PtdInsPs or PtdInsP$_2$s are used as substrates. PIP5Ks can also be assayed simultaneously with diacylglycerol kinases (DGKs) by adding 500 μM diacylglycerol (DG) and 1 mM deoxycholate (DOC) or with phosphoinositide 3-kinases (PI3Ks) by replacing PtdIns-4-P and PS with 40 μM PtdIns, PtdIns-4-P, and PtdIns-4,5-P$_2$ and 80 μM PS. As a control for PIP5K activity, kinase-inactive PIP5K mutants can be used. Among the mutants we have tested, we have found the PIP5Kα D227A, PIP5Kβ D307A, and PIP5Kγ D342A are the most impaired in kinase activity.

[29] A. Moritz, G. P. De, W. H. Gispen, and K. W. Wirtz, $J.$ $Biol.$ $Chem.$ **267,** 7207 (1992).

Analysis of Lipid Kinase Products

Two methods exist for analyzing the lipid products generated in the kinase assay. Of the two procedures, thin-layer chromatography is quicker and easier to use. TLC separation allows for the identification of PtdIns-P, PtdIns-P$_2$, and PtdIns-P$_3$. However, it cannot reliably differentiate between the different isomers such as PtdIns-4,5-P$_2$ and PtdIns-3,4-P$_2$. For this reason, anion-exchange HPLC analysis is the preferable method to use to distinguish between the various isomers.

Thin-Layer Chromatography

Two main solvent systems are typically used for TLC separation. The first solvent contains chloroform, methanol, water, and ammonium hydroxide (60 : 47 : 11 : 1.6) and is used to separate PtdIns-P from PtdIns-P$_2$. Both ATP and PtdInsP$_3$ remain at the origin. This system requires only one-half of a normal TLC plate and takes just 30 min to develop. The second solvent system, which contains n-propanol and 2 M acetic acid (65 : 35), separates PtdInsP, PtdInsP$_2$, and PtdInsP$_3$. This system requires a full TLC plate and 6 hr to develop (it may also be left to develop overnight if necessary).

To analyze the products from the lipid kinase assay, the lipids (contained in chloroform) are spotted directly onto a foil-backed, silica gel 60 TLC plate, about 1.5 cm from the bottom. The origin may be marked in pencil directly on the plate. TLC plates are prepared by precoating the plates with 1% potassium oxalate [in 50% (w/v) methanol]. Oxalate chelates divalent cations, which can cause phosphoinositides to aggregate. The plates are air dried and stored. Immediately before use, they are activated by heating at 100° for 10 min. A gentle stream of air can be used to help evaporate the chloroform while loading the samples onto the TLC plate. It is important to confine the sample to an area of about 5 mm in diameter. When the samples are dry, the TLC plate is placed in a TLC tank containing solvent. The solvents are first mixed in a flask and then added to the tank. The TLC tank should also contain a vertical piece of Whatman (Clifton, NJ) paper at the back that is saturated when the solvent is added. We generally prepare the solvents fresh for each run. When the solvent has reached the top of the plate, the TLC plate is removed from the tank and allowed to dry. Phosphorylated lipids are visualized by autoradiography and quantified using a Bio-Rad (Hercules, CA) Molecular Analyst. To identify the position of the various phosphoinositides on the TLC plate, phosphoinositide standards (2–5 μg each) are included and then visualized after the run by exposing the plate to iodine vapor for 2–5 min. Examples of the lipid

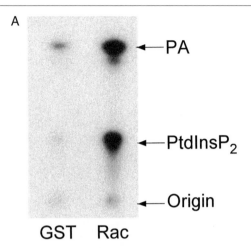

A

←—PA

←—PtdInsP₂

←—Origin

GST   Rac

FIG. 2. (A) GST and GST–Rac were incubated with rat brain homogenate, washed, and assayed simultaneously for PIP5K and DGK activity. The lipid products were then separated by TLC using chloroform, methanol, water, and ammonium hydroxide (60:40:13:1.5). The migration positions of the lipid standards are indicated. Adapted from Tolias et al.[22] (B) The activity of bacterially expressed recombinant PIP5Kα was assayed using PtdIns, PtdIns-4-P, PtdIns-3-P, and PtdIns-3,4-P₂ as substrates. The products of the reactions were separated by TLC using n-propanol and 2 M acetic acid (65:35). Adapted from Tolias et al.[19]

products of a Rac-associated PIP5K from rat brain homogenate and recombinant PIP5K separated by TLC are shown in Figs. 2A and 2B, respectively.

High-Performance Liquid Chromatography

As mentioned previously, HPLC analysis is used to precisely identify the various isomers generated in the lipid kinase assays. To prepare the phosphoinositide products for analysis by HPLC, they are deacylated with methylamine reagent.[30] Methylamine cleaves the fatty acid chains from the glycerol backbone, resulting in glycerophosphorylinositol phosphates (GroPInsPs), which contain only the inositol head group and the glycerol side chain. The generation of GroPInsPs is useful because these molecules can be separated easily by HPLC using anion-exchange chromatography. To deacylate lipids, the chloroform extract from the kinase reactions is dried under nitrogen and then resuspended in 1 ml methylamine reagent (26.8 ml 40% methylamine in water, 16 ml water, 43.7 ml methanol, 11.4

---

[30] K. R. Auger, C. L. Carpenter, L. C. Cantley, and L. Varticovski, *J. Biol. Chem.* **264,** 20181 (1989).

B

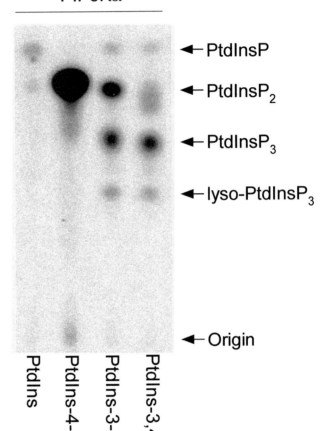

FIG. 2. (*Continued*)

ml *n*-butanol). We find glass scintillation vials to be convenient containers for this reaction. The vials are placed in a styrofoam float in a water bath. After a 50-min incubation at 53°, samples are dried in a Speed-Vac using a cold sulfuric acid trap to prevent methylamine vapor from damaging the vacuum pump. Once dry, the samples are resuspended in 1 ml water and transferred to a glass test tube, and the fatty acids are extracted twice with 1 ml extraction solution [*n*-butanol, petroleum ether, ethyl formate (20:4:1)]. The lower aqueous phase is transferred to an Eppendorf tube

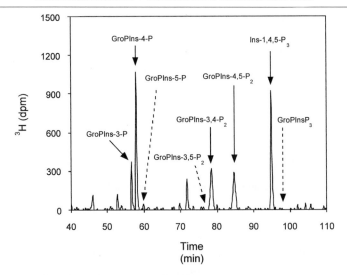

FIG. 3. HPLC analysis of deacylated, ^{3}H-labeled standards PtdIns-4-P, PtdIns-3,4-P$_2$, PtdIns-4,5-P$_2$, and Ins-1,4,5-P$_3$. Expected migration positions of deacylated PtdIns-5-P, PtdIns-3,5-P$_2$, and PtdIns-3,4,5-P$_3$ are also indicated.

and dried in a Speed-Vac (the sulfuric acid trap is no longer necessary). The deacylated lipids are resuspended in water, mixed with ^{3}H-labeled standards, filtered, and then loaded onto a Partisphere SAX anion-exchange HPLC column (Whatman), as described previously.[31]

After the sample is loaded, the column is washed for 5 min and the bound material is eluted at a flow rate of 1 ml/min with 1 M (NH$_4$)$_2$HPO$_4$, pH 3.8, and water with the following gradients: 0 to 15% 1 M (NH$_4$)$_2$HPO$_4$ for 55 min, 15% 1 M (NH$_4$)$_2$HPO$_4$ for 15 min, and 15 to 65% 1 M (NH$_4$)$_2$HPO$_4$ for 25 min. The column is regenerated by washing for 15 min with 1 M (NH$_4$)$_2$HPO$_4$, followed by a water wash for 10 min. The eluate from the HPLC column flows into an on-line continuous flow scintillation detector (Radiomatic Instruments, Downers Grove, IL) that can monitor and quantify two different radioisotopes simultaneously. Using this gradient, GroPIns-3-P is expected to elute from the column at approximately 24 min, GroPIns-4-P at 26 min, GroPIns-3,4-P$_2$ at 60 min, GroPIns-4,5-P$_2$ at 62 min, and GroPIns-3,4,5-P$_3$ at 90 min. Elution times often vary between runs, but the separation between lipids usually remains constant. To increase the separation of PtdIns-4-P from PtdIns-5-P and PtdIns-3,4-P$_2$ from PtdIns-3,5-P$_2$, the following modified ammonium phosphate gradient can

[31] L. A. Serunian, K. R. Auger, and L. C. Cantley, *Methods Enzymol.* **198**, 78 (1991).

be used: 0% 1 M $(NH_4)_2HPO_4$ for 5 min, 0 to 1% 1 M $(NH_4)_2HPO_4$ for 5 min, 1 to 3% 1 M $(NH_4)_2HPO_4$ for 40 min, 3 to 10% 1 M $(NH_4)_2HPO_4$ for 10 min, 10 to 13% 1 M $(NH_4)_2HPO_4$ for 25 min, 13 to 65% 1 M $(NH_4)_2HPO_4$ for 25 min, and 65 to 0% 1 M $(NH_4)_2HPO_4$ in 1 min, followed by a salt and water wash.[19] Figure 3 illustrates the elution profiles of ^{3}H-labeled GroPIns standards separated on an HPLC column using this modified ammonium phosphate gradient.

Acknowledgment

This work was supported by Grant GM 54389 from the NIH to CLC.

[19] Activity and Regulation of p35/Cdk5 Kinase Complex

By MARGARETA NIKOLIC and LI-HUEI TSAI

Introduction

To function as a serine threonine kinase, cyclin-dependent kinase 5 (Cdk5) has to associate with a regulatory partner.[1–4] The first known regulator of Cdk5, p35, is predominantly expressed in the central nervous system (CNS),[1,5,6] although recent evidence suggests that low levels may be present in the differentiating myotome.[7] In mammals, there has been no evidence of endogenous p35/Cdk5 kinase activity in cells of nonneuronal origin or proliferating neuronal precursors.[8,9] The onset of p35/Cdk5 kinase activity

---

[1] L. H. Tsai, I. Delalle, V. S. J. Caviness, T. Chae, and E. Harlow, *Nature* **371,** 419 (1994).

[2] J. Lew, Q. Q. Huang, Z. Qi, R. J. Winkfein, R. Aebersold, and T. Hunt, *Nature* **371,** 423 (1994).

[3] J. Lew and J. H. Wang, *Trends Biochem. Sci.* **20,** 33 (1995).

[4] D. Tang, J. Yeung, K. Y. Lee, M. Matsushita, H. Matsui, and K. Tomizawa, *J. Biol. Chem.* **270,** 26897 (1995).

[5] I. Delalle, P. G. Bhide, V. S. J. Caviness, and L. H. Tsai, *J. Neurocytol.* **26,** 283 (1997).

[6] K. Tomizawa, H. Matsui, M. Matsushita, J. Lew, M. Tokuda, and T. Itano, *Neuroscience* **74,** 519 (1996).

[7] M. Zheng, C. L. Leung, and R. K. Liem, *J. Neurobiol.* **35,** 141 (1998).

[8] L. H. Tsai, T. Takahashi, V. S. Caviness, Jr., and E. Harlow, *Development* **119,** 1029 (1993).

[9] M. H. Lee, M. Nikolic, C. A. Baptista, E. Lai, L. H. Tsai, and J. Massagué, *Proc. Natl. Acad. Sci. U.S.A.* **93,** 3259 (1996).

parallels the timing of p35 expression, which is evident as neuronal precursors exit the cell cycle.[1,5,6] The highest levels of this kinase are seen in the developing mammalian cerebral cortex when the young neurons are migrating out of the ventricular zone to form layers of the future cortex.[5,8,9]

The importance of the p35/Cdk5 kinase in cortical development has been shown by mouse knockout experiments. Mice that lack Cdk5 exhibit embryonic and perinatal lethality with severe defects of the CNS.[10–12] The absence of p35 is not lethal; however, these animals have a severely disrupted laminar configuration of the cortex and display a predisposition to lethal seizures.[13,14]

In neurons, both p35 and Cdk5 localize to the actin-rich peripheral lamellipodia of growing axons.[15–18] Disruption of the active kinase results in reduced neurite outgrowth in cultures of cortical neurons[15–17] and neuronal cell lines.[19] We have shown that the p35/Cdk5 kinase is an effector of the Rho GTPase Rac and regulates another Rac effector, the Pak1 (p21 activated kinase) kinase by phosphorylation.[18] These findings place the p35/Cdk5 kinase as an important component of the Rac signaling pathway in neurons.

To date, the p35/Cdk5 kinase is the only known neuronal-specific effector of Rho GTPases. Furthermore, by inhibiting the activity of Pak1, the p35/Cdk5 kinase provides insight into novel signaling cascades that regulate the actin cytoskeleton in neurons. Because the p35/Cdk5 kinase is fundamental for the correct development of the mammalian CNS, it is important to know how its activity is regulated. We have therefore chosen to describe in this article several methods used to isolate p35/Cdk5 kinase and determine its level of activity.

[10] T. Ohshima, J. M. Ward, C. G. Huh, G. Longenecker, Veeranna, and H. C. Pant, *Proc. Natl. Acad. Sci. U.S.A.* **93**, 11173 (1996).

[11] E. C. Gilmore, T. Ohshima, A. M. Goffinet, A. B. Kulkarni, and K. Herrup, *J. Neurosci.* **18**, 6370 (1998).

[12] T. Ohshima, E. C. Gilmore, G. Longenecker, D. M. Jacobowitz, R. O. Brady, and K. Herrup, *J. Neurosci.* **19**, 6017 (1999).

[13] T. Chae, Y. T. Kwon, R. Bronson, P. Dikkes, E. Li, and L. H. Tsai, *Neuron* **18**, 29 (1997).

[14] Y. T. Kwon and L. H. Tsai, *J. Comp. Neurol.* **395**, 510 (1998).

[15] M. Nikolic, H. Dudek, Y. T. Kwon, Y. F. Ramos, and L. H. Tsai, *Genes Dev.* **10**, 816 (1996).

[16] G. Pigino, G. Paglini, L. Ulloa, J. Avila, and A. Cáceres, *J. Cell Sci.* **110** (Pt.2), 257 (1997).

[17] G. Paglini, G. Pigino, P. Kunda, G. Morfini, R. Maccioni, and S. Quiroga, *J. Neurosci.* **18**, 9858 (1998).

[18] M. Nikolic, M. M. Chou, W. Lu, B. J. Mayer, and L. H. Tsai, *Nature* **395**, 194 (1998).

[19] W. Xiong, R. Pestell, and M. R. Rosner, *Mol. Cell. Biol.* **17**, 6585 (1997).

Preparation of Tissues and Cells for Isolation of
p35/Cdk5 Kinase

Preparation of Cortical Tissues

We use time pregnant rats of the Long Evans strain, as the embryos are easier to work with due to eye pigmentation. Usually 17-day-old embryos (E17) to newborns (P0) are harvested, as this is the time when the p35/Cdk5 kinase activity peaks; however, the same method can be used to obtain tissues from younger embryos or older rodents.

1. Pregnant mothers are killed rapidly in a CO_2 chamber. The embryos are immediately surgically removed, decapitated, and the heads placed in prechilled Hanks' buffered saline solution (GIBCO-BRL, Gaithersburg, MD) and kept continuously on ice. Newborns are decapitated rapidly using a sharp pair of surgical scissors.
2. Using a dissection microscope and fine forceps the brains are freed from surrounding tissue. After removal of as many meningies as possible, the two cortical hemispheres are separated from the remaining CNS tissue. During dissection, care is taken to continuously place the tissues in fresh, ice-cold Hanks' salt solution.
3. As each cortex is freed, it is placed into a prechilled microcentrifuge tube and either snap-frozen in liquid nitrogen or lysed in the appropriate buffer (see subsequent sections for lysis buffers and protocols). Frozen samples can be stored at $-80°$ for a long period of time prior to lysis. Alternatively, the tissues can be placed into prechilled dissociation media and processed for culturing.

Establishing Cultures of Cortical Neurons

Cortical neurons are usually obtained from E17–P0 rats. Neurons from older cortices are less viable in culture, while the establishment of neuronal precursors from younger cortices requires a different procedure.

1. The tissues are removed as described in the previous section and placed into ice-cold dissociation media [DM; 82 mM Na$_2$SO$_4$, 30 mM K$_2$SO$_4$, 5.8 mM MgCl$_2$, 0.25 mM CaCl$_2$, 1 mM HEPES, pH 7.3–7.4, 20 mM glucose, 0.001% (v/v) phenol red, and 0.2 mM NaOH]. Care must be taken to maintain sterility of all tools and solutions used in this procedure.
2. Using forceps, the tissues are cut up into smaller pieces, placed into 5 ml of prewarmed enzyme solution (pH ~ 7.4; 9.7 ml DM, 3.2 mg cysteine hydrochloride, and 100–200 units papain), and incubated at $37°$ for 20 min. This is repeated one more time. After two rinses in

light inhibitor solution (1 ml of heavy inhibitor solution and 6 ml of DM), the cortical tissues are incubated for 2 min at 37° in heavy inhibitor [pH ~ 7.4; 6 ml DM, 60 mg bovine serum albumin (BSA), and 60 mg trypsin inhibitor]. The tissues are subsequently rinsed and triturated in Optimem (GIBCO-BRL) containing 20 mM glucose. The neurons are counted and plated onto poly(D-lysine) and laminin-coated tissue culture plates or glass coverslips in glucose-supplemented Optimem. Usually, 12 × 10^6 neurons are plated onto a 10-cm plate, 4 × 10^6 onto a 6-cm plate, and 2 × 10^5 on a coverslip.

3. The neuronal cultures are incubated at 37°, 5% (v/v) CO_2 for 1–3 hr to allow attachment, followed by a careful media change. For the densities described earlier, the growth medium is BGM [90 ml BME media (Sigma, St. Louis, MO) 5% fetal calf serum, 35 mM glucose, 1 mM L-glutamine, 0.45 ml SVM (3 mg/ml L-proline, 3 mg/ml L-cysteine, 1 mg/ml p-aminobenzoic acid, 12.5 μg/ml vitamin B$_{12}$, 2 mg/ml myo-inositol, 2 mg/ml choline chloride, 5 mg/ml fumaric acid, and 80 μg/ml coenzyme A), 0.8 mg putrescine, 0.25 mg human transferrin, and 0.25 μg sodium selenite]. However, if neurons are to be plated at a lower density, we use Neurobasal media (GIBCO-BRL) containing 1 mM sodium pyruvate, 2 mM L-glutamine, 0.06 mg/ml L-cysteine, and 2% supplement B-27 (GIBCO-BRL).

4. At the time of plating the levels of p35/Cdk5 kinase are greatly reduced in these neuronal cultures, when compared to the tissues they originated from. We speculate that this is due to the mechanical stress the cells have gone through during the process of dissociation and plating. p35/Cdk5 kinase activity increases with the maturation of the cultures *in vitro,* peaking on the third and fourth day from culturing.[1]

Nonneuronal Expression of p35/Cdk5

For most experiments, Cos-7 cells are used due to their high transfectability. The conditions described in this section are determined for them. Many cell lines can be transfected in a similar manner, but will require additional optimization to achieve maximum efficiencies.

1. On the day of transfection, Cos-7 cells are plated out evenly on 10-cm plates from a pool of trypsinized cells (usually a 1:5 split from a confluent 10-cm plate). This is done in the morning, allowing at least 8 hr for the cells to attach before transfection.

2. Five hundred-microliter aliquots of 2× HBS (dissolve 16.4g NaCl, 11.9 g HEPES, and 0.21 g Na$_2$HPO$_4$ in 1 liter distilled H$_2$O, bring

pH to 7.1, and filter sterilize) are made for each transfection mix in 5-ml snap cap polypropylene Falcon tubes. In addition, a sterile microcentrifuge tube containing 450 μl sterile water, 50 μl 2.5 M CaCl$_2$, and a total of 20 μg plasmid DNA previously purified on a CsCl gradient is prepared for each transfection. The DNA mix is transferred to HBS with constant bubbling to avoid formation of large precipitates that will not be taken up by the cells and are more toxic.

3. The precipitate is allowed to form for approximately 10–15 min and is subsequently transferred onto a plate of cells. The cells are incubated overnight (between 12 and 16 hr) at 37° and 5% CO$_2$.

4. The following morning, all the plates are washed three times in sterile PBS (140 mM NaCl, 1.6 mM NaH$_2$PO$_4$, and 8.4 mM Na$_2$HPO$_4$, pH 7.4–7.6). The cells are subsequently incubated for 24 hr in fresh Dulbecco's modified Eagle's medium (DMEM) containing 10% fetal calf serum (FBS) at 37° and 5% CO$_2$. It is important not to culture the transfected cells for longer, as the p35/Cdk5 kinase causes cell death on expression.

Lysis Conditions and Immunoprecipitations

Lysis Conditions

The association between p35 and Cdk5 is strong enough to withstand harsh lysis in the presence of 0.1% SDS such as RIPA buffer (see Table I) and high concentrations of NaCl. However, these conditions do not allow for the interaction of other proteins with the p35/Cdk5 kinase. Therefore, the lysis buffer should vary depending on the experimental requirements. For instance, it is necessary to use a buffer containing MgCl$_2$ and no more than 150 mM NaCl when isolating p35/Cdk5 complexed to Rac and Pak1. A list of several buffers we routinely use and their applications is given in Table I. The buffers are stored at 4°. Prior to use, protease and phosphatase inhibitors are always added [5 mM NaF, 1 mM NaVO$_4$, 100 μg/ml phenylmethylsulfonyl fluoride (PMSF), 2 μg/ml aprotinin, 1 μg/ml pepstatin A, and 2 μg/ml leupeptin].

Lysis is always carried out on ice for 20 min. It is sufficient to pipette embryonic cortices approximately 20 times in lysis buffer using a p1000; other tissues are mashed in a Dounce homogenizer, whereas cultured neurons and transfected Cos-7 cells are incubated in 0.5–1 ml of buffer per 10-cm plate and placed on trays containing ice. Lysates are collected by centrifugation (see Table I for required speed), and the protein concentra-

TABLE I
MOST COMMONLY USED BUFFERS FOR CELL LYSIS, CONDITIONS, AND APPLICATIONS

Buffer name	Recipe	Centrifugation	Application
ELB	50 mM Tris–HCl, pH 7.4, 250 mM NaCl, 1 mM EDTA, 0.5% NP-40, 1 mM DTT	4°, 8 min, maximum speed in microcentrifuge	p35/Cdk5 kinase assays from membrane and cytoplasm
Buffer A	20 mM Tris–HCl, pH 7.2, 2 mM MgCl$_2$, 0.5% NP-40, 150 mM NaCl, 1 mM DTT	4°, 8 min, maximum speed in microcentrifuge	p35/Cdk5 kinase assays from membrane and cytoplasm, p35/Cdk5-associated proteins
STM	10 mM Tris–HCl, pH 8, 0.25 M sucrose, 10 mM MgCl$_2$, 1 mM DTT	4°, 15 min, 600g followed by 4°, 15 min, 100,000	Cytoplasmic p35/Cdk5 kinase and associated proteins, lower stringency
STM + NP-40	STM with 0.5% NP-40, starting material is pellets from STM lysis	4°, 15 min, 600g	Membrane-associated p35/Cdk5 kinase and associated proteins
RIPA	50 mM Tris–HCl, pH 8, 150 mM NaCl, 1% NP-40, 0.5% DOC, 0.1% SDS, 1 mM DTT	4°, 8 min, maximum speed in microcentrifuge	Whole cell lysate, stringent p35/Cdk5 kinase assays

tion is determined using a Bio-Rad (Hercules, CA) protein assay (as described by the manufacturer) and a spectrophotometer. Total protein concentrations are calculated by reference to a BSA standard curve.

Immunoprecipitations

In brain Cdk5, levels are in vast excess over p35. Anti-Cdk5 immunoprecipitates will contain no p35 if the DC-17 antibody is used (available from Santa Cruz), or a mixture of Cdk5/p35 complexes and Cdk5 free from p35, if the Santa Cruz antibody (C-8) is used. To obtain high levels of the p35/Cdk5 kinase complex it is necessary to carry out p35 immunoprecipitations. We commonly use Santa Cruz antibodies raised against the C (C-19) or N (N-20) terminus of human p35.

Because p35 is a very unstable protein,[20] which is present at low concentrations in brain, immunoprecipitations are carried out with large amounts

[20] G. N. Patrick, P. Zhou, Y. T. Kwon, P. M. Howley, and L. H. Tsai, *J. Biol. Chem.* **273,** 24057 (1998).

of starting material. Generally, a minimum of 1 mg of total protein from cortical tissue is used per immunoprecipitation; however, for unstable or weak interactions, such as association of p35 with Rac, this can be scaled up to 5–10 mg per immunoprecipitation. When using large amounts of starting material, 5–7 μl of p35 or Cdk5 antibody is used per immunoprecipitation, whereas less is sufficient for kinase assays (see following section). The immunoprecipitates are incubated for at least 1 hr with constant rocking at 4°, followed by the addition of 100 μl of a 10% protein A-Sepharose bead solution in the appropriate lysis buffer. The samples are incubated for another hour at 4°, and the beads are washed at least three times with buffer containing 150 mM NaCl and 0.5% Triton X-100 or NP-40 and resuspended in 30 μl of 1× sample buffer [50 mM Tris–HCl, pH6.8, 2% sodium dodecyl sulfate (SDS), 0.1% bromphenol blue, 10% glycerol, and 100 mM DTT]. These samples can be electrophoresed immediately on polyacrylamide gels or stored at −80° for future analysis. Using these conditions, association between Rac and p35/Cdk5 can be detected in mouse cortical lysates (see Fig. 1).

Determination of p35/Cdk5 Kinase Activity

p35 antibodies from Santa Cruz (N-20 and C-19) are both able to immunoprecipitate p35/Cdk5 kinase activity. However, only one commer-

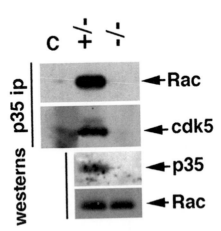

FIG. 1. In neurons the p35/Cdk5 kinase associates with Rac. Mouse cortices obtained from animals harboring a homozygous deletion of p35 (−/−) or heterozygous for p35 expression (+/−) were lysed and immunoprecipitated for p35 (C-19, Santa Cruz) followed by Western blotting for Rac (Upstate Biotechnology) and Cdk5 (DC-17). Note that Rac and Cdk5 coprecipitate with p35 only from +/− lysates. Western blots for Rac and p35 show equal distribution of the former in both samples and the absence of the latter from p35−/− mice. C, control immunoprecipitation using nonspecific antibody. From Nikolic[18] et al., with permission.

A   B

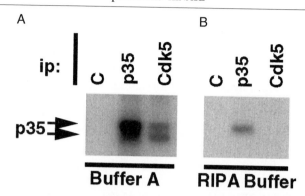

Buffer A   RIPA Buffer

FIG. 2. Comparison of p35/Cdk5 kinase activity levels. Buffer A lysates (A) or RIPA (B) from transfected Cos-7 cells immunoprecipiated with control (lane C), anti-p35 (antibody C-19; lane p35), or anti-Cdk5 (antibody C-8; lane Cdk5) antibodies and subjected to kinase assays in the absence of added substrate. The amounts of protein used per immunoprecipitation and kinase sample electrophoresed were equal in A and B. The length of film exposure was also the same. Arrows point to phosphorylated p35. Note the lower levels of phosphorylated p35 in anti-Cdk5 immunoprecipitations (A), whereas in B they are undetectable at this exposure time.

cially available antibody to Cdk5 can immunoprecipitate an active p35/ Cdk5 kinase, the Santa Cruz C-8 antibody. Others, such as the monoclonal antibody DC-17, interfere with the ability of Cdk5 to associate with p35. It is also important to note that anti-Cdk5 immunoprecipitates contain fewer kinase complexes than anti-p35 immunoprecipitates (see Fig. 2A) due to the excess of Cdk5 in cells. To date, all lysis conditions used have yielded detectable p35/Cdk5 kinase activity from both brain tissue or transfected cells. However, the levels of activity vary depending on the buffers used. For instance, lower levels of activity are detectable from RIPA buffer cell lysates than most other buffers, as illustrated in Fig. 2B.

We generally isolate p35/Cdk5 kinase by immunoprecipitation from embryonic cortical tissues or transfected cells. The latter provides a cleaner kinase immunoprecipitate, presumably due to the overexpression of both p35 and Cdk5 and the presence of fewer interacting proteins/kinases. It is generally sufficient to use as little as 50 μg of total protein from cell lysates to detect the active kinase, although in most experiments, 100–500 μg is our standard amount. To detect p35/Cdk5 kinase activity, the source of the complex must be appropriate, thus E17–adult rat or E15–adult mouse cortical tissues or transfected cells.

In most cases, histone H1 is used as a good substrate that is highly phosphorylated by the p35/Cdk5 kinase.[1,21] Alternatively, neurofilament

[21] J. Lew, R. J. Winkfein, H. K. Paudel, and J. H. Wang, *J. Biol. Chem.* **267**, 25922 (1992).

proteins, microtubule-associated proteins, Munc-18, Pak1, and other reported substrates can also be used.[3,18,22–26] In general, 1 μg of recombinant protein per kinase reaction is more than sufficient to detect phoshphorylation. Kinase reactions can also be carried out in the absence of any added substrate, as p35 rapidly autophosphorylates,[20] the 35kDa band is obscured when histone H1 is added to the reactions.

Immunoprecipitations for kinase assays are carried out as described in the previous section. If high background kinase activity is anticipated, the lysates are precleared in zysorbin (Zymed) for 1 hr at 4° prior to immunoprecipitation. The immunoprecipitates are washed twice in the appropriate lysis buffer that must contain at least 150 mM NaCl and twice in kinase buffer (50 mM HEPES, pH 7.0, 10 mM MgCl$_2$, and 5 mM MnCl$_2$). Each kinase reaction is carried out in 50 μl of kinase buffer containing the substrate of choice, 1 mM DTT, and 1 μCi [γ-^{32}P]ATP. Fifty microliters of this mix is added to the washed beads and the reactions are incubated at room temperature for 20 min when they are terminated with 50 μl of 2× sample buffer (100 mM Tris–HCl, pH 6.8, 4% SDS, 0.2% bromphenol blue, 20% glycerol, and 200 mM DTT). The samples can be analyzed immediately by SDS–PAGE or stored at −20°. Only one-tenth of the kinase reaction is generally loaded onto a 10–12% polyacrylamide minigel (e.g., the Bio-Rad minigel systems) or a third onto a larger gel system. After electrophoresis the gel is dried and exposed to autoradiography film for not more than 1 hr at room temperature, especially if histone H1 was the substrate. Comparisons can be made between different reactions for kinase activity levels using scanned images and computer programs such as NIH image analysis to examine the autoradiography films or a phosphoimager to directly analyze the gel.

Measuring Instability of p35/Cdk5 Kinase

The p35/Cdk5 kinase is very unstable. This is thought to be due, at least in part, to the rapid autophosphorylation of p35 on cdk5 activation, resulting in ubiquitin-mediated p35 proteolysis.[20] The p35 protein has a very short half-life, which we have measured using two different methods. The [^{35}S]methionine labeling and pulse chase method reveals the amount

[22] K. T. Shetty, W. T. Link, and H. C. Pant, *Proc. Natl. Acad. Sci. U.S.A.* **90**, 6844 (1993).
[23] M. R. Hellmich, J. A. Kennison, L. L. Hampton, and J. F. Battey, *FEBS Lett.* **356**, 317 (1994).
[24] K. Ishiguro, S. Kobayashi, A. Omori, M. Takamatsu, S. Yonekura, and K. Anzai, *FEBS Lett.* **342**, 203 (1994).
[25] D. Sun, C. L. Leung, and R. K. H. Liem, *J. Biol. Chem.* **271**, 14245 (1996).
[26] R. Shuang, L. Zhang, A. Fletcher, G. E. Groblewski, J. Pevsner, and E. L. Stuenkel, *J. Biol. Chem.* **273**, 4957 (1998).

of p35 that is translated during the time cells are exposed to radioactive methionine and remains after the amino acid is removed. Cyclohexamide treatment shows the amount of total p35 remaining in the cells following the complete inhibition of protein synthesis. Similar results are obtained using both approaches in either cultures of rat cortical neurons or cell lines overexpressing p35 and Cdk5.[20]

Experimental Procedure

To measure the half-life of p35 by radioactively labeling p35, cultures of primary cortical neurons or transfected cells (prepared as described in previous sections) are carefully washed twice in PBS and once with methionine-free DMEM. The cells are subsequently incubated for 1 hr in methionine-free DMEM containing 10% fetal bovine serum and 110 μCi/ml [^{35}S]Met at 37° and 5% (v/v) CO_2 followed by extensive washing with PBS. The medium is replaced with DMEM supplemented with 10% fetal bovine serum, and cell lysates are made at appropriate times (usually 10, 20, 40, 60, and 80 min from beginning of cold chase) in ELB buffer (see Table I). The lysates are routinely precleared with zysorbin (Zymed) for 1 hr at 4°, and p35 is immunoprecipitated with either anti-p35 or appropriate anti-Cdk5 antibodies as described in previous sections.

The immunoprecipitates are separated by electrophoresis on 10–12% polyacrylamide gels. It is advisable to also include a lane of *in vitro*-translated, ^{35}S-labeled p35 as a marker control to allow easy identification of p35 in the immunoprecipitates. Alternatively, a negative control of labeled lysates from cells that lack p35 expression can also be used. Prior to drying the ^{35}S signal is amplified either by using one of several commercially available amplification solutions (produced by companies such as Amersham or NEN) or amplification in diphenyloxazole (PPO). The latter method in general gives a better level of amplification, retaining the clarity and sharpness of the bands, although it is considered far more toxic. When amplifying in PPO, the polyacrylamide gels are gently rocked in destain [7% (v/v) glacial acetic acid and 25% (v/v) methanol] for 30 min and then rinsed in DMSO. The gels are incubated in DMSO with gentle shaking for 30 min at room temperature, and the process is repeated once in fresh DMSO. The gels are then placed in a PPO solution [22% (w/v) diphenyloxazole in DMSO] for 30 min, which must be carried out in a fume hood and with great care. Following a 30-min wash in running tap water (the gels noticeably turn white and stiffen as the PPO precipitates), the gels are dried and exposed to radiography at −80°. DMSO and PPO can be stored and reused several times. The p35 protein is identifiable in the immunoprecipitates as a clear 35-kDa labeled protein.

If the half-life is determined using the cyclohexamide method, neuronal cultures, or transfected cells are incubated in cyclohexamide (30 mg/ml) for required periods of time, we usually take 10-, 20-, 40-, 60-, and 80-min time points. After removal of cyclohexamide-containing media, the cells are lysed in ELB buffer (Table I), separated on polyacrylamide gels, and transferred to polyvinylidene difluoride (PVDF) or nitrocellulose membranes. The levels of p35 are determined by Western blot analysis using the Santa Cruz (N-20 or C-19) antibodies.

To obtain accurate comparisons of p35 levels, it is advisable to analyze the gels with a phosphoimager. Alternatively, the NIH image analysis program or equivalent computer software can be used to examine scanned autoradiographs, especially if solutions such as the new Amersham ECL, which is optimized for quantification, are used to develop the Western blots.

Inhibition of p35/Cdk5 Kinase Activity

There are several ways to inhibit the activity of the p35/Cdk5 kinase. We have successfully used Cdk5 mutants defective in the phosphotransfer reaction, which, when introduced into neurons, form a stabilized complex with p35, in this way outcompeting endogenous Cdk5.[20,27] Cortical neurons into which the Cdk5 mutants (Cdk5 T[33] and Cdk5N[144]) have been introduced have shorter neurites than controls,[15] revealing that kinase activity is important for neurite outgrowth. Despite the fact that Cdk5 mutants work really well, a limitation to this approach has been the difficulty of introducing DNA into primary neurons at a high enough efficiency to do biochemical experiments.

In recent years a family of chemical inhibitors has been developed with strong specificity to the Cdk family of kinases.[28,29] One of these reagents, roscovitine [2-(1-ethyl-2-hydroxyethylamino)-6-benzylamino-9-isopropyl-purine], displays the highest selectivity toward the p35/Cdk5 complex (IC_{50} value of 0.2 μM).[29] We have successfully used roscovitine to inhibit the p35/Cdk5 kinase both *in vitro* and *in vivo*,[18,20] the methods for which are described in this section.

[27] S. van den Heuvel and E. Harlow, *Science* **262,** 2050 (1993).
[28] W. F. De Azevedo, S. Leclerc, L. Meijer, L. Havlicek, M. Strnad, and S. H. Kim, *Eur. J. Biochem* **243,** 518 (1997).
[29] L. Meijer, A. Borgne, O. Mulner, J. P. Chong, J. J. Blow, and N. Inagaki, *Eur. J. Biochem.* **243,** 527 (1997).

Experimental Procedure

Roscovitine (Calbiochem, La Jolla, CA) is stored in powdered form at $-20°$. Just prior to use, a stock solution (10 mg/ml) is made up in DMSO from which further dilutions can be made depending on the final amount used and concentration required. We find that if care is taken when thawing and freezing, the stock solution can be used for *in vitro* experiments several times. However, for *in vivo* use, a fresh stock must be made for each experiment.

To inhibit the p35/Cdk5 kinase *in vitro*, p35 or Cdk5 immunoprecipitates are carried out as described previously. Selected immunoprecipitates are incubated in kinase reaction buffer that also contains roscovitine. We usually apply a range of concentrations between 0 and 20 μM roscovitine. Because DMSO is the solvent, it is used as the negative control (0) at the maximum volume applied in the experiment. The samples are subsequently treated as described in the kinase assay section.

The great advantage of roscovitine is that it is internalized by cells *in vivo*, allowing inhibition of the p35/Cdk5 kinase in neuronal cultures. For this purpose, primary neurons are cultured as described earlier. At a time when the p35/Cdk5 kinase is peaking (usually third day from plating) they are exposed to roscovitine (0–30 μM). The effects of roscovitine *in vivo* are most apparent at higher concentrations than *in vitro*, presumably due to some restraints of drug internalization. The cells are incubated with inhibitor for 1 hr at $37°$ and 5% CO_2, when the medium is removed and they are lysed. Because the p35/Cdk5 kinase is rapidly turned over, it is important not to exceed the 1-hr incubation time as the newly synthesized kinase will be free of inhibitor and fully active, therefore masking the result of the experiment. Figure 3A illustrates roscovitine inhibition of the p35/Cdk5 kinase when either histone H1 or Pak1 is used as the substrate.

Unfortunately, it is not possible to carry out a kinase reaction to show an *in vivo* decrease in p35/Cdk5 kinase activity as we find that roscovitine is removed from the kinase complex when the immunoprecipitates are washed, restoring the kinase activity to normal. However, when active, the p35/Cdk5 kinase autophosphorylates and downregulates itself *in vivo* by induced degradation of p35. Inhibition of p35/Cdk5 kinase activity results in a prolonged half-life of p35 and a consequent increase in overall p35 levels. Therefore, a good indicator of kinase inhibition is in fact an increase in p35 levels as determined by Western blot analysis.[20] In addition, we have shown that the p35/Cdk5 kinase phosphorylates Pak1 in neurons, inhibiting the activity of the Pak1 kinase as determined by the phosphorylation of histone H4.[18] Neuronal cultures exposed to increasing amounts of roscovi-

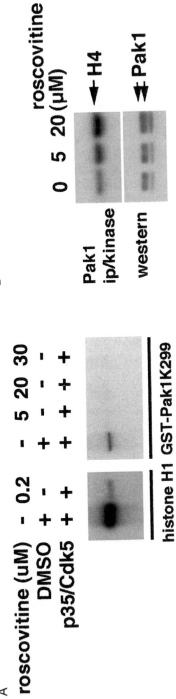

A

roscovitine (uM)   -   0.2   -   5   20   30   -   -
DMSO   +   -   +   +   +   +   -   -
p35/Cdk5   +   +   +   +   +   +   +   +

histone H1   GST-Pak1K299

B

roscovitine
0 5 20 (µM)

Pak1
ip/kinase   ←H4

western   ≠Pak1

Fig. 3. Inhibition of p35/Cdk5 kinase activity using the purine analog roscovitine. (A) The addition of roscovitine to kinase reactions effectively inhibits the p35/Cdk5 kinase activity when histone H1 or a kinase dead mutant of Pak1 (Pak1K299) is used as a substrate. (B) Anti-Pak1 immunoprecipitates and kinase assays were carried out using lysates obtained from neuronal cultures that had been incubated with roscovitine. Pak1 kinase activity increased with the amount of rosovitine used, as determined by histone H4 phosphorylation, whereas overall protein levels remained constant. The rise in Pak1 kinase activity demonstrated inhibition of the p35/Cdk5 kinase. From Nikolic[18] et al., with permission.

tine show a paralleling increase in Pak1 kinase activity as the p35/Cdk5 inhibition is alleviated (see Fig. 3B).

Concluding Remarks

The fascination of working with the p35/Cdk5 kinase is that it provides a unique aspect to thinking about downstream signaling from Rho GTPases, specifically Rac. The kinase dramatically affects the actin cytoskeleton in nonneuronal cells, which is manifested by rapid changes in cell shape. This was the first indication that at least part of the p35/Cdk5 signaling pathway is the same in both cells of neuronal and nonneuronal origin. However, in proliferating cells the effects of p35/Cdk5 kinase activity cannot be tolerated and its expression of causes rapid cell death (Nikolic and Tsai, unpublished data).

During migration, neurite outgrowth, and subsequent pathfinding, neurons require the constant ability to rapidly alter their direction of movement and cell shape. Often, different ends of an axonal growth cone are simultaneously subject to opposite effects on the actin cytoskeleton. Controling these processes requires tightly regulated rapid signaling bursts. Thus, activated Rac (RacGTP) has a fast turnover to RacGDP (nonsignaling), the subsequently activated Pak1 kinase is downregulated by p35/Cdk5 phosphorylation, and the p35/Cdk5 kinase itself has a very short half-life. Further understanding of the regulation of these proteins and how they integrate into signaling cascades should help us understand these complex processes.

Acknowledgments

This work was partly supported by NSF and NIH grants to L.-H.T. and the Medical Foundation Charles King Trust to M.N. M.N. is an independent research fellow supported by the Wellcome Trust. L.-H.T. is an assistant investigator of the Howard Hughes Medical Institute, a Rita Allen Foundation Scholar, and a recipient of an Esther A. and Joseph Klingenstein fund.

[20] Actin Assembly Mediated by Arp2/3 Complex and WASP Family Proteins

By R. Dyche Mullins and Laura M. Machesky

Introduction

The majority of actin cytoskeletal proteins were discovered and first characterized using biochemical techniques. Many powerful assays and analysis tools have been developed to study actin assembly and dynamics. Actin and many actin binding proteins are ubiquitous, abundant, and relatively easy to purify and study *in vitro*. Combined with cell biology and genetics, models developed from this *in vitro* work can be tested *in vivo,* resulting in a picture of how proteins can work together in complicated processes such as cell motility. For example, the Arp2/3 complex was discovered based on a biochemical interaction[1] found to have actin nucleating, cross-linking, and capping activities *in vitro,*[2] and has subsequently been shown to be essential for the assembly of lamellipodia *in vivo* and the motility of the intracellular pathogenic bacteria *Shigella flexnerii* and *Listeria monocytogenes.*[3–5] The Arp2/3 complex is now known to be regulated in cells by a direct interaction with WASP family proteins,[4] and *in vitro,* these proteins have been shown biochemically to stimulate the nucleation activity of the complex greatly.[6–8]

We have reconstituted a signal-dependent actin assembly mechanism from purified components.[6] Based on genetic and cell biological data, this mechanism appears to represent a cellular signaling pathway for *de novo* nucleation of actin filaments. There has been an explosion of interest in regulation of the actin cytoskeleton, and the simple, reconstituted systems

---

[1] L. M. Machesky, S. J. Atkinson, C. Ampe, J. Vandekerckhove, and T. D. Pollard, *J. Cell Biol.* **127,** 107 (1994).

[2] R. D. Mullins, J. A. Heuser, and T. D. Pollard, *Proc. Natl. Acad. Sci. U.S.A.* **95,** 6181 (1998).

[3] T. P. Loisel, R. Boujemaa, D. Pantaloni, and M. F. Carlier, *Nature* **401,** 613 (1999).

[4] L. M. Machesky and R. H. Insall, *Curr. Biol.* **8,** 1347 (1998).

[5] R. C. May, M. E. Hall, H. N. Higgs, T. D. Pollard, T. Chakraborty, J. Wehland, L. M. Machesky, and A. S. Sechi, *Curr. Biol.* **9,** 759 (1999).

[6] L. M. Machesky, R. D. Mullins, H. N. Higgs, D. A. Kaiser, L. Blanchoin, R. C. May, M. E. Hall, and T. D. Pollard, *Proc. Natl. Acad. Sci. U.S.A.* **96,** 3739 (1999).

[7] R. Rohatgi, L. Ma, H. Miki, M. Lopez, T. Kirchhausen, T. Takenawa, and M. W. Kirschner, *Cell* **97,** 221 (1999).

[8] D. Yarar, W. To, A. Abo, and M. D. Welch, *Curr. Biol.* **20,** 555 (1999).

FIG. 1. Spontaneous assembly of actin filaments from monomers. The first few interactions (nucleation) are extremely unfavorable and occur very slowly. Once a stable nucleus is formed, however, it elongates rapidly (elongation).

described here are extremely powerful tools for studying signal-dependent actin assembly.

The technical challenges in studying such a reconstituted system are to obtain highly purified proteins from a homogeneous tissue source; to collect meaningful data; and to analyze data correctly. We describe methods that have been used in our laboratories to characterize the nucleation activity of the Arp2/3 complex and its regulation by WASP family proteins. These methods can be applied to many systems for the study of actin dynamics.

Purification of Proteins

We focus mainly on the reconstitution of signal-dependent actin polymerization from purified components, but in many cases it is sufficient or even desirable to study the process in crude cell extracts. Zigmond[9] describes the preparation of such extracts elsewhere in this volume. For highly purified systems, it is desirable to use actin from the same organism and the same tissue as the regulatory proteins being studied. Mixed systems can be used for many purposes, but because differences have been found,[10,11] results must be interpreted with caution.

Data Collection

Actin assembly can be monitored *in vitro* by several methods, including sedimentation of filaments, light scattering, and fluorescence of labeled actin derivatives.[12] This article focuses on the time-dependent measurement of fluorescence of pyrene-labeled actin derivatives.

Data Analysis

Spontaneous actin assembly is a nucleation–condensation reaction[13] (Fig. 1), meaning that it occurs in two distinct steps. Nucleation is a slow

[9] S. H. Zigmond, *Methods Enzymol.* **325** [21] (2000) (this volume).
[10] P. C.-H. Tseng and T. D. Pollard, *J. Cell Biol.* **94**, 213 (1982).
[11] L. Blanchoin and T. D. Pollard, *J. Biol. Chem.* **274**, 15538 (1999).
[12] J. A. Cooper and T. D. Pollard, *Methods Enzymol.* **85**, 182 (1982).
[13] F. Oosawa and M. Kasai, *J. Mol. Biol.* **4**, 10 (1962).

process during which actin monomers assemble into a stable oligomer. Once formed, this nucleus elongates rapidly into a stable filament. Nucleation factors and associated signaling molecules do not alter the amount of actin assembled, but rather they change the rate of assembly by speeding up the nucleation step. Therefore, the activities of these proteins must be studied kinetically.

Proteins Required for Assays

Purified Actin

Sources of purified actin are rabbit muscle—easy to make, but not as physiologically correct (e.g., see Tseng and Pollard[10] and Blanchoin and Pollard[11]). Actin can also be purified from other sources, including human platelets[14,15] and erythrocytes or amebas.[16] Actin should be gel-filtered to remove oligomers.[16] For these assays, pyrene-labeled actin is required in trace amounts. Pyrene labeling, originally described by Pollard,[17] is described in detail by Zigmond.[9]

Arp2/3 Complex

The Arp2/3 complex can be purified using conventional methods by following the protein with a specific antibody. Various sources have been used, as the Arp2/3 complex appears to be ubiquitous in its tissue and organism distribution. Human platelets[18] or human neutrophils[19] can often be obtained from a local blood bank. It may be easier to use *Acanthamoeba,* which are rich in Arp2/3 complex and easy to grow in large quantities in the laboratory.[20]

Loisel *et al.*[3] described a simple method for purification of the Arp2/3 complex from bovine brain based on its affinity for the C terminus of N-WASP. Briefly, a glulathione *S*-transferase (GST) fusion protein of the N-WASP C terminus is linked to a glutathione-agarose column, and cell extracts can be passed over this column. Contaminants can be washed off the column with high salt (0.2 *M* KCl) and then the Arp2/3 complex can

[14] P. J. Goldschmidt-Clermont, L. M. Machesky, S. K. Doberstein, and T. D. Pollard, *J. Cell Biol.* **113,** 1081 (1991).

[15] S. R. Schaier, *Methods Enzymol.* **215,** 58 (1992).

[16] S. MacLean-Fletcher and T. D. Pollard, *J. Cell Biol.* **85,** 414 (1980).

[17] T. D. Pollard, *Anal. Biochem.* **134,** 406 (1983).

[18] M. D. Welch, A. H. DePace, S. Verma, A. Iwamatsu, and T. J. Mitchison, *J. Cell Biol.* **138,** 375 (1997).

[19] L. M. Machesky, E. Reeves, F. Wientjes, F. J. Mattheyse, A. Grogan, N. F. Totty, A. L. Burlingame, J. J. Hsuan, and A. W. Segal, *Biochem. J.* **328,** 105 (1997).

[20] J. F. Kelleher, R. D. Mullins, and T. D. Pollard, *Methods Enzymol.* **298,** 42 (1998).

be eluted using 0.2 M MgCl$_2$. A potential danger with this method is the possibility of contamination of the preparations with recombinant WASP family protein fragments. This should be detectable in nucleation assays (described later) or with suitable antibodies.

WASP Family Proteins

It has been difficult to obtain full-length active WASP family proteins. This appears to be due to multiple factors. First, WASP, N-WASP, and Scar 1 are all difficult to express in large quantities in *Escherichia coli*. This could be due to problems with folding or posttranslational modifications. Second, WASP and, to a lesser extent, N-WASP and Scar1 appear to aggregate at high concentrations (L. M., 1999, unpublished observations). This can be seen in mammalian cells and appears to result in the "actin clusters" that have been described previously[21] and it also occurs during purification (S. Kellie, personal communication). Third, there is currently no suitable assay for measuring the activity of WASP family proteins; the role of Cdc42 binding to WASP is unclear, as are the roles of numerous other binding proteins.

The best available methods to make recombinant WASP or N-WASP appear to be with the baculovirus system.[7,9] Scar1 can be made in relatively small quantities in *E. coli* using standard GST fusion protein purification methods.[4] Fragments of WASP family proteins lacking the N-terminal portions of the proteins are much simpler to express and isolate from bacteria as GST fusion proteins.[4,22] These can be very useful for actin polymerization studies, as they appear to be fully active in enhancing the nucleation activities of the Arp2/3 complex.[6,7] Briefly, the fragment of interest, e.g., Scar-PWA, is cloned into the pGEX GST fusion vector and transformed into BL21-DE pLysS cells containing a lysozyme expression plasmid under chloramphenicol selection. Lysis is carried out by a freeze-thaw cycle followed by sonication or two passes through a microfluidizer. Standard GST fusion protein purification methods can then be employed with the lysates following centrifugation to remove cell debris.[6]

Pyrene Actin Polymerization Assay

Buffer Solutions

Actin requires ATP and divalent cations for stability. The critical concentration for polymerization of Ca^{2+}–actin in low salt is approximately

---

[21] M. Symons, J. M. Derry, B. Karlak, S. Jiang, V. Lemahieu, F. McCormick, U. Francke, and A. Abo, *Cell* **84,** 723 (1996).

[22] H. Miki, K. Miura, and T. Takenawa, *EMBO J.* **15,** 5326 (1996).

$150~\mu M$, more than two orders of magnitude higher than that of Mg^{2+}–actin. Therefore, monomeric actin is most stable when stored at 0–4° in the presence of Ca^{2+} at pH 8.0. The buffer solution should be changed every 1–2 days to keep the actin in fresh DTT and ATP.[23] The easiest way to do this is to store the actin in a 4° cold room in dialysis tubing immersed in storage buffer. Actin used for kinetic studies should never be frozen and should not be stored in monomeric form for more than 2 weeks after gel filtration. Pyrene-labeled actin may be stored as aliquots frozen at −80°. For each reaction, pyrene-labeled actin should be included at about 5–15% of total actin.

Actin polymerization is best studied in solutions approximating intracellular ionic conditions. Spontaneous nucleation and elongation proceed rapidly in millimolar concentrations of KCl and $MgCl_2$ buffered to pH 7.0 with imidazole.[24] To ensure that the actin binds Mg^{2+} rather than Ca^{2+}, the polymerization buffer also includes millimolar EGTA. We make polymerization buffer as a 10× stock and supplement it with 1 mM dithiothreitol (DTT) and 0.2 mM ATP on dilution. Prior to assay, all proteins other than actin should be dialyzed or gel filtered into 1× polymerization buffer. The dialyzate or gel filtration column fractions that elute before the void volume should be saved and used as buffer controls.

Before beginning polymerization assays, actin should be converted from the Ca^{2+}-ATP form in buffer G into the physiological Mg^{2+}-ATP form. If the divalent cation is not preexchanged, the exchange reaction must be added to the kinetic model describing the polymerization reaction.[25,26] Mg^{2+}-ATP actin can be stored up to 4 hr on ice. Exchange the divalent cation by adding 0.1 volume of a 10× stock of Mg^{2+}-exchange solution and incubating for 2 min at 25°.

Buffers

Actin storage buffer (buffer A): 2 mM Tris–Cl (pH 8), 0.5 mM DTT, 0.2 mM ATP, 0.1 mM $CaCl_2$, 3 mM NaN_3

10× polymerization buffer: 500 mM KCl, 10 mM $MgCl_2$, 10 mM EGTA, 100 mM imidazole, pH 7.0. On dilution, supplement with 1 mM DTT and 0.2 mM ATP

10× Mg-exchange solution: 0.5 mM $MgCl_2$ and mM EGTA

[23] J. Xu, W. H. Schwarz, J. A. Kas, T. P. Stossel, P. A. Janmey, and T. D. Pollard, *Biophys J.* **74,** 2731 (1998).

[24] D. Drenckhahn and T. D. Pollard, *J. Biol. Chem.* **261,** 12754 (1986).

[25] J. A. Cooper, E. L. Buhle, Jr., S. B. Walker, T. Y. Tsong, and T. D. Pollard, *Biochemistry* **22,** 2193 (1983).

[26] C. Frieden, *Proc. Natl. Acad. Sci. U.S.A.* **80,** 6513 (1983).

Fluorimeter Setup and Assay

Before collecting data, the assay conditions and fluorimeter settings should be optimized. The three major concerns in fluorimetric assays are photobleaching, signal-to-noise ratio, and light scattering. For best results, set up the system in the following order.

1. Eliminate photobleaching. Photobleaching occurs when intense excitation light disrupts the chemical structure of the fluorophore. This causes a time-dependent decrease in measured fluorescence. Photobleaching also releases reactive free radicals that can affect the activity of the proteins in solution. For this reason, the effects of photobleaching cannot simply be corrected mathematically.

To eliminate photobleaching, first polymerize pyrene-labeled actin at a high concentration (20 μM) by adding 0.1 volume of 10× polymerization buffer and incubating for approximately 30 min at room temperature. Dilute polymerized actin to the concentration needed in the experiment and place in fluorimeter. Select a data collection time of 2000–3000 sec and measure fluorescence versus time. If the fluorescent signal decreases detectably over several hundred seconds, the excitation light is photobleaching the fluorescent probe. To decrease photobleaching, decrease the excitation light by (1) decreasing the slit widths on the excitation monochromator (Fig. 2, S1 and S2), (2) placing a neutral density filter (or closing a diaphragm) in the excitation light path (Fig. 2, F1), or (3) decreasing current to the lamp. Continue decreasing the excitation light until the fluorescent signal is stable

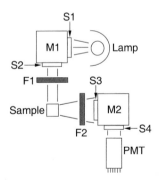

FIG. 2. Schematic diagram of a standard spectrofluorimeter setup. Light from the excitation light source passes through an excitation monochromator (M1) and perhaps an excitation filter (F1) before falling on the sample. The emitted fluorescent light passes through a low-fluorescence emission filter (F2) and an emission monochromator (M2) before falling onto a photomultiplier tube (PMT). The selectivity and light throughput of the monochromators are controlled by entrance and exit slits (S1, S2, S3, and S4). Rejection of stray light can be enhanced by the correct choice of excitation and emission filters (F1 and F2).

and does not decrease detectably for 200–300 sec. Once this point is found, do not alter the excitation light again.

2. Optimize the signal-to-noise ratio. After eliminating photobleaching, the fluorescence signal may need to be boosted to discriminate it from noise. Using the pyrene actin solution from step 1, boost the signal in one of two ways: (1) increase the width of the slits on the emission monochromator (Fig. 2, S3 and S4) or (2) increase the voltage and/or gain of the photomultiplier tube (PMT). If the signal is still too weak, increase the concentration of pyrene-labeled actin. After boosting the signal, photobleaching may be noted that was not previously detectable. If so, return to step 1. Repeat steps 1 and 2 until an acceptable signal is achieved with no detectable photobleaching.

3. Light scattering. Light scattered from air bubbles, dust particles, lipid vesicles, protein aggregates, or actin filaments can corrupt fluorescence data. Two factors contribute to light-scattering artifacts: sample clarity and monochromator selectivity. Remove particles from buffer solutions by filtering and degassing under vacuum. Do not filter protein-containing solutions, but, if necessary, spin them at a high speed in a tabletop ultracentrifuge or airfuge and degas them under vacuum.

Many samples, especially crude cell extracts or solutions containing lipid vesicles, are quite turbid and it may be impossible to decrease turbidity without affecting biochemical activity. In such cases the effects of light scattering can be decreased by optimizing the monochromator performance. Scattered light affects fluorescence measurements primarily because the excitation and emission monochromator are not perfectly selective. Some scattered light at the excitation wavelength passes through the emission monochromator and contributes to the signal. Increasing the selectivity of both monochromators decreases the contribution from scattered light. Each monochromator has two slits, input and output (Fig. 2, S1, S2, S3, and S4). Decreasing the slit widths increases selectivity. For narrow and symmetric wavelength profiles, the input and output slits should be set to the same width (i.e., S1 = S2 and S3 = S4).

Commonly available fluorimeter systems vary widely in their rejection of stray light. Monochromators on the SLM 8000 series (SLM Aminco, Champaign, IL) are specially designed for maximum stray light rejection and, in the authors' experience, are the best for use with turbid samples. Monochromators on other commonly used systems such as the PTI alpha-scan (Photon Technology International, NJ) or ISS PC1 (ISS, Champaign, IL) are designed for maximum sensitivity and do not reject stray light as well. The performance of these systems with turbid samples can be improved greatly by using inexpensive long-pass filters. For pyrene fluorescence, we insert a low-fluorescence 389- or 399-nm long-pass filter (KV-389, KV-399, Schott Glass Co., Duryea, PA) in front of the emission mono-chromator (Fig. 2, F2). If necessary, a 400-nm short-pass filter (03 SWP 602, Melles Griot, Irvine, CA) can also be inserted between the excitation monochromator and the sample (Fig. 2, F1).

Collecting Data

Essential Controls

If working with factors that may regulate or modify the nucleation machinery, determine if these factors themselves alter actin assembly in the absence of Arp2/3 and WASP family proteins. Proteins that bundle, sever, or cap actin filaments or bind monomers or oligomers can alter actin assembly. Three experiments are usually sufficient to control for these effects: measurement of spontaneous assembly, critical concentration, and barbed-end elongation. Because it may be necessary on occasion to deter-mine the effect of a protein on pointed-end elongation or on the ability of Arp2/3 to cap the pointed end of a filament, we include a protocol for this as well.

This list of controls is not intended to be exhaustive. Other control experiments may be necessary. Those discussed are the absolute minimal set that must be performed before making any claims about the effect of a factor on actin nucleation or assembly. The simple interpretations are provided as a guide and are no substitute for a general understanding of the kinetics of macromolecular assembly.

Spontaneous Assembly. The simplest control is to determine the effect of a given factor on spontaneous actin assembly. For all of the control experiments, the following materials are needed.

1. The factor to be tested, dialyzed, or gel filtered into $1\times$ polymeriza-tion buffer (see earlier discussion) along with the dialyzate or gel filtration column fractions that elute before the void to use as a buffer control (this is absolutely essential!).

2. A stock solution of monomeric actin (preferably at $>20~\mu M$) doped with $\leq15\%$ pyrene-labeled actin.

3. $10\times$ polymerization buffer.

Sample Protocol (Assumes 100 μl Final Volume)

1. Turn on fluorimeter, set up, and optimize for measurement of py-rene fluorescence.

2. Into a clean tube, pipette enough actin for a final concentration of 5 μM at 100 μl (e.g., if the stock solution is 20 μM, pipette 20 μl). This concentration is arbitrary but convenient because it polymerizes to equilibrium in about 1 hr.

3. Into a second tube, pipette enough 10× polymerization buffer to bring the actin solution to 1× (e.g., for 20 μl of actin stock solution, pipette 2 μl) and then add enough buffer control solution to bring the total (including the actin) up to 100 μl.

4. Set a pipetter to 110 μl and draw up the second solution. Add it to the first, triturate to mix, and transfer immediately to a quartz cuvette.

5. Place the cuvette in the fluorimeter and collect data until the pyrene fluorescence reaches a stable plateau value. If the actin is good, this should take about 1 hr.

6. Now, repeat the procedure, but in place of the buffer control solution, use a mixture of the control solution and the factor to be tested. If possible, use equimolar concentrations of the factor and actin. If not, use the maximum concentration of the factor to be used in other assays.

Repeat each condition a few times to determine the variability of the results. Compare the curves directly by overlaying on the same axes. If the fluorimeter control software does not allow this, export data to a spreadsheet or graphics program. If, within experimental variation, the curves are identical, you can assume that differences observed in the presence of Arp2/3 and WASP family proteins are caused by the interaction of the factor with these proteins.

If the shapes or plateau values of the curves are different, the factor alters actin assembly, which must be taken into account in any experiment containing nucleation factors. If polymerization is accelerated, the factor probably nucleates or severs actin filaments (creating new ends). If polymerization is retarded, the factor may cap[27] or tightly bundle[28] actin filaments, sequester actin monomers,[29] or interfere with spontaneous nucleation.[30] If the baseline values are different, the factor may bind actin monomers and alter pyrene fluorescence. Sequestering, capping, and bundling proteins can all alter the plateau fluorescence. The following controls can help discriminate between these possibilities.

Critical Concentration. The critical concentration for actin polymerization can be defined as the concentration above which filaments exist in stable equilibrium with monomers and below which filaments are unstable.

[27] G. H. Isenberg, U. Aebi, and T. D. Pollard, *Nature* **288**, 455 (1980).
[28] J. W. Murray, B. T. Edmonds, G. Liu, and J. Condeelis, *Cell Biol.* **135**, 1309 (1996).
[29] D. Safer, R. Golla, and V. Nachmias, *Proc. Natl. Acad. Sci. U.S.A.* **87**, 2536 (1990).
[30] T. D. Pollard and J. A. Cooper, *Biochemistry* **23**, 6631 (1984).

The simplest way to determine the critical concentration is to dilute actin under polymerizing conditions to a range of concentrations and assay for the presence of filaments (Fig. 3).

In addition to the solutions noted earlier, the following are also needed.

1. Actin monomer storage buffer (buffer A).
2. A stock solution of the factor of interest made up or dialyzed into buffer A.

Protocol (Assumes 100 μl Sample Volumes)

1. Start by polymerizing actin at a high concentration. Add 50 μl of 10× polymerization buffer to 450 μl of actin stock solution and incubate for 1 hr at room temperature to induce polymerization. It is much easier to dilute filamentous actin than to polymerize dilute monomeric actin.

2. Set up four groups of 11 tubes each. Label them with the concentrations of actin to be used in the experiment. We suggest the following (in μM): 0.0, 0.05, 0.075, 0.1, 0.2, 0.4, 0.6, 0.8, 1.0, 1.5, and 2.0.

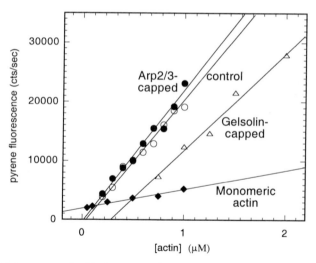

FIG. 3. Measurement of critical concentration of actin polymerization. The fluorescence of monomeric actin changes linearly with concentration (closed diamonds). Intersection of the filamentous actin fluorescence with this line determines the critical concentration. The critical concentration of actin alone (open circles, control) is 0.1 μM, approximately equal to the critical concentration of the fast-growing barbed end of the filament. Capping the barbed end with gelsolin increases the critical concentration to 0.6 μM (open triangles, gelsolin capped), the critical concentration of the pointed end. Capping the slow-growing pointed end with the Arp2/3 complex, however, has only a small effect on the critical concentration (closed circles, Arp2/3 capped).

3. Into two sets of tubes carefully pipette the appropriate amount of polymerized actin stock solution. This is the critical step in the assay. A critical concentration experiment is essentially an exercise in the pipetting technique. To the first set of tubes, add enough $1\times$ polymerization buffer to bring up to final volume. To each tube in the second set, add the same amount of the factor of interest and then enough polymerization buffer to bring to final volume.

4. Into the other two sets of tubes, carefully pipette monomeric actin in buffer A. Bring one set up to final volume with buffer A. To each tube in the other set, add the same amount of the factor of interest and bring up to final volume with buffer A.

5. Incubate all solutions for 4 hr to overnight in the dark at room temperature. An empty desk drawer works well.

6. Measure the steady-state pyrene fluorescence of each solution. For each of the four sets, plot fluorescence versus concentration.

Much information can be extracted from this experiment. Compare the two sets prepared in buffer A (nonpolymerizing conditions). Each should fall on a straight line. If the slopes are different, the factor may bind monomeric actin and alter pyrene fluorescence.

Now compare the two sets containing actin alone. Under polymerizing conditions the first few points should fall on a line identical to nonpolymerizing conditions. The higher concentrations should fall on a different line with a higher slope. The intersection of the two lines is the critical concentration. If pipetting is careful and the actin is okay, the intersection should occur around 0.1 μM. Make the same comparison to determine the critical concentration in the presence of the factor of interest.

Here are several possible results and the simplest interpretations.

1. The critical concentration is lower in the presence of the added factor. This indicates that the factor stabilizes filaments, probably by decreasing the rate of monomer dissociation from the barbed end. Several factors, including the protein tropomyosin[31] and the mushroom toxin phalloidin, have this effect.[32]

2. The critical concentration is higher in the presence of the factor. In this case, the factor probably either caps the barbed end of the filament or sequesters actin monomers. If high concentrations of the factor cause the critical concentration to keep increasing above 0.6 μM, the answer is probably sequestering. If the factor increases the critical concentration to

[31] A. A. Lal and E. D. Korn, *Biochemistry* **25**, 1154 (1986).
[32] P. Sampath and T. D. Pollard, *Biochemistry* **30**, 1973 (1991).

0.6 μM, even at high concentrations, it may cap the barbed ends of actin filaments (see Fig. 3).

3. At the highest actin concentrations the slopes of the fluorescence versus concentration plots are different in the presence and absence of the factor. In this case the factor probably binds actin filaments and alters pyrene fluorescence.

Barbed-End Elongation. Measurement of barbed-end elongation in the presence and absence of the factor of interest is essential in discriminating between nucleation and severing or between capping and sequestration.

In addition to the reagents listed earlier, the following items are needed.

1. A stock solution of unlabeled (dark) actin at a concentration $\leq 20\ \mu M$.
2. Cut pipette tips.

Protocol (Assuming 100 μl Final Volume)

1. Polymerize dark actin at a high concentration. Mix 50 μl of 10× polymerization buffer with 450 μl of actin stock solution and incubate for 1 hr at room temperature. Use the filaments in this solution as seeds for elongation.

2. Place three clean tubes in a rack. Into the first, pipette enough pyrene-doped monomeric actin to obtain 2 μM at final dilution. Into the second, pipette 5–10 μl of the polymerized dark actin. Use a cut pipette tip to avoid breaking the filaments. Into the third add enough 10× polymerization buffer to bring the actin in the first tube up to 1× along with enough 1× polymerization buffer to bring the total volume of the three tubes up to 100 μl.

3. Set a pipetter to 110 μl and use a cut tip. Draw up the solution from tube 3 and mix with tube 2. Draw up the mixture and add to tube 1. Mix by trituration, but be gentle to avoid breaking filaments. Quickly transfer to a quartz cuvette and measure fluorescence. Do this several times until the results are consistently reproducible.

4. Repeat steps 2 and 3, substituting the factor of interest for the 1× polymerization buffer in the third tube. Use several different final concentrations.

The barbed end of an actin filament elongates 10 times more rapidly than the pointed end.[33] Therefore, to a good approximation, the initial rate of fluorescence increase is proportional to the rate of barbed-end elongation. If the factor increases the rate of fluorescence change, it probably severs the filaments and creates new barbed ends. If, at concentrations

[33] T. D. Pollard, *J. Cell Biol.* **103**, 2747 (1986).

well below 2 μM, it decreases the rate of fluorescence change by 10-fold, it probably caps the barbed end of the filament. If it decreases the rate to near zero but only at concentrations of 2 μM or higher, it probably sequesters actin monomers.

Pointed-End Elongation. To measure pointed-end elongation, make actin seeds with capped barbed ends. The simplest way to do this is to begin with gelsolin–actin dimers.

In addition to reagents listed previously, the following items as needed.

1. Gelsolin. Gelsolin may be expressed and purified from *E. coli* or purified from horse or bovine serum (not heat inactivated serum!) using the method of Bryan and Coluccio.[34] We have used both sources with identical results. Plasma gelsolin may also be purchased from Sigma (St. Louis, MO).

2. 10 mM CaCl$_2$ in water.

Protocol (Assumes 100 μl Final Volume)

1. First, make gelsolin–actin dimers. Mix gelsolin with a twofold molar excess of unlabeled monomeric actin in buffer A. Add CaCl$_2$ to a final concentration of 100 μM. Incubate at room temperature for 2–4 hr and then on ice overnight. Gelsolin–actin dimers are slow to form.

2. Add 0.1 volume of 10× polymerization buffer to the gelsolin–actin mixture and store on ice. Once made, gelsolin–actin dimers are good for about 5 days. Do not freeze.

3. Before assay, allow the gelsolin–actin dimers and the 1× polymerization buffer to warm to room temperature. If the solutions are cold at the beginning of the assay, there is a significant artifactual change in the slope caused by warming.

4. The elongation assay is similar to barbed-end elongation. Place three clean tubes in a rack. Into the first, pipette enough pyrene-labeled monomeric actin for a final concentration of 2 μM. Into the second, pipette 5–20 μl gelsolin–actin dimers. Into the third, add enough 10× polymerization buffer to bring the actin in tube 1 up to 1× and enough 1× polymerization buffer to bring the total volume in the tubes up to 100 μl.

5. Set a pipetter to 110 μl. Withdraw the solution in tube 3 and mix with tube 2. Withdraw the mixture and add to tube 1. Mix by trituration and transfer to a quartz cuvette. Measure pyrene fluorescence. Repeat several times until the results are consistently reproducible.

---

[34] J. Bryan and L. M. Coluccio, *J. Cell Biol.* **101**, 1236 (1985).

Experimental Design

To extract the maximum amount of information from kinetic analysis, data should be collected across a range of protein concentrations. For each condition, data must be collected until the system reaches steady state and the concentration of polymerized actin comes to a stable plateau. In most experiments there will be several proteins whose concentrations can be varied, but generally the most useful information comes from varying only two of the components: actin and/or the most upstream component of the signaling pathway used in the experiment. Collecting data sets in which every component is held constant except the actin concentration gives information on the mechanism of nucleation (at this time still a poorly understood process). The change in polymerization kinetics as a function of actin concentration is related to several important unknowns, including the number of kinetically distinguishable steps in nucleation and the number of actin-binding sites in the pathway that contribute to the formation of a stable nucleus (see Data Analysis). Varying the most upstream component in the pathway (e.g., an active fragment of Scar1, Fig. 4) provides informa-

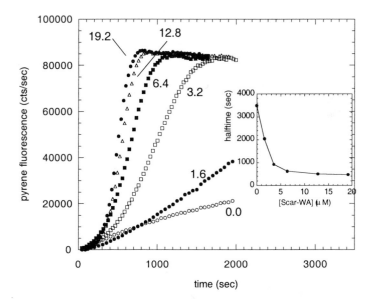

FIG. 4. Enhancement of Arp2/3 nucleation activity by the C-terminal portion of human Scar1. Conditions: 3.3 μM *Acanthamoeba* actin, 10% pyrene-labeled, 0.9 μM *Acanthamoeba* Arp2/3, and various concentrations of human Scar1-WA. Concentrations in micromolar are noted beside each curve. (Inset) Plot of the time to half-maximal polymerization vs concentration of Scar1-WA.

tion about the affinity of this component for its effectors and its efficiency as an activator or inhibitor of nucleation.

In each assay, monomeric actin should be added last. Place three clean tubes in a rack. Into the first tube, pipette monomeric actin, in the second 0.1 volume of 10× polymerization buffer, and in the third mix all other reagents. To start the polymerization reaction, simply mix the contents of the three tubes. Work quickly but avoid injecting air bubbles. Withdraw the liquid from the third tube and mix with 10× polymerization buffer. Withdraw this mixture and add to the tube containing actin. Mix by trituration, pipette into a quartz cuvette, place the cuvette in the fluorimeter, and begin data collection. With this method there is a significant delay or dead time between the start of the reaction and the beginning of data collection (typically 20–30 sec), which may affect data analysis. Correct for this in one of two ways: (1) trigger data collection to begin as soon as actin is added to the reaction mix but open the shutter only after the cuvette is placed in the fluorimeter or (2) start a stopwatch the instant actin is added and measure the time until the first data point is recorded. Both methods require alteration of data after collection. In the first, data points collected with the shutter closed must be removed and in the second the measured dead time must be added to all the time values.

Sample Assay

Measure the effect of increasing concentrations of the C-terminal (WA) fragment of Scar1 Arp2/3 complex nucleation activity (Fig. 4). Given 100 μl final sample volume and initial stock solutions as follows: actin 34 μM, Pyrene–actin 20 μM, Arp2/3 2 μM, and Scar-WA 320 μM.

Buffer control (μl)	10× salts (μl)	Dark actin (μl)	Pyrene–actin (μl)	Arp2/3 (μl)	Scar-WA (μl)	[Scar-WA] (μM)
39.5	1	8	1.5	50	0	0
39.0	1	8	1.5	50	0.5	1.6
38.5	1	8	1.5	50	1.0	3.2
37.5	1	8	1.5	50	2.0	6.4
35.5	1	8	1.5	50	4.0	12.8
33.5	1	8	1.5	50	6.0	19.2

Data Analysis

The kinetics of multistep processes can be analyzed by either model-dependent or model-independent methods. In a model-dependent method, we propose a set of possible mechanisms to explain the assembly process. Next, we run a computer simulation of each model and vary the unknown rate constants until we obtain a best fit to experimental data. Finally, we compare all the best fits of all the models. Models that predict unrealistic rate constants are thrown out and the model whose best fit most closely matches experimental data is chosen.

When an assembly process involves multiple steps with unknown rate constants, the determination of a mechanism from model-dependent analysis can become an ill-conditioned problem. That is, the more steps in a proposed mechanism, the more easily it can be made to fit the data, even if it is not the actual assembly mechanism. In such situations, model-independent analysis techniques are extremely useful. A model-independent analysis technique makes few or no assumptions about the mechanism underlying the process.

Model-Dependent Analysis Using KINSIM

There are many excellent kinetic simulation programs available, both commercially and as shareware/freeware, but the most readily available and user friendly is KINSIM, which was originally written to study the mechanism of actin polymerization. Versions of KINSIM compiled for several different computer-operating systems can be downloaded from a server at Washington University School of Medicine (http://biochem. wustl.edu/cflab). We do not cover the details of using KINSIM here. The manual (also available for download from the Washington University site) describes in detail how to compile and run mechanisms and how to compare simulated curves to experimental data. Additionally, an excellent tutorial on kinetic analysis using KINSIM was published previously.[34a]

Simple Models for Actin Polymerization

The kinetics of actin polymerization are dominated by the first few steps in the assembly reaction. The simplest model for actin polymerization that reproducibly fits experimental data is

$$A + A \rightleftharpoons 2A \qquad (1)$$
$$2A + A \rightleftharpoons 3A \qquad (2)$$
$$3A + A \rightleftharpoons 4A \qquad (3)$$
$$4A + A \rightleftharpoons F \qquad (4)$$
$$F + A \rightleftharpoons F \qquad (5)$$

where A is an actin monomer, 2A is a dimer, etc., and F is a filament.

[34a] D. H. Wachsstock and T. D. Pollard, *Biophys. J.* **67**, 1260 (1994).

In this model there are four steps to formation of a stable filament. Because stable filaments rarely depolymerize back to monomers, step 4 is modeled as an irreversible reaction. If it were reversible, all filaments would be modeled as 5-mers and would disassemble back to monomers at an unrealistically high rate. Step 5 does not look right but it reflects the fact that the elongation of filaments does not change the number concentration of filaments. The kinetics of this step also determine the critical concentration of the simulation. Some authors have added fragmentation and annealing steps to this reaction.[25,35,36] For highly purified actin, however, the model provides excellent fits of experimental data. Only the rate constants of the first two reactions need to be varied. All others should be set to the rate constants for the elongation and depolymerization of ATP-Mg-actin from the barbed end ($k_+ = 11.3\ \mu M^{-1}\ sec^{-1}$, $k_- = 1.5\ sec^{-1}$).[33]

To describe actin polymerization in the presence of nucleation factors, additional reactions must be added. The unactivated Arp2/3 complex appears to weakly nucleate actin polymerization by stabilizing actin dimers. Polymerization data in the presence of the unactivated Arp2/3 complex are well fit by adding the following reactions to the scheme described earlier:

$$P + 2A \rightleftharpoons P2A \tag{6}$$

$$P2A \rightleftharpoons F \tag{7}$$

Nucleation in the presence of activated Arp2/3 and/or other nucleation factors requires additional reactions. Several plausible schemes should be proposed, simulated, and compared.

Preparation of Data for Analysis

The easiest way to compare simulations of actin polymerization to experimental data is to plot the decrease in monomeric actin rather than the increase in polymerized actin. If the proteins used do not affect the critical concentration for polymerization, the monomeric actin concentration can be calculated from pyrene fluorescence. This is done most conveniently using a spreadsheet program such as Lotus 1-2-3 or Microsoft Excel or a graphics package such as Kaleidagraph, Cricket Graph, or Microcal Origin. After loading data into the program, the following steps are needed.

1. Subtract out the baseline fluorescence (F_0). This is the fluorescence of unpolymerized actin at time zero in the experiment. It can be estimated by either measuring fluorescence of a solution of monomeric actin at the

[35] A. Wegner and P. Savko, *Biochemistry* **21**, 1909 (1982).
[36] C. T. Zimmerle and C. Frieden, *Biochem.* **25**, 6432 (1986).

appropriate concentration or, if the dead time of the measurement is short, from the fluorescence of early time points (Fig. 5A).

2. After subtracting the baseline, determine the fluorescence at the steady-state plateau (F_{ss}). If there is noise in data, pick a value that reflects the average fluorescence at the plateau (Fig. 5A). Normalize data by dividing all data points by F_{ss} (Fig. 5B). Because this normalization step is absolutely required for meaningful simulation, it is important to collect complete data sets in which polymerization reaches a steady-state plateau.

3. Multiply all data points by the amount of polymerized actin present at steady state. This is just the concentration of actin added to the assay minus the critical concentration ($[A]_0 - A_{cc}$). If proteins added to the assay do not alter the critical concentration, $A_{cc} = 0.1 \ \mu M$. The first three steps convert pyrene fluorescence (f) to the concentration of polymerized actin ($[P]$).

4. Convert polymer concentration to monomer concentration by subtracting each data point from the initial concentration of actin used in the assay ($[A]_0 - [P]$). These data should be saved with the time information in the ASCII two-column text format and then converted to a KINSIM-readable file (Fig. 5C).

Optimizing Model Parameters

Unknown rate constants in a kinetic model can be optimized either by hand or by nonlinear least-squares alogrithms. In many situations, particularly when only one or two unknown rate constants must be varied, it is possible to obtain a best fit using KINSIM by varying the rate constants, performing the simulation, and judging the fit by eye from the graph. Any proposed best fit should reasonably fit several curves collected at different protein concentrations.

With a complicated model, several unknowns must be fit to a series of data sets collected at different concentrations of actin or nucleation factor and it is very difficult to obtain an unambiguous best fit by hand. In such cases, an automated routine should be used to perform nonlinear least-squares fitting. FITSIM is an extension of KINSIM that performs iterative nonlinear least-squares fitting using the modified Gauss's method.[37] The program and its operating manual may also be downloaded from the Washington University server. Because it uses an iterative method, the ability of FITSIM to find a best fit is sensitive to the initial guesses for unknown rate constants. The most efficient procedure is to load data into KINSIM and obtain an approximate fit by hand and then use these values as initial guesses for the FITSIM algorithm.

---

[37] C. T. Zimmerle and C. Frieden, *Biochem. J.* **258**, 381 (1989).

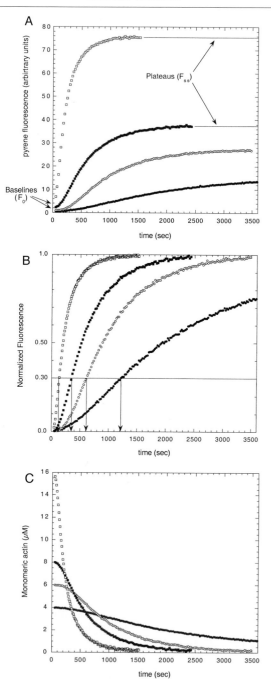

A

B

C

Physically Meaningful Rate Constants

One test of a mathematical model is whether it predicts plausible values for unknown parameters. For this reason it is useful to have a sense of the upper bound on rate constants for molecular association. The fastest bimolecular reactions are limited only by the rate of molecular diffusion and are described by the Debye–Smoluchowski equation:

$$k_+ = 4\pi k f_{elec} b (D_1 + D_2) N_0 10^{-3}$$

where D_1 and D_2 are the diffusion constants of the two molecules, b is the interaction radius (in cm), N_0 is Avogadro's number, k is the fraction of collisions that result in binding, and f_{elec} is an electrostatic factor. In most cases, f_{elec} is approximately 1.0, for globular proteins in dilute aqueous solutions $D \le 10^{-6}$ cm^2 sec^{-1}, a good approximation for b is 2×10^{-7} and an upper bound for k is 10^{-1}. This gives an upper bound for rate constants of approximately 300 μM^{-1} sec^{-1}. Association rate constants significantly higher than this are unlikely to have any physical meaning and may indicate that the model is incorrect.

Model-Independent Analysis of Kinetic Data

In many cases we can use kinetic data to make strong statements about a molecular mechanism without constructing a detailed kinetic model. Making only a few assumptions, we can calculate parameters, such as the instantaneous concentration of free barbed ends, or we can simply relate the overall time scales of actin assembly reactions to the concentrations of various components without making any assumptions at all.

Calculation of Free Barbed Ends

If we assume that all of the filament growth takes place at the barbed end (a safe assumption in the presence of Arp2/3, which caps the pointed end of the filament) and that the proteins being tested do not affect actin elongation or critical concentration (assumptions that should be directly

---

FIG. 5. Preparation of actin polymerization data for analysis by KINSIM. Raw data are collected in fluorescence units and must be converted to actin concentration (A). The first step is to normalize data by first subtracting the baseline fluorescence and then dividing by the steady-state plateau fluorescence (B). Data are then converted to polymeric actin concentration by multiplying the actin concentration minus the critical concentration. This is then subtracted from the initial actin monomer concentration to obtain the concentration of monomeric actin (C).

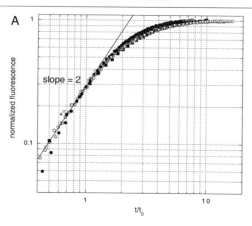

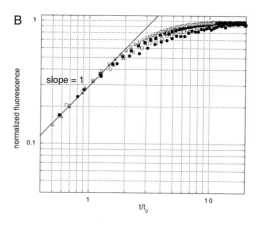

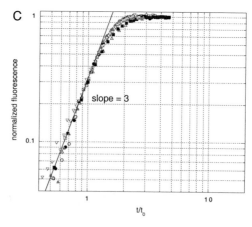

tested, see Essential Controls), then the concentration of free barbed ends at a given time is

$$
[e] = \frac{[A]_0 - cc}{F_p} \left[\frac{\dfrac{\Delta F}{\Delta t}}{k_+\left(1 - \dfrac{F}{F_p}\right)[A]_0 - k_-} \right]
$$

where $[e]$ is the concentration of free barbed ends and k_+ and k_- are the association and dissociation rate constants for barbed-end elongation of Mg–actin ($11.6\ \mu M^{-1}\ \sec^{-1}$ and $1.3\ \sec^{-1}$ [33]). F and F_p are the instantaneous and plateau fluorescence values. $[A]_0$ and cc are the initial actin concentration and the critical concentration, and $\Delta F/\Delta t$ is an approximation of the instantaneous slope of the fluorescence curve. The most critical parameter in this equation is the slope, $\Delta F/\Delta t$. There are many ways to calculate $\Delta F/\Delta t$. One simple calculation is to estimate $\Delta F/\Delta t$ at a given time, t, by

$$
\frac{F(t + \tau) - F(t - \tau)}{2\tau}
$$

where F is the instantaneous fluorescence and τ is a small time increment. Noise in data is amplified by this calculation, but increasing the value of τ can decrease the noise. More sophisticated methods of calculating derivatives can be found in any standard text on numerical methods (e.g., "Numerical Recipes in C," by Press, Flannery, Teukolsky, and Vetterling, Cambridge University Press).

Comparing Time Scales of Polymerization

One simple but useful test to compare the effectiveness of two nucleation factors is to measure the half-time to maximum actin polymerization as a

---

Fig. 6. Time scaling of actin polymerization data. (A) Time scales of the spontaneous actin polymerization data collected at multiple concentrations (4, 6, 8, 10, 14, 16, and 20 μM) were normalized by dividing each time scale by the time it took to polymerize to 30% of plateau level (see arrows in Fig. 5B). Data are plotted on a log–log scale. (B) Similarly scaled data for actin assembly from preformed actin seeds. Actin concentrations were 1, 1.5, 2.0, 3.0, 7.0, and 10.0 μM. (C) Scaled polymerization data for actin assembly in the presence of Arp2/3 and Scar1-WA. In each case, data fall onto a single curve, suggesting that a single assembly pathway dominates each set of conditions. The slope at early time points is related to the number of steps required to form a stable nucleus.[38] Nucleation in the presence of Arp2/3 and Scar1-WA appears to require more steps than either spontaneous or seeded polymerization.

function of nucleator concentration. We varied the concentrations of two Arp2/3 activators (fragments WA and PWA from human Scar1[16]) (Fig. 4) and plotted half-time versus concentration. From the concentration dependence of the half-time, we estimated the relative affinities of the activators. High concentrations of activator should saturate the Arp2/3 complex, so by comparing the half-times at high concentration we can compare the efficiency with which different molecules stimulate Arp2/3 nucleation. By comparing different deletion mutants of the same molecule, we can gain insight into the molecular mechanism of Arp2/3 activation.

The molecular assembly events required for Arp2/3-mediated nucleation are not well understood. Without knowing the mechanisms, however, we can compare assembly kinetics in the presence of different nucleation factors and determine whether the nucleation mechanisms are the same or different. In addition to normalizing the fluorescence as described earlier (see Fig. 5B), normalizing the time scales of polymerization curves can also provide useful information on assembly mechanisms. One use of time scaling is to test for the presence of multiple assembly pathways. First, collect data sets at several different concentrations of actin, keeping all of the other components constant. Normalize data by subtracting the baseline fluorescence and dividing by the plateau, steady-state fluorescence. Next, choose an arbitrary amplitude value between 0 and 1 and determine the time (t_0) that each data set crosses this value. Finally, for each data set, divide all the times by t_0 and plot on a log–log scale (Fig. 6). If to a good approximation all data collapse onto a single curve, this is evidence that actin assembles via a single pathway across all these concentrations.[38] If data do not collapse to a single curve, particularly if the slope of data at early time points varies continuously with concentration, this is evidence for multiple, distinct assembly pathways.

When normalized and scaled, spontaneous actin polymerization data collapse onto a single curve (Fig. 6A). Assembly from existing seeds collapses onto a very different looking curve (Fig. 6B) and assembly in the presence of a constant amount of Arp2/3 and ScarWA onto yet another curve. At early time points, the slopes of these curves on log–log plots are related to the number of kinetically resolvable steps in the nucleation mechanism. (Flyvbjerg et al.[38]) investigated the general case of nucleation–condensation reactions and showed that, for an integer slope k, the number of kinetically resolvable steps required to form a nucleus is $k - 1$. Our actin assembly data are consistent with this. For assembly from seeds, the slope is approximately 1, reflecting the fact that there are nuclei present at $t = 0$ and no steps required to create them. For spontaneous assembly,

[38] H. Flyvbjerg, E. Jobs, and S. Leibler, Proc. Natl. Acad. Sci. U.S.A. 93, 5975 (1996).

the slope is approximately 2, consistent with a single, kinetically dominant step to nucleation.[13] For assembly in the presence of Arp2/3 and ScarWA, the slope is 3, possibly indicating a two-step mechanism (Mullins *et al.*, manuscript in preparation). The mechanism of actin filament nucleation in the presence of activated Arp2/3 appears to be kinetically more complex than either spontaneous or seeded actin polymerization. Note that even though we do not understand the details of the mechanism, we can use model-independent methods to compare data sets and make statements about differences in underlying mechanisms.

A further study of reconstituted, signal-dependent actin polymerization will be required to uncover the details of Arp2/3-mediated actin assembly and to decipher its regulation by cellular signaling systems.

Acknowledgments

RDM is supported by HHMI Research Resource Grant 57895. Special thanks to J. Zalevsky for helpful discussions and for collecting data presented in the figures. LMM is supported by the Medical Research Council, UK.

[21] *In Vitro* Actin Polymerization Using Polymorphonuclear Leukocyte Extracts

By Sally H. Zigmond

Introduction

Actin polymerization is essential for pseudopod extension. During cell locomotion, the amount of F-actin is constant but in rapid flux with polymerization occurring at the barbed (high affinity) end of an actin filament and depolymerization occurring at the pointed (low affinity) end. In the absence of stimulation, polymorphonuclear leukocytes (PMN) are round and immotile; addition of chemoattractant causes a transient twofold increase in the amount polymerized actin (filamentous or F-actin)[1] and in the number of actin filaments.[2] These large chemoattractant-induced changes make PMN useful for studies of the signaling pathways leading to actin polymerization.

Most chemoattractants (chemokines) initiate actin polymerization in leukocytes by binding a serpentine receptor that activates a pertussis toxin-sensitive heterotrimeric G-protein. Some chemoattractants, e.g., colony

---

[1] P. N. Devreotes and S. H. Zigmond, *Annu. Rev. Cell Biol.* **4**, 649 (1988).
[2] M. Cano, D. A. Lauffenburger, and S. H. Zigmond, *J. Cell Biol.* **115**, 677 (1991).

stimulating factor (CSF), utilize a tyrosine kinase receptor. In both cases, small GTPases of the Rho family are activated.[3-6] Both Rac and Cdc42 appear to contribute to actin polymerization. In leukocytes from a Rac2 knockout in mouse, chemoattractant-induced migration and actin polymerization are inhibited.[7] In RAW 264.7 macrophages, expression of either dominant-negative Rac or Cdc42 inhibits chemoattractant-induced actin polymerization.[8] In Bac1.2F5 macrophages, injection of dominant-negative Rac blocks migration, whereas dominant-negative Cdc42 allows migration but inhibits chemotaxis toward CSF.[9] The downstream effectors of Rac and Cdc42 that lead to actin polymerization are under investigation by many laboratories.

Cell extracts that retain functional responses to agonists are useful for dissecting signaling pathways. It is important to first determine if the agonist produces the response of interest in the extract. Simplifying the composition of the extract can helpful, but it can also result in loss of response to an agonist (Table I). In an intact cell, chemoattractant stimulates a spatially organized actin polymerization. A permeabilized cell responds to chemoattractant as long as GTP is included but most spatial organization is lost. After mechanical lysis, GTPγS and GTPγS-charged Cdc42 induce polymerization; chemoattractant does not.[10] The ability of exogenous Rac to induce polymerization is retained in permeabilized platelets,[11] but is lost following lysis of most cells (Table I). In the high-speed supernatant (HSS) of lysed cells, GTPγS-charged Cdc42 stimulates actin polymerization, but GTPγS alone no longer works, presumably due to the loss of guanine nucleotide exchange activity.[10,12,13] The addition of acidic lipids restores the ability of GTPγS to induce polymerization in the high-speed supernatant. The stability of HSS makes it particularly useful for fractionation studies.

[3] G. M. Bokoch, B. P. Bohl, and T.-H. Chuang, J. Biol. Chem. **269**, 31674 (1994).
[4] M. R. Phillips, A. Feoktistov, M. H. Pillinger, and S. B. Abramson, J. Biol. Chem. **270**, 11514 (1995).
[5] C. Laudanna, J. J. Campbell, and E. C. Butcher, Science **271**, 981 (1996).
[6] V. Bernard, B. P. Bohl, and G. M. Bokoch, J. Biol. Chem. **274**, 13198 (1999).
[7] A. W. Roberts, C. Kim, L. Zhen, J. B. Lowe, R. Kapur, B. Petryniak, A. Spaetti, J. D. Pollock, J. B. Borneo, G. B. Bradford, S. J. Atkinson, M. C. Dinauer, and D. A. Williams, Immunity **10**, 183 (1999).
[8] D. Cox, P. Chang, Q. Zhang, P. G. Reddy, G. M. Bokoch, and S. Greenberg, J. Exp. Med. **186**, 1487 (1997).
[9] W. E. Allen, D. Zicha, A. J. Ridley, and G. E. Jones, J. Cell Biol. **141**, 1147 (1998).
[10] S. H. Zigmond, M. Joyce, J. Borleis, G. M. Bokoch, and P. N. Devreotes, J. Cell Biol. **138**, 363 (1997).
[11] J. H. Hartwig, G. M. Bokoch, C. L. Carpenter, P. A. Janmey, L. A. Taylor, A. Toker, and T. P. Stossel, Cell **82**, 643 (1995).
[12] L. Ma, L. C. Cantley, P. A. Janmey, and M. W. Kirschner, J. Cell Biol. **140**, 1125 (1998).
[13] V. L. Katanaev and M. P. Wymann, J. Cell Sci. **111**, 1583 (1998).

TABLE I
INDUCTION OF ACTIN POLYMERIZATION IN POLYMORPHONUCLEAR CELLS AND CELL EXTRACTS

Cell form	Effective agonist	Ineffective agonist	Ref.
Intact	Chemoattractant	GTP, GTPγS	
Permeabilized	Chemoattractant + GTP	Chemoattractant,	a, b
(streptolysin O)	GTPγS	GTP	
Broken	GTPγS	Chemoattractant + GTP	c
(bombed or	GTPγS-Cdc42	GTPγS-Rac	
sonicated)		GTPγS-Rho	
Supernatant	GTPγS-Cdc42	GTPγS	
(high speed)	GTPγS + active GEF	Active GEF	
of broken cells	GTPγS + PIP	PIP, PIP2	

[a] T. Redmond, M. Tardif, and S. H. Zigmond, *J. Biol. Chem.* **269**, 21657 (1994).
[b] M. Tardif, S. Huang, T. Redmond, D. Safer, M. Pring, and S. H. Zigmond, *J. Biol. Chem.* **270**, 28075 (1995).
[c] S. H. Zigmond, M. Joyce, J. Borleis, G. M. Bokoch, and P. N. Devreotes, *J. Cell Biol.* **138**, 363 (1997).

The delicate balance required for monomer flux through F-actin depends on small differences in the affinity of G-actin for the barbed and pointed ends of actin filaments and for various cytoplasmic components. Because biochemical assays often involve ~10-fold dilution of the cytosol, it is surprising that agonists can still induce actin changes. The function is maintained because the concentration of free G-actin* is little changed following dilution. In the whole cell lysate, F-actin can maintain a constant concentration of free G-actin concentration by depolymerizing when the concentration falls or polymerizing when it rises. In high-speed supernatants of PMN, where there is little F-actin, the free G-actin concentration is "buffered" largely by thymosin β_4 (Tβ_4) (Table II). The presence of Tβ_4 helps maintain a relatively constant–free G-actin concentration following dilution. Thus, a supernatant containing 175 μM Tβ_4 and 100 μM actin would have a free G-actin from 0.8 μM (given a K_d of ATP-G-actin for Tβ_4 of 0.6 μM); following a 10-fold dilution, the free G-actin would decrease only to 0.7 μM. However, because Tβ_4-bound actin also serves as a reservoir of G-actin, the G-actin released to maintain the free concentration decreases the amount in this reservoir. Thus, the maximal amount of actin that can polymerize decreases roughly in parallel with the dilution.

The consequences of dilution on other components depend on their concentrations and affinities. In the case of capping protein, the concentration in the intact cell is sufficient high (2 μM) that given its affinity for a

---

*The free G-actin concentration in cytoplasm remains unknown because it is regulated by multiple factors and varies with the state of cell stimulation.

TABLE II
POLYMORPHONUCLEAR CELLS AND EXTRACTS

		range	K_d for ATP-actin
Cytoplasmic volume	2×10^{-13} liter/cell[b]		
Protein per cell	1×10^{-10} g		
Actin/cell	$5.6 \pm 0.8 \times 10^{-12}$ g[c]		
Protein concentrations in high-speed supernatants			
After Triton lysis (10^8 cells/ml)	~3 mg/ml		
After N_2 bombing (3×10^8 cell/ml)	3.7 ± 0.7 mg/ml (mean $\pm$ SD; $n = 20$)		
After sonication (4×10^8 cells/ml)	6.6 ± 1.6 mg/ml ($n = 20$)		
Cytoplasmic concentrations used in this article			
Actin (total)	$200 \ \mu M$[d]	200–600[c]	
G-actin (resting)	$120 \ \mu M$[d]	120–400	
G-actin (stimulated)	$40 \ \mu M$		
Thymosin β_4	$175 \ \mu M$[d]	50–420[e]	0.6[f]
Profilin	$40 \ \mu M$[g]		0.1[h]
Capping protein	$2 \ \mu M$[i]	1–5	0.001[i]
Filament number[j,k]			
Resting cell	$1.4 \ \mu M$ ($1.7 \pm 0.4 \times 10^5$ filaments/cell)		
Stimulated cell	2.2 to $3.3 \ \mu M$ (2.7 to 4×10^5 filaments/cell)		

[a] Determined by (i) the diameter of a round PMN of ~10 μm and (ii) when pelleted by centrifugation, there are ~10^9 PMN/ml of pellet; the volume of this pellet is ~50% buffer (determined from cell-impermeable radiolabel).
[b] D. Roos, A. A. Voetman, and L. J. Meerhof, *J. Cell Biol.* **97**, 368 (1983).
[c] J. R. White, P. H. Naccache, and R. I. Sha'afi, *J. Biol. Chem.* **258**, 14041 (1983).
[d] L. Cassimeris, D. Safer, V. T. Nachmias, and S. H. Zigmond, *J. Cell Biol.* **119**, 1261 (1992).
[e] E. Hannapel and M. V. Kampen, *J. Chromatogr.* **397**, 279 (1987).
[f] A. Weber, V. T. Nachmias, C. R. Pennise, M. Pring, and D. Safer, *Biochemistry* **31**, 6179 (1992).
[g] F. S. Southwick and C. L. Young, *J. Cell Biol.* **110**, 1965 (1990).
[h] I. Perelroizen, J.-B. Marchand, L. Blanchoin, D. Didry, and M.-F. Carlier, *Biochemistry* **33**, 8472 (1994).
[i] M. J. DiNubile, L. U. Cassimeris, M. Joyce, and S. H. Zigmond, *Mol. Biol. Cell,* **12**, 1659 (1995).
[j] M. Carson, A. Weber, and S. H. Zigmond, *J. Cell Biol.* **103**, 2707 (1986).
[k] M. Cano, D. A. Lauffenburger, and S. H. Zigmond, *J. Cell Biol.* **115**, 677 (1991).

barbed end ($K_d \sim 1$ nM), only a small fraction of the filaments (~1 μM) in an intact PMN should have free barbed ends[2,14] (Table II). Even after a thousandfold dilution (as in the pyrenylactin assay, see later), the capping protein concentration is still above its K_d for capping. However, because the rate of capping is proportional to the capper concentration, capping becomes slow. In intact cytosol, the half-time of capping a barbed end is about 0.3 sec ($k_{on} \sim 1 \times 10^6 \ M^{-1} \ sec^{-1}$)[14]; after a 20-fold dilution, the half-time is about 6 sec, and after 1000-fold dilution, about 300 sec. The rate of actin polymerization at a barbed end and depolymerization at a pointed end are enhanced by profilin and cofilin, respectively. Because the actions

[14] M. J. DiNubile, L. U. Cassimeris, M. Joyce, and S. H. Zigmond, *Mol. Biol. Cell* **12**, 1659 (1995).

of profilin and cofilin depend on their concentrations, they are decreased by dilution.[15]

This section, describes the preparation of a HSS of lysed PMN that responds to GTPγS-charged Cdc42 with increases in both F-actin level and filament number. Assays for F-actin and actin filament ends are described.

Lysate Preparation

Cells

Extracts of rabbit peritoneal exudate PMN (~95% neutrophils),[16] human PMNs obtained by venipuncture or by apheresis of volunteers, or undifferentiated HL60 cells all exhibit GTPγS-induced actin polymerization. In each case, the isolated cells are suspended at $3-6 \times 10^8$ cells/ml in saline and incubated for 5 min on ice with 1 mM diisopropyl fluorophosphate (DFP) (Sigma, St. Louis, MO; stored at 4° as a 1 M stock in propylene glycol). After pelleting the cells (1000 rpm for 9 min), the DFP in the supernatant can be inactivated by decanting the supernatant into 5 N NaOH. The cells are washed two times with cold saline before resuspending at $3-6 \times 10^8$ cells/ml in intracellular physiological buffer (IP; 135 mM KCl, 10 mM NaCl, 2 mM MgCl$_2$, 2 mM EGTA, and 10 mM HEPES, pH 7.1) plus protease inhibitors: 1 μg/ml leupeptin, 1 μg/ml benzamidine, 10 μg/ml aprotinin, and 10 μg/ml TAME.

Cell Lysis

Cells are lysed mechanically by nitrogen bombing in a Parr bomb (350 lb/in^2 for 15 min on ice)[10] or by the minimal sonication required to break the cells (3 × 1 sec pulse on setting 40 of a Dynatech probe sonicator), again on ice. F-actin levels, number of nucleation sites, and responses to GTPγS and Cdc42 are similar in cells broken by bombing and sonication. Detergent lysis interferes with the ability of both Cdc42 and GTPγS to induce polymerization.

The fresh lysate is unstable, with actin responses declining within 30 min. Thus, the lysate is immediately spun at 14,000 rpm in a microfuge for 5 min at 4°. The supernatant of this spin (low-speed supernatant) is immediately spun at 80,000 rpm for 20 min in a Beckman TL 100, using a 100.2 or 100.3 rotor, $\sim 7.2 \times 10^6 g$ min to obtain a high-speed supernatant. The HSS can be used immediately (stable at 4° for several hours) or frozen in liquid nitrogen in 200-μl aliquots and stored at $-80°$ (stable for months).

[15] D. Didry, M.-F. Carlier, and D. Pantaloni, *J. Biol. Chem.* **273,** 25602 (1998).
[16] S. J. Sullivan and S. H. Zigmond, *J. Cell Biol.* **85,** 703 (1980).

The protocol for making HSS is shown schematically in Fig. 1. On thawing, the supernatant is respun at high speed to remove proteins precipitated by freezing and thawing. It is important to avoid warming the HSS as the response to GTPγS plus lipid is decreased by 5 min at room temperature.

The composition of supernatants can be manipulated by fractionation, affinity adsorption, or addition of selected substances. The protein concentration of treated supernatants can be restored to starting values by spin-filtration in a Centricon-10 (Amicon, Danvers, MA).

Induction of Actin Changes

An aliquot of supernatant (known protein concentration) is mixed with agonist (e.g., GTPγS-charged Cdc42, GTPγS + lipids[10]) and warmed at 37° for various periods of time. The sample is then diluted ~20-fold into fluorescently labeled phalloidin for assays of F-actin or ~100-fold into pyrenylactin for assays of filament ends. A control sample is warmed without agonist.

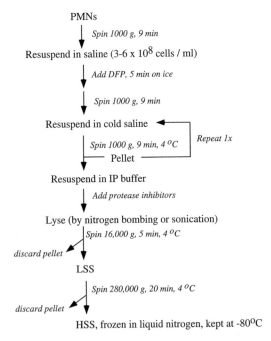

PMNs

Spin 1000 g, 9 min

Resuspend in saline (3-6 x 10^8 cells / ml)

Add DFP, 5 min on ice

Spin 1000 g, 9 min

Resuspend in cold saline

Spin 1000 g, 9 min, 4 °C | Repeat 1x

Pellet

Resuspend in IP buffer

Add protease inhibitors

Lyse (by nitrogen bombing or sonication)

Spin 16,000 g, 5 min, 4 °C

discard pellet

LSS

Spin 280,000 g, 20 min, 4 °C

discard pellet

HSS, frozen in liquid nitrogen, kept at -80°C

FIG. 1. Protocol for preparation of high-speed supernatant from PMNS.

Assay of F-Actin Level by Phalloidin Binding

Principle

The F-actin level is measured by binding of fluorescently labeled phalloidin (we use tetramethylrhodamine isothiocyanate, TRITC–phalloidin but phalloidin labeled with other fluorophores is available). Phalloidin binds tightly ($K_d \sim 10$ nM) and stoichiometrically to F-actin (one phalloidin per actin molecule in the filament). The bound phalloidin is pelleted with the F-actin and then extracted with methanol. The fluorescence in methanol is proportional to the F-actin. Some F-actin binding proteins, such as cofilin interfere with phalloidin binding. Nevertheless, estimates of actin by DNase assays, analysis of SDS gels, and phalloidin binding usually give similar values, within the error of the methods.

Method

This method is modified from the original description.[17]

1. Incubate a 20- to 40-μl aliquot of HSS (at 2–4 mg/ml protein) with agonist or buffer at room temperature or 37°.

2. Dilute into 860 μl IP buffer containing 0.4 μM TRITC–phalloidin (Sigma). Incubate for 1 hr (phalloidin binding plateaus after about 30 min) and then pellet the F-actin at $7 \times 10^6 g$ min (80,000 rpm for 20 min in a Beckman 100.2 rotor in a Beckman TL100 ultracentrifuge).

The concentration of phalloidin should be in excess of the concentration of F-actin. At 0.4 μM, TRITC–phalloidin is usually $\sim$0.2 μM in excess of that needed to saturate the F-actin induced in 40 μl of PMN HSS at 3 mg/ml. To test if phalloidin is limiting, determine if the fluorescence is a linear function of the amount of stimulated HSS added. If fluorescence plateaus with large aliquots of HSS, reduce aliquot size or increase phalloidin concentration.

Dilution of HSS in the absence of phalloidin allows F-actin to depolymerize,[2] whereas prolonged incubation in phalloidin drives polymerization. If it is necessary to postpone phalloidin addition, add glutaraldehyde or formaldehyde to stabilize the F-actin and denature the G-actin. Controls that include aldehyde are required as aldehydes alter the fluorescent signal.

3. Decant, or remove with a pipette, the supernatant, taking care not to lose the small pink pellet (because you may not even see the pellet, it is helpful to mark on the centrifuge tube where the pellet is expected). Add 1 ml methanol to the tube, seal tube with Parafilm, and extract TRITC–

[17] T. H. Howard and C. O. Oresajo, *Cell Motil.* **5,** 545 (1985).

phalloidin from the pellet overnight. (For a small pellet or a pellet disrupted by sonication, extraction is complete in 1 hr.)

4. Remove methanol and measure its fluorescence in a fluorimeter (for TRITC–phalloidin, excitation 540 nm/emission 575 nm). Any small pieces of pellet present can be repelleted; however, in most cases, they do not interfere.

The amount of agonist-induced actin polymerization is proportional to the difference between the fluorescence of a control sample and a sample incubated with agonist. Using this difference corrects for most nonspecific effects. Nonsaturable signals can be determined by including a 10-fold excess unlabeled phalloidin (reducing saturable staining by 90%). The fluorescence values can be converted into micromolar F-actin by preparing a standard curve with known concentrations of F-actin. We find that optimal stimulation with Cdc42 or GTPγS plus lipid induces about 30% of the actin in a HSS at 3 mg/ml to polymerize to F-actin. Phalloidin fluorescence and quantification of pelleted actin examined on Coomassie blue-stained SDS gels give similar results.

The maximum amount of F-actin that can be polymerized is, of course, limited by the amount of actin in the extract. If exogenous actin is added, the final concentrations of F-actin can be higher. However, the addition of $\geq 2~\mu M$ actin to HSS at 3 mg/ml often causes an agonist independent increase in the F-actin level.

Assay of Free Filament Ends by Pyrenylactin Polymerization

Principle

The rate-limiting step in the polymerization of pure G-actin is the formation of small actin oligomers or "nuclei" that then elongate to form filaments. The rate of nucleation is a function of the concentration of G-actin raised to the fourth power. At low concentrations, $<2~\mu M$, there is a sufficient lag period before polymerization begins (several minutes) that if actin filaments are added at the beginning of the lag period, the rate of polymerization is described by Eq. (1).

$$dF/dt = [\text{ends}_b][\text{G-actin} - G_{\text{crit} \cdot b}]\,k_{\text{on} \cdot b} + [\text{ends}_p][\text{G-actin-}G_{\text{crit} \cdot p}]k_{\text{on} \cdot p} \quad (1)$$

where b and p subscripts refer to barbed and pointed ends of the filament, respectively. Because the "on" rate at the barbed end ($k_{\text{on} \cdot b}$) is about 10 times that at the pointed end ($k_{\text{on} \cdot p}$), when both ends of the filament are free (i.e., uncapped), barbed-end elongation dominates the rate of polymerization.

The rate of actin polymerization is followed easily by using G-actin covalently labeled with pyrene[18-20]; pyrenylactin increases its fluorescence ~20-fold on polymerization. Although many proteins, including actin itself, bind pyrenylactin as well as native actin, profilin has a decreased affinity for pyrenylactin. Both skeletal muscle and nonmuscle actin can be labeled with pyrene. Skeletal muscle actin can be extracted from "acetone powder"[21] as modified by Murray et al.[22] and labeled by a modification[23] of the method of Kouyama and Mihashi[18] as described in the Appendix. Both native and pyrenylactin are gel filtered prior to use to remove capping proteins and actin oligomers that serve as nucleation sites.[24]

Characterization of Pyrenylactin

Before starting, the pyrenylactin should be characterized to determine its extent of labeling, critical concentration, specific fluorescence (change in fluorescence/μM actin polymerized), and spontaneous nucleation activity. Assays for these properties are as follows.

Fraction of Actin Labeled with Pyrene. Determine the OD_{290} (protein) and OD_{344} (pyrene) for the actin solution (without DTT which interferes with the OD_{290}) and buffer. The actin concentration (μM) is corrected for pyrene fluorescence using Eq. (2):

$$[\text{Actin}] = (\text{actin solution } OD_{290} - 1/3 \text{ actin solution } OD_{344})$$
$$- (\text{buffer } OD_{290} - 1/3 \text{ buffer } OD_{344})/0.0249 \quad (2)$$

The pyrenylactin concentration (μM) equals OD_{344} (4.5×10^{-5}). Usually (using the method described in the Appendix) between 60 and 90% of the actin is labeled. If the actin is used at ≤ 1.5 μM, there is little internal quench of fluorescence. To measure F-actin at higher concentrations, dilute the labeled actin with unlabeled actin to avoid internal quench (adsoption by the molecules in the cuvette decreasing the fluorescence).

Critical Concentration of Actin. Incubate different concentrations of pyrenylactin (e.g., 0.1, 0.15, 0.2, 0.25, 0.3, 0.5, 0.75, and 1 μM) in polymerizing (see earlier discussion) and nonpolymerizing buffer (e.g., column buffer, see Appendix). The pyrene fluorescence is read after the fluorescence in polymerizing buffer has reached steady state (~48 hr; the time can be decreased if, after some polymerization has occurred, the solution is pi-

[18] T. Kouyama and K. Mihashi, *Eur. J. Biochem.* **114,** 33 (1981).
[19] J. A. Cooper, S. B. Walker, and T. D. Pollard, *J. Muscle. Res. Cell Motil.* **4,** 253 (1983).
[20] T. D. Pollard, *Anal. Biochem.* **134,** 406 (1983).
[21] J. A. Spudich and S. Watt, *J. Biol. Chem.* **246,** 4866 (1971).
[22] J. M. Murray, A. Weber, and M. K. Knox, *Biochemistry* **20,** 641 (1981).
[23] M. Carson, A. Weber, and S. H. Zigmond, *J. Cell Biol.* **103,** 2707 (1986).
[24] S. MacLean-Fletcher and T. D. Pollard, *Biochem. Biophys. Res. Commun.* **96,** 18 (1980).

petted vigorously to shear filaments and increase the number of ends). The critical concentration is the actin concentration at the intercept of a line drawn through the fluorescence in polymerizing buffer with that in nonpolymerizing buffer (Fig. 2).

The critical concentration of actin is a function of the composition of the polymerization buffer. The values of our preparations of rabbit skeletal muscle actin are usually between 0.1 and 0.15 μM in 0.1 M KCl, 2 mM MgCl$_2$, 1 mM ATP, 25 mM Tris, pH 7.4. A high critical concentration indicates that (i) the actin is not pure (i.e., some of the protein is not actin), (ii) some of the actin is denatured, or (iii) a capping factor is present.

Pyrenylactin can be used to determine the critical concentration of unlabeled actin merely by including the pyrenylactin at a ratio such that the concentration of pyrenylactin is always below the critical concentration (~1/20 pyrenyl to unlabeled actin).

Specific fluorescence is the fluorescence change due to polymerization of given concentration of actin, e.g., 1 μM. The slope of the polymerized actin on the critical concentration graph (Fig. 2) gives the fluorescence change per 1 μM actin polymerized.

Spontaneous (de novo) Nucleation Activity of Actin. The actin should be monitored to check for existing nuclei and nucleation activity: determine the time required to reach 2% of the final steady state. Dilute the actin (from column buffer at 20–30 μM) to 1.5 μM into assay buffer (see earlier discussion) and follow the time course of polymerization in the fluorimeter. Determine the end point of polymerization. The mean time for 1.5 μM gel-filtered pyrenylactin to reach 2% polymerization is 8 ± 3 min.

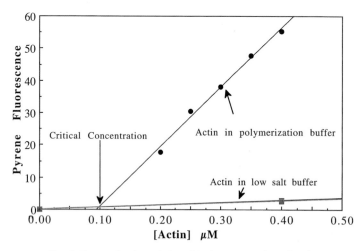

FIG. 2. Determination of the critical concentration of actin.

The absence of a lag, or decrease in the lag period, indicates that some actin oligomers or nucleating factors are present. This is often the case if the actin has not been gel filtered, as the fractions just preceding the main actin peak frequently have high levels of nucleation.[24] An excessively prolonged lag may indicate that the actin concentration is lower than expected or that a barbed-end capping protein is present.

Conversion of Ca–Actin to Mg–Actin. Dilute G-actin stock stored in column buffer to ~3 μM into water buffered with 1 mM Tris, pH 7.4 or 8, containing enough $MgCl_2$ to give 50 μM Mg over final ATP (0.7 mM provided by the column buffer) and enough EGTA to give 10–20 μM EGTA over the calcium present (0.3 mM from column buffer). Many investigators use the actin after 4–6 min (although conversion is not complete). It is important not to convert much ahead of time as magnesium actin is not stable.

Storage of Actin. Pyrenylactin is more stable than unlabeled actin because the pyrenyl group on cysteine-373 inhibits oxidation, a major cause of actin denaturation. Pyrenylactin stored at 20 to 30 μM in column buffer (because calcium–actin is much more stable than magnesium actin) at 4° shows little increase in critical concentration or change in *de novo* nucleation even after 2 months. Unlabeled actin stored at 20 to 30 μM in calcium and ATP in the presence of DTT is stable for a week to 10 days. For longer periods, small aliquots ($\leq$250 μl) at 20 to 40 μM actin in column buffer containing 1 mM DTT are flash frozen in liquid nitrogen and stored at −80°. For use, thaw quickly, sonicate briefly on ice to break any filaments, store on ice for ~30 min to allow depolymerization, and then spin at high speed to pellet any denatured actin aggregates. Some investigators store actin as G-actin at 4° by continually changing the DTT and ATP.[25]

Pyrenylactin Assay of Filament Free Ends

1. Because fluorescence is sensitive to temperature (decreasing the temperature increases the fluorescence), it is important to prewarm everything, e.g., empty quartz cuvettes, pyrenylactin as G-actin (either straight pyrenylactin or at a decreased fraction labeled by dilution with unlabeled G-actin), and polymerization buffer (140 mM KCl, 2 mM $MgCl_2$, 1 mM ATP, 1 mM EGTA, and 25 mM Tris–HCl, pH 7.4), to the temperature of the fluorimeter, usually room temperature, before each assay.

2. Incubate an aliquot of HSS with agonist or buffer for the desired period of time as for F-actin assays. With HSS at 3 mg/ml, we use an

[25] J. Xu, W. H. Schwarz, J. A. Kas, T. P. Stossel, P. A. Janmey, and T. D. Pollard, *Biophys. J.* **74,** 2731 (1998).

incubation volume of 6 μl and incubation times varying between 15 sec and 5 min.

3. To transfer this small sample efficiently, add 994 μl polymerization buffer containing pyrenylactin (final concentration of 1.5 μM added as G-actin for less than 30 sec) to the tube with HSS and then transfer the solution immediately to the prewarmed cuvette (this transfer provides mixing). Follow the pyrene fluorescence (excitation 370/emission 410) for 4 or 6 min. With time the rate will decrease as the G-actin concentration is decreased by polymerization. Include a control with just pyrenylactin in the assay buffer to detect nucleation by the actin itself. Because G-actin is stored as calcium–actin, which has a different k_{on} than G-actin bound by magnesium, for accurate measure of early initial rates ($\sim$60 sec), convert to magnesium G-actin just before use.

Because of the $\sim$100-fold dilution, the HSS contributes little G-actin or G-actin sequestering factors: 10 μl cytosol 1 $\times$ 10^8 PMN equivalents diluted to 1 ml contributes $\sim$0.1 μM actin, 0.2 μM Tβ_4, and 0.03 μM profilin. Thus, the free G-actin concentration is that of the pyrenylactin actin.

4. After measuring the initial rate, save each sample (transfer to a microfuge tube and store in the dark) and measure the final (end point) fluorescence again at 5 hr. Nucleated samples usually reach end point by 5 hr; however, samples with few nucleation sites may not reach their end point for 12 hr or longer. It is best to read at several times as (unlike pure F-actin which can be read at 72 hr) in HSS the end point values can decrease between 5 and 18 hr. If the end point in the presence of supernatant is lower than that in its absence (pure actin), consider the possibility that the supernatant has quenched the fluorescence or that a barbed-end capping protein or actin sequestering factors are present.

To determine if HSS *quenches* pyrenylactin fluorescence, polymerize several milliliters of pyrenylactin in polymerization buffer to steady state. Read the fluorescence before and immediately after adding a small aliquot ($\sim$6 μl) of HSS. Determine if the the HSS causes (rapid, i.e., seconds) a decrease in fluorescence (compared another sample with 6 μl buffer). Capping proteins and monomer-binding proteins in the HSS also decrease fluorescence; however, the change due to monomer binders and capping proteins requires depolymerization and is slower (occurs over minutes or hours).

Barbed-end capping factors can be detected by their ability to slow the rate of seeded polymerization (initiate polymerization by adding a known volume of spectrin–actin seeds or sheared F-actin). Because the capping protein contributed by HSS is probably dilute, it will slowly cap growing ends and slowly decrease the rate of polymerization relative to a control with only seeds added. To confirm that the slowing of polymerization is due to barbed-end capping, repeat the experiment in the presence of a barbed-end capper cytochalasin; if the ends are already capped, an additional capper from the HSS will have no effect. However, if an actin-sequestering factor is present, it will decrease the rate in the presence and absence of cytochalasin.

One can also examine the effect of a given aliquot of HSS on the critical concentration of the pyrenylactin. A factor that quenches the fluorescence will decrease the specific fluorescence (change in fluorescence per micromolar polymerized) without shifting the critical concentration. G-actin sequestering factors and cappers will increase the critical concentration, but at concentrations above the new critical concentration, the increase in fluorescence per micromolar actin added will be the same as in the control.

5. Convert the initial rate, measured in fluorescence units, to micromolar pyrenylactin polymerized using the specific fluorescence determined in the critical concentration assay (see earlier discussion). The specific fluorescence was determined without HSS, thus if HSS quenches, one needs to correct the specific fluorescence for the quench.

6. To determine the relative contribution of barbed and pointed-end elongation, compare the rate of pyrenylactin polymerization in the presence and absence of a factor that preferentially inhibits barbed-end elongation. If both ends are free, the presence of enough cytochalasin in the polymerization buffer to cap >90% of the barbed ends ($\sim$2 μM cytochalasin B or 0.2 μM cytochalasin D) decreases the polymerization rate 8- to 10-fold.† If only pointed ends are elongating, e.g., with filaments capped by gelsolin, these concentrations of cytochalasin cause less than a 10% decrease in the rate of polymerization. Because cytochalasin, especially cytochalasin D in the presence of HSS, stimulates *de novo* nucleation (detected as an increase rate of polymerization over time), the change in nucleation of control supernatant +/− cytochalasin must be measured before assuming that the rate in the presence of cytochalasin represents pointed-end elongation.

7. It is possible to estimate the number concentration of ends if one knows the actin rate constants for the conditions being used. The number of actin nucleation sites can be estimated from Eq. (1) or rewritten as in Eq. (3):

---

† The contribution of pyrenylactin elongation at barbed and pointed ends can be estimated given the following conditions: G-actin concentration 1.5 μM; $k_{on \cdot p}/k_{on \cdot b}$ = 1/10; critical concentration p = 0.5 μM; and critical concentration b = 0.05 μM. The rate of elongation at each end equals ([G-actin]- [G$_{crit.}$ at that end])k_{on} for that end. Thus, when both filament ends are free, the rate of elongation at the barbed end is 14.5 times that at the pointed end.

$$dF/dt = -dG/dt = [end_b](k_{on \cdot b}[\text{G-actin}] - k_{off \cdot b})$$
$$+ [end_p](k_{on \cdot p}[\text{G-actin}] - k_{off \cdot p}) \qquad (3)$$

The polymerization rate, expressed as micromolar actin polymerized/sec, varies dramatically with the nucleotide bound to G-actin (ATP versus ADP) and the ionic conditions.[20] For net polymerization of ATP–G-actin in 0.1 M KCl and 1 mM MgCl$_2$ at pH 8: ATP–G-actin $k_{on \cdot b}$ ~1 × 10^7 M^{-1} sec^{-1}; ATP–G-actin $k_{off \cdot b}$ ~0.8 sec^{-1}; the ATP–G-actin $k_{on \cdot p}$ ~0.6 × 10^6 M^{-1} sec^{-1}; and ADP–G-actin $k_{off \cdot p}$ ~0.4 sec^{-1}.[15] Using these values, we estimate that warming HSS at 3 mg/ml for 3 min with optimal concentrations of Cdc42 or GTPγS plus lipid induces 1–4 nM free barbed ends.

8. To ensure that the pyrenylactin assay is functioning as expected, determine the rate of polymerization at different concentration of G-actin (the rate is expected to be proportional to the concentration of G-actin-G$_{crit.}$) and the rate using different amounts of HSS (if the high-speed supernatant provides only ends, the rate should be proportional to the amount of HSS added).

Agonists

Recombinant GTPases

Recombinant Cdc42 is expressed in a baculovirus insect cell expression system as GST fusion proteins as described.[10] The GTPase is extracted with cholate from the membrane fraction of Sf9 cells (to optimize recovery of geranylgeranylated forms) and retained on glutathione beads. The beads are washed with detergent and buffer and then cleaved with thrombin. After removal of the thrombin with benzamidine beads, proteins are stored frozen. At the time of use, the GTPase is charged with GTPγS by incubating for 10 min at 30° with 100 μM GTPγS in EDTA/Mg to give a final free Mg concentration between 100 and 1000 nM. The free Mg concentration is then increased to 2 mM in excess of the EDTA present and the samples are stored on ice until use.

GTPγS plus Lipid

In HSS, GTPγS does not work unless an active GEF or an acidic lipid (presumably to activate an endogenous GEF) is added. A variety of pure lipids suffice, with PIP being one of the best. However, mixed brain lipids, which are inexpensive, work well when the exact lipid composition is not critical. For use, the lipids are suspended and sonicated in water at 10 mg/ml diluted to 2 mg/ml and stored frozen. The sample is resonicated just before use at 50 μg/ml.

Microscopic Examination of Actin Filaments

Electron Microscopy of Negatively Stained Actin Filaments

HSS is incubated with buffer or agonist on a freshly carbon-coated EM grid. Alternatively, the samples are incubated in a tube and transferred to the grid (transfer can increase filament bundling). The grid is washed with two water drops (to remove salt) but containing 1 μM phalloidin (to protect F-actin). The grid is then placed on a drop of 1% uranyl acetate, drained, and dried. The samples are viewed at magnifications between 10,000 and 50,000×.

High-speed supernatants contain particulate material, which makes it difficult to identify unambiguously filaments less than 0.25 μm long (4.5 mm on the photo). With pure actin, shorter filaments can be recognized.

Fluorescence Microscopy

Agonist and TRITC–phalloidin are mixed briefly with HSS (final protein concentration ~4 mg/ml), immediately 5–8 μl is placed on the slide, and a 22 × 40-mm coverslip is lowered gently to minimize bubble formation. To reduce fluorescence quenching, 1/10 dilutions of glucose (1%) and 1/10 dilution of catalase and glucose oxidase (2.5 mg/ml catalase and 7.5 mg/ml glucose oxidase in 3 ml distilled water; keeps for 2–3 months) can be added to the HSS. Polymerization occurs *in situ* at room temperature or at 37°. At early times (1–2 min), fine TRITC-stained filaments are observed with a 63 or 100× lens. Because individual filaments in solution are moving, only filaments attached to the glass can be photographed. Adsorption of nitrocellulose[26] or NEM-treated myosin (E. DeLaCruz, personal communication, 1999) to glass can help capture and immobilize F-actin. With extended incubation (10 min), the filaments form bundles and networks.

Because the glass surface adsorbs protein, the protein concentration of the HSS needs to be higher for polymerization on the slide than in a tube. Samples can be incubated in a tube, stained with TRITC–phalloidin, and then spread between a slide and coverslip, although the transfer results in filament bundling. The TRITC–phalloidin, needed to visualize the filaments, can itself with time induce polymerization. Phalloidin-induced filaments are short and bundle rapidly into short rods.

[26] J. Xu, J. F. Casella, and T. D. Pollard, *Cell Motil. Cytoskel.* **42**, 73 (1999).

Appendix: Preparation of Pyrenylactin from Muscle Acetone Powder

Solutions

Calcium depolymerizing solution (CDS)

10 mM Tris–HCl	10 ml of 1 M stock, pH 8.0
0.2 mM CaCl$_2$	2 ml of 0.1 M stock
0.5 mM ATP	5 ml of 0.1 M stock, pH 8.0
50 μM CaEDTA	5 ml of 10 mM stock
1 mM DTT	1 ml of 1 M stock
1 mM sodium azide	32 mg (or 10 ml of 0.1 M)

Make volume up to 1 liter; store at 4°.

Column buffer

TEA	1.33 ml
0.3 mM CaCl$_2$	6 ml of 0.1 M stock
0.1 mM EDTA	2 ml of 0.1 M stock, pH 7.0
0.7 mM ATP	0.75 g dry
Sodium azide	0.4 g (or 125 ml of 0.1 M)

Make volume up to 2 liters and adjust pH to 7.5 (use concentrated HCl) Store at 4°.

Procedure

1. Rinse a 400- and a 600-ml beaker, a stir bar, mortar and pestle, and double-layered cheesecloth with 1 mM CaEDTA, pH 7.4 (use stock 10 mM CaCl$_2$ and 10 mM EDTA, pH 7.4). Repeat rinse two times and then rinse with deionized water.
2. Grind 10 g of muscle acetone power with the mortar and pestle.
3. Measure out 200 ml of cold CDS. Use a little to wet and mash up the powder with the mortar and pestle, then transfer this mash into the 400-ml beaker, and add the rest of CDS. Stir for 25 min in the cold room.
4. Filter the extracted powder through cheesecloth into a 600-ml beaker and squeeze out as much liquid as possible. Reextract the solid mass with 100 ml CDS for 5 min. Filter and squeeze out liquid and combine with original filtrate.
5. Centrifuge the liquid (~210 ml) in a cold rotor at 4° for 10 min at 14,000.
6. Filter the supernatant from step 5 through a fluted Whatman (Clifton, NJ) #1 or #41 filter (15 or 18.5 cm) into the preweighed 500-ml Erlenmeyer flask. Reweigh and calculate supernatant volume (using 1 g = 1 ml). Remove a 100-μl sample for a protein determination and go on to step 7. During the polymerization in step 7, dilute this

sample to 1 ml (10-fold) with CDS and read OD at 290 and 310 nm (the 310 reading allows a background correction). Read CDS buffer at both wavelengths. The amount of protein extracted is:

(Actin OD_{290} − actin OD_{310}) − (buffer OD_{290} − buffer OD_{310})/
0.00249 = actin concentration (μM)

already corrected for the 10-fold dilution. To determine the concentration in milligrams per milliliter, multiply the micromolar concentration by 0.042 (molecular mass of actin is 42,000 kDa).

7. Induce polymerization by making the solution 2 mM $MgCl_2$ and 0.1 M KCl (will require <0.5 ml of 1 M $MgCl_2$ and slightly more than 7 ml of 3 M KCl). Let stand for 30 min at room temperature and then in a 15° water bath for another 30 min.

8. Return flask to ice bucket. When the solution has reached 4°, add dry KCl to give a final concentration of 0.8 M (remember the solution is already 0.1 M KCl). Stir in cold until the salts are dissolved. If necessary, adjust to pH 8.3–8.5 with 1 N KOH.

9. Spin the solution at 4° for 100 min at ~150,000g (35 K in a Beckman 50.2 Ti rotor.

10. Discard supernatant and rinse pellets once with distilled water and scrape off pellets into about 20 ml cold CDS. Homogenize in glass Dounce homogenizer until the solution is clear of any solid pieces (it will be viscous.).

11. Dialyze against CDS (using 1/2 inch tubing, dialyze for 12 to 36 hr at 4° on a stir plate).

End of day 1.

12. Change dialysis solution to CDS without DTT and continue to dialyze (~another 12 hr with 1/2 inch tubing).

13. Remove actin and spin for 85 min at ~150,000g at 4°.

14. Remove supernatants to preweighed flask and again calculate the volume. Read OD values as described earlier and recalculate the actin concentration. Put the actin in an ice bucket with water/ice at a temperature of 15° and set this on a stir plate.

15. For labeling the actin needs to be at ~40 μM in 2 mM $MgCl_2$, 0.1 M KCl, and 45 μM pyrene [N-(1-pyrene)iodoacetamide]. Calculate the amount of each solution needed and add in the following order: CDS (without DTT), $MgCl_2$, from 1 M stock, pyrene (drop by drop), from stock of N-(1-pyrene)iodoacetamide dissolved in dimethyl sulfoxide (DMSO) at 1 mg/ml = 2.6 mM, and KCl, dry or from 3 M stock. From now on, protect the actin from exposure to light by wrapping in foil. Allow to polymerize in 15° water bath for 90 min. Then spin it for 2.5 hr at ~150,000g at 4°.

16. Homogenize pellets the same as described earlier except use column buffer. Dialyze. (Using 1/2 inch tubing, dialyze 2 nights and 1 day in column buffer.)

17. Remove solution and spin for 2.5 hr at 150,000g at 4°. Save the supernatants; calculate the volume by weight and read a sample on the spectrophotometer as described earlier to determine the actin concentration. This time, to correct for pyrene fluorescence at 290 nm (extinction coefficient of 29,400 $M^{-1}cm^{-1}$ [19]), read OD_{344} (rather than OD_{310}) and OD_{290}. The extinction coefficient of pyrene actin at 344 is 22,000 M^{-1} cm^{-1}.[18] Substitute 1/3 times the 344-nm reading for the 310-nm readings:

$$(\text{Actin } OD_{290} - 1/3 \text{ actin } OD_{344}) - (\text{buffer } OD_{290} - 1/3 \text{ buffer } OD_{344})/0.00249 = \text{actin concentration } (\mu M)$$

18. Load solution at ~80 to 100 μM onto 450- to 500-ml Sephacryl S-300 column equilibrated with column buffer.

19. Collect ~4-ml fractions. The actin elutes at about 200 ml. Read fractions from the actin peak, without dilution, at 290 and 344 nm. Read the OD values of the column buffer.

$$\text{Actin concentration in each fraction } (\mu M) = (\text{actin } OD_{290} - 1/3 \text{ actin } OD_{344}) - (\text{buffer } OD_{290} - 1/3 \text{ buffer } OD_{344})/0.0249$$

We recover about 90% of the protein loaded on the column. Actin yields vary between 6 and 87 mg actin (mean 30 ± 19) from 10 g acetone powder. The variation in yield depends in part on how much connective tissue was dissected from the muscle before extraction. Even with careful dissection, only a fraction of the total actin present is extracted.

20. Wash the column with 500 ml of 0.5 M KCl and store with 0.02% sodium azide (w/v) until the next use. Run at least 3 bed volumes of low salt with ATP through the column immediately before the next actin run.

21. Assay the "spontaneous" nucleation activity of each fraction in the actin peak (see assay described earlier).

22. Pool fractions with low nucleating activity and determine the critical concentration (see assay described earlier).

Acknowledgments

I am extremely grateful to Annemarie Weber for sharing various actin protocols. I thank Drs. A. Weber, E. DeLaCruz, C. Yang, and M. Huang and Mr. M. Joyce for reading a draft of the method. SHZ was supported by NIH Grant AI19883.

Section III

Analysis of Rho GTPase Function

[22] Determination of GTP Loading on Rac and Cdc42 in Platelets and Fibroblasts

By ANSER C. AZIM, KURT L. BARKALOW, and JOHN H. HARTWIG

Introduction

The Rho family of small Gtpases are guanine nucleotide-binding proteins of 20–50 kDa that modulate the organization of the actin cytoskeleton, cell growth, vesicular cell trafficking, and transcriptional regulation by linking receptors to cytoplasmic targets. They function as molecular switches that cycle between inactive GDP-bound forms and active GTP-bound forms.[1] To transduce signals initiated by ligated receptors, the GTP-bound forms of these small Gtpases bind and activate target proteins that then mediate biological responses or propagate the signal to other effector molecules. GTP–Rac causes cells to spread and ruffle. GTP–Cdc42 mediates filopodial growth from the cell surface in a process that involves the Arp2/3 complex and N-WASP, WASP, or SCAR.[2–5] Filopodial growth can also be mediated by the Ras family GTPase, RalA[6]; GTP–Rho promotes stress fiber formation and the formation of adhesion sites.[7–9]

The temporal relationship between Rac and Cdc42 activation and cell functions, in many cases, remains to be established. As an example, for Rac and Cdc42 to function in actin assembly as proposed, or in other cellular processes for that matter, they must activate prior to the barbed-end nucleation reaction that starts the actin assembly reaction. In order to gain insight into the mechanics of receptor-mediated activation of Cdc42 and Rac, we use the GTPase-binding domain of p21-activated kinase 1 (PAK-1), expressed as a glutathione S-transferase (GST) fusion protein, then bound to Sepharose beads as a means to selectively collect the active

[1] A. Hall, *Annu. Rev. Cell Biol.* **10**, 31 (1994).
[2] P. Aspenstrom, U. Lindberg, and A. Hall, *Curr. Biol.* **6**, 70 (1996).
[3] M. Symons, J. Derry, B. Karlak, S. Jiang, V. Lemahieu, F. McCormick, U. Francke, and A. Abo, *Cell* **84**, 723 (1996).
[4] L. Machesky and K. Gould, *Curr. Opin. Cell Biol.* **11**, 117 (1999).
[5] L. Machesky, R. Mullins, H. Higgs, D. Kaiser, L. Blanchoin, R. May, M. Hall, and T. Pollard, *Proc. Natl. Acad. Sci. U.S.A* **96** (1999).
[6] Y. Ohta, N. Suxuki, S. Nakamura, J. Hartwig, and T. Stossel, *Proc. Natl. Acad. Sci. U.S.A.* **96**, 2122 (1999).
[7] A. Ridley, H. Paterson, C. Johnston, D. Diekmann, and A. Hall, *Cell* **70**, 401 (1992).
[8] A. Ridley and A. Hall, *Cell* **70**, 389 (1992).
[9] A. Hall, *Curr. Opin. Cell Biol.* **5**, 265 (1993).

GTP-bound form of the GTPase.[10] This domain of PAK-1 has been used to assay GTP–Rac2 from neutrophils.[11] PAK1 kinase has a CRIB (Cdc42/Rac interactive binding) motif and the binding of GTP–Rac or GTP–Cdc42 activates its kinase activity. This GTP-dependent trapping assay is straightforward if the GTPases are soluble in cell lysates, allowing them to be collected rapidly using the derivatized Sepharose beads and bacterially expressed protein constructs. When the GTPases are forced into a GTP-bound state, this assay will trap 100% of these proteins (Fig. 1). In platelets, all of the Rac protein is soluble following lysis with Triton X-100 in physiological mimic buffers containing millimolar EGTA. GTP–Cdc42, however moves into the cytoskeletal fraction following cell activation, adding to the complexity of the assay by requiring experiments that follow Cdc42 in both the soluble and the cytoskeletal fraction.

This article details the quantitation of the GTP and GDP forms of Rac and Cdc42 from platelets after PAR-1 receptor ligation with thrombin receptor-activating peptide (TRAP). We also show that the GTP-bound forms of Rac and Cdc42 can be quantitated from fibroblasts after stimulation with growth factors.

Methods

Solutions

Platelet lysis buffer (10×): 500 mM Tris–HCl, pH 7.2, 50 mM EGTA, 50 mM EDTA, 10% (v/v) Triton X-100, 52 nM leupeptin, 10 nM benzamidine, 123 nM aprotinin, and 10 μM phallacidin. Wash buffer (1×): 20 mM Tris–HCl, pH 7.5, 150 mM NaCl, and 0.1% (v/v) Tween 20.

Experimental Procedures

Expression of GST-PAK

The GTPase binding domain of PAK-1 (aa 67–150) is expressed as a (GST) fusion protein in *Escherichia coli* and affinity purified from cell lysates using a glutathione-Sepharose column. Cells are induced to express the protein construct with 1 mM of isopropylthiogalactoside (IPTG) after 4 hr of growth in the log phase. GST-PAK1 is purified from frozen cell lysates and freshly bound to Sepharose beads for all experiments.

---

[10] A. Azim, K. Barkalow, J. Chou, and J. Hartwig, *Blood* **95**, 959 (2000).
[11] V. Benard, B. Bohl, and G. Bokoch, *J. Biol. Chem.* **274**, 13198 (1999).

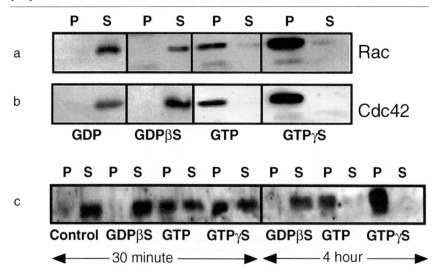

FIG. 1. Immunoblot analysis demonstrating that GST-PAK1-glutathione-Sepharose 4B bead conjugates bind only to the GTP-bound forms of Rac and Cdc42. (a) Anti-Rac immunoblot showing the effect of exogenous loading of Rac in a 100,000g supernatant of Triton X-100-lysed human platelets with 100 μM GDP, GTP, GDβS, or GTPγS on the amount of Rac collected with 10 μg of GST-PAK1 beads. (b) Anti-Cdc42 immunoblots showing the amount of Cdc42 collected after exogenous loading with guanine nucleotides as described in the text. (c) Effect of incubation time on the amount of Rac collected following forced loading with different guanine nucleotides in the trapping assay.

Preparation and Treatment of Platelets

Human blood from healthy individuals is drawn into 1/10 volume of Aster-Jandl anticoagulant and centrifuged at 110g for 10 min at room temperature. The platelet-rich plasma is removed (the upper phase in the centrifuged blood), and 10–15 ml of platelet-rich plasma is passed through a small Sepharose 2B column (60-ml syringe with an ~50-ml bed volume) equilibrated with 145 mM NaCl, 10 mM HEPES, 10 mM glucose, 0.2 mM Na$_2$HPO$_4$, and 5 mM KCl, pH 7.4, containing 0.3% (w/v) bovine serum albumin (bsa) and 1.8 g/liter glucose at room temperature. This column separates the platelets from the plasma proteins. Platelets elute in the void as milky-colored fractions. Plasma proteins are retained on the column.[12] The gel-filtered platelets are allowed to rest for 30 min at 37° before use. Platelets were activated using 25 μM TRAP. Resting or activated platelets are lysed by the addition of 1/10 volume of lysis buffer. The cell solution

[12] J. Hartwig, in "The Cytoskeleton: A Practical Approach" (K. Carraway and C. Carraway, eds.), p. 23. Oxford Univ. Press, Oxford, 1992.

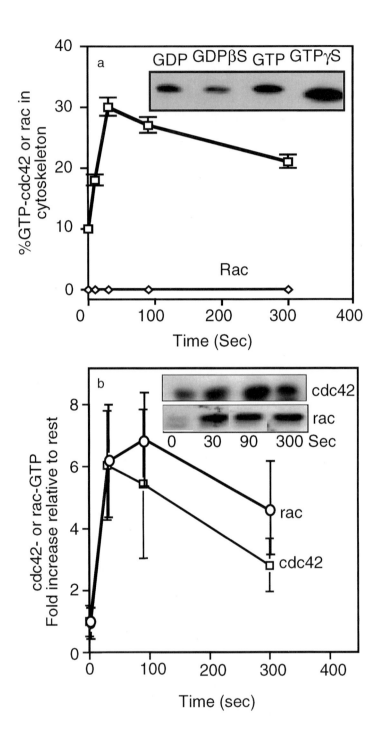

is mixed and the tubes placed on ice. Cytoskeletal and soluble fractions are separated by centrifugation of this lysate at 100,000g for 30 min at 4°.

Fibroblasts

NIH 3T3 mouse fibroblasts are seeded in 10 × 2-cm plastic petri dishes and grown to confluency. The cells are serum starved for 12 hr before treatment with 20 nM of epidermal growth factor (EGF). Resting or EGF-treated cells are released from plates with 5 mM EDTA in Hanks' balanced salt solution and lysed as described previously. Cytoskeletal and soluble fluid fractions are separated by centrifugation of the lysate at 100,000g for 30 min.

Trapping Assay

The soluble supernatant, prepared as described from lysed cells, is carefully removed and transferred to Eppendorf tubes. The supernatant (0.5 ml), derived from ~10^8 platelets, which have a total protein concentration of ~150 μg, is mixed and incubated with Sepharose 4B-glutathione bead conjugated with 5–10 μg of recombinant GST-PAK. The sample is incubated for 4–12 hr at 4°C with agitation. After the incubation period, beads are collected by centrifugation at 10,000g for 1 min. The supernatant is removed and 1/4 volume of 5× SDS–PAGE buffer is added. The pellet is washed five to six times with chilled wash buffer by centrifugation. The pellet is then suspended in 1× SDS–PAGE buffer such that the final volume is equal to that of the supernatant fluid removed after the first centrifugation plus the dilution factor resulting from the addition of 5× SDS–PAGE sample buffer. Rac and Cdc42 are separated on 12% SDS–PAGE gels[13] and transferred electrophoretically to a polyvinylidene difluoride (PVDF)

---

[13] U. Laemmli, *Nature (Lond.)* **227**, 680 (1970).

Fig. 2. (a) Cdc42, but not Rac, moves into the cytoskeleton after stimulation of platelets with 25 μM TRAP. Activated GTP–Rac is not collected in the 100,000g pellet. Approximately 30% of the total Cdc42 becomes associated with the actin cytoskeleton following platelet activation. (Inset) Immunoblot showing that GTP and GTPγS induce the translocation of Cdc42 into actin cytoskeleton. (b) Comparison of the time courses of Rac and Cdc42 GTP loading following platelet activation with TRAP. Quantitation of GTP–Rac and GTP–Cdc42 in the soluble fraction prepared from platelet lysates. Each point is an average of a minimum of four experiments (mean ± SD). Resting platelets have 2–10% of their total Rac and Cdc42 in GTP forms. Activation increases the amount of GTP–Rac and GTP–Cdc42 by six- to sevenfold in the lysates. (Inset) Immunoblots showing trapped GTP-Rac and GTP–Cdc42 in resting and activated platelets.

Immobilion P membrane (PVDF, 0.45 μm: Millipore Corp., Bedford, MA) at 15 V for 30 min in a semidry blotting apparatus. Rac and Cdc42 are quantitated from the pellet and supernatants by immunoblotting using a mouse anti-Rac monoclonal antibody and an anti-Cdc42 antibody (BD Transduction Laboratories, Lexington, KY) followed by peroxidase-labeled goat antimouse immunoglobulin G (IgG, Bio-Rad, Hercules, CA). Linearity of the immunoblot signal is confirmed using recombinant GTPase standards.

Figure 1 shows the validity of this assay in platelet lysates. Soluble platelet proteins released by detergent permeabilization of resting platelets (100,000g supernatant) are incubated with 100 μM of GDPβS, GTPγS, GDP, or GTP in lysis buffer containing 1 mM dithiothreitol and 1 mM MgCl$_2$ to load the endogenous Rac and Cdc42 with the various guanine nucleotides. Ten micrograms of PAK1-GST-Sepharose 4B beads is added to 100 μg of lysates loaded with nucleotide and incubated for 4 hr at 4°. The beads are collected by centrifugation at 10,000g for 1 min at 4°. Beads are washed five times and bound protein is denatured with SDS. Samples are electrophoresed, transferred to PVDF membrane, and immunoblotted using mouse monoclonal anti-Rac (Fig. 1a) and Cdc42 (Fig. 1b) IgG. Incubation with GTP or GTPγS causes all of the Rac and Cdc42 protein in the lysate to become bound to the beads. None of Rac or Cdc42 is bound, however, after loading with GDP or GDPβS. This demonstrates that the binding of Rac and Cdc42 to the GBD domain of PAK1 requires them to have GTP bound. Figure 1c shows that the maximal trapping of the GTP form of Rac requires incubation with the beads for >30 min.

Detergent permeabilization releases soluble proteins into the supernatant while leaving an insoluble actin-based cytoskeleton, which we separate

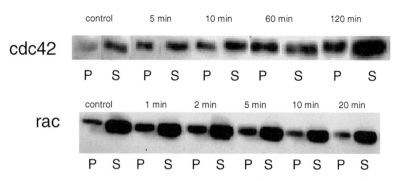

FIG. 3. Immunoblots showing the amount of Rac and Cdc42 trapped by PAK1 bead conjugates in resting and EGF-treated NIH 3T3 cells. Cells were serum starved for 12 hr, stimulated with 20 nM EGF for the indicated times, and lysed as described using Triton X-100.

by centrifugation at 100,000g for 20 min. Figure 2a shows that Triton X-100-containing buffers release all of the platelet Rac into the soluble phase independent of the activity state of the cells. Cdc42, however, distributes in both soluble and cytoskeletal fractions. GTP loading promotes Cdc42 movement into the cytoskeletal fraction, indicating that it is the GTP-bound form, which binds to the cytoskeleton. Quantification of GTP–Rac is straightforward because the soluble protein fraction contains all the Rac protein of the platelet. Figure 2b shows that there is ~sixfold increase in the relative amount of GTP–Rac in platelets after PAR1 receptor ligation with 25 μM TRAP. The increase in GTP–Rac content is rapid and peaks 30–60 sec after the addition of TRAP to platelets. Although the extent of Rac activation varies among platelet preparation, the kinetics of GTP loading is similar in all experiments. In comparison, actin assembly is maximal after 60 sec.[14] TRAP activation of platelets leads not only to increased cytoskeletal association of Cdc42 (Fig. 2a) but also to a rapid increase in the amount of GTP–Cdc42 measured in the soluble protein supernatant (Fig. 2b). Like Rac, the increase in GTP–Cdc42 was rapid and increased six- to sevenfold in comparison to its levels in resting platelets. The kinetics of Cdc42 charging (Fig. 2b) and its movement into the cytoskeleton (Fig. 2a) follow the same time course.

Fibroblasts stimulated with growth factors spread and ruffle in a Rac-dependent fashion.[7,15] Using the trapping assay, we analyzed the GTP status of Rac and Cdc42 in serum-starved and EGF-stimulated fibroblasts. Figure 3 shows that EGF induces GTP binding by both Cdc42 and Rac. Rac activation is maximal 2–5 min following exposure to EGF. Cdc42 activation occurred somewhat more slowly and peaked between 10 and 60 min.

Conclusion

We have described a simple assay that can be used to quantify GTP–Rac and GTP–Cdc42 in cells. In platelets, there is a six- to sevenfold increase in GTP binding within 30 sec of PAR-1 receptor ligation. In fibroblasts, GTP charging induced by EGF is also measured easily. This assay provides a simple but powerful quantitative measure of GTPase activation that can be employed in most cell-based systems.

Acknowledgments

This work was supported by NIH Grants HL HL56252 and HL56949 and a Scientist Development Grant (KLB) from the American Heart Association.

[14] J. Hartwig, *J. Cell Biol.* **118**, 1421 (1992).
[15] A. Ridley, *Bioessays* **16**, 321 (1994).

[23] Determination of GTP Loading on Rho

By XIANG-DONG REN and MARTIN ALEXANDER SCHWARTZ

Introduction

Rho is a member of the Ras superfamily of low molecular weight GTPases that is implicated in the regulation of actin cytoskeleton organization,[1,2] formation of focal adhesions,[2] transcriptional activation, and cell cycle progression.[3,4] Since the discovery of its basic biological functions in the early 1990s, there have been substantial studies on how Rho executes its biological functions. Less, however, is known about its regulation by upstream elements initiated by interactions of cells with soluble factors, other cells, or extracellular matrix components. The slower progress in this area may be in part because a reliable assay for Rho activity was not available. We have established an affinity precipitation assay for endogenous GTP-loaded (active) Rho.[5] This method has proved to be successful in analyzing cellular Rho activities in a variety of systems. Since its publication, there have been numerous requests for the assay. This article presents our experience with this assay in the hope that it may answer most of the common technical questions and provide some additional information to facilitate further optimization.

Technical Difficulties

Like Ras, Rho cycles between a GTP-bound active form and a GDP-bound inactive form, thus serving as a molecular switch for cellular signals. The bound GTP can be hydrolyzed to GDP by the intrinsic GTPase activity of Rho. Although the intrinsic GTPase activity is low, it can be accelerated greatly (by orders of magnitude) by interaction with GTPase-activating proteins (GAPs).[6,7] The classical way to detect the activity of a GTPase

[1] H. F. Paterson, A. J. Self, M. D. Garrett, I. Just, K. Aktories, and A. Hall, *J. Cell Biol.* **111**, 1001 (1990).
[2] A. J. Ridley and A. Hall, *Cell* **70**, 389 (1992).
[3] C. S. Hill, J. Wynne, and R. Treisman, *Cell* **81**, 1159 (1995).
[4] M. F. Olson, A. Ashworth, and A. Hall, *Science* **269**, 1270 (1995).
[5] X.-D. Ren, W. B. Kiosses, and M. A. Schwartz, *EMBO J.* **18**, 578 (1999).
[6] N. Morii, K. Kawano, A. Sekine, T. Yamada, and S. Narumiya, *J. Biol. Chem.* **266**, 7646 (1991).
[7] K. Rittinger, P. A. Walker, J. F. Eccleston, S. J. Smerdon, and S. J. Gamblin, *Nature* **389**, 758 (1997).

(e.g., Ras) is to immunoprecipitate the protein from lysates of cells that are labeled metabolically with [^{32}P]phosphate and then analyze the associated radioactive nucleotides by thin-layer chromatography.[8] The activity of the GTPase is usually expressed as the ratio of the amount of GTP versus total nucleotides (GTP+GDP). In the case of Ras, the antibody Y13-259 used for immunoprecipitation inhibits the intrinsic and GAP-stimulated GTPase activity of Ras so that the bound GTP could be preserved for chromatographic analysis. However, such an antibody is not available for Rho. Furthermore, the corresponding Y13-259 recognition site in Rho (amino acids 65–75) is identical to other Rho family members, such as Rac and Cdc42; therefore, even if a suitable antibody were available, it would be unlikely to distinguish Rho from Rac and Cdc42.

As the intrinsic GTPase activity of Rho is fairly low, we tried to immunoprecipitate Rho from cells labeled with [^{32}P]phosphate. We tested antibodies against Rho-specific sequences near the C terminus to precipitate endogenous Rho and anti-HA antibody with cells expressing epitope-tagged Rho. Those efforts were unsuccessful, either because the anti-Rho antibody did not immunoprecipitate endogenous Rho efficiently or the tagged Rho could be precipitated but the bound nucleotide was entirely GDP. To determine whether Rho-bound GTP might be hydrolyzed during the immunoprecipitation, we loaded recombinant Rho proteins (on agarose beads) with [α-^{32}P]GTP *in vitro* and incubated the beads with cell lysate at different temperatures. At the end of the incubation, the bound nucleotides were eluted and analyzed by chromatography (Fig. 1). We found that the total amount of radioactive nucleotides stayed largely unchanged; however, the fraction of Rho-bound GTP dropped dramatically in the presence of cellular extracts. Even under these conditions where Rho protein is in excess, much of the Rho-bound GTP was lost after a 1-hr incubation. These results indicate that GAP activity is a major obstacle for the Rho–GTP-loading assay. The results also indicate that low temperature and a high concentration of NaCl (500 mM) help to minimize GAP activity.

Expression and Purification of GST–TRBD

The Rho-binding domain (RBD) from Rhotekin (one of the Rho effectors) was reported to inhibit both the intrinsic and the GAP-enhanced GTPase activity of Rho.[9] Because Rho effectors interact only with GTP–

[8] J. Downward, J. D. Graves, P. H. Warne, S. Rayter, and D. A. Cantrell, *Nature* **346,** 719 (1990).

[9] T. Reid, T. Furuyashiki, T. Ishizaki, G. Watanabe, N. Watanabe, K. Fujisawa, N. Morii, P. Madaule, and S. Narumiya, *J. Biol. Chem.* **271,** 13556 (1996).

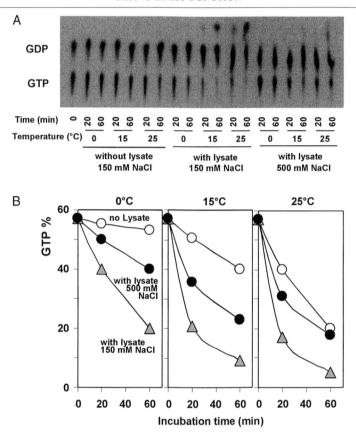

FIG. 1. Influence of Rho GTPase activity by cellular proteins and temperature. Recombinant GST–Rho on glutathione beads was loaded with [α-^{32}P]GTP (Amersham) at 30° for 10 min in 50 mM Tris, 2 mM EDTA, and 1 mM DTT. Loading was stopped by adding 20 mM MgCl$_2$, and the free [^{32}P]GTP was washed away. The Swiss 3T3 cell lysate (1 mg/ml) was prepared in Tris buffer containing 1% Triton X-100, 0.5% sodium deoxycholate, 0.1% SDS, 150 mM NaCl, 10 mM MgCl$_2$, 10 μg/ml each of leupeptin and aprotinin, and 1 mM PMSF. Half of the lysate was adjusted to 500 mM NaCl. Cell lysates (with either 150 or 500 mM NaCl) or the lysis buffer alone was added to aliquots of Rho beads preloaded with [^{32}P]GTP and incubated on ice, 15°, or 25° for 20 or 60 min. After incubation, the beads were washed twice with lysis buffer, and Rho-associated nucleotides were released by heating at 70° for 10 min in 40 mM EDTA, 2% SDS, 4 mM DTT, and 0.5 mM each of GTP and GDP. Nucleotides were resolved by thin-layer chromatography as described.[11] The autoradiograph (A) was digitized and quantified (B) in the AlphaEase system (Alpha Innotech, CA).

Rhotekin sequence

```
    7                               89
    I    L    E    D       E    K    T    G   Stop
GGTGGATCC ATC CTG GAG GAC... ...GAG AAG ACA GGC TAG AATTCAT
CCACCTAGG TAG GAC CTC CTG... ...CTC TTC TGT CCG ATCTTAAGTA
   BamHI                                        EcoRI
```

FIG. 2. A map of the pGEX-2T-TRBD construct.

Rho, we used the Rhotekin Rho-binding domain (TRBD) to affinity precipitate endogenous GTP–Rho from cell lysates.[5] This domain should bind only GTP–Rho and inhibit its conversion to the GDP form, thus the amount of Rho retained on TRBD beads should provide a measure of Rho–GTP loading. Similar strategies had been employed to assay other Ras family proteins.[10]

The coding sequence for the Rhotekin Rho-binding domain (amino acids 7–89) was amplified by polymerase chain reaction from a mouse testicle cDNA preparation (Clontech, Palo Alto, CA) and cloned into the pGEX-2T vector (Pharmacia, Piscataway, NJ) at the BamHI and EcoRI sites.[5] Sequences flanking the cloning sites are shown in Fig. 2. The DNA was transformed into the DH5α strain of Escherichia coli, and its sequence was confirmed to be identical to the published sequence.[9] The resulting recombinant protein is a fusion protein with the glutathione S-transferase (GST) at the N terminus.

For expression and purification of GST–TRBD, bacteria expressing GST–TRBD [stored at $-70°$ in 20% (v/v) glycerol/L-broth] are inoculated from the glycerol stock into 400 ml L-broth containing 100 μg/ml ampicillin. After overnight incubation with shaking at 37°, the culture is diluted 1 : 10 into 3600 ml LB with ampicillin in four 2-liter flasks. Incubation is continued at 37° until the OD_{600} reaches 0.8, which usually takes 1.5–2 hr. Protein expression is then induced with 0.5 mM isopropyl-β-D-thiogalactopyranoside (IPTG; 0.5 M stock in H_2O) for 2 hr. We noticed that when induction was carried out at 37°, most of the proteins (70% or more) were insoluble; however, the Rho-binding activity of the residual soluble material was generally high. When induced at room temperature the yield of soluble protein was higher but the binding activity was lower. In practice we find that results are satisfactory at 30–37°.

Bacteria are collected by centrifugation at 4000g for 15 min and resuspended in 40 ml cold lysis buffer [50 mM Tris, pH 7.5, 1% Triton X-100, 150 mM NaCl, 5 mM $MgCl_2$, 1 mM dithiothreitol (DTT), 10 μg/ml aprotinin, 10 μg/ml leupeptin, and 1 mM phenylmethylsulfonyl fluoride (PMSF)]. It is

[10] J. de Rooij and J. L. Bos, Oncogene **14,** 623 (1997).

important that bacteria be completely resuspended before sonication. The bacteria suspension (about 45 ml in total) is transferred into two prechilled 50-ml centrifugation tubes (for JA17 rotor) and sonicated on ice with a microtip at setting 3 for six to eight times, 15 sec each. It is cooled on ice for 1–2 min between sonications. The lysate is clarified by centrifugation in a prechilled JA17 rotor at 14,000 rpm (27,000g) for 30 min at 4°. The lysate is mixed with 0.6 ml (wet volume; preswelled with water) glutathione beads (Sigma) and rotated at 4° for 60 min. The beads are collected by brief centrifugation (2000 rpm, 1 min at 4° in a Beckman bench-top centrifuge). They are washed six times with 12 ml wash buffer (50 mM Tris, pH 7.5, 0.5% Triton X-100, 150 mM NaCl, 5 mM MgCl$_2$, 1 mM DTT, 1 μg/ml aprotinin, 1 μg/ml leupeptin, and 0.1 mM PMSF) and once with the same buffer supplemented with 10% (v/v) glycerol. When resuspending the beads, invert the tube gently to avoid foaming. After the last wash, the beads are resuspended gently in 8 ml wash buffer supplemented with 10% glycerol, aliquoted into prechilled tubes, and stored at −70°. To determine the amount of purified protein, run 20 μl (out of 8 ml) bead suspension in a 13% reduced gel and include bovine serum albumin (BSA) standards (2.5, 5, and 10 μg) for comparison. GST–TRBD in 20 μl bead suspension should contain about 4 μg. We often see two additional bands right below the GST–TRBD band, but they are usually less than 1/10 of the major TRBD band.

GST–RBD is a labile protein, and the binding activity of the TRBD to GTP–Rho may vary significantly among different batches of purification. This is the most common problem when the assay is not working well. A "quality-control" Rho assay is therefore recommended for each new batch of TRBD beads (see later). Cautions should be taken to avoid anything that can possibly lead to protein denaturation (such as freeze-thawing, temperature higher than 4°, foaming of protein solution) during purification. It is recommended that the preparation be completed in 1 day (which takes about 9.5 hr) to avoid unnecessary freeze-thawing.

Preparation of Cell Lysates

Immediately after stimulation or other treatment, cells are washed twice with ice-cold Tris-buffered saline (TBS). The dishes are placed on ice at an angle to drain (30–60 sec) and as much TBS as possible is removed. Cold lysis buffer (50 mM Tris, pH 7.2, 1% Triton X-100, 0.5% sodium deoxycholate, 0.1% SDS, 500 mM NaCl, 10 mM MgCl$_2$, 10 μg/ml each of leupeptin and aprotinin, and 1 mM PMSF) is added. The cells are scraped rapidly and the crude lysates are transferred into 1.5-ml tubes and spun at 13,000g for 10 min at 4°. The cleared lysates are then transferred quickly into tubes with GST–TRBD (20–30 μg) beads and rotated at 4° or on ice for 45–60 min. The beads are washed four times (in a microfuge at 5000 rpm for 15–20 sec), each with 600 μl cold Tris buffer containing 1% Triton X-100, 150 mM NaCl, 10 mM MgCl$_2$, 10 μg/ml each of leupeptin and aprotinin, and 0.1 mM PMSF. After the last wash, the wash buffer is removed carefully, SDS sample buffer (containing 40 mM DTT) is added to the beads, and they are heated at 95° for 10 min. The samples are run on 13% gels and transferred to a nitrocellulose membrane for immunodetection of GTP–Rho pulled down by TRBD.

As discussed in the introduction, GAP activity is the major obstacle of the assay. GTP-bound Rho in the lysate is not protected from GAPs before being bound by TRBD. In some initial experiments we tested adding soluble GST–TRBD in the lysis buffer to begin binding the GTP–Rho at the time of lysis and then adding the glutathione beads after centrifugation. However, the retention of GTP–Rho was lower. Rebinding of the GST–TRBD to the glutathione beads was not 100% efficient under these conditions, which may partially account for the low binding. We therefore prefer to lyse the cells in the absence of TRBD, clear the lysate rapidly by a brief centrifugation, and then mix the cleared lysate with GST–TRBD beads, as described earlier.

As indicated in Fig. 1, GTP hydrolysis increases with time and temperature. It is therefore critical to lyse the cells and process the samples in a timely fashion and keep the samples cold the entire time. The assay is best carried out in a cold room. Avoid touching the lower portion of the tube as this may warm up the sample and increase GTP hydrolysis. We noticed that the temperature of the samples increased significantly after centrifugation for a short period of time (Fig. 3) even when the microfuge was placed in a cold room. This raised the question as to how long the lysates should

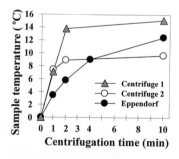

FIG. 3. Warming of samples during centrifugation. A 1.5-ml Eppendorf tube containing 1 ml of ice-cold water was placed in the centrifuge and spun at full speed for the indicated time in a cold room. After spinning, the temperature of the water was measured. The rotor was then cooled down on ice for 30 min before the next sample was centrifuged.

be centrifuged. While longer times at a higher temperature should increase GTP hydrolysis, in practice we found that net retention of GTP–Rho was similar or slightly higher when the lysate was spun for 10 min compared with 1 min (data not shown). We therefore recommend centrifugation for 10 min for the clarification of lysates.

When doing time course experiments, however, a shorter centrifugation time (2 min) can make multiple samples much easier to handle. This is because centrifugation is needed for clearing lysates as well as washing the beads at different times; additionally, it is necessary to keep the centrifuge temperature more or less constant for different samples. We therefore reduced the centrifugation time to 2 min and kept the rotor on ice after centrifugation during time course experiments. We also incubated the lysate with GST–TRBD beads on ice to avoid effects of temperature fluctuation in the cold room during experiments.

For the lysis buffer, it is important to avoid phosphate-based buffer, as phosphate forms precipitates with magnesium, a cation that is required to stabilize GTP-bound Rho. Use of SDS, deoxycholate, and a high concentration ($500\ mM$) of NaCl in the lysis buffer helps minimize the interaction of GTP–Rho with GAPs and other Rho effectors, but does not noticeably decrease binding to TRBD. Therefore it is preferable to include SDS, deoxycholate, Triton X-100, and $500\ mM$ NaCl in the lysis buffer. We have noticed that some cell lines have nuclear membranes that are sensitive to these harsher detergents, such that nuclear lysis occurs and the cell extracts become viscous due to released DNA. Under these conditions, a Triton X-100-based lysis buffer may be used instead. Although the assay becomes less sensitive, possibly because GTP–Rho dissociates less efficiently from complexes formed with cellular proteins, changes of Rho activity may still be detected (Fig. 4).

The binding of TRBD to GTP–Rho is relatively rapid, reaching maximal levels after about 45–60 min (Fig. 5). A longer incubation time is not necessary, as the free GTP–Rho may have already been converted to

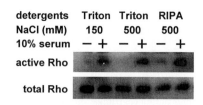

FIG. 4. Effect of different lysis buffers on the detection of GTP–Rho. Serum-starved Swiss 3T3 cells were stimulated with 10% serum for 3 min, and the Rho activity was measured in different lysis buffers.

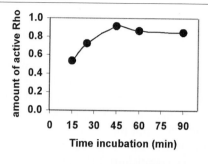

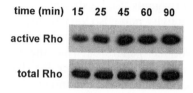

FIG. 5. Optimizing incubation time of TRBD with cell lysate. The Swiss 3T3 cell lysate was aliquoted into tubes containing equal amounts of GST–TRBD beads and incubated for indicated times. The beads were washed four times at the end of incubation, and the bound Rho proteins were analyzed by Western blotting. The image on the X-ray film (bottom) was digitized and quantified (top) with the AlphaEase system.

GDP–Rho by the action of GAPs. Additionally, GAPs may act on GTP–Rho molecules that transiently dissociate from the TRBD. It is essential to keep the incubation time consistent when samples have to be processed at different time points. Because the portion of Rho that can be pulled down by TRBD also appears to vary with the protein concentration of the lysate, it is important to keep the protein concentration of the lysates reasonably constant so that Rho activities can be compared accurately.

Finally, because GAP activities may vary in different cell lines, the optimal conditions may vary accordingly.

Detection of Rho by Western Immunoblotting

After Western transfer, the nitrocellulose membrane is blocked in 5% (w/v) nonfat dried milk for 60 min at room temperature or overnight at 4°. The RhoA antibody (Santa Cruz Biotechnology, Santa Cruz, CA) is diluted 1 : 1000 in 5% (w/v) milk and incubated for 2 hr at room temperature. Incubation overnight at 4° appears to decrease the signal/background ratio. After washing four times with TBS/0.1% Tween 20, the transfers are incubated in HRP-labeled secondary antibody (BioSource International, Camarillo, CA; 1 : 1000 in milk) for 60 min. The membrane is washed six times and is then developed with ECL (enhanced chemiluminescence) reagents (Amersham, Piscataway, NJ). The amount of TRBD-bound Rho is normalized to the total amount of Rho in cell lysates for the comparison of Rho activity (level of GTP-bound Rho) in different samples.

Typically, fibroblasts from one 10-cm dish are extracted with 800 μl lysis buffer. From that volume, 650 μl is incubated with TRBD and 20 μl is kept for Western blotting as a "total Rho" control. The lysate concentration is usually around 0.8–1 mg/ml (from a confluent dish), and the total Rho signals from cell lysates can be detected after a 1-min exposure. The exposure time may vary significantly, however, depending on the specific blotting material and reagents used, especially ECL reagents. In addition the Rho antibody reactivity may also vary from batch to batch.

Western blotting results are often recorded by exposing X-ray films to the light-emitting products generated from the ECL reagents. The film images are then scanned into a computer for densitometric analysis. It should be kept in mind that the linear range of the X-ray film is fairly narrow and that scanning of the film image sets a further limit on the linearity of the signals. Thus, care is required to avoid inaccuracies. A digital camera system [such as the ChemiImager4400 (Alpha Innotech, San Leandro, CA) or ChemiDoc (Bio-Rad, Hercules, CA)] can record the Western immunoblot results directly into a computer for densitometry analysis. Because it has a wider linear range and can identify saturated signals, it is less prone to errors.

Quality Control of TRBD and Positive Control for Rho Assay

When the assay does not work, the first thing to check is whether the purified TRBD has binding activity to GTP–Rho. This can be done by incubating TRBD beads with lysate prepared from cells transfected with immunotagged V14-Rho (a mutant of Rho that stays in GTP-bound form) to see whether the TRBD binds V14-Rho. In our laboratory, we test the quality of the TRBD preparations by assaying Rho activity in Swiss 3T3 cells growing in serum. With a reasonably good preparation of TRBD, the Rho signal pulled down by the GST–TRBD from 650 μl of lysate is about the same as the total Rho signal from 20 μl lysate (see Fig. 5) in the experimental scheme described earlier. A good quality TRBD preparation does not, however, guarantee detection of changes of Rho activity. A positive control for performing the assay with endogenous Rho is suggested. We recommend using serum-starved and serum-stimulated (2–3 min) Swiss 3T3 cells. In this system, Rho activity is reliably stimulated three- to sixfold.[5]

[24] Use and Properties of ROCK-Specific Inhibitor Y-27632

By SHUH NARUMIYA, TOSHIMASA ISHIZAKI, and MASAYOSHI UEHATA

Introduction

The small GTPase Rho functions as a molecular switch to control various cellular processes such as stimulus-induced cell to substrate adhesion and motility, neurite retraction, cytokinesis, calcium sensitization of smooth muscle contraction, the G_1 to S progression of the cell cycle, cell transformation, and nuclear transcription by serum response factor.[1,2] Rho acts on downstream effectors to exert these actions. Rho effectors include the ROCK family of protein serine/threonine kinases composed of p160ROCK (ROCK-I)[3] and ROKα/Rho kinase/ROCK-II,[4–6] protein kinase (PK)N,[7,8] citron and citron kinase,[9] and mammalian diaphanous homologs, mDia1[10] and mDia2.[11] Expression of various mutants of these effectors, as well as analysis of their activities in an *in vitro* cell-free system, indicates that the ROCK family of kinases is involved in the increase of actomyosin-based contractility through phosphorylation and down regulation of myosin phosphatase[12] and in Rho-induced bundling of actin stress fibers and clustering

---

[1] S. Narumiya, *J. Biochem.* **120**, 215 (1996).

[2] L. van Aelst and C. D'Souza-Schorey, *Genes Dev.* **11**, 2295 (1997).

[3] T. Ishizaki, M. Maekawa, K. Fujisawa, K. Okawa, A. Iwamatsu, A. Fujita, N. Watanabe, Y. Saito, A. Kakizuka, N. Morii, and S. Narumiya, *EMBO J.* **15**, 1885 (1996).

[4] T. Leung, E. Manser, L. Tan, and L. Lim, *J. Biol. Chem.* **270**, 29051 (1995).

[5] T. Matsui, M. Amano, T. Yamamoto, K. Chihara, M. Nakafuku, M. Ito, T. Nakano, K. Okawa, A. Iwamatsu, and K. Kaibuchi, *EMBO J.* **15**, 2208 (1996).

[6] O. Nakagawa, K. Fujisawa, T. Ishizaki, Y. Saito, K. Nakao, and S. Narumiya, *FEBS Lett.* **392**, 189 (1996).

[7] G. Watanabe, Y. Saito, P. Madaule, T. Ishizaki, K. Fujisawa, N. Morii, H. Mukai, Y. Ono, A. Kakizuka, and S. Narumiya, *Science* **271**, 645 (1996).

[8] M. Amano, H. Mukai, Y. Ono, K. Chihara, T. Matsui, Y. Hamajima, K. Okawa, A. Iwamatsu, and K. Kaibuchi, *Science* **271**, 648 (1996).

[9] P. Madaule, M. Eda, N. Watanabe, K. Fujisawa, T. Matsuoka, H. Bito, T. Ishizaki, and S. Narumiya, *Nature* **394**, 491 (1998).

[10] N. Watanabe, P. Madaule, T. Reid, T. Ishizaki, G. Watanabe, A. Kakizuka, Y. Saito, K. Nakao, B. M. Jockusch, and S. Narumiya, *EMBO J.* **16**, 3044 (1997).

[11] A. S. Alberts, N. Bouquin, L. H. Johnston, and R. Treisman, *J. Biol. Chem.* **273**, 8616 (1998).

[12] K. Kimura, M. Ito, M. Amano, K. Chihara, Y. Fukata, M. Nakafuku, B. Yamamori, J. Feng, T. Nakano, K. Okawa, A. Iwamatsu, and K. Kaibuchi, *Science* **273**, 245 (1996).

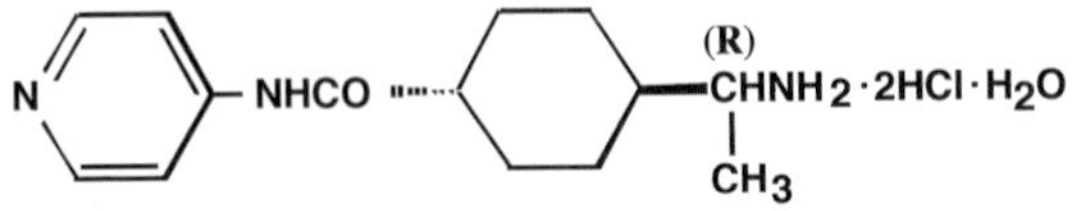

FIG. 1. Structure of Y-27632.

of focal adhesions[13–15] and that the mDia family of proteins is involved in Rho-induced actin polymerization and alignment of ROCK-induced actin stress fibers.[10,16] While these molecular biological and biochemical approaches have been valid in elucidating the actions of Rho effectors, they also have limitations in that they cannot be applied easily in elucidation of the roles of the Rho effector-mediated pathways in cells, tissues, and intact animals in response to physiological stimuli. We have reported a new synthetic compound named Y-27632 as a specific inhibitor of the ROCK family of kinases.[17] Since this report, Y-27632 has been widely used as a ROCK inhibitor to identify and evaluate the involvement and roles of the ROCK family of kinases in a variety of systems. This article describes general properties of this compound and summarizes its applications in cultured cells, isolated tissues, and intact animals.

Properties of Y-27632

Chemical Properties

Y-27632, (R)-$(+)$-$trans$-4-(1-aminoethyl)-N-(4-pyridyl)cyclohexanecarboxamide dihydrochloride monohydrate ($C_{14}H_{21}N_3O \cdot 2HCl \cdot H_2O$, molecular weight 338.3) (Fig. 1) is a white powder that melts at 262.5° with decomposition. It can be stored at room temperature in a lightproof desiccated container for at least 1 year. It is more than 10% (w/v) soluble in distilled water and is stable in a 0.5% water solution either at room temperature or at 4° for at least 4 weeks. Y-27632 shows an ultraviolet absorption maximum at 269.6 nm. Purity can be examined by high-performance liquid chromatography (HPLC) on a reversed-phase ODS column with a solvent of 0.05 M sodium perchlorate (adjusted with perchloric acid to pH 2.5)–acetonitrile

[13] T. Leung, X. Q. Chen, E. Manser, and L. Lim, *Mol. Cell. Biol.* **16,** 5313 (1996).
[14] T. Ishizaki, M. Naito, K. Fujisawa, M. Maekawa, N. Watanabe, Y. Saito, and S. Narumiya, *FEBS Lett.* **404,** 118 (1997).
[15] M. Amano, K. Chihara, K. Kimura, Y. Fukata, N. Nakamura, Y. Matsuura, and K. Kaibuchi, *Science* **275,** 1308 (1997).
[16] N. Watanabe, T. Kato, A. Fujita, T. Ishizaki, and S. Narumiya, *Nature Cell Biol.* **1,** 136 (1999).
[17] M. Uehata, T. Ishizaki, T. Satoh, T. Ono, T. Kawahara, T. Morishita, H. Tamakawa, Yamagami, J. Inui, M. Maekawa, and S. Narumiya, *Nature* **389,** 990 (1997).

(7/1, v/v). Although the metabolic stability of Y-27632 in intact animals has not yet been reported, this compound loses its inhibitory activity on serum-induced stress fiber induction after incubation with cultured cells for 24 hr (see later).

Biochemical Properties

Y-27632 and related compounds specifically bind to the ROCK family of kinases and inhibit their kinase activity.[17] Kinetic analysis using a purified enzyme revealed that Y-27632 inhibits ROCK by competing with ATP for binding to the catalytic site.[18] However, in contrast to general ATP-competitive inhibitors where kinase inhibition can be reversed by the milli-molar physiological concentrations of ATP seen in the cell, Y-27632 can potently inhibit ROCK-mediated cellular responses at micromolar concentrations. This potent inhibition is also observed when Y-27632 is added together with millimolar concentrations of ATP to permeabilized tissues.[17] These findings indicate that although Y-27632 is directed to the ATP-binding site of the kinase, this compound inhibits ROCK, not by simply preventing ATP from binding, but in a more complex manner. The affinities of Y-27632 to various kinases determined as K_i values are 0.22, 0.30, 5.3, 3.1, 73, 25, and more than 250 μM for ROCK-I, ROCK-II, citron kinase, PKN, PKCα, PKA, and myosin light chain-kinase (MLCK), respectively.[17,18] Thus, Y-27632 inhibits ROCK-I and ROCK-II equally well, and the affinities of Y-27632 for ROCK kinases are at least 10 times higher than those for citron kinase and PKN and about 100 more times or higher than those for PKC, PKA, and MLCK. Y-27632 does not appreciably inhibit the kinase activity of PAK at 100 μM concentration.

When added to intact cells, Y-27632 does not appear to freely permeate the cell membrane but must be taken up by the cell, apparently by carrier-mediated facilitated diffusion. The uptake of a Y-27632-related compound, [³H]Y-30141, occurs in a manner dependent on time and temperature. As shown in Fig. 2, the uptake reaches close to a plateau in 30 min at 37°, in which the intracellular concentration of the compound is comparable to the concentration outside the cell.[18] Therefore, cells or tissues should be incubated with Y-27632 for at least 30 min to obtain the maximal inhibition.

In Vitro Applications

Y-27632 has been used in many biological systems both *in vitro* and *in vivo* to examine the involvement of ROCK and to characterize its actions.

[18] T. Ishizaki, M. Uehata, I. Tamechika, J. Keel, K. Nonomura, M. Maekawa, and S. Narumiya, *Mol. Pharm.* **57,** 976 (2000).

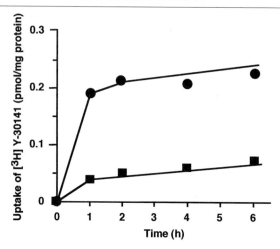

FIG. 2. Time- and temperature-dependent uptake of [^{3}H]Y-30141, a Y-27632-related compound, by rat lymphocytes. Rat lymphocytes were suspended in minimal essential medium at 2×10^8 cells/ml and were incubated with 1 μM [^{3}H]Y-30141. After indicated times, the cells were spun down and were washed twice with phosphate-buffered saline. The cells were then lysed in 1 N NaOH, and the radioactivity in the lysates was determined.

The results of these experiments are summarized in Table I. This section describes three examples of *in vitro* applications of Y-27632 in tissues and cultured cells.

Inhibition of Smooth Muscle Contraction

Y-27632 has been developed as a smooth muscle relaxant. It potently relaxes the contraction of various smooth muscles induced by a variety of agonists but to a less extent than by exposure to high concentrations of potassium ion.[17] An example of the concentration-dependent relaxation of smooth muscle contraction is shown in Fig. 3A. A coronary artery is dissected from a pig heart, and a ring is suspended in Krebs–Henseleit solution bubbled with 95% O_2–5% CO_2 (v/v). The preparation is incubated for about 1 hr and exposed to 50.9 mM KCl to obtain a standard contraction. After the tissue is washed and the tone returns to the basal level, histamine is added at final 3 μM, and the isometric contraction is monitored. When the contraction reaches a plateau, Y-27632 dissolved in distilled water at 20 mM is added to the medium at various concentrations, and resultant relaxation is observed. Y-27632 relaxes the histamine-induced contraction in a concentration-dependent manner from 10^{-8} to 10^{-5} M.

Smooth muscle contraction is evoked primarily by the Ca^{2+} calmodulin-dependent pathway, which activates MLCK to phosphorylate myosin light chain (MLC), thereby inducing contraction. Agonist stimulation also activates another pathway involving Rho and ROCK along with the primary pathway, and the activated ROCK augments the action of the first pathway by phosphorylating and inhibiting myosin phosphatase[12] and Y-27632 decreases the contraction by inhibiting ROCK in this reaction and interfering with the second pathway (Fig. 3B). Indeed, agonist-induced phosphorylation of the myosin-binding subunit of myosin phosphatase and inhibition of this phosphorylation by Y-27632 have been shown in blood platelets and Swiss 3T3 cells.[19,20] Therefore, the inhibitory effects of Y-27632 on agonist-induced contraction is dependent on the strength of agonist stimulation. Y-27632 may not inhibit the contraction of smooth muscle or other contractile processes in the cell when stimulation of the Ca calmodulin pathway by agonists is strong enough to induce the maximum contraction by itself or the activation of this pathway is induced without stimulation of the Rho–ROCK pathway, e.g., through opening of a voltage-gated calcium channel by exposure to high potassium ion. An example of the dependency of the Y-27632 effect on the strength of agonist stimulation is shown in Fig. 3C.[19] Y-27632 potently inhibits the increase in MLC phosphorylation in platelets stimulated with 0.05 U/ml thrombin but not in those with 0.1 U/ml.

Whereas Rho kinase (ROCK-II) has been shown to phosphorylate MLC directly *in vitro* in a cell-free system and has been suggested to induce a Ca^{2+}-independent contraction,[21,22] this role of ROCK has not been shown *in vivo*. Studies using Y-27632 show that the ROCK-mediated smooth muscle contraction does not work in the absence of Ca^{2+} ions[23] and that the calcium-independent phosphorylation of MLC evoked by myosin phosphatase inhibitors is catalyzed by a kinase insensitive to Y-27632.[23,24]

[19] Y. Suzuki, M. Yamamoto, H. Wada, M. Ito, T. Nakano, Y. Sasaki, S. Narumiya, H. Shiku, and M. Nishikawa, *Blood* **93**, 3408 (1999).

[20] J. Feng, M. Ito, K. Ichikawa, N. Isaka, M. Nishilawa, D. J. Hartshorne, and T. Nakano, *J. Biol. Chem.* **274**, 37385 (1999).

[21] M. Amano, M. Ito, K. Kimura, Y. Fukata, K. Chihara, T. Nakano, Y. Matsuura, and K. Kaibuchi, *J. Biol. Chem.* **271**, 20246 (1996).

[22] Y. Kureishi, S. Kobayashi, M. Amano, K. Kimura, H. Kanaide, T. Nakano, K. Kaibuchi, and M. Ito, *J. Biol. Chem.* **272**, 12257 (1997).

[23] K. Iizuka, A. Yoshii, K. Samizo, H. Tsukagoshi, T. Ishizuka, K. Dobashi, T. Nakazawa, and M. Mori, *Br. J. Pharmacol.* **128**, 925 (1999).

[24] Y. Kureishi, M. Ito, J. Feng, T. Okinaka, N. Isaka, and T. Nakano, *Eur. J. Pharmacol.* **376**, 315 (1999).

TABLE I
EFFECTS OF Y-27632 IN VARIOUS BIOLOGICAL SYSTEMS

Biological system	Concentrations of Y-27632 (μM)	Effects	Refs.
Cells			
Platelets	10–100	Inhibition of agonist-induced shape change and MLC phosphorylation	32, 33, 34
	0.1–30	Inhibition of agonist-induced phosphorylation of myosin phosphatase and MLC	19
Human neutrophils	1–10	Inhibition of chemoattractant-induced locomotion and shape change	39
Rat peritoneal mast cells	30	Inhibition of Rho-mediated cortical F-actin disassembly without affecting secretion	40
Hepatic stellate cells		Inhibition of migration and endothelin-induced cell contraction	41
Rat aortic smooth muscle cells	10	Inhibition of DNA synthesis and cell migration	37
Rat cardiac myocytes	1–30	Inhibition of endothelin-induced expression of natriuretic peptides and hypertrophy	30
Bovine endothelial cells	10	Inhibition of hypotonic shock-induced activation of volume-regulated anion channel	42
Human umbilical vein endothelial cells	10	Inhibition of endothelial contraction	28
N1E-115 neuroblastoma cells	0.01–100	Inhibition of LPA- and Rho-induced neurite retraction and MLC phosphorylation	25
	10	Inhibition of LIM-kinase activation and cofilm phosphorylation	31
NIH 3T3 fibroblasts	10	Inhibition of Rho-mediated stress fiber formation and cell transformation No effects on SRF activation and cell cycle progression	26
	10	No effect on ERM phosphorylation and microvillus formation	27

Inhibition of Stress Fiber Formation in Cultured Cells

Swiss 3T3 cells are seeded on a coverglass at a density of 3×10^4 cells per 3.5-cm dish and are cultured for 1 day in Dulbecco's modified Eagle's

TABLE I (*continued*)

Biological system	Concentrations of Y-27632 (μM)	Effects	Refs.
Swiss 3T3 fibroblasts	10	Inhibition of formation of stress fibers and induction of shape change	25
	10	Inhibition of LPA-induced phosphorylation of myosin phosphatase target subunit	20
HeLa cells	10	Inhibition of Rho-induced formation of stress fibers and focal adhesions	16, 17, 43
	30	No effects on cytokinesis	9
Rat embryonic fibroblasts	20–30	Acceleration of wound closure	29
CCL39 Chinese hamster lung fibroblasts	10–30	Inhibition of agonist-induced Na^+–H^+ exchanger type 1 activation	44
MM1 rat hepatoma cells	0.2–100	Inhibition of MLC phosphorylation and cell invasion	38
Tissues and organs			
Rabbit aorta and mesenteric artery	1–100	Inhibition of phenylephrine-induced contraction and Ca^{2+} sensitization of contraction	17
Rabbit pulmonary artery	1–30	Inhibition of GTPγS- but not PDBu-induced Ca^{2+} sensitization of contraction	45
Rabbit portal vein	10–100	No inhibition of Ca^{2+}-independent contraction	24
Rabbit trachea	1–300	Inhibition of Ca^{2+}-sensitization of contraction	23, 46
Isolated liver	20	Inhibition of endothelin-induced portal vein constriction	41
Rats	30 mg/kg body weight po	Lowering blood pressure in hypertensive animals	17
	0.13 mg/hr ip	Inhibition of tumor dissemination	38

medium (DMEM) containing 10% (v/v) fetal bovine serum (FBS). The cells are then serum starved for 1 day in DMEM. Y-27632 is dissolved in sterilized distilled water at 10 mM, and 1000 volume of the stock solution is added to the culture medium. After the cells are preincubated with or without 10 μM Y-27632 for 30 min, lysophosphatidic acid (LPA) dissolved in distilled water at 10 mM is added to the cells at 5 μM and incubation is performed in the continued presence or absence of Y-27632. The cells are fixed and subjected to double staining with Texas Red–phalloidin and

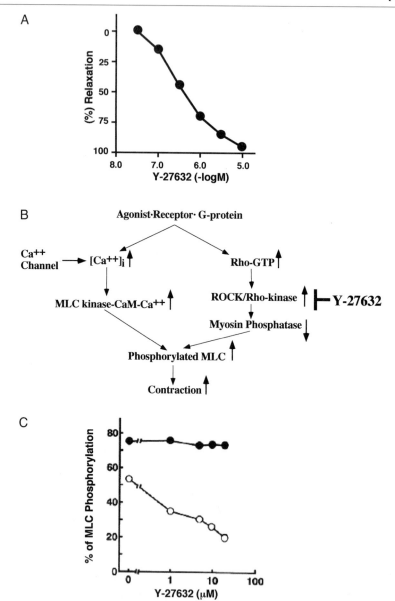

FIG. 3. (A) Concentration-dependent inhibition of smooth muscle contraction by Y-27632. A pig coronary artery was contracted with 3 μM histamine and then subjected to relaxation with various concentrations of Y-27632. (B) Site of action of Y-27632 in the signaling pathway of smooth muscle contraction. (C) Dependency of effect of Y-27632 on the strength of agonist stimulation. Washed platelets were preincubated with various concentrations of Y-27632 for

LPA treatment
(min):   0   2   10

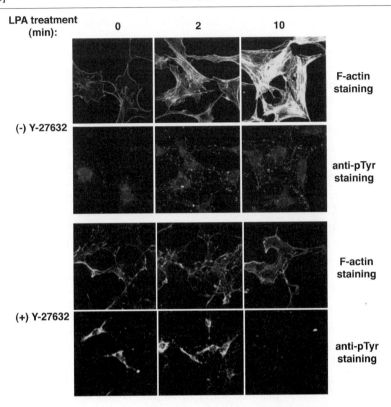

F-actin
staining

(-) Y-27632

anti-pTyr
staining

F-actin
staining

(+) Y-27632

anti-pTyr
staining

FIG. 4. Inhibition by Y-27632 of LPA-induced formation of stress fibers and focal adhesions in serum-starved Swiss 3T3 cells. Serum-starved Swiss 3T3 cells were stimulated with LPA for 0.2 and 10 min in the absence (top) and presence (bottom) of 10 μM Y-27632. Cells were fixed and stained with Texas Red–palloidin and antiphosphotyrosine (pTyr) antibody for F-actin and focal adhesions, respectively.

antiphosphotyrosine antibody 4G10 for F-actin and focal adhesions, respectively. As shown in Fig. 4, stress fibers and focal adhesions are detected 2 min after LPA stimulation and have matured by 10 min in control Swiss 3T3 cells without Y-27632. Y-27632 at 10 μM suppresses this response significantly. The Y-27632 treatment also causes outgrowth of long pro-

3 min and then stimulated with either 0.05 (○) or 0.1 (●) U/ml thrombin for 30 sec without stirring. Platelets were then lysed and subjected to SDS–PAGE for immunoblotting with anti-Ser[19]-phosphorylated MLC antibody. Reproduced from Suzuki et al.[19]

cesses in treated cells as seen in cells treated with botulinum C3 exo-enzyme.[25]

Inhibition of stress fiber formation by Y-27632 was also reported in HeLa cells expressing Val [14]Rho or an active form of ROCK,[17] cultured NIH 3T3 cells,[26,27] and human umbilical endothelial cells.[28] Y-27632 also abolished bundled actin fibers in migrating rat embryonic fibroblasts subjected to an *in vitro* wound-healing assay[29] and in attenuated striated actomyosin bundles in cultured rat cardiac myocytes.[30]

When added to cultured cells in the presence of serum, Y-27632 abolished stress fibers in 30 min after addition. However, after 24 hr of treatment, thin F-actin bundles reappeared in treated cells.[18,26] The resumed F-actin bundles were abolished by treating the cells with freshly prepared Y-27632. The Y-27632-containing medium recovered from 24 hr of culture with cells shows the decreased activity to inhibit stress fibers, whereas Y-27632 incubated with the culture medium alone retains the original potency, indicating that Y-27632 is inactivated during incubation with cultured cells.

Inhibition of LPA-Induced Neurite Retraction in N1E-115 Neuroblastoma Cells

Mouse N1E-115 neuroblastoma cells[25] are maintained in DMEM containing 10% FBS and are allowed to extend neurites by culture in serum-free DMEM for 24 hr. By this procedure, more than 90% of cells extend neurites. Neurite retraction is evoked by incubating cells with 1 μM LPA. Y-27632 dissolved in distilled water at 10 mM is added to the culture at various concentrations 30 min before the LPA addition, and neurite retraction is observed in the continued absence or presence of Y-27632. The cells are then lysed in Laemmli's SDS–PAGE sample buffer and are subjected to immunoblotting to detect phosphorylated MLC. Figure 5A shows the concentration-dependent inhibition of neurite retraction by Y-27632. The inhibition is detected at 10^{-7} M Y-27632 and is complete at 10^{-5} M with the IC_{50} value of 0.56 μM. Under this condition, Y-27632 also inhibits MLC phosphorylation in a similar concentration–inhibition manner (Fig. 5B).

[25] M. Hirose, T. Ishizaki, N. Watanabe, M. Uehata, O. Kranenburg, W. H. Moolenaar, F. Matsumura, M. Maekawa, H. Bito, and S. Narumiya, *J. Cell Biol.* **141,** 1625 (1998).

[26] E. Sahai, T. Ishizaki, S. Narumiya, and R. Treisman, *Curr. Biol.* **9,** 136 (1999).

[27] T. Matsui, S. Yonemura, S. Tsukita, and S. Tsukita, *Curr. Biol.* **9,** 1259 (1999).

[28] M. Essler, M. Retzer, M. Bauer, J. W. Heemskerk, M. Aepfelbacher, and W. Siess, *J. Biol. Chem.* **274,** 30361 (1999).

[29] C. D. Nobes and A. Hall, *J. Cell Biol.* **144,** 1235 (1999).

[30] K. Kuwahara, Y. Saito, O. Nakagawa, I. Kishimoto, M. Harada, E. Ogawa, Y. Miyamoto, I. Hamanaka, N. Kajiyama, N. Takahashi, T. Izumi, R. Kawakami, N. Tamura, Y. Ogawa, and K. Nakao, *FEBS Lett.* **452,** 314 (1999).

A

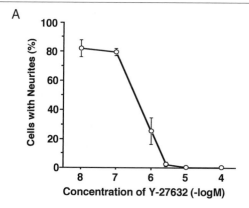

B

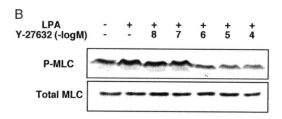

FIG. 5. (A) Concentration-dependent inhibition of neurite retraction of N1E-115 neuroblastoma cells by Y-27632. (B) Concentration-dependent inhibition of MLC phosphorylation by Y-27632 during neurite retraction of N1E-115 neuroblastoma cells. Note that the increase in the level of phosphorylated MLC by the LPA addition was decreased by Y-27632 in a concentration-dependent manner from 10^{-8} to 10^{-5} M. Reproduced from Sahai et al.[26]

This neuroblastoma cell system can be used in the identification of a novel ROCK substrate, phosphorylation of which is induced by the LPA addition and inhibited by pretreatment with Y-27632 (see, for example, Maekawa et al.[31]).

Specificity of Y-27632 Actions

In blood platelets, treatment with Y-27632 inhibits agonist-induced shape change without affecting calcium mobilization; this inhibition is accompanied by the decrease in phosphorylation of MLC as well as that of

[31] M. Maekawa, T. Ishizaki, S. Boku, N. Watanabe, A. Fujita, A. Iwamatsu, T. Obinata, K. Ohashi, K. Mizuno, and S. Narumiya, *Science* **285,** 895 (1999).

the myosin-binding subunit of myosin phosphatase.[19,32–34] Y-27632 does not inhibit the G_1–S progression of the cell cycle in Swiss 3T3 cells[18] and NIH 3T3 cells,[26] the growth of which is arrested in the G_1 phase by treatment with C3 exoenzyme.[35,36] However, inhibition of thrombin-induced DNA synthesis by Y-27632 is observed in rat aortic smooth muscle cells.[37] Among other Rho-mediated cellular processes, no inhibition by Y-27632 is found in the activation of serum response factor[26] and cytokinesis is resistant to Y-27632 treatment.[9,18] Y-27632 does not inhibit phosphorylation of ERK and JNK in NIH 3T3 cells stimulated with serum and either ultraviolet radiation or anisomysin, respectively.[26]

In Vivo Applications

Y-27632 can be administered at 30 mg/kg body weight orally per day for 10 days to intact animals without any acute toxicity. Oral administration of 30 mg/kg body weight of Y-27632 was reported to lower the blood pressure of hypertensive rats, whereas the same dose of the drug hardly affected the blood pressure of normal control rats.[17] Y-27632 was also administered continuously for 11 days by the aid of an osmotic pump at the rate of 0.13 mg per hour into the peritoneal cavity of rats without any toxicity. This treatment significantly prevented hepatoma cells implanted into the cavity from disseminating and producing tumors and nodules.[38]

[32] B. Klages, U. Brandt, M. I. Simon, G. Schultz, and S. Offermanns, J. Cell Biol. 144, 745 (1999).
[33] B. Z. Paul, J. L. Daniel, and S. P. Kunapuli, J. Biol. Chem. 274, 28293 (1999).
[34] M. Bauer, M. Retzer, J. I. Wilde, P. Maschberger, M. Essler, M. Aepfelbacher, S. P. Watson, and W. Siess, Blood 94, 1665 (1999).
[35] M. Yamamoto, N. Marui, T. Sakai, N. Morii, S. Kozaki, K. Ikai, S. Imamura, and S. Narumiya, Oncogene 8, 1449 (1993).
[36] M. F. Olson, A. Ashworth, and A. Hall, Science 269, 1270 (1995).
[37] T. M. Seasholtz, M. Majumdar, D. D. Kaplan, and J. H. Brown, Circ. Res. 84, 1186 (1999).
[38] K. Itoh, K. Yoshioka, H. Akedo, M. Uehata, T. Ishizaki, and S. Narumiya, Nature Med. 5, 221 (1999).
[39] V. Niggli, FEBS Lett. 445, 69 (1999).
[40] R. Sullivan, L. S. Price, and A. Koffer, J. Biol. Chem. 274, 38140 (1999).
[41] N. Kawada, S. Seki, T. Kuroki, and K. Kaneda, Biochem. Biophys. Res. Commun. 266, 296 (1999).
[42] B. Nilius, T. Voets, J. Prenen, H. Barth, K. Aktories, K. Kaibuchi, G. Droogmans, and J. Eggermont, J. Physiol. (Lond.) 516, 67 (1999).
[43] K. Fujisawa, P. Madaule, T. Ishizaki, G. Watanabe, H. Bito, Y. Saito, A. Hall, and S. Narumiya, J. Biol. Chem. 273, 18943 (1998).
[44] T. Tominaga, T. Ishizaki, S. Narumiya, and D. L. Barber, EMBO J. 17, 4712 (1998).
[45] X. Fu, M. C. Gong, T. Jia, A. V. Somlyo, and A. P. Somlyo, FEBS Lett. 440, 183 (1998).
[46] A. Yoshii, K. Iizuka, K. Dobashi, T. Horie, T. Harada, T. Nakazawa, and M. Mori, Am. J. Respir. Cell Mol. Biol. 20, 1190 (1999).

[25] Inducible Membrane Recruitment of Small GTP-Binding Proteins by Rapamycin-Based System in Living Cells

By FLAVIA CASTELLANO and PHILIPPE CHAVRIER

Introduction

The regulation of actin cytoskeleton dynamics in response to extracellular stimuli is mediated by members of the Rho GTP-binding protein family, including Rho, Cdc42, and Rac, which control stress fiber, filopodia, and lamellipodia, formation, respectively.[1] Activation of Rho GTP-binding proteins is promoted by guanine-nucleotide exchange factors (GEFs), which catalyze the replacement of bound GDP with GTP. It is generally assumed that signaling leads to GEF activation and the accumulation of activated GTP-bound Rho proteins at the membrane. Actin filament reorganization is then regulated by specific effector proteins that bind to and are activated by GTP-bound Rho proteins.[1]

Most of the studies analyzing the role of Rho proteins in the regulation of actin reorganization have used either expression of dominant inhibitory or constitutively active Rho proteins on microinjection/transfection into recipient cells[2,3] or, more recently, have attempted to reconstitute actin filament assembly on the addition of recombinant Rho GTP-binding proteins in cell-free assays.[4] While the first experimental approach relies on proteins that may be abnormally localized or temporally regulated, the second may oversimplify the analysis, perhaps omitting important components required for actin filament homeostasis.

An approach has been developed to induce the translocation of various proteins to the plasma membrane, to activate specific signal transduction pathways, and to induce protein–protein interactions in vivo. This system exploits the ability of certain natural cell-permeable compounds to induce

---

[1] L. Van Aelst and C. D'Souza-Schorey, Genes. Dev. 11, 2295 (1997).
[2] A. J. Ridley and A. Hall, Cell 70, 389 (1992).
[3] A. J. Ridley, H. F. Paterson, C. L. Johnston, D. Diekmann, and A. Hall, Cell 70, 401 (1992).
[4] S. H. Zigmond, M. Joyce, J. Borleis, G. M. Bokoch, and P. N. Devreotes, J. Cell Biol. 138, 363 (1997).

protein dimerization.[5,6] One such compound, the immunosuppressant rapamycin, binds with high affinity to intracellular receptors comprising the FK506-binding protein family, including FKBP12.[7] FKBP12-bound rapamycin interacts with another protein called FRAP through its 11-kDa domain called FRB[8] and inhibits the kinase activity of FRAP, resulting in cell cycle arrest.[9] In this tripartite complex, rapamycin acts as an adaptor to join FKBP to FRB.[6]

We have modified this system to permit the controlled and local recruitment of activated Rho proteins to discrete sites underneath the plasma membrane mimicking the activation step of Rho GTP-binding proteins by guanine exchange factors. We have constructed chimeric Rho proteins containing the FRB domain that, on the addition of rapamycin, are recruited to FKBP domains expressed as the cytoplasmic region of an engineered transmembrane receptor, which includes the extracellular domain of CD25. A further element of this system has been to analyze the effect of the local enrichment of the Rho chimeras at discrete sites of the plasma membrane on FKBP receptor clustering using anti-CD25 antibodies. Using this approach, we have found that the recruitment of Cdc42 to the plasma membrane elicits the formation of actin-based membrane protrusions.[10]

Construction of Chimeric Proteins

Rat basophil leukemia (RBL-2H3) cells are transfected with a vector encoding the membrane receptor CD25-FKBP$_2$ that consists of two tandemly repeated rapamycin-binding FKBP domains as a cytoplasmic region, which are fused with the extracellular and transmembrane regions of human CD25 (Fig. 1a). After selection, individual clones are screened for the expression of CD25-FKBP$_2$ by fluorescence activated cell sorting (FACS) analysis with anti-CD25 antibodies (not shown). One of the selected clones (clone 15B[10]) is subsequently transfected with a construct expressing consti-

[5] D. M. Spencer, T. J. Wandless, S. L. Schreiber, and G. R. Crabtree, Science **262**, 1019 (1993).

[6] V. M. Rivera, T. Clackson, S. Natesan, R. Pollock, J. F. Amara, T. Keenan, S. R. Magari, T. Phillips, N. L. Courage, F. Cerasoli, Jr., D. A. Holt, and M. Gilman, Nature Med. **2**, 1028 (1996).

[7] B. E. Bierer, P. S. Mattila, R. F. Standaert, L. A. Herzenberg, S. J. Burakoff, G. Crabtree, and S. L. Schreiber, Proc. Natl. Acad. Sci. U.S.A. **87**, 9231 (1990).

[8] J. Chen, X. F. Zheng, E. J. Brown, and S. L. Schreiber, Proc. Natl. Acad. Sci. U.S.A. **92**, 4947 (1995).

[9] E. J. Brown, P. A. Beal, C. T. Keith, J. Chen, T. B. Shin, and S. L. Schreiber, Nature **377**, 441 (1995).

[10] F. Castellano, P. Montcourrier, J. C. Guillemot, E. Gouin, L. M. Machesky, P. Cossart, and P. Chavrier, Curr. Biol. **9**, 351 (1999).

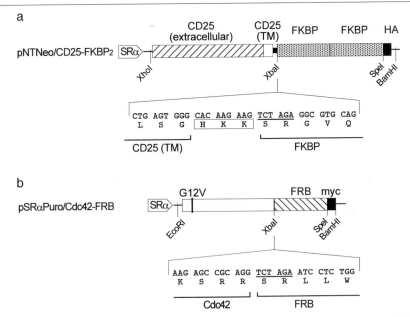

FIG. 1. Construction of FKBP and FRB domain-containing chimeras. (a) CD25-FKBP$_2$ consists of CD25 extracellular and transmembrane (TM) domains fused to two tandemly repeated copies of the rapamycin-binding protein FKBP12 (FKBP) and the HA epitope as the cytoplasmic region. The sequence HKK (single letter code), which has been added at the junction between the CD25 TM domain and FKBP, is boxed. (b) In Cdc42V12-FRB, the Cdc42V12-encoding sequence ends at R185 and the carboxy-terminal membrane-anchoring motif (CAAX box) of Cdc42 is replaced by the FKBP/rapamycin-binding domain (FRB) followed by a myc epitope. *Xba*I restriction sites used during the construction of the chimeras are underlined. Srα, Srα promoter region.

tutively active Cdc42V12 fused to the FRB domain (Fig. 1b). The FRB domain is inserted in place of the carboxy-terminal membrane-anchoring motif of Cdc42 in order to generate a chimeric protein that is expected to accumulate in the cytosol as a nonfunctional product.

Chimera Preparation

Oligonucleotides 5'-acagcggaattcctcgaggccaccatggattcatacctgctg-3' and 5'-ggtacctctagacttcttgtgcccactcaggaggag-3' are used to amplify the extracellular and transmembrane domains of human CD25 adding the sequence HKK (single letter code) at the intracellular domain junction. The polymerase chain reaction (PCR) product is digested with *Eco*RI and *Xba*I (sites are underlined in the oligo nucleotides) and ligated in a three-way reaction with an *Xba*I–*Bam*HI fragment derived from plasmid pCMF2E (kindly provided by ARIAD Pharmaceuticals Inc., Boston, MA) containing two tandem copies of FKBP C-terminally tagged with an HA epitope (FKBP$_2$-HA) into pGEM1 restricted with EcoRI and BamHI (Promega, Madison, WI). The CD25-FKBP$_2$ insert is removed as an XhoI–BamHI fragment and is subcloned into the mammalian expression vector pNTNeo, which bears a neomycin resistance gene (a gift of Dr. B. Malissen, CIML, Marseille, France) (Fig. 1a). The primers 5'-acacgcgaattcgccaccatggagacaat-tagtgc-3' (EcoRI site is underlined) and 5'-caggcgtctagacctgcggctcttcttcgg-3' (XbaI site is underlined) are used to PCR amplify the GTPase-defective V12 allele of Cdc42 while removing the C-terminal CAAX box. The FRB chimera is assembled in pGEM1 restricted with EcoRI and BamHI by ligating the EcoRI–XbaI Cdc42V12 PCR fragment in a three-way reaction with an XbaI–BamHI FRB fragment [from plasmid pCGNN-FRB(B), provided by ARIAD Pharmaceutical Inc.] into the vector. A myc tag obtained by hybridizing the two oligonucleotides—5'ctagtgaacaaaaactcatct-cagaagaggatctgtaatgag-3' and 5'-gatccacattacagatcctcttctgagatgagttttttgttca-3'(SpeI and BamHI compatible ends are underlined)—is inserted as an SpeI–BamHI fragment at the carboxy terminus of the Cdc42-FRB construct. The sequence encoding the Cdc42V12-FRBmyc chimera is subcloned as an EcoRI–BamHI fragment into the mammalian expression vector pSRαPuro (a gift of Dr. C. Bonnerot, Institut Curie, Paris, France) bearing a puromycin resistance gene. A schematic representation of the Cdc42V12-FRB vector is shown in Fig. 1b.

Cell Culture and Derivation of Stable Transformants

RBL-2H3 cells (a gift of Dr. M. Daëron, Institut Curie, Paris, France) are cultured in Dulbecco's modified Eagle's medium (DMEM) supplemented with 10% (v/v) fetal calf serum (FCS) at 37° in a 6% (v/v) CO$_2$ atmosphere. Electroporation is carried out as follows. Cells are cultured for 24 hr in DMEM/10% FCS, trypsinized, and resuspended at a concentration of 4×10^7 cells/ml in DMEM supplemented with 20% FCS and 20 mM HEPES, pH 7.2. Cells (10^7) are incubated with 20 μg of ScaI-linearized pNTneo-CD25-FKBP$_2$ vector for 10 min at room temperature and pulsed at 250 V and 500 μF with a Bio-Rad (Hercules, CA) GenePulser. After a further 10-min incubation at room temperature, cells are seeded in a 75-cm^2 flask in the same medium as described earlier. After 24 hr, cells are subjected to selection in medium containing 1 mg/ml G418 (GIBCO-BRL Gaithersburg, MD). Resistant clones are picked following trypsinization, expanded, and tested for CD25-FKBP$_2$ expression by immunofluorescence and immunoblot analysis with an anti-HA antibody (clone 3F10, Boehringer

Mannheim, Meylan, France) and FACS analysis with a fluorescein isothiocyanate (FITC)-conjugated antihuman CD25 antibody (Coulter-Immunotech). Clone 15B[10] expressing CD25-FKBP$_2$ is chosen for further transfection of the FRB chimera following the same transfection protocol. Doubly transfected clones are selected in DMEM containing both 1 mg/ml G418 and 1 μg/ml puromycin (Sigma, St. Louis, MO) and are tested for expression of the FRB chimera by immunofluorescence and immunoblot analysis using an anti-myc tag antibody (clone 9E10). A clone expressing both CD25-FKBP$_2$ and Cdc42V12-FRB (clone 15BH7) is used for further studies.[10]

Regulation of Protein–Protein Interaction by Rapamycin

The physical interaction between FKBP- and FRB-containing chimeras is investigated by surface immunoprecipitation of CD25-FKBP$_2$ receptors using anti-CD25 antibodies. Cells (7.5×10^5) in a 9-cm^2 well (six-well plate) are incubated overnight in medium supplemented with 100 nM rapamycin [from a 10 μM stock solution in dimethyl sulfoxide (DMSO)] or an equivalent volume of DMSO as a control, followed by a 1-hr incubation in medium containing antihuman CD25 rat monoclonal antibody (5 μg/ml, clone 33B3.1, Coulter-Immunotech, Marseille, France) at 37°. After extensive washing, 4.5-μm-diameter magnetic beads coated with sheep antirat immunoglobulin G (IgG) (1.5×10^7 beads/well in 0.5 ml medium) (Dynal, Oslo, Norway) are added to the cells, followed by an incubation of 20 min on ice and then 30 min at 37°. After a wash in phosphate-buffered saline (PBS), cells are lysed in 0.5 ml ice-cold PINT buffer (50 mM Tris–HCl, pH 7.5, 150 mM NaCl, 2 mM EDTA) supplemented with 0.5% Triton X-100 and protease inhibitors (complete cocktail, Boehringer Mannheim). The magnetic beads are separated from the lysate using a magnet (Magnetic Particle Concentrator MPC-2, Dynal). The beads are washed by three cycles of resuspension into ice-cold PINT buffer and magnetic separation. Proteins in the bead fraction are eluted by resuspension in sample buffer, separated by SDS–PAGE, and transferred to a polyvinylidene difluoride (PVDF) membrane (Millipore, Bedford, MA) for Western blotting. CD25-FKBP$_2$ is detected with an anti-HA rat monoclonal antibody (clone 3F10), and Cdc42V12-FRB is detected using an anti-myc tag mouse monoclonal antibody (clone 9E10). Western blots are developed using HRP-conjugated antibodies (Sigma) and the ECL (enhanced chemiluminescence) revelation system (Amersham, Buckinghamshire, UK). As shown in Fig. 2, association is detected only on treatment of 15BH7 cells with 100 nM rapamycin (lane 4), demonstrating that rapamycin triggered the translocation of cytosolic Cdc42V12-FRB to membrane-expressed CD25-FKBP$_2$ receptors.

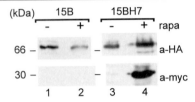

FIG. 2. Rapamycin-mediated association of Cdc42V12-FRB with CD25-FKBP$_2$. Intracellular association of Cdc42V12-FRB with CD25-FKBP$_2$ was tested in an anti-CD25 surface coimmunoprecipitation assay using stably transformed RBL-2H3 cells cultured in the absence (lanes 1 and 3) or presence of rapamycin (lanes 2 and 4). Immunoprecipitates were analyzed by immunoblotting with anti-HA and anti-myc antibodies to detect HA-tagged CD25-FKBP$_2$ and myc-tagged Cdc42V12-FRB, respectively. 15B, CD25-FKBP$_2$-expressing cells. 15BH7, CD25-FKBP$_2$- and Cdc42V12-FRB-coexpressing cells. Modified from Castellano et al.[10] with permission from Elsevier Science.

Cytoskeletal Reorganization on Rapamycin-Mediated Membrane Recruitment of Activated Cdc42

15BH7 cells grown on glass coverslips are treated for 16 hr with 100 nM rapamycin in conditions sufficient to induce protein–protein interaction, as shown earlier, and actin distribution is analyzed by immunofluorescence microscopy. Under these conditions, we did not observe substantial changes in F-actin distribution (data not shown).[10] Therefore, experimental conditions aimed at increasing the local concentration of activated Cdc42 at discrete sites of the plasma membrane by CD25-FKBP$_2$ receptor clustering are tested. Clustering of the CD25-FKBP chimera is obtained either by incubating cells with anti-CD25 antibody-coated beads or by plating cells on anti-CD25 antibody-coated glass coverslips.

Bead-Clustering System

On clustering of FKBP receptors using anti-CD25 antibody-coated beads and the addition of rapamycin to the medium, we anticipated that we could trigger the recruitment and concentration of activated Cdc42 to the FKBP receptor. As depicted in Fig. 3, the rapamycin-controlled protein–protein interaction would then be expected to result in a local enrichment of activated Cdc42 at the plasma membrane at the bead–membrane interface. 15BH7 cells (2×10^5) are seeded onto 12-mm-diameter glass coverslips the day prior to the experiment. Cells are incubated for different times (2 to 16 hr) in medium supplemented with DMSO (as control) or with increasing concentrations of rapamycin (0.1 to 100 nM) and then for an additional 1 hr at 37° in 0.5 ml of medium supplemented with 2.5 μg/ml of biotinylated antihuman CD25 mouse monoclonal anti-

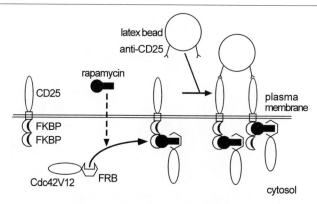

FIG. 3. Model of inducible recruitment of soluble Cdc42V12 to a membrane-localized docking protein by rapamycin. On the addition of membrane-permeable rapamycin, which acts as an heterodimerizer due to its ability to bind both FKBP and FRB domains, Cdc42V12 translocates from the cytosol to the plasma membrane protein CD25-FKBP$_2$. The concentration of CD25-FKBP$_2$, at the plasma membrane is increased locally by aggregation with anti-CD25-coated beads. Modified from Castellano et al.[10] with permission from Elsevier Science.

body (clone B.1.49.9, Coulter-Immunotech). After extensive washing to remove unbound antibody, 1-μm green fluorescent streptavidin-labeled latex beads (Sigma) are added to the coverslips (approximately 10 beads per cell) and spun onto the cells at 400g for 2 min at 4° followed by an incubation on ice for 20 min. Excess beads are eliminated by washing at 4° and prewarmed medium at 37° is added to the cells. The coverslips are incubated at 37° for 1 hr before immunofluorescence analysis. Cells are then washed once with PBS and fixed in 3% (v/v) paraformaldehyde in PBS for 20 min. After quenching the free aldehyde groups with 50 mM NH$_4$Cl in PBS for 10 min, cells are permeabilized with 0.1% Triton X-100 for 4 min. After blocking in 10% horse serum in PBS, F-actin is stained with Texas Red (TR)-conjugated phalloidin (Molecular Probes, Eugene, OR). Confocal microscopy is performed with a Leica TCS 4D confocal microscope equipped with a mixed gas argon/krypton laser (Leica Laser Teknik, Heidelberg, Germany). As shown in Fig. 4, when 15BH7 cells treated for 16 hr with 100 nM rapamycin are incubated with anti-CD25 antibody-coated beads, actin filaments appeared to polymerize within ~5-μm long membrane protrusions with a bead at their tip (Fig. 4e). Shorter incubation times such as 2 or 8 hr in rapamycin-supplemented medium are sufficient for the induction of membrane protrusions (Figs. 4c and 4d, respectively). Induction of membrane protrusions by rapamycin is dose dependent, requiring concentrations of rapamycin as low as 1 nM (Fig. 4b). These structures could also be induced on clustering of CD25-FKBP$_2$

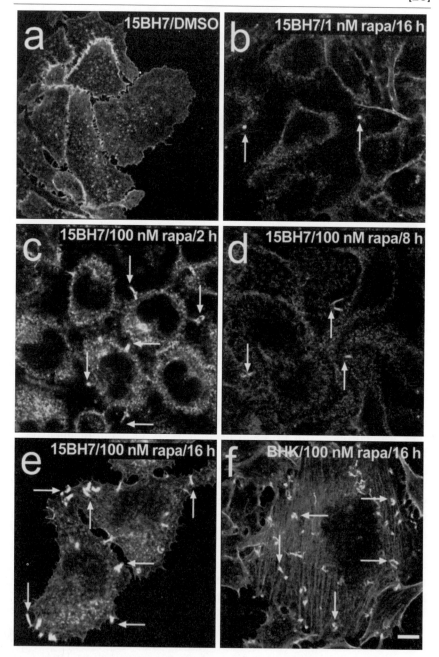

receptors prior to the recruitment of Cdc42-FRB by rapamycin (not shown). Finally, we investigated whether actin-based membrane protrusions could also be obtained in transiently transfected cells. Baby hamster kidney (BHK) cells (10^5 cells plated on a 12-mm-diameter coverslip in a 2-cm^2 well of a 24-well plate) are transfected with pNTNeo/CD25-FKBP$_2$ and pSRα/Cdc42V12-FRB (0.25 μg each) using 1.5 μl of Lipofectamine (GIBCO-BRL) following the manufacturer's conditions. After 4 hr, transfected cells are incubated with 100 nM rapamycin for 16 hr, and CD25-FKBP$_2$ receptors are clustered using anti-CD25 antibody-coated beads as described earlier. Figure 4f shows the induction of actin-rich membrane protrusions at the bead–membrane interface of transfected BHK cells, demonstrating that this system may also be employed under transient transfection conditions.

Anti-CD25 Antibody-Coating Experiments

Engagement of the high-affinity IgE receptor (FcεRI) on RBL-2H3 cells induces membrane ruffling at the dorsal surface of the cells, whereas it triggers actin-plaque assembly at the cell–substratum interface.[11] We have demonstrated that the former response is controlled by Rac1, whereas the latter is regulated by Cdc42.[12] Therefore, we investigated whether increasing the local concentration of Cdc42V12-FRB at the cell–substratum interface may be sufficient to induce cytoskeletal reorganization. Glass coverslips are incubated for 2–4 hr in a solution of antihuman CD25 mouse monoclonal antibody (clone B.1.49.9) in PBS (10 μg/ml) and then washed once in PBS to remove unbound antibody. 15BH7 cells (2 × 10^5) are seeded onto anti-CD25 antibody-coated coverslips and, after 4–24 hr at 37°, cells

[11] J. R. Pfeiffer, G. G. Deanin, J. C. Seagrave, B. H. Davis, and J. M. Oliver, *J. Cell Biol.* **101**, 2145 (1985).

[12] J. C. Guillemot, P. Montcourrier, E. Vivier, J. Davoust, and P. Chavrier, *J. Cell Sci.* **110**, 2215 (1997).

FIG. 4. Rapamycin-mediated recruitment of Cdc42V12 at the plasma membrane promotes the formation of actin-based membrane protrusions. 15BH7 cells grown on glass coverslips (a–e) were treated with DMSO as control (a) or with either 1 (b) or 100 (c–e) nM rapamycin for 2 (c), 8 (d), or 16 (b and e) hr. Cells were then incubated with biotinylated anti-CD25 antibody, and 1-μm green fluorescent streptavidin-labeled latex beads were added, followed by another incubation of 1 hr at 37°. Finally, cells were processed for immunofluorescence analysis and F-actin labeling with Texas Red-conjugated phalloidin. (f) BHK cells were cotransfected with the CD25-FKBP$_2$ and Cdc42V12-FRB constructs under transient conditions, incubated with 100 nM rapamycin for 16 hr, and CD25-FKBP$_2$ receptors were clustered with anti-CD25 antibodies and latex beads as described. For clarity, the position of beads visualized in the green channel is indicated by arrows. Bar: 10 μm.

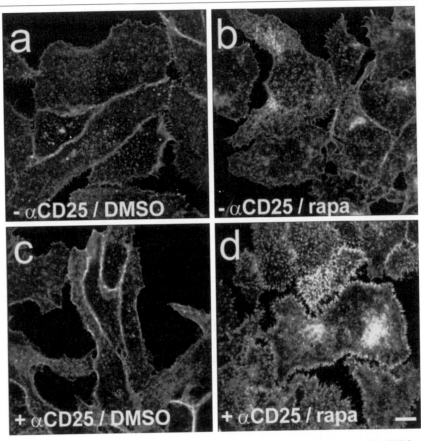

FIG. 5. Actin polymerization induced by rapamycin-mediated recuitment of Cdc42V12 to the cell–substratum interface. 15BH7 cells were plated on naked glass coverslips (a and b) or on glass coverslips coated with antihuman CD25 antibodies (c and d). Cells were then treated with rapamycin for 16 hr (b and d) or DMSO as a control (a and c), fixed, and labeled with Texas Red-conjugated phalloidin to visualize F-actin distribution by confocal microscopy. Optical sections were recorded at the ventral surface of the cells. Bar: 10 μm.

are treated with 100 nM rapamycin (or DMSO) for 16 hr before staining for F-actin as described earlier. Anti-CD25 antibody coating did not visibly alter F-actin distribution at the ventral surface of control (DMSO-treated) cells (compare Figs. 5a and 5c). In contrast, rapamycin treatment of 15BH7 cells plated on anti-CD25 antibody-coated coverslips triggered the formation of numerous F-actin-rich foci spread throughout the ventral surface as well as microspikes at the cell periphery resembling actin-plaque assembly (Fig. 5d). These structures are dependent on the expression of the

Cdc42V12-FRB chimera as they were not observed in 15B cells under the same conditions (not shown).

Conclusion

The system described in this article provides an alternative and powerful approach for assessing the role of Rho family GTPases in intact cells. Here, the rapamycin system was applied to the study of the role of Cdc42 in actin polymerization. This system should be easily applicable to any protein, such as members of the Ras superfamily or their downstream effectors,[10] whose function depends on plasma membrane localization. Indeed, the application of this approach to the study of Rac1 function has revealed an important role for this protein in phagocytic uptake (F. Castellano, P. Montcourrier, and P. Chavrier, unpublished).

Acknowledgments

We are indebted to Dr. V. M. Rivera and ARIAD Pharmaceuticals, Inc. for providing for FKBP and FRB encoding cDNAs. F.C. was a recipient of a NATO fellowship. This work was supported by INSERM and CNRS institutional fundings.

[26] Expression of Rho GTPases Using Retroviral Vectors

By Frits Michiels, Rob A. van der Kammen, Lennert Janssen, Garry Nolan, and John G. Collard

Introduction

The introduction of wild-type or mutant Rho-like GTPases into cells or tissues has facilitated the study of these proteins in diverse cellular processes. Proteins can be transiently administered by direct injection of purified proteins or by injection or transfection of constructs that encode these proteins. Stable expression can be achieved by transfection of constructs, followed by selection of positive cells using a selection marker. Each of these methods has its limitations, and the method of choice will strongly depend on the cells and processes that one wants to study.

Several groups have applied viral vectors for efficient transduction of Rho-like GTPases.[1,2] Two different types of viral vectors are frequently

---

[1] D. J. Sulciner, K. Irani, Z. X. Yu, V. J. Ferrans, P. Goldschmidt-Clermont, and T. Finkel, *J. Cell Biol.* **16,** 7115 (1996).
[2] P. L. Hordijk, J. P. ten Klooster, R. A. van der Kammen, F. Michiels, L. C. Oomen, and J. G. Collard, *Science* **278,** 1464 (1997).

employed: adenoviruses and retroviruses. The high transduction efficiencies that can routinely be obtained using these vectors compensate for the somewhat more work that is required for their construction. Advantages of adenovirus-based vectors are that they can infect almost any cell type with high efficiency and that they infect proliferating as well as resting cells. Disadvantages are that they can have an influence on the processes that one wants to study and their expression is always transient. Moloney murine leukemia virus-derived retroviruses can similarly infect a wide variety of cell lines, leading to stable expression, because the proviral DNA is integrated into the genome. However, the latter only occurs in dividing cells, making this system not applicable for nonproliferating primary cells.

In our laboratory, we have successfully implemented second-generation retroviral vectors for the establishment of stable cell lines of different cell types.[2–7] The main advantage is that virtually all cell types can be infected. Furthermore, large amounts of retroviral stocks can be made and used to infect the cell lines of interest. These novel retroviral plasmids facilitate the cloning of cDNA constructs and allow the generation of directionally cloned cDNA libraries. Selection markers that are embedded in the retroviral constructs can be used to generate stable expressing cell lines. Different selectable genes allow the introduction of different genes into the same cells. This article describes the features of these plasmids and describes in detail how they are used for the generation of stable expressing cells.

Design of Retroviral Plasmids

The basis of our retroviral plasmids is the LZRSpBMN-LacZ vector.[8] In this Moloney murine leukemia virus (MMLV)-based plasmid, the env-pol sequences are replaced by the LacZ gene. Furthermore, the EBNA-1 gene and oriP origin of replication of Epstein–Barr virus are present in the plasmid backbone. These allow episomal replication of the plasmid in the packaging cells. In this way, a high-titer supernatant can be generated

[3] E. E. Sander, S. van Delft, J. P. ten Klooster, T. Reid, R. A. van der Kammen, F. Michiels, and J. G. Collard, *J. Cell Biol.* **30,** 1385 (1998).

[4] F. N. van Leeuwen, H. E. Kain, R. A. van der Kammen, F. Michiels, O. W. Kranenburg, and J. G. Collard, *J. Cell Biol.* **139,** 797 (1997).

[5] J. C. Stam, F. Michiels, R. A. van der Kammen, W. H. Moolenaar, and J. G. Collard, *EMBO J.* **15,** 4066 (1998).

[6] E. E. Sander, J.-P. ten Klooster, S. van Delft, R. A. van der Kammen, and J. G. Collard, *J. Cell Biol.* **147,** 1009 (1999).

[7] F. N. van Leeuwen, S. van Delft, H. E. Kain, R. A. van der Kammen, and J. G. Collard, *Nature Cell Biol.* **1,** 242 (1999).

[8] T. M. Kinsella and G. P. Nolan, *Hum. Gene Ther.* **7,** 1405 (1996).

after transient transfection of the packaging cells. Transfected cells can be selected because a puromycin resistance gene, under the control of the phosphoglycerol kinase-1 promoter, is also present in the plasmid backbone. Culturing of the transfected packaging cells in the presence of puromycin usually generates a high-titer supernatant. However, some constructs could not be enriched in this way as the expressed proteins appear to be toxic for the packaging cells.

The vector was adapted by replacing the LacZ gene for a sequence that contains a multiple cloning site (see Fig. 1A), followed by an internal ribosome entry site (IRES) sequence and the neomycin resistance gene. Because the Ψ-packaging signal, which contains ATGs, is not properly spliced out in most cell lines, we also inserted multiple stop codons in all three reading frames in front of the polylinker. The resulting plasmid, pLZRS-IRES-NEO, can be used for the insertion of cDNAs in the poly-linker, provided that the cDNA contains its own ATG translation start codon. The maximum length of cDNA that we have inserted in this vector is 4.8 kb, but presumably larger cDNAs can also be cloned.

To allow selection of double-infected cells, we replaced the neomycin resistance gene in pLZRS-IRES-NEO with the zeocin resistance marker, resulting in pLZRS-IRES-ZEO. Both pLZRS-IRES-NEO and pLZRS-IRES-ZEO function well in expressing the desired cDNA in the target cells. We have also constructed plasmids that contain the LacZ gene or the enhanced green fluorescent protein (eGFP) gene instead of the neomycin or zeocin gene. However, the LacZ vector gave a strongly reduced expression level of the cDNA of interest in target cells compared to pLZRS-IRES-ZEO/NEO vectors. Similarly, pLZRS-IRES-GFP also showed reduced expression levels for some, but not all, constructs. The reason for this is not known at this time.

Because some of the constructs that we generated over the past were unstable in *Escherichia coli,* we also replaced the high-copy number origin of replication in these plasmids for the ColE1 origin of pBR 322, resulting in pLZRS-IRES-NEO/PBR (see Fig. 1B) and pLZRS-IRES-ZEO/PBR (see Fig. 1C). The latter appeared indeed to be more stable in *E. coli.*

Transfection and Virus Collection

Retroviral plasmids are transfected in Phoenix (φNX) packaging cells.[9] These cells express the gag/pol and env proteins from two different con-structs, thus minimizing the chance of generating replication-competent retroviruses. Furthermore, the presence of a selection marker on each

---

[9] G. P. Nolan and A. R. Shatzman, *Curr. Opin. Biotechnol.* **9,** 447 (1998).

A

BamHI   SwaI   EcoRI   XhoI
      BstBI   SfiI   AscI   NotI   SnaBI

t a a g t a a g t a g g a t c c a t t t a a a t t c g a a t t c c t g c a g g c c t c g a g g g g c c g c g g c c g c g g c c g c t a c g t a

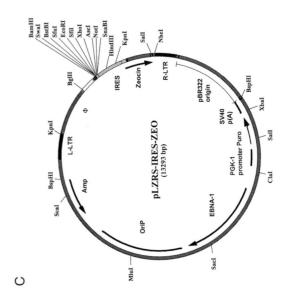

C

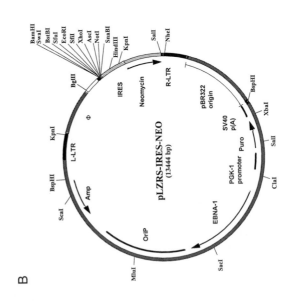

B

of these constructs allows the selection of cells that have retained both (see later).

Ecotrophic packaging cells (ϕNX-E) produce retrovirus that can infect only rodent cells, whereas viruses produced in amphotrophic (ϕNX-A) packaging cells can infect cells from different origins. A few rules should be taken into account when culturing ϕNX cells. First, every few months the cells should be cultured in the presence of diphtheria toxin (2μg/ml) and hygromycin (200 μg/ml) to select for the env and gag-pol plasmids, respectively. Second, because gag-pol is expressed as an IRES-CD8 surface marker, cells can be monitored by fluorescence activated cell sorting (FACS) analysis using anti-CD8 antibodies. Third, ϕNX cells do not adhere very well to tissue petri dishes. Care must be taken that trypsinization is as short as possible and that the cells do not grow under confluent conditions.

Transfection of packaging cells is routinely performed using standard calcium phosphate coprecipitation. For this, 2.5×10^6 ϕNX cells are seeded in 6.0 cm dishes in Dulbecco's modified Eagle's medium (DMEM) supplemented with 10% (v/v) fetal calf serum (FCS), 2 mM L-glutamine, and 100 U each of penicillin and streptomycin (culture medium). It is important that the cells are not clumped and that the dishes are approximately 80% confluent the day of transfection. The next day, medium is replaced by 3 ml of culture medium containing 25 μM chloroquine. Ten micrograms of plasmid is mixed with 62 μl of 2.0 M CaCl$_2$ and water to a final volume of 500 μl. Next, 500 μl of 2× HBS (50 mM HEPES, pH 7.5, 10 mM KCl, 12 mM dextrose, 280 mM NaCl, 1.5 mM Na$_2$HPO$_4$) is added dropwise by bubbling, and the resulting mixture is immediately layered onto the cells. Transfection is allowed to proceed for at least 6 hr, with a maximum of 10 hr, after which the medium is withdrawn. Cells are gently washed with phosphate-buffered saline (PBS) and cultured further in the presence of 5 ml of culture medium.

Twenty-four hours after transfection, one can start to collect the retrovirus-containing supernatant by replacing the medium with 3 ml of fresh

---

FIG. 1. Retroviral plasmids. (A) The polylinker is shown that was inserted into all of the pLZRS-derived plasmids. Recognition sequences that are unique to most of the retroviral plasmids are indicated. The only exception is *Eco*RI, which is also present in the LacZ sequences of pLZRS-IRES-LacZ. The polylinker is preceded by stop codons in all three reading frames to prevent translational read through from the gag sequences. (B and C) Maps of the pLZRS-IRES-NEO/pBR and pLZRS-IRES-ZEO/pBR plasmids, respectively. The indicated plasmids contain the pBR322-derived bacterial origin of replication, which ensures better stability in *E. coli.* pLZRS-IRES-NEO and pLZRS-IRES-ZEO plasmids containing a high copy number origin of replication have also been generated. pLZRS-IRES-LacZ and pLZRS-IRES-eGFP have only been constructed as high-copy number plasmids.

culture medium, followed by a 24-hr collection period, which is usually sufficient for a high-titer supernatant. Subsequently, the culture medium is gently withdrawn from the dish, centrifuged at 3000g for 10 min at 4°, and aliquoted in portions of 1.0 ml. Aliquots can be stored at −80° and are stable for at least 6 months.

Alternatively, one can start to select for transfected cells by trypsinizing the cells and transferring them to a T25 culture dish 24 hr after transfection. Culture medium, containing 1 μg/ml of puromycin, is added and selection is allowed to proceed for at least 8 days with a maximum of 30 days while regularly changing the medium. Take care that the cells do not become confluent. If this is the case, transfer them to a T75 culture dish. After this period, cells are plated in 6-cm dishes at 2.5 × 10^6 cells per dish in culture medium without puromycin. The next day, the medium is replaced with 3 ml fresh culture medium. Twenty-four hours later, one can again collect the viral supernatants as described earlier. We routinely collect the viruses twice from these plates using intervals of 24 hr. Supernatants are centrifuged at 3000g for 10 min at 4° and stored in aliquots at −80°. Please note that some constructs, such as (N43)R-RAS and N17Rac1 and V12Cdc42, are rather toxic for the packaging cells. In that case, retrovirus-containing supernatants can be obtained only from transiently transfected packaging cells.

Because the packaging cells express the transfected cDNA, some of the constructs will morphologically affect the ϕNX cells. Both active Tiam1 and V12Rac1 will induce the generation of cell–cell contacts, leading to the formation of islands of cells. Alternatively, expression can be monitored in the packaging cells by Western blot analysis, using protein- or tag-specific antibodies.

Infection Protocol

Prior to infection, the desired number of viral aliquots is thawed on ice. Subsequently, 1% (v/v) of Dotap (N-(1-[2,3-dioleoyloxy]propyl)-N,N,N-trimethylammonium; Roche, Mannheim, Germany) is added to the viral supernatants and the mixture is left on ice for 10 min. Alternatively, Polybrene (hexadimethrine bromide; Sigma, St. Louis, MO) can be added from a stock solution of 1 M to a final concentration of 3 mM. Polybrene, however, is more toxic for cells than Dotap. This, however, depends on the cell lines used and has to be determined in advance whether a particular cell line is sensitive to polybrene.

Infection is performed by adding the retrovirus-containing supernatants/Dotap mixture to the target cells. Adhering target cells should be seeded 1 day in advance in a 6-well dish at 20,000 cells per well. Supernatant (0.5–1.0 ml) is added to the cells and incubated for 1–4 hr. After this period, the supernatant is removed and replaced by fresh culture medium. Suspension cells are either added directly to the supernatant in a 24-well plate at a density of 20,000–100,000 cells per well or centrifuged for 3 hr at 32° at 1600 rpm in the viral supernatants. After infection, cells are centrifuged again and resuspended in fresh culture medium. We routinely take the empty vector along as a control for further experiments.

The efficiency of retroviral transduction has to be determined empirically. NIH 3T3 cells can be infected very well, and normally 0.5 ml of retrovirus-containing supernatant is sufficient to infect 20,000 NIH 3T3 cells with an efficiency of nearly 100%, allowing the use of this system for transient assays. Madin–Darby canine kidney (MDCK) cells are more difficult to infect, and 1 ml of supernatant for 20,000 cells will give an efficiency of about 40%. Mouse BW5147 T lymphoma cells are even more difficult to infect, leading to an efficiency of roughly 20% using the indicated amounts of viruses and cells (see Table I). Transduction efficiencies can be determined by infecting target cells with a retrovirus carrying the LacZ

TABLE I

EFFICIENCY OF RETROVIRAL TRANSDUCTION OF GENES IN DIFFERENT CELL TYPES

Cells	Source	Efficiency (%)	Reference
NIH 3T3	Mouse fibroblasts	~100	Sander et al.[6]
Swiss 3T3	Mouse fibroblasts	~100	R. van der Kammen, unpublished results
BW5147	Mouse T lymphoma	~20	Stam et al.[5]
Jurkat	Acute T-cell leukemia	~50	R. van der Kammen, unpublished results
K562	Chronic erythroleukemia	~25–50	R. van der Kammen, unpublished results
MDCK	Madin–Darby kidney cells	~40	Hordijk et al.[2] and Sander et al.[3]
T47D	Human ductal breast carcinoma	~50	R. van der Kammen, unpublished results
A431	Human vulvar epidermoid carcinoma	~50–100	R. van der Kammen, unpublished results
Colo320	Human colon adenoma carcinoma	~50–100	R. van der Kammen, unpublished results
CHO	Chinese hamster ovary cells	~50	R. van der Kammen, unpublished results
N1E115	Mouse neuroblastoma	~50	van Leeuwen et al.[4,7]
PC12	Rat pheochromocytoma	~50	van Leeuwen et al.[7]
1F6	Human melanoma	~50	R. van der Kammen, unpublished results
M14	Human melanoma	~50	R. van der Kammen, unpublished results

gene or the egFP gene followed by FACS analysis of the cells 2 days after infection. Alternatively, a rough estimation of the efficiency can be obtained by selection of the transduced cells (see later) and determining the cell number after 10 days. Efficiencies can differ considerably between different constructs. The main reason for a low-virus titer is the effect of the expressed protein of the ϕNX cells. In general, (N43)R-Ras, V12Rac, V12Cdc42, and N17Rac give rather low titers.

Selection of Infected Cells

Twenty-four hours after infection, the medium is replaced by medium containing the desired selection drug (neomycin or zeocin). The amount of selection drug required for a particular cell line should be determined previously. For NIH 3T3 cells, we use 400 μg/ml of active G418 (neomycin; GIBCO-BRL, Gaithersburg, MD) and 500 μg/ml of zeocin (Cayla, France). For BW5147 cells, we use 1 mg/ml of active G418 and 0.5 mg/ml of zeocin. For both antibiotics, selection has to proceed for at least 10 days. Although the cells are quite stable compared to stable cell lines generated from transfected cells, we noted that growing the cells on normal culture medium after this period may lead to a gradual loss of expression, particularly for genes that affect cell growth. In our laboratory, we use pools of infected cells for further analyses. There is no need for subcloning, as the efficiency is rather high and one can compare independent pools generated from different wells of the infection dish.

In general, expression levels from the retroviral constructs in the stable lines are rather low compared to stable cells that have been generated by other means. All of the proteins that we have studied, e.g., Tiam1 and other GEFs, GAPs, Rho-like GTPases, PI3-kinase, Ras, Raf, integrin subunits, catenins, and mutants thereof, do not show more than threefold overexpression compared to endogenous proteins. In this regard, we feel quite confident with this system, knowing that the effects we observe are obtained under near-physiological conditions.

Acknowledgment

This research was supported by grants from The Netherlands Organization for Scientific Research and the Dutch Cancer Society to JGC.

[27] Expression of Rho Gtpases Using Adenovirus Vectors

By DANIEL M. SULLIVAN and TOREN FINKEL

Introduction

Much of what is known about the function of Rho family Gtpases in mammalian cells comes from data obtained using recombinant dominant negative or constitutively active Gtpases. Using these proteins it is possible to evaluate Rho involvement in physiological processes through selective activation or inhibition of signaling downstream of individual Rho family members. The primary challenge associated with using Rho mutants, as with any recombinant reagent, is delivering the protein into a sufficient number of cells in sufficient quantity to elicit a measurable change. There are, of course, a number of vehicles available for the expression of recombinant proteins in cells, ranging from transfection, an option accessible to almost any laboratory, to microinjection, a procedure requiring both special skill and equipment. Another option that has gained popularity in recent years involves the employment of recombinant adenovirus as the gene transfer vehicle. The primary advantage of adenovirus over other vectors or microinjection is that highly efficient gene transfer, greater than 90% in many cases, can be achieved in a wide range of cell types and tissues.[1-7] For instance, we have used recombinant adenovirus to transfer genes into primary cultures of cardiac myocytes. These nondividing cells can neither be infected with retrovirus nor transfected with high efficiency. Indeed, the

[1] H. Yamamoto, N. Atsuchi, H. Tanaka, W. Ogawa, M. Abe, A. Takeshita, and H. Ueno, *Eur. J. Biochem.* **264,** 110 (1999).

[2] M. Hoshijima, V. P. Sah, Y. Wang, K. R. Chien, and J. H. Brown, *J. Biol. Chem.* **273,** 7725 (1998).

[3] D. Kalman, S. N. Gomperts, S. Hardy, M. Kitamura, and J. M. Bishop, *Mol. Biol. Cell* **10,** 1665 (1999).

[4] D. J. Sulciner, K. Irani, Z. X. Yu, V. J. Ferrans, P. Goldschmidt-Clermont, and T. Finkel, *Mol. Cell Biol.* **16,** 7115 (1996).

[5] K. S. Kim, K. Takeda, R. Sethi, J. B. Pracyk, K. Tanaka, Y. F. Zhou, Z. X. Yu, V. J. Ferrans, J. T. Bruder, I. Kovesdi, K. Irani, P. Goldschmidt-Clermont, and T. Finkel, *J. Clin. Invest.* **101,** 1821 (1998).

[6] J. B. Pracyk, K. Tanaka, D. D. Hegland, K. S. Kim, R. Sethi, I. I. Rovira, D. R. Blazina, L. Lee, J. T. Bruder, I. Kovesdi, P. J. Goldschmidt-Clermont, K. Irani, and T. Finkel, *J. Clin. Invest.* **102,** 929 (1998).

[7] Y. Kitamura, T. Kitamura, H. Sakaue, T. Maeda, H. Ueno, S. Nishio, S. Ohno, M. Sakaue, W. Ogawa, and M. Kasuga, *Biochem. J.* **322,** 873 (1997).

efficiency obtained with both plasmid and retrovirus vectors is highly cell type specific, and generally only effective in actively dividing cells. The high efficiency of gene transfer obtained with adenovirus simplifies analysis of transgene effects, as a population of cells uniformly expressing the transgene can be obtained without the need to identify and/or sort cells based on expression of the transgene or a secondary marker gene. This allows for direct biochemical analysis of transgene effects following infection. Another advantage related to the high efficiency of adenovirus-mediated gene transfer was demonstrated by Kalman and co-workers,[3] who used a variety of dominant negative and constitutively active Ras family mutants in combination to dissect the GTPase cascade leading from basic fibroblast growth factor (bFGF) to actin reorganization.

Although somewhat more difficult to make than plasmid or retrovirus constructs, once an adenovirus construct has been established it is easier to work with than either of the aforementioned vectors. There is no special equipment required beyond an ultracentrifuge, no special procedures or additives needed to obtain efficient gene transfer, and, because the virus can be stored indefinitely in highly concentrated form, a single large-scale preparation of virus can supply dozens of experiments. Therefore, once the initial hurdle of establishing a recombinant adenovirus is cleared, maintenance and use of the construct require little effort. There have been several Rho-adenoviruses described in the literature[1–4,6,7]; however, all of these constructs contain constitutively active or dominant negative mutants of the RhoA or Rac1 isoforms and therefore only two of the seven distinct Rho family proteins are represented. In addition to these, Hoshijima and co-workers[2] have described adenovirus constructs carrying dominant negative mutants of Rho kinase and demonstrate their usefulness in defining a role for this RhoA target in myofibrile formation. It is anticipated that a more complete library of Rho–adenovirus constructs will develop as more laboratories begin to use adenovirus as a gene transfer vehicle and that such a library will be a powerful tool for understanding the function of Rho GTPases.

In our experience, the growth characteristics of adenovirus vectors carrying Rho transgenes do not differ in any significant way from any of the other constructs in our laboratory. It is worth noting that Rac and Cdc42 are involved in transducing the signal for adenovirus endocytosis.[8] This finding suggests that an adenovirus carrying a dominant negative Rac or Cdc42 transgene could interfere with a second round of infection. This hypothetical limitation is not relevant to establishment and propagation of adenovirus constructs, as reinfection is not required and would only rarely

[8] E. Li, D. Stupack, G. M. Bokoch, and G. R. Nemerow, *J. Virol.* **72**, 8806 (1998).

impose a limitation on the experimental use of a dominant negative construct.

This article describes the techniques for construction, propagation, and use of adenovirus vectors as these techniques have developed in our laboratory, with the hope that they will aid researchers interested in using adenovirus vectors to study the function of Rho GTPases.

Safety

Adenovirus vectors are considered a level 2 biohazard. Researchers should therefore consult with their institution's safety office regarding the proper procedure for handling and disposing of adenovirus before beginning work. Under ordinary circumstances, adenovirus vectors can be considered quite safe as they are both replication deficient and based on a virus with a good laboratory safety record in its wild-type, replication-competent form. It should be pointed out, however, that the use of adenovirus vectors to express potent oncogenic proteins such as V12Ras is potentially dangerous.

Maintenance of 293 Cells

The most commonly used adenovirus vectors contain deletions in their E1 genes that render them replication deficient. These deletions serve the dual purpose of preventing the expression of cytopathogenic adenovirus genes and increasing the size of the insert DNA that the virus can carry. Because E1 genes are necessary for adenovirus replication, E1-deleted vectors must be propagated in 293 cells that have been transformed with and express the left end of the adenovirus genome, including E1 genes.[9] In our experience, careful maintenance of 293 cells is crucial for success in the establishment and propagation of recombinant adenoviruses, and cells that have not been properly maintained rarely produce satisfactory results.

293 cells should be grown in Eagle's minimum essential medium (MEM) (Life-Technologies, Gaithersburg, MD) supplemented with 2 mM glutamine and 10% heat-inactivated certified fetal bovine serum (FBS) (Life-Technologies) at 37° and 5% (v/v) CO_2. The cells must never be allowed to become more than 80% confluent, and a healthy culture should require splitting at $1:10$ (cm^2/cm^2) every 2–3 days. The media should be changed every second day if the cells are split at a 3-day interval. Because better results are generally obtained with low (<40) passage 293 cells, it is impor-

---

[9] F. L. Graham, J. Smiley, W. C. Russell, and R. Nairn, *J. Gen. Virol.* **36,** 59 (1977).

tant to establish a stock of frozen cells as early as possible. In addition, it is advisable that the cultures used to create frozen stocks be established from commercially available low passage 293 cells (Microbix Biosystems, Toronto, Canada). Cells can be frozen in growth medium containing 10% (v/v) dimethyl sulfoxide (DMSO).

Construction of Recombinant Adenoviruses

Establishing a recombinant adenovirus is by far the most trouble-prone aspect of working with adenovirus vectors. In our experience, most problems can be attributed to the low efficiency of transfection arising either from improperly maintained cells or poor quality plasmid DNA. Therefore, careful attention should be paid to caring for 293 cells prior to transfection and only good quality DNA should be used. The procedure described here is based on the cotransfection technique developed by Graham and co-workers.[10–14] Briefly, the approach utilizes two plasmids: one which contains map units 0 to 16 of the adenovirus genome with the E1 region replaced by a promoter and polylinker and the other containing the adenovirus genome expanded with a stuffer DNA to beyond the packaging limit of the virus. Using this approach, the transgene can be inserted easily into the shuttle vector (the plasmid containing the promoter and polylinker), and the recombinant adenovirus can be obtained through homologous recombination in 293 cells with the second, adenovirus genome containing, plasmid (rescue plasmid).

Preparation of Plasmid DNA

The rescue plasmids pJM17 or pBHG10 and a variety of shuttle vectors can be purchased from Microbix Biosystems. In general the shuttle vectors can be handled using standard protocols for expansion and purification of plasmid DNA and good results have been obtained from commercial kits. The rescue plasmids are, however, very large (~40 kb) and a number of problems can arise if they are not handled carefully. A very good protocol for purifying large plasmids is available by request from Microbix Biosys-

---

[10] F. L. Graham, *EMBO J.* **3**, 2917 (1984).
[11] W. J. McGrory, D. S. Bautista, and F. L. Graham, *Virology* **163**, 614 (1998).
[12] A. J. Bett, L. Prevec, and F. L. Graham, *J. Virol.* **67**, 5911 (1993).
[13] A. J. Bett, W. Haddara, L. Prevec, and F. L. Graham, *Proc. Natl. Acad. Sci. U.S.A.* **91**, 8802 (1994).
[14] F. L. Graham and L. Prevec, *in* "Methods in Molecular Biology" (E. J. Murray, ed.), Vol. 7, p. 109. Humana Press, Clifton, NJ, 1991.

tems (http://www.microbix.com/). It is particularly important to start cultures from fresh colonies and, if NaOH is used to lyse bacteria, to keep the incubation in lysis buffer as short as possible to avoid nicking the DNA. In addition, we prefer cesium chloride banding[15] to silica adsorption in the preparation of these large plasmids. Good quality rescue plasmid DNA should run as a single large band (no downward smearing) and produce the following bands when digested with *Hin*dIII (0.7, 2.1, 2.9, 3.4, 3.8, 4.3, 4.6, 5.3, 5.4, and 8.0 kb). The rescue plasmid should be stored in TE buffer (10 mM Tris-HCl, 1 mM EDTA, pH 8.0) at 4°.

Cotransfection

293 cells can be transfected efficiently using standard calcium phosphate precipitation protocols. For laboratories that do not already have an established protocol, we recommend the calcium phosphate transfection system kit (Life-Technologies).

Method

Day 0. Plate 293 cells in 60-mm dishes at several different densities (e.g., 0.3, 0.4, and 0.5 × 10^6 cells per dish). We have found that cell density at the time of cotransfection is a critical determinant and therefore feel that chances of success are improved by transfecting over a range of different densities. In addition, the cells should be split from a fast-growing culture, preferably at a passage earlier than 40. The cells should be 60–80% confluent on day 1.

Day 1. Perform calcium phosphate transfection according to standard laboratory procedure or the protocol included with the Life-Technologies kit using 10 μg of transgene-containing shuttle vector, 5 μg of rescue plasmid, and 5 μg of carrier DNA per dish. The cells should be left in the precipitate for 12–24 hr.

Day 2. Replace the transfection medium with fresh growth medium and allow the cells to come to confluence. As soon as the cells reach confluence, 24–48 hr after transfection, the monolayer should be overlaid with a solid medium so that individual recombinants can be isolated and screened for expression.

---

[15] J. Sambrook, E. F. Fritsch, and T. Maniatis (eds.), "Molecular Cloning: A Laboratory Manual." Cold Spring Harbor Laboratory Press, Cold Spring Harbor, NY, 1989.

Agarose Overlay

Materials

Low-melting temperature agarose (SeaPlaque; FMC, Rockland, ME)
100× Pen–Strep (penicillin–streptomycin) solution (Biofluids, Rockville, MD)
100× Glutamine solution (Biofluids)
2× MEM (Life-Technologies)
Heat-inactivated certified FBS
42° and 37° water baths

Method

1. Make up 100 ml of a 2% agarose suspension in tissue culture grade
water and melt the agarose in a microwave. The resulting solution can be
stored indefinitely at room temperature. For use, melt the agarose and cool
to 42° in a water bath.

2. Make up a 2× media stock consisting of 120 ml 2× MEM, 30 ml
FBS, 3 ml 100× Pen–Strep, and 6 ml 100× glutamine and store at 4°. For
use, transfer 2 ml of media per 60-mm dish to be overlaid into a sterile
50-ml tube and warm to 37°.

3. Add an equal volume of the 42° agarose to the 2× media and mix thoroughly.

4. Remove the media from the transfected cells and gently add 4 ml
of the overlay mixture per 60-mm dish, taking care not to create bubbles.

5. Allow the overlay to solidify for 15 min at room temperature before
returning the cultures to the incubator. To obtain well-formed plaques, it
is important to allow the agarose to solidify completely before returning
the culture to the incubator.

6. A feed overlay can be performed every 4–5 days by repeating the
procedure just described and placing the fresh overlay on top of the existing one.

Harvesting and Expanding Plaques

Plaques from recombinant adenovirus might appear as early as 5 days
after cotransfection or can take up to 2 weeks to appear. The plaques are
quite obvious and can be seen easily with the naked eye. It should be noted
that foci appearing somewhat similar to the early stages of adenovirus
cytopathic effect (CPE) often develop in the 293 monolayer after an extended time at confluence (7–10 days). These foci can be distinguished

from adenovirus plaques based on their tendency to remain small (<1 mm) with densely populated centers, as opposed to adenovirus plaques, which should grow to several millimeters and have few or no living cells at the center.

Materials

Sterile disposable 1-ml pipettes
Sterile Eppendorf tubes
293 growth medium
293 cells at ~80% confluence in a 12-well plate
Dry ice
37° water bath
4° microcentrifuge

Method

1. Draw a circle on the bottom of the plate around plaques to be picked.

2. Aspirate 300 μl of MEM into a 1-ml pipette and then approximately 100 μl of air to create an air cushion between the medium and the tip of the pipette.

3. Force the tip of the pipette through the agarose overlay at the center of the plaque and gently aspirate the agarose plug into the pipette.

4. Express the plug and the medium into a sterile Eppendorf tube, vortex, and place the tube in dry ice. Repeat the procedure for all of the plaques to be harvested.

5. Thaw the plaques in a 37° water bath and vortex vigorously for 30 sec. Repeat four more cycles of freeze-thaw and vortex.

6. Pellet debris for 5 min at 12,000g and 4°, and transfer the supernatant containing the crude viral lysate (CVL) to a fresh Eppendorf tube. This CVL can be stored at −80°, although it is recommended that the virus be expanded as soon as possible.

7. To expand the CVL, dilute 200 μl into 2 ml of growth medium and infect 293 cells in 1 well of a 12-well dish by replacing the growth medium with the diluted CVL. Incubate overnight and then replace the virus medium with 2 ml of fresh 293 growth medium. Adenovirus CPE should be apparent within 24 to 48 hr after infection (depending on the titer of the plaque CVL).

8. When 80–90% of the cells have detached from the plate, harvest the cells and medium into a polypropylene tube (it should be possible to detach any adherent cells by pipetting up and down). Prepare a freeze-thaw CVL as described in steps 5 and 6.

The viruses obtained can be screened at this point for expression of the transgene. The simplest way to do this is to infect a fresh culture of 293 cells as described in step 7 and harvest protein for Western blot analysis at 24–48 hr after infection. Positive isolates can be expanded further by preparing CVL, as described earlier, from larger cultures.

Large-Scale Purification of Adenovirus

One reason that it is possible to achieve very high efficiency of gene transfer using adenovirus is that the virus is stable at high concentrations and therefore can be delivered at a high multiplicity of infection [MOI; 100 plaque-forming units (pfu) per cell if necessary]. Even if a high MOI is not necessary, it is important to purify constructs before they are used for experiments as CVL is toxic to many cells, possibly due to the presence of large amounts of unincorporated adenovirus capsid proteins.

Materials

293 cells (100–200 × 150-mm plates at ~80% confluence; as always, cells should be plated from a good quality culture the day before infection)

CVL or purified virus from an earlier lot (use 80 μl of CVL prepared as described earlier or 1–2 MOI of purified virus per 150-mm plate; it is a good idea to screen CVL for wild-type contamination, as described later, before using it to inoculate a large-scale culture)

293 growth medium

Tissue culture centrifuge and sterile centrifuge bottles (the type of centrifuge and bottles is not important as they will only be used to pellet cells at low speed; however, their volume capacity should be as large as possible)

High-speed centrifuge (Sorvall, Newtown, CT, or equivalent)

50-ml polypropylene centrifuge tubes (Beckman, Fullerton, CA)

Beckman ultracentrifuge and SW41 rotor or equivalent

Beckman 14 × 89-mm ultracentrifuge tubes

Cesium chloride solutions (ρ equals 1.25 and 1.4 g/ml in TD buffer (137 mM NaCl, 5 mM KCl, 5.5 mM glucose, 0.7 mM Na$_2$HPO$_4$, 25 mM Tris–HCl, pH 7.4)

3 cm^3 syringes fitted with 19-gauge needles

Slide-A-Lyzer dialysis cassettes (Pierce, Rockford, IL)

Dialysis solution [4 liter; 10 mM Tris–HCl (pH 7.0), 1 mM MgCl$_2$, 10% (v/v) glycerol, autoclaved and cooled to 4°]

37° water bath

Dry ice

Method

1. Dilute adenovirus in 15 ml of 293 cell growth medium per 150-mm plate to be inoculated.

2. Replace the growth medium on each dish of 293 cells with 15 ml of the virus-containing medium and incubate overnight at 37° in a tissue culture incubator.

3. Replace the virus-containing media with fresh growth medium and continue the incubation until >90% of the cells have detached from the plates (48–72 hr).

4. Detach the remaining cells by pipetting up and down and transfer the cells and media into centrifuge bottles.

5. Pellet the cells for 10 min at 2000g and 4°.

6. Remove 10 ml of the supernatant to a sterile tube and discard the remaining supernatant according to the institutional guidelines for a level 2 biohazard.

7. Resuspend the cells in 6 ml of the supernatant that was saved and transfer them to a 50-ml polypropylene centrifuge tube. Prepare CVL as described earlier.

8. Prepare cesium chloride step gradients in two 14 × 89-mm ultracentrifuge tubes by overlaying 2.5 ml of the 1.25-g/ml solution onto 2.5 ml of the 1.4-g/ml solution.

9. Divide the CVL in half and carefully overlay it onto the cesium chloride step gradients.

10. Use MEM to fill the tubes to within 0.5 cm of the top and transfer them to the ultracentrifuge buckets. Centrifuge for 1 hr at 35,000 rpm and 20°. Following ultracentrifugation, three bands should be apparent: the densest band, resolved at the step interface, contains the fully assembled virus; the next band above the virus band contains capsid proteins and should be less than one-half the size of the first; and the largest and least dense of the three bands, running at the aqueous–cesium chloride interface, contains cellular debris.

11. Pull the virus band by inserting a 19-gauge needle attached to a 3 cm³ syringe at a point 3–5 mm below the band and aspirating as much of the band as possible into the syringe. Inserting the needle without disrupting the gradient can be tricky; therefore it is a good idea to practice this technique before attempting to pull the virus.

12. Mix the virus bands pulled from each of the step gradients and transfer them to a sterile 15-ml conical tube. Add enough of a 1:1 mixture of the 1.25- and 1.4-g/ml cesium chloride solutions (ρ equals 1.325 g/ml) to give ~10 ml final volume.

13. Mix the virus-containing cesium chloride solution and transfer it to a fresh 14 × 89-mm ultracentrifuge tube. Fill the tube to within 0.5 cm of

the top with 1.325 g/ml cesium chloride and prepare a balance tube containing 1.325 g/ml cesium chloride.

14. Centrifuge for 18 hr at 35,000g and 20°.

15. Pull the virus band as described in step 12. The virus will again run as the densest band approximately halfway down the equilibrium gradient.

16. Transfer the virus to a dialysis cassette and dialyze (4× 1 liter changes of dialysis solution at 4°, 1 hr per change). The virus can be stored in small (≤50 μl) aliquots at −80°.

Plaque Titration

It is important to obtain an estimate of the virus concentration for each stock prepared so that the MOI does not vary from one experiment to the next. Although several methods have been devised to estimate the number of infectious units in a virus stock, the most commonly used is plaque titration. To obtain a titer, plate 293 cells in two six-well dishes so that they are approximately 70% confluent the following day; when the cells are ready, obtain 6 × 10-fold serial dilutions of the virus stock in MEM from 1×10^{-7} to 10^{-12} μl/ml; infect each well of a six-well dish with 1 ml of the diluted virus so that all of the dilutions are represented (do a duplicate titration in the second six-well plate); infect for 90 min; and then perform an agarose overlay as described previously. The plates should be kept in a tissue culture incubator at 37° for 7 to 10 days, with feed overlays every 3–4 days, before counting plaques. The titer is usually expressed as plaque-forming units per milliliter of stock solution and should be between 1×10^{10} and 10^{12} pfu/ml.

Wild-Type Contamination

It is not uncommon for recombinant adenovirus stocks to become contaminated with wild-type virus. Because infection with wild-type virus is likely to elicit cellular responses that could be misinterpreted as transgene effects, it is important to rule out the possibility of significant contamination before using a virus stock. This can be accomplished most reliably by assaying for virus replication in A549 cells, which are capable of replicating wild-type but not E1-deleted adenovirus.

Materials

A549 cell culture (CCL-185; ATCC, Manassas, VA)
A549 culture medium [Dulbecco's modified Eagle's medium (DMEM), 10% heat-inactivated FBS]

Method

1. Plate A549 cells in a 150-mm dish so that they are ~70% confluent the following day.

2. Dilute 1×10^9 pfu from the recombinant adenovirus stock into 15 ml of culture medium.

3. Infect A549 cells by replacing the growth medium with the diluted virus and incubating overnight in a tissue culture incubator at 37°.

4. Replace the virus medium with fresh growth medium and incubate the culture for an additional 5 days. (It is not uncommon for the cells to show signs of CPE in this first round, even in the absence of wild-type virus, as some components of the adenovirus capsid are toxic to A549 cells.)

5. Harvest the cells and culture medium by scraping and prepare a freeze-thaw lysate as described under "Harvesting and Expanding Plaques."

6. Use the lysate to infect another 150-mm dish of A549 cells by incubating the cells in the undiluted lysate for 90 min in a tissue culture incubator at 37°.

7. Replace the lysate with fresh growth medium and maintain the infected culture along with a control dish of cells with media changes every 3 days. If wild-type virus is present, the infected culture should show signs of CPE and die within 14 days.

An adenovirus construct that has been contaminated with wild-type virus can be repurified by plaquing as described for plaque titration and by expanding and testing well-isolated plaques as described under "Harvesting and Expanding Plaques." This method of plaque purifying and Western blot analysis of transgene expression can also be used in lieu of the A549 cell assay to detect wild-type contamination. Although screening plaques is significantly less sensitive than the A549 cell assay, it does have the advantages of being faster and providing purified CVL for expansion should contamination be found.

Transgene Expression Using Adenovirus Vectors

As mentioned earlier, it is easy to achieve efficient transgene expression in a wide range of cell types and tissues using adenovirus vectors. In general, almost any cultured cell type can be infected using a slight modification of the protocol described earlier for infecting A549 cells. When doing experiments, it is important to keep in mind that adenovirus infection can, by itself, activate signal transduction pathways and induce cytokine secretion in some cell types.[16] Also, some cells will support low-level expres-

---

[16] J. T. Bruder and I. Kovesdi, *Virology* **71,** 398 (1997).

sion of adenovirus genes in the absence of E1. Any of these phenomena can produce artifacts in experimental results; therefore, it is important to have a vector control, a parallel culture infected with an adenovirus vector carrying either no transgene or an irrelevant transgene, for all of the assays performed. In addition, the potential for vector-induced effects can be reduced by infecting cells at the minimum effective MOI. This can be determined by infecting with the adenovirus construct at various MOIs and immunohistochemically determining transgene expression. The lowest MOI producing transgene expression in nearly 100% of the cells, usually between 1 and 100 pfu/cell, should be used for experiments. If an antibody is not available for immunohistochemistry, another option is to determine the minimum effective MOI using an adenovirus vector carrying a reporter gene such as GFP or β-galactosidase.

The protocol described here is for adherent cultures, but can be modified for suspended cells by simply performing the infection in a sterile centrifuge tube and pelleting the cells for media changes. Adenovirus vectors can also be used to deliver transgenes to cells *in vivo*[17]; however, the procedure for accomplishing this varies significantly, depending on the target tissue, and a detailed discussion is beyond the scope of this article.

Materials

Adenovirus constructs (experimental and control)
Cell culture medium
Cultured cells (adenovirus will infect both actively growing and quiescent cells)

Method

1. Trypsinize a replicate dish of the cells to be infected and determine the number of cells in the dish.
2. Calculate the amount of virus needed to infect at the minimum effective MOI and dilute the virus into as small a volume of culture medium as possible, depending on the size of the dish.
3. Replace the cell growth medium with the virus medium and incubate overnight in a tissue culture incubator at 37°.
4. Replace the virus medium with growth medium.

Transgene expression is evident by Western blot from approximately 6 hr after infection and usually peaks by 72 hr after infection.

[17] S. L. Brody and R. G. Crystal, *Ann. N.Y. Acad. Sci.* **716,** 90 (1994).

[28] *In Vivo* Activity of Wild-Type and Mutant PAKs

By Charles C. King, Luraynne C. Sanders, and Gary M. Bokoch

Introduction

The PAK family of serine/threonine kinases serves as effector molecules for members of the Rho family of GTPases. Four PAK isoforms have been described in human tissue to date, and all have been shown to bind to Rac, Cdc42, and/or TC10 in a GTP-dependent manner.[1] PAKs have been implicated in the regulation of various cellular processes, including proliferation,[2,3] oxidant production,[4] apoptosis,[5–7] and actin cytoskeletal reorganization.[8–11]

Our understanding of the *in vivo* activity of PAKs has been facilitated by the use of mutations in key functional and regulatory motifs of PAK. PAKs contain a carboxy-terminal kinase domain and an amino-terminal regulatory domain, which includes the p21-binding domain (PBD), the autoinhibitory domain, several proline-rich motifs, and an acidic region. Binding of Cdc42–GTP, Rac–GTP, or biologically active lipids, such as sphingosine, disrupts the intramolecular interactions of the autoinhibitory domain and the kinase domain, resulting in PAK1 autophosphorylation and an increase in the phosphorylation of PAK1 substrates.[12] Commonly

[1] R. H. Daniels and G. M. Bokoch, *Trends Biochem. Sci.* **24,** 350 (1999).
[2] J. P. Mira, V. Benard, J. Groffen, L. C. Sanders, and U. G. Knaus, *Proc. Natl. Acad. Sci. U.S.A.* **97,** 185 (2000).
[3] Y. Tang, Z. Chen, D. Ambrose, J. Liu, J. B. Gibbs, J. Chernoff, and J. Field, *Mol. Cell. Biol.* **17,** 4454 (1997).
[4] U. G. Knaus, S. Morris, H.-J. Dong, J. Chernoff, and G. M. Bokoch, *Science* **269,** 221 (1995).
[5] T. Rudel and G. M. Bokoch, *Science* **276,** 1571 (1997).
[6] N. Lee, H. MacDonald, C. Reinhard, R. Halenbeck, A. Roulston, T. Shi, and L. T. Williams, *Proc. Natl. Acad. Sci. U.S.A.* **94,** 13642 (1997).
[7] A. Schürmann, A. F. Mooney, L. C. Sanders, M. A. Sells, H-G, Wang, J. C. Reed, and G. M. Bokoch, *Mol. Cell. Biol.* **20,** 453 (2000).
[8] S. Dharmawardhane, L. C. Sanders, S. S. Martin, R. H. Daniels, and G. M. Bokoch, *J. Cell Biol.* **138,** 1265 (1997).
[9] Z. S. Zhao, E. Manser, Z. Q. Chen, C. Chong, T. Leung, and L. LIM, *Mol. Cell. Biol.* **18,** 2153 (1998).
[10] E. Manser, H. Y. Huang, T. H. Loo, X. Q. Chen, J. M. Dong, T. Leung, and L. LIM, *Mol. Cell. Biol.* **17,** 1129 (1997).
[11] M. A. Sells, U. G. Knaus, S. Bagrodia, D. M. Ambrose, G. M. Bokoch, and J. Chernoff, *Curr. Biol.* **7,** 202 (1997).
[12] F. Zenke, C. King, B. Bohl, and G. Bokoch, *J. Biol. Chem.* **274** 32565 (1999).

TABLE I
PAK1 WILD-TYPE AND MUTANT CONSTRUCTS AND PHENOTYPES

PAK1 construct	Phenotype and use
Wild type	Not constitutively active, little activity *in vivo* in the absence of activating stimuli
T423E	Mimics phosphorylation of activation loop Thr; highly constitutively active as kinase[a]
Y107E	Mutation disrupts binding of autoinhibitory domain to catalytic domain; highly constitutively active as a kinase[b,c]
K299R K299A	Kinase-dead variants due to mutation of the ATP-binding region; still binds GTPases, have dominant negative activity[a]
H83,86L	Double mutation in p21-binding and autoinhibitory domains, does not bind GTPases; weakly kinase active but highly cytoskeletally active[d]
H83,86L, K299R	Triple mutant that does not bind GTPases and is kinase dead; has dominant negative activity[a,d]
P13A P42A P210A P193,194A	Mutations in N-terminal PXXP motifs prevent binding of SH3-containing regulatory proteins[e,f]
aa 1–74 aa 67–150 aa 169–205	Fragments of PAK1 N terminus shown to exert dominant negative effects on PAK-dependent neuronal growth factor-induced neurite outgrowth in PC12 cells[g]

[a] E. Manser, H. Y. Huang, T. H. Loo, X. Q. Chen, J. M. Dong, T. Leung, and L. Lim, *Mol. Cell Biol.* **17,** 1129 (1997).

[b] J. L. Brown, L. Stowers, M. Baer, J. A. Trejo, S. Coughlin, and J. Chant, *Curr. Biol.* **6,** 598 (1996).

[c] J. A. Frost, A. Khokhlatchev, S. Steppec, M. A. White, and M. H. Cobb, *J. Biol. Chem.* **273,** 28191 (1998).

[d] M. A. Sells, U. G. Knaus, S. Bagrodia, D. M. Ambrose, G. M. Bokoch, and J. Chernoff, *Curr. Biol.* **7,** 202 (1997).

[e] G. M. Bokoch, Y. Wang, B. P. Bohl, M. A. Sells, L. A. Quilliam, and U. G. Knaus, *J. Biol. Chem.* **271,** 25746 (1996).

[f] R. H. Daniels, F. T. Zenke, and G. M. Bokoch, *J. Biol. Chem.* **274,** 6047 (1999).

[g] R. H. Daniels, P. Hall, and G. M. Bokoch, *EMBO J.* **17,** 754 (1998).

used PAK1 mutants and their characteristics are described in Table I. For example, mutation of histidine-83 and histidine-86 to leucine (PAK1 H83,86L) in the p21-binding domain of PAK1 abolishes the interaction of GTPases with PAK1 and partially disrupts the interaction of the N-terminal autoinhibitory domain with the C-terminal catalytic domain. Although only modestly catalytically active, this PAK1 mutation has profound effects on the actin cytoskeleton.[11] In contrast, mutation of the activation loop threonine of PAK1 (aa 423) to glutamic acid (PAK1 T423E), mimicking a critical phosphorylation at this site,[12] results in a highly constitutively active enzyme.[10]

This article focuses on the use and *in vivo* effects of wild-type PAK1 and selected PAK1 mutants that have proven useful for *in vivo* analysis of PAK function. We present methods to analyze select biological activities of endogenous and transfected PAK1 *in vivo*. We also describe a virus-mediated transfection system for cellular analysis of the effects of PAKs on downstream targets and the actin cytoskeleton.

In Vivo Activity of PAKs as Measured by In-Gel Kinase Assay

The ability to measure the activity of PAK kinases after renaturation on polyacrylamide gels provides a valuable tool for assessing stimulus-induced kinase activation *in vivo*. The in-gel kinase assay is based on the ability of radiolabeled ATP to diffuse into polyacrylamide gels containing immobilized kinase substrates and renatured kinases. For *in vivo* analysis of endogenous protein kinase activity, an unstimulated or stimulated cell lysate is prepared in the presence of phosphatase inhibitors, such as okadaic acid or microcystin-LR, and separated on SDS–polyacrylamide gels into which a specific kinase substrate has been polymerized. The endogenous kinase, run under denaturing conditions, is subsequently treated to remove the SDS and then renatured *in situ*. The gel containing the renatured protein is subsequently incubated with $[\gamma\text{-}^{32}\text{P}]$ATP. After extensive washing to remove the $[\gamma\text{-}^{32}\text{P}]$ATP, the gel is stained for protein, destained, and dried, and the radioactive band(s) corresponding to the action of the renatured kinases on the immobilized substrate is detected by autoradiography.

The *in vivo* kinase assay has been used routinely to measure the activity of endgenous PAK isoforms in neutrophils and other cell lines.[4,13,14] Additionally, this technique can be used readily to examine the *in vivo* activities of PAK mutants in stimulated cells after transient transfection. In situations where the endogenous levels of PAK are very low, or if problems with backgrounds from renaturable non-PAK kinases present in the cellular lysate are encountered, the selectivity of the in-gel assay can be improved by prior immunoprecipitation of PAK with specific antibodies now available commercially from a variety of sources. Indeed, it is recommended that immunoprecipitations with PAK antibodies be performed at an early stage in order to verify that the kinases detected in the 62- to 68-kDa region are in fact PAKs.

[13] J. Ding, U. G. Knaus, J. P. Lian, G. M. Bokoch, and J. A. Badwey, *J. Biol. Chem.* **271**, 24869 (1996)

[14] G. M. Bokoch, A. M. Reilly, R. H. Daniels, C. C. King, A. Olivera, S. Spiegal, and U. G. Knaus, *J. Biol. Chem.* **273**, 8137 (1998).

We have found the in-gel PAK kinase assay of particular use in neutrophils for the analysis of the rapid PAK activation that is seen after stimulation with chemoattractants such as fMLP (fMet-Leu-Phe) (Fig. 1, top). Resting neutrophils exhibit little to no PAK activity, but after 30 sec to 1 min stimulation with the chemoattractant fMLP, substantial PAK activation is detected. PAK activity was decreased after 5 min and returned to basal levels by 30 min. In control experiments, PAK expression levels were shown to be constant as determined by immunoblotting with isoform-specific PAK antibodies (Fig. 1, bottom).

Methods

Preliminary immunoblotting experiments should be performed to determine whether significant amounts of endogenous PAK exist prior to performing an in-gel kinase assay. In the method described herein, freshly isolated human neutrophils are used because they contain high levels of two PAK isoforms: PAK1 and PAK2.[4] Neutrophils are stimulated with 1 μM fMLP for 0, 1, 5, 15, 30, or 60 min, and then the incubation is stopped by the addition of ice-cold lysis buffer.[4] The amount of protein in each sample is quantitated by Bicinchoninic Acid (BCA) assay (Pierce, Rockford, IL) according to the manufacturer's protocol, and 140 μg of protein is used for each sample. Additionally, 10 μg of lysate protein is used to verify PAK expression by immunoblotting. Two control samples are also used. First, a catalytically active PAK1 (PAK1 T423E) overexpressed in COS-7 cells is used as a positive control to ensure that the assay is working

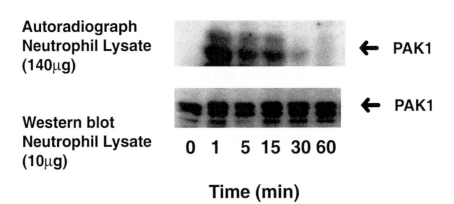

**Autoradiograph
Neutrophil Lysate
(140µg)**   **← PAK1**

**Western blot
Neutrophil Lysate
(10µg)**   **0   1   5 15 30 60**   **← PAK1**

Time (min)

FIG. 1. PAK1 activation in neutrophils after fMLP stimulation. Human neutrophils, incubated with $1 \times 10^{-6} M$ fMLP for the indicated times, were lysed and subjected to an in-gel kinase assay (top) using p47[phox] peptide substrate, as described in the text. PAK1 protein levels were compared by immunoblotting equal amounts of cell lysate (bottom).

properly. The PAK T423E preparation is described in this series.[15] Second, 140 μg of unstimulated neutrophil lysate is incubated with 1–2 μg of the PAK activator Cdc42–GTPγS in kinase buffer [50 mM HEPES, pH 7.5, 10 mM MgCl$_2$, 2 mM MnCl$_2$, 0.2 mM dithiothreitol (DTT)] with 20 μM ATP for 30 min at 30° in order to fully activate the endogenous PAK present. The guanine nucleotide loading of Rho family GTPases has been described previously.[16–18] Because the latter positive control can be prepared from any cellular lysate of interest, it is highly recommended as a routine practice. Load each sample on a 7% SDS–polyacrylamide gel containing either the carboxyl-terminal peptide (aa 297–331) of p47[phox] (0.5 mg/ml) or histone H4 (1 μg/ml) dissolved in the gel mixture. Although other substrates can be used (e.g., myelin basic protein, histone H1), we have found that these two substrates give relatively low backgrounds from non-PAK kinases present in the cellular lysates. Prestained molecular weight markers (GIBCO, Grand Island, NY) are loaded in the outer lanes for orientation. After electrophoresis, soak the gel in buffer 1 [20% (v/v) propanol, 50 mM Tris–HCl, pH 7.5; make 100 ml total] at room temperature for 40 min, changing the buffer after 20 min. Transfer the gel to buffer 2 (50 mM Tris–HCl, pH 7.5, 5 mM 2-mercaptoethanol; make 100 ml total) at room temperature for 2 hr, changing the buffer after 1 hr. Transfer the gel to buffer 3 (6 M guanidinium, 50 mM Tris–HCl, pH 7.5; make 100 ml total) at room temperature for 2 hr, changing the buffer after 1 hr. Because the guanidinium takes about 1 hr to completely go into solution, this buffer should be made well in advance. Soak the gel overnight at 4° in buffer 4 (50 mM Tris–HCl, pH 7.5, 5 mM 2-mercaptoethanol, 0.04% Tween 40; make 250 ml total). Both the 50 mM Tris–HCl, pH 7.5, and the Tween 40 should be heated at 37° for at least 1 hr prior to making this buffer. The solution should be allowed to mix for 30 min at room temperature before the addition of 2-mercaptoethanol. The following morning, soak the gel in buffer 4 for 1.5 hr, changing the buffer every 30 min. Transfer the gel to buffer 5 (10 mM HEPES, pH 8.0, 1 mM DTT, 0.1 M EGTA, 1 mM EDTA, 5 mM MgCl$_2$; make 100 ml total) at room temperature for 30 min. Transfer the gel to a 30° shaking bath and wash for another 30 min with buffer 5. To perform the in-gel kinase assay, add 25 μM ATP and 10 μl [^{32}P]ATP (4500 Ci/mmol) to kinase buffer for 30 min at 30° with gentle agitation. After the kinase assay, incubate the gel at room temperature in the wash

---

[15] C. C. King, A. M. Reilly, and U. G. Knaus, *Methods Enzymol.* **325** [15] (2000) (this volume).
[16] A. J. Self and A. Hall, *Methods Enzymol.* **256,** 3 (1995).
[17] A. J. Self and A. Hall, *Methods Enzymol.* **256,** 67 (1995).
[18] E. Manser, T. Leung, and L. LIM, *Methods Mol. Biol.* **84,** 295 (1998).

buffer (5% TCA, 1% sodium pyrophosphate; make 250 ml total) for 6 hr, changing the solution frequently. Put the gel in wash buffer overnight at room temperature. The next morning, stain the gel for 10 min in Coomassie Blue and then destain in a solution containing 20% methanol and 30% acetic acid. Dry the gel and expose to Kodak (Rochester, NY) AR film overnight.

Transient Expression of PAKs *in Vivo*

Traditional transient protein expression methods, including microinjection or lipid-mediated transfection, have proven useful for evaluating the biology of PAK at the individual cell level. However, many cell types do not lend themselves well to traditional transfection methods, and the typical low transfection efficiencies obtained make it difficult to study the effects of PAK at a biochemical level. We have employed a Semliki Forest virus (SFV) gene expression system to overcome these problems. This system has allowed us to transiently express PAKs (and a variety of GTPases) with an efficiency greater than 95% in baby hamster kidney (BHK)-21 cells, enabling us to study a population of cells placed in a bioassay. Virus-transfected BHK-21 cells have been used to investigate the role of PAK1 in cell spreading.[19] BHK-21 cells infected with PAK1 T423E, PAK1 wild-type, and LacZ (used as a control) were allowed to express protein for 6 hr. Cells were harvested, placed on fibronectin-coated coverslips, and allowed to attach and spread for 2 hr. PAK1 wild-type and LacZ cells attached and spread normally. In marked contrast, PAK1 T423E cells attached normally, but spreading was inhibited by ~80%. The higher transfection efficiency of the SFV expression system makes biochemical analysis of downstream targets for PAK easier (see section on cytoskeletal regulation). Furthermore, this virus has a broad host range and has allowed us to study PAK effects on cells normally difficult to transfect (i.e., many different breast cancer cell lines).[2]

Methods

cDNAs encoding PAK1 wild-type and mutation T423E are expressed in cells using the Semliki Forest virus gene expression system (Life Technologies, Gaithersburg, MD). The cDNAs are polymerase chain reaction (PCR) amplified using primers that contained a *Bam*HI restriction enzyme site and a myc tag at the 5' end. These constructs are subcloned into the *Bam*HI site of Semliki Forest vector pSFV3. *In vitro* transcription of

---

[19] L. C. Sanders, F. Matsumura, G. M. Bokoch, and P. de Lanerolle, *Science* **283**, 2083 (1999).

linearized pSFV3 constructs and pSFV-Helper2 is performed using SP6 RNA polymerase. RNA transfection of BHK-21 cells is done by electroporation as described previously,[20] yielding recombinant viral stocks of approximately 10^7 plaque-forming units (pfu)/ml. Viral stocks are stored at $-80°$ until use. The virus is activated per manufacturer's instruction, and BHK-21 cells are infected in serum-free media. Transfection efficiency of recombinant virus is routinely greater than 95% in BHK-21 cells. Cells are allowed to express protein for 6 to 8 hr after infection in serum-free media before use in experiments.

Antiapoptotic Activity of PAK *in Vivo*

PAKs have been shown to exert both pro- and antiapoptotic effects in a variety of cells.[5–7,21] The proapoptotic effects of PAK2 have been well documented.[5,6,22–24] On activation of apoptotic pathways by tumor necrosis factor α (TNF-α), ceramide, stress, UV light, or occupancy of the Fas receptor by the Fas ligand, PAK2 is cleaved at aspartate-212 by caspases in a region linking the amino-terminal regulatory domain and the carboxy-terminal catalytic domain. Without the amino-terminal regulatory domain, the catalytic domain of PAK2 becomes constitutively activated and phosphorylates a variety of substrates. These may result in the activation of JNK and p38, two important proapoptotic signaling molecules.[22] PAK2 cleavage by this mechanism contributes to the cytoskeletal changes associated with the progression of apoptosis as well.[5,6] The proapoptotic events associated with PAK2 cleavage and activation occur late in the apoptotic cascade.

Data suggest that PAK1 exerts antiapoptotic effects in FL5.12 cells and NIH 3T3 cells.[7] Bad is a member of the Bcl2 protein family which induces apoptosis when complexed with either Bcl-XL or Bcl2. Phosphorylation of Bad on two physiologically relevant sites, serine-112 and serine-136, disrupts the Bad/Bcl complexes and promotes the binding of Bad to 14-3-3τ. A number of kinases have been reported to phosphorylate these sites and to

[20] K. Lundstrom, A. Mills, G. Buell, E. Allet, N. Adami, and P. Liljestrom, *Eur. J. Biochem.* **224**, 917 (1994).
[21] U. G. Knaus and G. M. Bokoch, *Int. J. Biochem. Cell Biol.* **30**, 857 (1998).
[22] T. Rudel, F. T. Zenke, T.-H. Chuang, and G. M. Bokoch, *J. Immunol.* **160**, 7 (1998).
[23] B. N. Walter, Z. Huang, R. Jakobi, P. T. Tuazon, E. S. Alnemri, G. Litwack, and J. A. Traugh, *J. Biol. Chem.* **273**, 28733 (1998).
[24] T. K. Tang, W. C. Chen, W. H. Chen, S. D. Yang, M. H. Ni, and J. S. Yu, *J. Cell Biochem.* **70**, 442 (1998).

vector PAK1 T423E PAK1 K299R

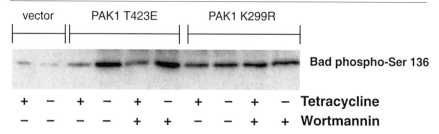

Bad phospho-Ser 136

| + | − | + | − | + | − | + | − | + | − | Tetracycline |
| − | − | − | − | + | + | − | − | + | + | Wortmannin |

FIG. 2. PAK1 T423E, but not PAK1 K299R, phosphorylates the antiapoptotic protein Bad *in vivo*. NIH 3T3 cells containing the pTetSplice control, PAK1 T423E, or PAK1 K299R vector were transfected with 5 μg Bad cDNA and allowed to express for 36 hr in the presence (+; PAK expression repressed) or absence (−; PAK expression induced) of tetracycline. When added, wortmannin was at 20 nM for 24 hr prior to cell lysis. Proteins were separated on 12% SDS–polyacrylamide gels and immunoblotted with a rabbit polyclonal phospho-specific Bad Ser-136 antibody (New England BioLabs, Beverly, MA).

exert antiapoptotic effects, the most well known of which is Akt.[25–32] We have shown that expression of constitutively activated PAK1 T423E protected FL5.12 cells from apoptosis due to interleukin 3 (IL3) withdrawal, and NIH 3T3 cells from Bad- and ceramide-induced apoptosis.[7] Activated PAK1 phosphorylated Bad at both serine-112 and serine-136 *in vitro,* which led to a dissociation of Bad from Bcl-XL/Bcl-2 and the association of Bad with 14-3-3. To determine whether PAK1 phosphorylated these sites in intact cells, we transfected hemagglutinin-tagged murine Bad (HA-mBad) into NIH 3T3 cells containing stably transfected vector alone or vector containing catalytically inactive PAK1 (PAK1 K299R) or constitutively active PAK1 (PAK1 T423E) under the control of an inducible tetracycline promoter.[33] A phospho-specific Bad Ser-136 antibody was used to immuno-precipitate Bad from these cells, and the precipitated protein was detected with an antibody to the Bad N terminus (N20 antibody). Figure 2 shows the results of such an experiment. Vector control cells or noninduced cells

[25] J. Zha, H. Harada, E. Yang, J. Jockel, and S. J. Korsmeyer, *Cell* **87,** 619 (1996).
[26] S. R. Datta, H. Dudek, X. Tao, S. Masters H. Fu, Y. Gotoh, and M. E. Greenberg, *Cell* **91,** 231 (1997).
[27] L. del Peso, M. Gonzalez-Garcia, C. Page, R. Herrera, and G. Nunez, *Science* **278,** 687 (1997).
[28] T. F. Franke, D. R. Kaplin, and L. C. Cantley, *Cell* **88,** 435 (1997).
[29] A. Rukenstein, R. E. Rydel, and L. A. Greene, *J. Neurosci.* **11,** 2552 (1991).
[30] P. Salomoni, M. A. Wasik, R. F. Riedel, K. Reiss, J. K. Choi, T. Skorski, and B. Calabretta, *J. Exp. Med.* **187,** 1995 (1998).
[31] H-G. Wang, U. R. Rapp, and J. C. Reed, *Cell* **87,** 629 (1996).
[32] S. Yano, H. Tokumitsu, and T. R. Soderling, *Nature* **396,** 584 (1998).
[33] M. A. Sells, J. T. Boyd, and J. Chernoff, *J. Cell Biol.* **145,** 837 (1999).

did not have a significant amount of phosphoserine-136. Withdrawal of tetracycline induced the expression of various PAK1 proteins. Expression of constitutively active PAK1 T423E significantly increased the amount of Bad phosphorylation at serine-136. The effect was seen even in the presence of wortmannin, indicating that this effect is independent of the phosphatidylinositol 3-kinase/Akt pathway. Conversely, a constant low level of Bad phosphorylation at serine-136 was observed in NIH 3T3 cells expressing catalytically inactive PAK1 K299R. Correlating with the phosphorylation of Bad by PAK1 T423E, changes in the levels of Bad complexed with Bcl2 were observed.[7]

Methods

NIH 3T3 cells containing the pTet-Splice vector alone, pTet-Splice PAK1 T423E, or pTet-Splice PAK1 K299R[33] are cultured in Dulbecco's modified Eagle's medium with 5% (v/v) calf serum, 5% (v/v) newborn calf serum, 2.5 mM histidinol, 2 μg/ml puromycin, 100 U/ml penicillin, 100 μg/ml streptomycin, 2 mM glutamine, and 10 mM HEPES, pH 7.5, in the presence or absence of 1 μg/ml filter-sterilized tetracycline in 10% (v/v) CO_2. For experiments, NIH 3T3 cells are seeded at 1×10^6 cells and transfected with 10 μg of pcDNA3 HA-mBad using LipofectAMINE (Life Technology, Inc., Gaithersburg, MD) according to the manufacturer's protocol and grown in the presence or absence of tetracycline for an additional 36 hr. Transfected cells are washed three times with phosphate-buffered saline and scraped into lysis buffer 1 [20 mM Tris, pH 8.0, 40 mM NaCl, 1.5 mM MgCl$_2$, 1 mM EDTA, 50 mM NaF, 0.5% Nonidet P-40 (NP-40) with 0.15 U/ml aprotinin, 20 mM leupeptin, and 1 mM phenylmethylsulfonyl fluoride (PMSF)]. Lysates are incubated with a 1:200 dilution of phosphoserine-136 Bad antibody prebound to protein A beads (50 μl 1:1 slurry/immunoprecipitation; Repligen Corp.) for 2 hr at 4°. The immunoprecipitates are washed three times in wash buffer containing 100 mM KCl, 3 mM NaCl, 3.5 mM MgCl$_2$, 10 mM PIPES (piperazine-N,N'-bis-[2-ethanesulfonic acid]), pH 7.3, 1.0 mM dithiothreitol, 1 mM PMSF, and 1% NP-40 and then three times in wash buffer without detergent. The samples are separated on 12% SDS–polyacrylamide gels, transferred to nitrocellulose, and blotted with the Bad (N-20) antibody (Santa Cruz Biotechnology, Santa Cruz, CA) at 1:250 dilution.

In Vivo Analysis of PAK Regulatory Effects on
  Actin–Myosin Cytoskeleton

Endogenous PAK has been shown to colocalize with F-actin in membrane ruffles and other cytoskeletal structures in stimulated cells.[8] Transient overexpression of PAK1 H83,86L in Swiss 3T3 cells induced cell polarization, with accompanying lamellipodia and edge and dorsal ruffles.[11] Expression of catalytically active PAK1 (T423E) resulted in losses of actin stress fibers and focal contacts.[10] Additionally, overexpression of membrane-targeted PAK wild-type (which has constitutive kinase activity[14]) could stimulate neurite outgrowth, a highly specialized type of cytoskeletal rearrangement, in PC12 cells.[34] Studies in *Drosophila* have shown that PAK is required for directed growth of photoreceptor axons,[35] and mutations in PAK3 have been found in a nonsyndromic X-linked form of mental retardation.[36] Consistent with the dramatic cytoskeletal regulatory effects of PAK1 observed in these studies, it has been established that PAK is required for normal motility of fibroblasts and endothelial cells.[33,37] While these findings clearly establish a critical role for PAK1 in cytoskeletal dynamics, the mechanism(s) by which this occurs remains unclear. We have described two novel PAK substrates, myosin light chain kinase (MLCK) and Lim-kinase, that are important regulators of cytoskeletal dynamics.

PAK Regulation of Myosin Light-Chain Kinase

Myosin light-chain kinase mediates phosphorylation of the 20-kDa regulatory myosin light chain (MLC) and thus controls contractility and tension of the cytoskeleton. PAK1 was shown to phosphorylate MLCK both *in vitro* and *in vivo*.[19] This phosphorylation results in a decreased V_{max} of MLCK and, hence, decreased phosphorylation of MLC on Ser-19. Because phosphorylation of this site is necessary for promoting the force-generating ability of myosin II, the net effect of active PAK1 is to decrease cell contractility and tension. In order to examine the effects of PAK1 on myosin function *in vivo*, BHK-21 cells were transfected with PAK1 mutants, and the effects on phosphorylation of Ser-19 on MLC during cell attachment and spreading were analyzed using an antibody that detects phosphorylation at Ser-19.[38] Cells overexpressing PAK1 T423E, PAK1 wild-type, LacZ, or control cells (nontransfected) were allowed to attach and spread on a fibronectin matrix. The cells were lysed at various times during spreading, and immunoblot analysis was performed using the phospho-specific anti-

[34] R. H. Daniels, P. Hall, and G. M. Bokoch, *EMBO J.* **17**, 754 (1998).
[35] J. Hing, J. Xiao, N. Harden, L. LIM, and S. L. Zipursky, *Cell* **97**, 853 (1999).
[36] K. M. Allen, J. G. Gleeson, S. Bagrodia, M. W. Partington, J. C. MacMillan, R. A. Cerione, J. C. Mulley, and C. A. Walsh, *Nature Genet.* **20**, 25 (1998).
[37] W. B. Kiosses, R. H. Daniels, C. Otey, G. M. Bokoch, and M. A. Schwartz, *J. Cell Biol.* **147**, 831 (1999).
[38] F. Matsumura, S. Ono, Y. Yamakita, G. Totsukawa, and S. Yamashiro, *J. Cell Biol.* **140**, 119 (1998).

body. During cell spreading, control cells showed a gradual increase in MLC phosphorylation, with the maximum at the 45-min time point. Cells expressing PAK1 T423E, however, showed a substantially reduced phosphorylation of MLC at each time point, consistent with the inability of these cells to spread (Fig. 3).

Methods

Approximately 1×10^6 cells are lysed in 100 μl of RIPA buffer (100 mM Tris, pH 7.2, 150 mM NaCl, 1% cholate, 1% NP-40, 0.1% SDS; just before use, add 2 mM PMSF, 1% aprotinin, 1 mM vanadate, 50 mM NaF). When blotting for a phosphorylated protein, samples are boiled immediately in Laemmli sample buffer. Thirty microliters of the cell lysate is run on 15% SDS–PAGE gels to resolve proteins less than 20 kDa. Immunoblotting of small molecular weight proteins can be difficult, as proteins <20 kDa often migrate through the transfer membrane or do not bind to the membranes efficiently. The 20-kDa MLC is difficult to detect by conventional transfer methods. We recommend transfer using a semidry transfer apparatus onto Immobilon P^{SQ} (Millipore, Bedford, MA), a type of polyvinylidene fluoride (PVDF) membrane modified to increase protein binding. To transfer, the gel and membrane are sandwiched between six pieces of blotting paper, three on each side. The two pieces of blotting paper closest to the anode are equilibrated in anode buffer I (300 mM Tris, 40% methanol), while the next piece of blotting paper, the gel, and the membrane are all equilibrated in anode buffer II [25 mM Tris, 40% (v/v) methanol] for 10 to 20 min. The three pieces of blotting paper closest to the cathode are equilibrated in cathode buffer III [25 mM Tris, 40 mM glycine, 10% (v/v) methanol]. To enhance binding of small molecular weight proteins, the anode buffers contain increased amounts of methanol, and the applied current is reduced during electroblotting (the milliamperes depend on the size of the gel). After transfer is complete, the membrane is blocked with buffer containing 10% goat serum, 3% bovine serum albumin (BSA) in 10 mM HEPES, pH 7.4. Blots are incubated with antiphosphoserine-19 MLC

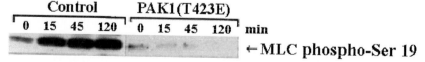

FIG. 3. Overexpression of activated PAK1 (T423E) inhibits MLC phosphorylation. BHK-21 cells overexpressing PAK1 T423E and control cells were lysed at various times after adhesion on fibronectin. Cell lysates were run on a 15% SDS–polyacrylamide gel, transferred, and immunoblotted with phospho-MLC Ser-19.

antibody[38] for 1 hr and washed three times for 20 min with TBS-T [20 mM Tris, pH 7.6, 137 mM NaCl, 0.1% (w/v) Tween 20]. Protein bands are visualized with horseradish peroxidase-conjugated antirabbit IgG (Pierce, Rockford, IL) diluted 1:20,000 and chemiluminescence (Pierce).

PAK Regulation of LIM-kinase

We have found that PAK1 phosphorylates and activates another enzyme important in controlling cytoskeleton dynamics, LIM-kinase. LIM-kinase phosphorylates and inactivates the small actin binding and depolymerizing protein cofilin/actin depolymerizing factor (ADF). Studies have established that LIM-kinase acts downstream of Rac to regulate cortical actin assembly.[39,40] We showed that PAK1 phosphorylates LIM-kinase at threonine-508 in the activation loop and increases LIM-kinase-mediated phosphorylation of cofilin 10-fold *in vitro*.[41] To establish whether LIM-kinase was required for PAK1-induced cytoskeletal rearrangements *in vivo*, we cotransfected BHK-21 cells with the activated PAK1 mutant H83,86L and the catalytically inactive LIM-kinase mutant D460N, which acts as a dominant negative inhibitor. After protein expression, cells were fixed and stained with anti-myc to detect PAK1, anti-LIM-kinase, and rhodamine–phalloidin to assess changes in cytoskeletal structures. PAK1 H83,86L induced dorsal ruffling in BHK-21 cells, and this phenotype was completely inhibited by the catalytically inactive LIM-kinase (Fig. 4, see color insert), indicating that LIM-kinase is acting downstream of PAK1 itself. Conversely, a PAK-specific inhibitor corresponding to the PAK1 autoinhibitory domain (aa 83–149) blocked LIM-kinase-induced cytoskeletal changes, suggesting that phosphorylation and activation of LIM-kinase require PAK1 catalytic activity.

Methods

Cells are plated, grown, and transfected on glass coverslips. After transfection, the cells are fixed in 4% paraformaldehyde for 20 min, permeabilized with 0.5% Triton X-100 in PBS for 20 min, and blocked with 1% BSA in phosphate-buffered saline (PBS) for 1 hr. Coverslips are incubated with anti-myc (9E10) at 1:1000 (v/v) to detect overexpressed PAK1 and anti-LIM-kinase at 1:1200 (v/v) for 1 hr and then washed three times with PBS.

[39] S. Aber, F. A. Barbayannis, H. Hanser, C. Scnieder, C. A. Stanyon, O. Bernard, and P. Caroni, *Nature* **393**, 805 (1998).
[40] N. Yang, O. Higuchi, K. Ohashi, K. Nagata, A. Wada, K Kangawa, E. Nishida, and K. Mizuno, *Nature* **393**, 809 (1998).
[41] D. C. Edwards, L. C. Sanders, G. M. Bokoch, and G. N. Gill, *Nature Cell Biol.* **1**, 253 (1999).

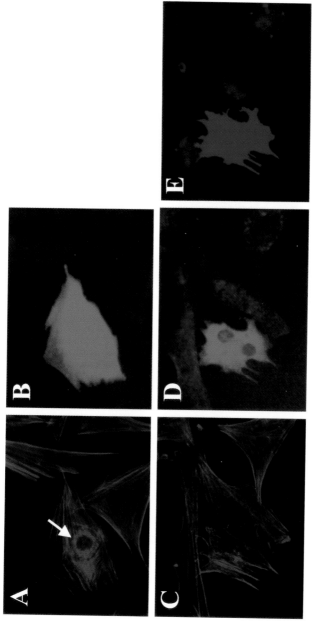

Fig. 4. Inhibition of PAK-induced cytoskeleton changes by kinase-dead LIM-kinase. (A and C) Phalloidin-stained F-actin (red). (B and D) Anti-myc staining to detect PAK1 H83,86L-expressing cells (green). (E) Anti-LIM-kinase staining to detect overexpression of LIM-kinase D460N (blue). (A and B) BHK-21 cell overexpressing PAK1 H83,86L. (A) Phalloidin-stained actin cytoskeleton shows dorsal ruffle (arrow) commonly observed with the PAK1 H83,86L mutation. (C–E) BHK-21 cell expressing PAK1 H83,86L and dominant negative LIM-kinase (D460N). (C) Inhibition of dorsal ruffles by coexpression of LIM-kinase D460N with PAK1 H83,86L. All micrographs were taken with a 40× objective.

Cells are incubated for 1 hr with rhodamine–phalloidin [1:500 (v/v)] (Sigma, St. Louis, MO), fluorescein isothiocyanate (FITC)-conjugated anti-mouse IgG 1:300 (Cappel Laboratories, Cochranville, PA), and Alexa 350 antirabbit IgG (Molecular Probes, Eugene, OR), PBS washed, and mounted with Pro-Long mounting media (Molecular Probes, Eugene, OR). Micrographs are taken using an Olympus IX 70 inverted microscope system equipped with ISee analytical imaging software (Innovision, Raleigh, NC).

Acknowledgments

The authors thank Drs. F. Matsumura for use of the phosphoserine-19 MLC antibody and Drs. M. A. Sells, V. Benard, C. Clougherty, and A. Schürmann for collaborative studies. This work was supported by the NIH (G.M.B.), a National Arthritis Foundation fellowship (C.C.K.), and a U.S. Army Breast Cancer Program award (L.C.S.).

[29] Single Cell Assays for Rac Activity

By Laura J. Taylor, Amy B. Walsh, Patrick Hearing, and Dafna Bar-Sagi

Introduction

The use of the needle microinjection technique to investigate the function of small GTPases offers a number of unique advantages. These include the ability to introduce into cells a controlled amount of macromolecules in a relatively synchronized fashion as well as the ability to analyze short-term cellular responses. Both features of the technique are particularly beneficial for the study of ectopically expressed Rho GTPases because of their broad-spectrum effects on the cellular architecture, which often yields a complex cellular phenotype. The degree to which microinjection studies can provide mechanistic insights is strictly dependent on the deployment of function-based single cell assays. This article describes several single cell assays that can be exploited effectively to study the biological role of Rac proteins.

Microinjection

The following parameters should be considered in preparing cells for injection as well as in analyzing the injected cells.

1. Gridded glass coverslips (Bellco, Vineland, NJ) are used to locate injected cell areas. Coverslips should be acid washed [2:1 $HNO_3:HCl$

(v/v)] for 2 hr in a fume hood followed by 2 hr to overnight rinsing with water. Keep coverslips separated throughout this process. Rinse several times in distilled, deionized water and store in an acid-washed glass container in ethanol.

2. Cells should be plated for 48 hr prior to injection to allow proper adhesion to occur. Shorter intervals often give rise to larger experimental variations.

3. Injection time under ambient conditions should be kept to a minimum (not to exceed a 30-min span). For cells that are sensitive to pH fluctuations, the use of HEPES-buffered medium is recommended.

4. The most adjustable parameters of the actual injection are the sample concentration, injection time, and injection pressure. The following parameters work well for the Eppendorf microinjector 5242 and micromanipulator 5170 or 5171 systems using a micropipette with a tip diameter of 0.5–1 μm. For most plasmid DNA samples, a pressure of 50–60 hPa for 0.3 sec is sufficient with the upper limit of DNA concentration being 125 μg/ml. Protein samples require a higher pressure of 70–80 hPa for 0.3 sec with 10 mg/ml as an upper concentration limit. It is generally best to standardize the pressure and time for a specific cell type and vary the concentration of sample introduced.

5. Most cell types are amenable to microinjection of either DNA or protein solutions. Plasmid DNA for microinjection can be purified using a CsCl–ethidium bromide gradient, Qiagen (Valencia, CA) resin, or polyethylene glycol (PEG) precipitation. The purified DNA is diluted in DNA microinjection buffer [50 mM HEPES (pH 7.2), 100 mM KCl, 5 mM NaH$_2$PO$_4$ (pH 7.3)] to a concentration ranging between 1 and 125 μg/ml. All samples for a microinjection experiment should have equal amounts of total DNA. For microinjection of proteins, samples must be of the highest attainable purity. The protein is diluted in a protein microinjection buffer [20 mM Tris–acetate (pH 7.4), 20 mM NaCl, 1 mM MgCl$_2$, 0.1 mM EDTA, 5 mM 2-mercaptoethanol] or 0.5× phosphate-buffered saline (PBS) (68.5 mM NaCl, 1.35 mM KCl, 2.15 mM Na$_2$HPO$_4 \cdot$ 7H$_2$O, 0.7 mM KH$_2$PO$_4$, pH 7.4) to a concentration ranging between 2 and 10 mg/ml. Both DNA and protein samples need to be centrifuged for 10 min at 14,000g (4°) to remove aggregates and insoluble particles.

6. There are three common methods of detecting injected cells.

Indirect Immunofluorescence

This method is applicable when using epitope-tagged protein or when antibodies against the ectopically expressed protein are available. The following staining protocol has been applied successfully in a variety of cell types.

1. The injected cells are washed with PBS and fixed in 3.7% (v/v) formaldehyde in PBS for 1 hr at room temperature.

2. The cells are washed three times for 5 min each with PBS.

3. Cells are permeabilized with 0.1% (v/v) Triton X-100/PBS for 3 min at room temperature.

4. Cells are washed three times for 5 min each with PBS.

5. Nonspecific antibody binding is blocked by incubating for 5 min at room temperature in 1% bovine serum albumin (BSA)/PBS.

6. Cells are incubated with the primary antibody in 1% BSA/PBS for 1 hr at 37°. The concentration of the primary antibody needs to be determined empirically depending on the level of the ectopically expressed protein and the affinity of the antibody.

7. The cells are washed three times for 5 min each with PBS.

8. The fluorochrome-conjugated secondary antibody [tetramethylrhodamine isothiocyanate (TRITC), fluorescein 5-isothiocyanate (FITC), or Alexa 350 (Molecular Probes, Eugene, OR)] diluted in 1% BSA/PBS is incubated with the cells at 37° for 1 hr. It is often useful to stain the cells with the nuclear stain DAPI (4',6-diamidino-2-phenylindole, Sigma, St. Louis, MO). This is done following secondary antibody incubation. Cells are rinsed once with PBS, and 1–2 μg/ml DAPI (10 mg/ml stock prepared in Milli-Q water) is added in PBS for 10 min at room temperature under light-protected conditions.

9. The cells are then washed three times with PBS and once with distilled water and mounted with 10 μl per 1 cm^2 coverslip of 200 μl Immu-mount (Shandon, Pittsburgh, PA) supplemented with 5 μl of 2% (w/v) p-phenylenediamine (dissolved in distilled, deionized water).

Coinjection of Cytomegalovirus–Green Fluorescent Protein (CMV-GFP)

Injection solution is supplemented with 10 μg/ml of E-GFP (Clontech, Palo Alto, CA). GFP fluorescence can usually be detected within 1 hr of injection.

Coinjection of IgG

Injection solution is supplemented with dog immunoglobulin G (IgG) (Rockland, Gilbertsville, PA) to a final concentration of 4 mg/ml. Detection is done by indirect immunofluorescence with fluorochrome-conjugated goat antidog (Rockland) using the protocol described earlier.

Single-Cell Assays

The following section focuses on single-cell assays for monitoring the effects of Rac on the actin cytoskeleton, cell cycle progression, and superoxide generation.

Membrane Ruffling

One of the earliest responses to Rac activation is the reorganization of the cortical actin cytoskeleton, which results in the stimulation of membrane ruffling.[1] Membrane ruffling activity can be assayed as follows.

1. Prior to microinjection, confluent or subconfluent cells are serum starved for 24–48 hr to reduce background ruffling activity.

2. Microinjected cells are analyzed for the appearance of membrane ruffles as early as 15 min (for injected proteins) or 2 hr (for injected DNA) following injection.

3. Membrane ruffles are observed easily in living cells by phase-contrast microscopy. Because ruffling activity is asynchronous, scoring should be done at several time points after injection. Significant fluctuations in pH and temperature (often caused by prolonged microscopic inspection) inhibit membrane ruffling.

4. Membrane ruffles can also be visualized by staining fixed cells with fluorochrome-conjugated phalloidin (Molecular Probes). Although used frequently, this method might provide an underestimate of ruffling activity because some membrane ruffles are not sufficiently stable to withstand the fixation procedure.

Fluid-Phase Pinocytosis

Membrane ruffling is accompanied by fluid-phase pinocytosis.[2] Fluid-phase pinocytic vesicles form soon after the induction of membrane ruffling and can be detected easily by their highly refractile appearance under phase-contrast microscopy. Fluorescein-conjugated dextran (FITC-dextran) is often used to monitor fluid-phase pinocytosis. FITC-dextran (Molecular Probes) is stable and soluble in aqueous solution and does not bind to the plasma membrane.

1. Confluent or subconfluent cells are serum starved for 24 hr.

2. Cells are microinjected with protein or plasmid DNA.

3. As membrane ruffling commences, FITC-dextran (0.5–1 mg/ml) is added to the media and incubated for 30 min. Conjugated dextrans varying in molecular weight (3000–150,000) can be used to distinguish between micropinocytosis and macropinocytosis.[3]

4. The cells are washed with Dulbecco's modified Eagle's medium (DMEM) and visualized using fluorescence microscopy. Due to the high rate of recycling of pinocytic vesicles it is advisable to use fixable dextrans.

---

[1] A. J. Ridley, H. F. Paterson, C. L. Johnston, D. Diekmann, and A. Hall, *Cell* **70,** 401 (1992).
[2] D. Bar-Sagi and J. R. Feramisco, *Science* **233,** 1061 (1986).
[3] N. Araki, M. T. Johnson, and J. A. Swanson, *J. Cell Biol.* **135,** 1249 (1996).

5. The rates of FITC-dextran uptake are compared between experimental and control injected cells by counting the number of pinocytic vesicles per injected cell.

Wound Healing

Rac-mediated changes in the organization of the actin cytoskeleton contribute to the regulation of cell movement. To monitor actin-based cell motility, a wound-healing assay can be employed as follows.

1. Cells are grown to 80–90% confluency and then serum starved for 24–48 hr.

2. A wound diameter between 130 and 200 μm is introduced into the confluent monolayer with a pair of extra fine microforceps.

3. After 10 min, cells aligning the wound and the adjacent row of cells are microinjected with the desired expression plasmid or protein.

4. The wound diameter is measured at the time of wounding and various time points postinjection until the wound heals. The mean wound width is calculated by measuring the diameter of the wound every 80 μm over the length of the wound for approximately 500 μm.

5. The expression of the protein in the injected cells can be verified by indirect immunofluorescence.

Activation of Serum Response Element

Immediate early gene expression of the c-*fos* gene is dependent on transcription of the *cis*-acting serum response element (SRE) within its promoter. The transcriptional activity of an SRE reporter construct consisting of five copies of the SRE fused to chloramphenicol acetyltransferase (5× SRE-CAT) can thus be used to assess progression through the early G_1 phase of the cell cycle.[4]

1. Cells are grown to 80–90% confluency and subsequently serum starved for 36–48 hr. Serum-starved cells are injected with a solution containing 5× SRE-CAT (25 μg/ml), injection marker (10 μg/ml E-GFP or 4 mg/ml dog IgG), and a test plasmid.

2. Cells are incubated for 3–12 hr and then fixed with 3.7% formaldehyde.

3. CAT expression is detected by indirect immunofluorescence using anti-CAT antibody (Accurate Chemical and Scientific Corporation, Westbury, NY).

4. Mechanical stress results in a significant activation of the SRE. Therefore, injection conditions need to be adjusted with respect to positive pressure and duration such that no CAT expression is detected in the negative control. Serum stimulation (20%) can be used as a positive control.

[4] R. Graham and M. Gilman, *Science* **251**, 189 (1991).

Activation of E2F1

The E2F1 promoter is activated during mid-to-late G_1 phase of the cell cycle as a consequence of cyclin D/cdk 4/6-mediated phosphorylation of the retinoblastoma protein.[5] Therefore, an E2F1 promoter linked to green fluorescent protein can be used as a reporter construct to monitor Cdk4/6 activation in live or fixed cells. We have introduced GFP downstream of the $-176/+36$ portion of the E2F1 promoter, which has been demonstrated to undergo a 40-fold induction on serum stimulation.[6]

1. Cells are grown to 80–90% confluency and then serum starved for 36–48 hr. Serum-starved cells are injected with a solution containing E2F1-GFP (75 μg/ml), injection marker (4 mg/ml dog IgG), and the test plasmid.

2. Cells are incubated for 14–26 hr. The optimal time for GFP needs to be established empirically and is dependent on the length of G_1 in the cell type used.

3. GFP expression can be detected in either live or fixed cells. In live cells, detection of the GFP signal is improved by reducing the amount of growth medium covering the cells.

4. Serum stimulation (20%) can be used as a positive control, and injection of Cdk inhibitors (e.g., p21^{cip}) or serum starvation can be used as negative controls.

BrdU Incorporation

S-phase entry can be assayed by monitoring the incorporation of the nucleotide analog 5-bromo-2′-deoxyuridine (BrdU) in newly synthesized DNA.

1. Cells are grown to 80–90% confluency and then serum starved for 36–48 hr. Serum-starved cells are injected with expression plasmid or purified protein. Note that expression plasmid concentrations in excess of 10 μg/ml inhibit BrdU incorporation. Serum stimulation (20%) can be used as a positive control.

2. After approximately 1 hr, 10 μM BrdU (Boehringer Mannheim, Indianapolis, IN) is added to the medium and cells are incubated for 26–30 hr, depending on the length of G_1 and S phase of the cell type injected. Alternatively, BrdU can be added for a much shorter "pulse" of 30 min to 4 hr if the kinetics of cell cycle progression are well established in the cell type used.

3. Rinse cells twice with PBS and then fix in 90% ethanol/5% water/5% acetic acid (v/v) for 20 min at $-20°$. The fixation protocol for BrdU incorporation is incompatible with GFP detection.

[5] C. J. Sherr, *Science* **274**, 1672 (1996).
[6] K. M. Hsiao, S. L. McMahon, and P. J. Farnham, *Genes. Dev.* **8**, 1526 (1994).

4. Cells are washed with at least four changes of PBS for 1 hr at room temperature, rinsed in TBS-T [20 mM Tris–HCl (pH 7.6), 150 mM NaCl, 0.1% Tween 20], and then incubated for 10 min in 1% BSA/TBS-T.

5. Cells are then incubated sequentially at room temperature for 10 min each in 2 N HCl, 0.1 M sodium borate (pH 8.5), and 1% BSA/TBS-T.

6. Cells are incubated with anti-BrdU (Boehringer Mannheim) made in 1% BSA/PBS for 1 hr at 37°.

7. Cells are washed 3 × 5 min with 1% BSA/TBS-T.

8. Cells are incubated for 1 hr with secondary monoclonal antibody made in 1% BSA/PBS at 37°. If dog IgG was injected as a marker, the secondary antibody solution is supplemented with goat antidog IgG and the incubation time is extended to 1 hr and 20 min.

9. Cells are washed and mounted for indirect immunofluorescence as described earlier.

Superoxide Production

Rac proteins have been implicated in the regulation of the activity of a membrane-bound NADPH oxidase to produce superoxide radicals.[7] Thus, the production of superoxide can be used as a readout for Rac activity. At the single-cell level this can be achieved using the nitroblue tetrazolium (NBT) reduction assay as follows.

1. Microinjected cells are rinsed once with DMEM and then incubated with DMEM containing 0.5% (w/v) NBT (Sigma) for 1 hr at 37°.

2. Cells are fixed for 1 hr at room temperature with 3.7% formaldehyde.

3. Cells are processed for indirect immunofluorescence using the protocol described earlier.

4. The production of superoxide can be detected by the appearance of a purple precipitate over the entire cell. The precipitate is clearly visible by phase-contrast microscopy (40× objective), but has a more pronounced appearance using bright-field microscopy. The immunofluorescence signal is detected by fluorescence microscopy.

5. The specificity of staining is assessed by two criteria. The first is inhibition by superoxide dismutase (SOD). SOD (Sigma, St. Louis, MO) is added along with NBT to the incubation medium at a concentration of 300 units/ml. Under these conditions, specific NBT staining is completely abolished. The second is the correlation between NBT staining and protein expression. Typically 70–80% of cells microinjected with RacV12 expression plasmid are positive for NBT staining.

[7] A. Abo, E. Pick, A. Hall, N. Totty, C. Teahan, and A. Segal, *Nature* **353**, 668 (1991).

In principle, the assay is applicable for both phagocytic and nonphagocytic cells. However, among nonphagocytic cells the intensity of NBT staining is variable and feasibility should be established on a per cell type basis.
It should be noted that the formation of the NBT precipitate is accompanied
by morphological changes of the positively stained cells. The most notice-
able change is cell rounding and partial loss of adherence. Prolonged incubation with NBT can lead to a complete detachment of the stained cells.

Acknowledgment

Work discussed in this article was supported by National Institutes of Health Grants
CA55360 and CA28146.

[30] Effect of Rho GTPases on Na–H Exchanger in Mammalian Cells

By SHERYL P. DENKER, WEIHONG YAN, and DIANE L. BARBER

Introduction

Na–H exchangers comprise a family of ion exchangers that regulate
intracellular pH (pH_i) and cell volume by catalyzing the electroneutral
exchange of intracellular protons (H^+) for extracellular sodium ions (Na^+).
In mammalian cells, the family consists of five isoforms (NHE1–NHE5)
found at the plasma membrane[1,2] and one isoform (NHE6) found at the
mitochondrial membrane.[3] All the isoforms have a predicted topology of
12 transmembrane-spanning α-helical domains at the N terminus and a
cytoplasmic domain at the C terminus. The transmembrane segment translocates ions and modulates the pH sensitivity of ion translocation through
a modifier site that binds H^+.[4] The C-terminal cytoplasmic domain binds
regulatory proteins and modulates ion translocation through a postulated
conformational change. In addition, the cytoplasmic domain contains serine
residues that are phosphorylated when the exchanger is regulated by growth

[1] J. Noel and J. Pouyssegur, *Am. J. Physiol.* **268,** C283 (1995).
[2] J. Orlowski and S. Grinstein, *J. Biol. Chem.* **272,** 22373 (1997).
[3] M. Numata, K. Petrecca, N. Lake, and J. Orlowski, *J. Biol. Chem.* **273,** 6951 (1998).
[4] S. Wakabayashi, P. Fafournoux, C. Sardet, and J. Pouyssegur, *Proc. Natl. Acad. Sci. U.S.A.*
89, 2424 (1992).

factors.[5] Our studies on the regulation of NHE by Rho family GTPases[6–9] have focused on NHE1, which is the only ubiquitously expressed isoform of the exchanger.

We have taken three approaches to study the regulation of NHE1 by Rho family GTPases: measuring transport activity, identifying associations with regulatory proteins, and monitoring phosphorylation. We study transport activity primarily in fibroblasts that express only the NHE1 isoform[10] and transiently express mutationally active and inactive GTPases and kinases. We also use pharmacological reagents that selectively target signaling proteins in the Rho pathway, specifically the bacterial exotoxin C3 and the pyridine derivative Y-27632 to inhibit the activities of RhoA and the RhoA kinase, p160ROCK, respectively. Using this approach, we have determined that RhoA and p160ROCK mediate NHE1 activation by integrin receptors, lysophosphatidic acid receptors, and the GTPase $G\alpha_{13}$, and that p160ROCK directly phosphorylates NHE1.[7]

Measurement of NHE1 Activity

Ion translocation is generally determined by measuring the rate of either Na^+ influx or H^+ efflux. We determine H^+ efflux by monitoring pH_i with H^+-sensitive fluorescent dyes.[6,11] Steady-state and dynamic changes in pH_i are determined by the balance between acid-extruding and acid-loading processes. Hence, if HCO_3^- transport mechanisms are blocked by measuring pH_i in the nominal absence of HCO_3^- (using a HEPES buffer), the only substantial acid extrusion is due to NHE1 activity. The standard HEPES buffer we use contains (in mM): 125 NaCl, 3 KCl, 1 $CaCl_2$, 2 NaH_2PO_4, 10 glucose, and 32 HEPES titrated to pH 7.4 at 37° with NaOH.

Dyes, Instrumentation, and Calibration

We routinely use the dual-excitation pH-sensitive dye BCECF [2,6-biscarboxyethyl-5(6)-carboxyfluorescein; Molecular Probes, Eugene, OR] to measure pH_i, although the dual-emission pH-sensitive SNARF-1 and

[5] C. Sardet, P. Fafournoux, and J. Pouyssegur, *J. Biol. Chem.* **266**, 19166 (1991).
[6] R. Hooley, C. Y. Yu, M. Symons, and D. L. Barber, *J. Biol. Chem.* **271**, 6152 (1996).
[7] T. Tominaga, T. Ishizaki, S. Narumiya, and D. L. Barber, *EMBO J.* **17**, 4712 (1998).
[8] T. Tominaga and D. L. Barber, *Mol. Biol. Cell* **9**, 2287 (1998).
[9] Z. S. Vexler, M. Symons, and D. L. Barber, *J. Biol. Chem.* **271**, 22281 (1996).
[10] X. Lin, T. A. Voyno-Yasenetskaya, R. Hooley, C. Y. Lin, J. Orlowski, and D. L. Barber, *J. Biol. Chem.* **271**, 22604 (1996).
[11] T. A. Voyno-Yasenetskaya, M. P. Faure, N. G. Ahn, and H. R. Bourne, *J. Biol. Chem.* **271**, 21081 (1996).

SNAFL dyes (Molecular Probes) have been used with confocal imaging.[12] BCECF solutions are unstable, and prepackaged 50-μg aliquots can be purchased for use in a single day. We prepare a 100 mM stock solution by dissolving 50 μg in 100 μl dimethyl sulfoxide (DMSO), sonicating briefly, and then adding HEPES buffer. Cells are incubated with 1 μM of the membrane-permeant acetoxymethyl ester BCECF/AM for 10–15 min at 37° in the absence of CO_2. Intracellular esterases cleave the dye to a membrane-impermeant form that remains inside the cell. Because intracellular esterases vary among cell lines, incubation times and dye concentrations required to obtain efficient dye loading will differ among cell types and must be determined empirically. Dye concentrations greater than 5 μM increase noise and variability in the measurements.

We determine NHE1 activity primarily in cultured cell lines, although freshly isolated cells in culture[13] or in suspension[14] have also been used. Cultured cells are plated on coverslips 48 hr prior to determinations and used at ~80% confluency. The pH$_i$ in cell populations is determined using a spectrofluorometer by mounting coverslips in a cuvette attached with an inflow/outflow apparatus and placed in a thermostatically controlled cuvette holder. The pH$_i$ of single cells is determined on coverslips mounted in a thermostatically controlled perfusion chamber on the stage of a Zeiss Axiovert. Images are collected with an intensified charged-coupled device (CCD) and analyzed using the MetaFluor software program (Universal Imaging, West Chester, PA). Cells are perfused at a rate of 2 ml/min using a multichannel perfusion pump to achieve rapid delivery and exchange of solutions.

For measurements using BCECF, ratiometric determinations are made by monitoring emission at 530 nm while alternately exciting the dye at pH-sensitive (~500 nm) and pH-insensitive (~440 nm) wavelengths. Determining emission at a pH-insensitive excitation is necessary to normalize pH-sensitive measurements to the intracellular concentration of dye. Ratiometric determinations are calibrated to pH$_i$ by using the nigericin technique as described by Thomas et al.[15] Nigericin is a K–H exchanger, and if external and internal [K$^+$] are equal, the extracellular pH and pH$_i$ should be equilibrated. Experiments are terminated by perfusing cells with 10 μM

[12] B. J. Muller-Borer, H. Yang, S. A. M. Marzouk, J. J. Lemasters, and W. E. Cascio, Am. J. Physiol. 275, H1937 (1998).
[13] G. Boyarsky, M. B. Ganz, R. B. Sterzel, and W. F. Boron, Am. J. Physiol. 255, C844 (1988).
[14] N. Demaurex, G. P. Downey, T. K. Waddell, and S. Grinstein, J. Cell Biol. 133, 1391 (1996).
[15] J. A. Thomas, R. N. Buchsbaum, A. Zimniak, and E. Racker, Biochemistry 18, 2210 (1979).

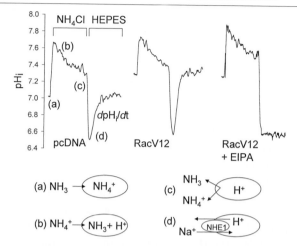

FIG. 1. Activated Rac increases pH_i recoveries from an NH_4Cl-induced acid load. Cells were exposed to 30 mM NH_4Cl, which results in a rapid increase in pH_i due to the entry of uncharged NH_3 (a), followed by a slow decline in pH_i as NH_4^+ diffuses into the cell and dissociates to NH_3 and H^+ (b). Replacing NH_4Cl with a HEPES buffer causes pH_i to acidify below the steady-state level as NH_3 diffuses out of the cell rapidly, trapping intracellular H^+ (c). The rate of pH_i recovery (dpH_i/dt) from this acid load is an index of Na–H exchange activity (d). The indicated tracings represent determinations made using cells transfected with pcDNA control vector (left tracing) or mutationally activated RacV12 (middle tracing). The contribution of NHE1 to pH_i recoveries was determined by treating cells expressing RacV12 with EIPA (25 mM) for 10 min before incubating with NH_4Cl (right tracing).

nigericin (Molecular Probes) in a KCl buffer containing (in mM): 105 KCl, 32 HEPES, and 1 $MgCl_2$. The emission ratio at specified extracellular pH values can therefore be used to calibrate pH_i. Stock nigericin solutions of 100 mM, prepared in DMSO and KCl buffer, are stable for 1 month, and working solutions of 10 μM should be prepared daily. The pH of working solutions is titrated to a specified pH using KOH. In most cells, the nigericin calibration curve is linear between pH 6.4 and 7.6, which allows us to use a two-point calibration (generally pH 6.8–6.9 and 7.1–7.2) at the end of each experiment.

NHE1 Activity

In intact cells, NHE1 activity is best studied by inducing an acid load within the cell and monitoring the rate of pH_i recovery (Fig. 1).[16] Cells are exposed for 5–10 min to external NH_4^+, which results in a rapid increase in pH_i as NH_3 diffuses into the cell and complexes with intracellular H^+

[16] W. F. Boron and P. De Weer, *J. Gen. Physiol.* **67**, 91 (1976).

(Fig. 1a). NH_4^+ diffuses more slowly, which causes a steady decline in pH_i as NH_4^+ dissociates to NH_3 and H^+ (Fig. 1b). Rapidly removing external NH_4^+ and replacing it with the HEPES buffer causes pH_i to quickly fall lower than its initial value as NH_3 diffuses out of the cell, trapping intracellular H^+ (Fig. 1c). The rate of pH_i recovery from this acid load (dpH_i/dt) is an index of NHE1 activity (Fig. 1d). This is determined by fitting the pH_i as a function of time to a third- or fourth-order polynomial and evaluating the derivative of the exponential fit at selected pH_i. NHE1 activity is modified by a H^+-binding site within the transmembrane domain, which renders the exchanger exquisitely pH sensitive. It is critical, therefore, to compare dpH_i/dt for different conditions at identical pH_i values. Moreover, to confirm that pH_i recovery is due to NHE1 activity, we make control measurements that include NHE1 inhibitors such as ethylisopropylamiloride (EIPA; 25 $\mu M;$ Research Biochemicals, Natick, MA) or the Hoechst compound HOE694 (10 μM; Frankfurt, Germany) or that replace extracellular Na^+ with choline (125 m$M;$ Sigma, St. Louis, MO) or N-methyl-D-glucamine ($NMDG^+$, 125 m$M;$ Sigma) (Fig. 1).

Determining H^+ Efflux

The H^+ efflux rate is determined as the product of dpH_i/dt and the intrinsic buffering power of the cell (β_1) at a given pH_i. By exposing the cell to a weak acid or weak base, we can determine the biochemical and organellar buffering as a function of pH_i, independently of CO_2 and transport processes. Because this determination requires that the ΔpH_i not be regulated by transport processes, it is performed in a nominally HCO_3^- free HEPES buffer to block HCO_3^- exchangers, and in the absence of external Na^+ (replaced by choline or $NMDG^+$) with added EIPA to block Na–H exchangers. We determine β_1 according to previously described methods[13,17] by exposing cells to a weak base of NH_4^+/NH_3. Because one intracellular H^+ is consumed for each NH_4^+ formed, the mean buffering power of all non-NH_4^+/NH_3 intracellular buffers is determined according to Eq. (1).

$$\beta = \frac{\Delta(NH_4^+)_i}{\Delta pH_i} \tag{1}$$

At a specified pH_i, achieved by using the NH_4Cl prepulse technique described earlier, cells are exposed to 1 mM NH_4^+/NH_3 for 1–2 min and the resulting pH changes on addition and removal are measured. The specified pH_i is determined as the midpoint of the ΔpH_i, and calculations for de-

[17] A. Roos and W. F. Boron, *Physiol. Rev.* **61**, 296 (1981).

termining unknown parameters such as the pK and the outside concentration of NH_4^+ can be found in Roos and Boron[17] and Boyarsky et al.[13] The buffering power of the clonal cell lines we have studied is not constant, but varies inversely as a function of pH_i (Fig. 2A). We have used cellular buffering powers and pH_i recovery rates from an acid load to determine that expression of constitutively active RhoAV14 or Cdc42V12 increases H^+ efflux rates and that coexpression of dominant negative p160ROCK selectively blocks the increase due to RhoAV14 but not that due to Cdc42V12 (Figs. 2B and 2C).

NHE1-Interacting Proteins

The C-terminal cytoplasmic domain of NHE1 directly binds a number of regulatory proteins. This binding results in conformational changes in NHE1 that modulate ion transport activity. Determining which proteins associate with NHE1, therefore, is important for understanding how Rho family GTPases regulate exchanger activity. To identify NHE1-interacting proteins that are regulated by GTPases, we have used biochemical methods and two cloning strategies, the yeast two-hybrid system and λgt11 expression library screening with a ^{32}P-labeled glutathione S-transferase (GST) NHE1 probe.[18] Proteins that interact with NHE1 include the calcineurin B homologous protein, CHP, whose phosphorylation status is regulated by Rho family GTPases[18]; the NCK-interacting kinase, NIK (W. Yan, submitted), which may act downstream of Rac[19]; and the ERM family of actin-binding proteins (S. P. Denker, submitted) (Fig. 3), whose membrane association is regulated by Rho family GTPases.[20,21] Additionally, Wakabayashi et al.[22] determined that direct association with calmodulin increases NHE1 activity. To establish direct association with NHE1 and its physiological relevance, we use in vitro-binding assays and coimmunoprecipitation approaches, respectively. This section focuses on methods used successfully in our laboratory that may be applied to other NHE isoforms.

In Vitro Binding Assays

In vitro binding using GST–NHE1 fusion proteins and [^{35}S]methionine-labeled in vitro-translated proteins is a quick method to screen candidate

[18] X. Lin and D. L. Barber, Proc. Natl. Acad. Sci. U.S.A. 93, 12631 (1996).

[19] Y. C. Su, J. Han, S. Xu, M. Cobb, and E. Y. Skolnik, EMBO J. 16, 1279 (1997).

[20] M. Hirao, N. Sato, T. Kondo, S. Yonemura, M. Monden, T. Sasaki, Y. Takai, and S. Tsukita, J. Cell Biol. 135, 37 (1996).

[21] D. J. Mackay, F. Esch, H. Furthmayr, and A. Hall, J. Cell Biol. 138, 927 (1997).

[22] S. Wakabayashi, T. Ikeda, T. Iwamoto, J. Pouyssegur, and M. Shigekawa, Biochemistry 36, 12854 (1997).

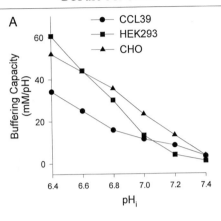

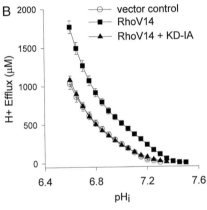

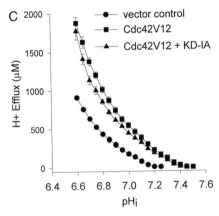

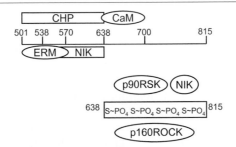

FIG. 3. Binding and interaction sites for the indicated structural and regulatory proteins on the carboxy-terminal cytoplasmic domain of NHE1. ERM (ezrin, radixin, and moesin), NIK (Nck-interacting kinase), CHP (calcineurin homologous protein), and CaM (calmodulin), directly bind to NHE1. The maximum possible overlaps in binding regions are shown. p160ROCK (Rho kinase), p90[RSK] (ribosomal S6 kinase), and NIK are serine/threonine kinases that phosphorylate the distal C terminus of the cytoplasmic domain. Of these kinases, only NIK directly binds to NHE1.

binding partners identified by sequence motifs in NHE1. Once a positive interaction has been established, truncations and deletions of the fusion protein can be used to determine the binding site more precisely.

Preparation of GST–NHE1 Fusion Proteins

The full-length cytoplasmic domain (amino acids 501–815, human se-quence) of NHE1 expressed as a GST fusion protein is relatively insoluble in BL21 cells. We have optimized the bacterial growth and induction condi-tions to increase the yield of GST-NHE1 full-length constructs in the soluble fraction. General transformation and purification protocols are found in the "GST Gene Fusion System Handbook." [23] A single transformed colony is inoculated into 20 ml 2XYTAG media and grown overnight at 37°. The next day the culture is transferred at 1/10 dilution into 200 ml 2XYTAG and grown at 30° until absorbance at 600 nm is 0.7–0.8 (approximately 3

---

[23] GST Gene Fusion System, 18-1123-20, Amersham Pharmacia Biotech, Piscataway, NJ (1997).

FIG. 2. Cell-specific buffering capacities and increases in H^+ efflux in response to activated Rho and Cdc42. (A) The intracellular buffering capacity decreases with increasing pH_i and varies in the clonal cell lines CCL39, hamster lung fibroblasts; HEK293, human embryonic kidney fibroblasts; and CHO, hamster ovary fibroblasts. (B and C) Increases in H^+ efflux as a function of pH_i in cells expressing mutationally activated RhoV14 or Cdc42V12. Coexpres-sion of a kinase-inactive p160ROCK (KD-IA) containing a $K^{105} \rightarrow A$ substitution in the kinase domain and an $I^{1009} \rightarrow A$ substitution in the Rho-binding domain inhibits increases in H^+ efflux in response to RhoV14, but not to Cdc42V12.

hr). Protein expression is induced in the presence of 0.25 mM isopropyl-β-D-thiogalactoside (IPTG) at 25–27° for 3 hr. If the bacteria are induced at a density greater than 0.8 or a temperature greater than 28°, the amount of soluble full-length GST-NHE1 obtained is reduced. We resuspend bacterial cell pellets in 10 ml of ice-cold phosphate-buffered saline (PBS) with 0.1 U/ml aprotinin, 20 μg/ml leupeptin, and 1 mM Pefabloc (Roche Molecular Biochemicals, Indianapolis, IN). Cells are lysed on ice by four cycles of sonication (15 sec each, output = 3 W) and solubilized in the presence of 1.0% Triton X-100 at 4° with rocking for 1 hr. Subsequent purification steps are as described in the GST handbook. Our typical yields from a 200-ml culture are 30–50 μg for full-length cytoplasmic domain NHE proteins and 10-fold more for the more soluble truncated proteins. We estimate the concentration of fusion protein on the glutathione-Sepharose beads by SDS–PAGE and Coomassie blue staining, using bovine serum albumin as a standard.

In Vitro Translation and Binding Assays

In vitro translation can be performed using commercially available kits. We use the Promega TNT[24] kit for coupled transcription and translation in the presence of [^{35}S]methionine. Because the specific labeling of each protein differs, the amount of labeled product used in binding assays should first be normalized by autoradiography of samples separated by SDS–PAGE (Fig. 4a).

We prewash fusion protein beads with binding buffer containing (in mM) 50 Tris, pH 7.5, 150 NaCl, 1 EDTA, and 0.1% NP-40. Fusion proteins (1–2 μg) and in vitro translation products are incubated in 0.5 ml of binding buffer with rocking (4° for 1 hr). Fusion protein complexes are pelleted (10,000g for 2 min at 4°), the supernatant is aspirated, and the beads are washed in 0.5 ml of buffer with a quick and gentle vortex. We set the vortex on 5 and pulse for 1–2 sec. This method of washing reduces nonspecific binding, especially for fusion proteins that may aggregate in vitro. Protein complexes are pelleted as described earlier, and the washes are repeated three times. To reduce background signals, the liquid remaining after the final wash is carefully removed with a narrow gel-loading tip. Samples are boiled in SDS–PAGE sample buffer with dithiothreitol (DTT) for 3 min, vortexed briefly, and pelleted. Standard fixation and fluorography techniques can then be used; film exposure times range from 8 to 16 hr. When available, positive controls for either NHE or the in vitro-translated product should be included because the buffer recipe is empirical and modifications

---

[24] TNT Coupled Reticulocyte Lysate System, TB126, Promega Corp., Madison, WI (1997).

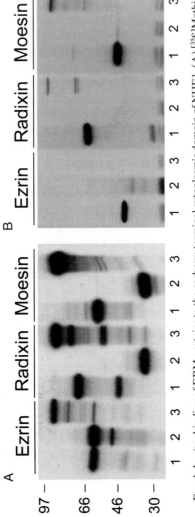

FIG. 4. *In vitro* binding of ERM proteins to the carboxy-terminal cytoplasmic domain of NHE1. (A) [^{35}S]Methionine-labeled *in vitro*-translated products of amino (lane 1), carboxyl (lane 2), and full-length (lane 3) ezrin, radixin, and moesin used in binding assays. (B) *In vitro* binding to GST–NHE1 (501–815). Only the amino termini (lane 1) of ERM proteins bind NHE1. Molecular masses (in kDa) are indicated on the left.

to the salt concentration and detergent type may be necessary. Useful negative controls include GST alone or, if available, a second fusion protein. Negative controls may also include truncations or deletions of the *in vitro*-translated product. Using this approach, we have determined that amino, but not carboxyl, termini of ERM proteins interact with the proximal 70 amino acids of the C terminus of NHE1 (Fig. 4b).

Coimmunoprecipitation

There are currently no commercially available antibodies that can immunoprecipitate endogenous NHE1. The model system we use, therefore, is an NHE1-deficient cell line stably expressing recombinant NHE1 tagged at the carboxyl terminus with an HA or an EE epitope. In this system it has been straightforward to demonstrate an association of NHE1 with CHP, as the pool of NHE1 associated with CHP is released by gentle solubilization conditions. The association of NHE1 and ERM proteins, however, has been more difficult to demonstrate consistently, probably because this complex is also associated with the cytoskeleton and harsher solubilization conditions are required (S. P. Denker, submitted). Thus, although we can detect ezrin in NHE1 immunoprecipitates from cells treated without cross-linker, we obtain more consistent results when a cross-linker is used (Fig. 5). The specific cross-linker to be used must be determined empirically. For example, using the *N*-hydroxysuccinimide ester dithiobis(succinimidyl propio-

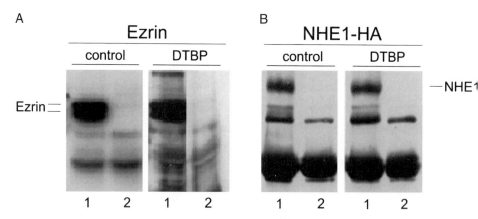

Fɪɢ. 5. Ezrin associates with NHE1 *in vivo*. Fibroblasts stably expressing an HA-tagged NHE1 (lane 1) or vector (lane 2) were treated without (control) or with cross-linker (DTBP; 2 m*M*), and NHE1 was immunoprecipitated from cell lysates. (A) Immunoblot for ezrin (doublet) shows the presence of ezrin in immunoprecipitates only from cells expressing NHE1. (B) Immunoblot for HA shows equivalent amounts of NHE1 are immunoprecipitated in the presence and absence of DTBP.

nate) (DSP; Pierce, Rockford, IL), we have detected NHE1 oligomerization but not ERM coimmunoprecipitation, whereas using the imido ester dimethyl 3,3'-dithiobispropionimidate-dihydrochloride (DTBP; Pierce), we have detected ERM coimmunoprecipitation but not NHE1 oligomerization.

Cells are grown to a maximum of 90% confluency, which maintains NHE1 expression. Monolayers are washed twice with PBS and incubated with 2 mM DTBP in an amine-free buffer, pH 8.5, at 4° for 1 hr or at room temperature for 30 min. We use the HEPES buffer described in the previous section for determining pH$_i$. The free cross-linker is quenched by a brief rinse, and monolayers are incubated (4° for 15 min) in the same buffer containing 25 mM glycine. Monolayers are solubilized on ice for 10 min (buffer as described earlier, with pH 7.0 and containing 1% NP-40, 0.5% deoxycholic acid, 0.5% SDS, and protease inhibitors), collected, and then solubilized at 4° for another hour. The longer solubilization time increases the efficiency of coimmunoprecipitation, presumably by increasing the amount of NHE1–ezrin complexes released from the cytoskeleton. NHE1 and associated proteins are immunoprecipitated from precleared postnuclear supernatants (850g, 3 min at 4°) by an antibody directed against the HA epitope (Roche Molecular Biochemicals).

NHE1 Phosphorylation

Activation of NHE1 by mitogenic stimuli is associated with the phosphorylation of a number of serine residues within the distal C-terminal cytoplasmic domain (amino acids 638–815 of human NHE1). The Rho kinase p160ROCK,[7] the Ste20-related kinase NIK (W. Yan, submitted), and the ERK-regulated kinase p90^{RSK} [25] directly phosphorylate NHE1. Transport activity, however, can be regulated by both phosphorylation-dependent and -independent mechanisms. Phosphorylation of NHE1 in response to integrin receptors and heptahelical receptors coupled to Gα_{13} is mediated by p160ROCK, and this phosphorylation is critical for increases in NHE1 activity induced by these pathways.[7,8] In contrast, abolishing NHE1 phosphorylation by deleting C-terminal serine residues inhibits increases in transport activity in response to growth factors by only 50%[4,26] and has no effect on activity in response to osmotic shrinkage.[27]

[25] E. Takahashi, J. Abe, B. Gallis, R. Aerbersold, D. J. Spring, E. G. Krebs, and B. C. Berk, *J. Biol. Chem.* **274**, 20206 (1999).

[26] S. Wakabayashi, B. Bertrand, M. Shigekawa, P. Fafournoux, and J. Pouyssegur, *J. Biol. Chem.* **269**, 5583 (1994).

[27] S. Grinstein, M. Woodside, C. Sardet, J. Pouyssegur, and D. Rotin, *J. Biol. Chem.* **267**, 23823 (1992).

A

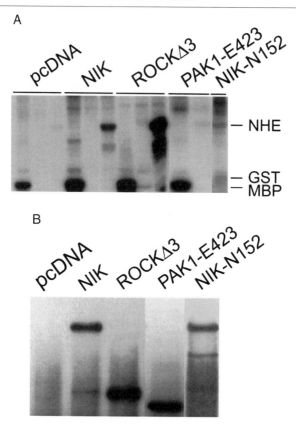

FIG. 6. The serine/threonine kinases NIK and p160ROCK, but not PAK1, phosphorylate NHE1 *in vitro*. (A) The indicated kinases, tagged with a myc epitope and transiently expressed in HEK293 cells, were immunoprecipitated and used for *in vitro* kinases assays. Phosphorylation of the substrates myelin basic protein (lane 1), GST (lane 2), and a truncation of the cytoplasmic domain of NHE1 containing amino acids 638–815 (lane 3) was determined by autoradiography. The mobility of each substrate is indicated on the right. (B) Immunoblot analysis to determine the abundance of the indicated kinases in the immunoprecipitates that were used for *in vitro* kinase assays.

In Vitro Kinase Assay

We use GST–NHE1 fusion proteins (full-length or a distal 638–815 segment) as substrates for *in vitro* kinase assays, and as positive controls we use nonspecific kinase substrates such as myelin basic protein and histone. Enzymes include kinases that are immunoprecipitated from cells using either antibodies to endogenously expressed proteins or antibodies to epitopes tagged on heterologously expressed proteins. The latter approach permits investigating the effects of wild-type and mutationally active or inactive kinases on NHE1 phosphorylation.[7] Cells are lysed in buffer containing (in mM): 20 Tris, pH 7.4, 150 NaCl, 1 EDTA, 1 Na$_3$VO$_4$, 1% NP-40, supplemented with protease inhibitors. Kinases are immunoprecipitated from precleared lysates by standard protocols. Immune complexes are collected and washed three times with lysis buffer and once with wash buffer containing (in mM): 20 Tris, pH 7.4, 1 EDTA, 0.1% NP-40, 1 Na$_3$VO$_4$, 5 2-mercaptoethanol, and 10% glycerol.

An aliquot (~25%) of the immune complex is removed for Western analysis to determine the abundance of kinases (Fig. 6B). The remainder of the immune complex (~75%) is used for *in vitro* kinase reactions in the presence of 5 μg of GST–NHE1 and 20 μl of kinase reaction buffer (25 mM HEPES, pH 7.5, 1 mM DTT, 10 mM MgCl$_2$, 3 mM MnCl$_2$, 1 μM Na$_3$VO$_4$, 10 μM ATP, and 0.5 μCi [^{32}P]ATP). The reaction mixture is incubated at 30° for 20 min and the reaction is stopped by adding SDS–PAGE sample buffer. Phosphorylated substrates are separated by SDS–PAGE and visualized by autoradiography (Fig. 6A). Parallel positive control assays are performed using nonspecific substrates such as myelin basic protein or histone.

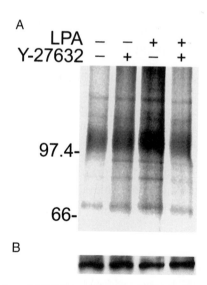

A

LPA   −   −   +   +
Y-27632   −   +   −   +

97.4-

66-

B

FIG. 7. Phosphorylation of NHE1 *in vivo* by lysophosphatidic acid (LPA) is mediated by p160ROCK. (A) Fibroblasts stably expressing an EE-tagged NHE1 were labeled with ortho[^{32}P]phosphate, and NHE1 was immunoprecipitated from untreated and treated cells (LPA; 10 μM; 10 min) in the absence or presence of the p160ROCK inhibitor Y-27632 (30 μM; 30 min). (B) Immunoblot with anti-EE antibodies to determine the abundance of NHE1 in the immunoprecipitates used in (A). Adapted with permission from Tominaga *et al.*[7]

In Vivo Phosphorylation of NHE1

Determining the *in vivo* phosphorylation of endogenously expressed NHE1 was first described by Sardet and colleagues.[4,27,28] We use a modification of these techniques to measure the phosphorylation of heterologously expressed NHE1 tagged at the C terminus with either an HA or an EE epitope.[7] In most cell types, basal phosphorylation of NHE1 is relatively high (Fig. 7), and thus the most critical step in determining increases in NHE1 phosphorylation is to obtain quiescent cells. We plate cells in 10-cm dishes 48 hr prior to experiments at a density that will reach 80% confluency at the time of the experiment. Twenty-four hours prior to experiments, cells are incubated in low-serum (0.5%) medium; 3 hr prior to experiments, cells are incubated in phosphate-free low-serum medium. Proteins are labeled in the presence of 200 μCi/ml ortho[^{32}P]phosphate for 2–3 hr. Cells are washed and collected in buffer A containing (in mM): 50 HEPES, pH 7.4, 150 NaCl, 3 KCl, 12.5 sodium pyrophosphate, 10 ATP, 5 EDTA, and protease inhibitors) and lysed using a probe sonicator by applying three 10-sec pulses at 3 V. Lysates are centrifuged (100,000g for 15 min at 4°), and pellets are resuspended in buffer A containing 1% Brij 96, the most effective detergent for solubilizing NHE1 (Sigma Chemicals, St. Louis, MO), and sonicated for 40 sec. NHE1 is immunoprecipitated from precleared (100,000g for 30 min at 4°) sonicates by standard techniques. Phosphorylated proteins are separated by 7.5% SDS–PAGE and detected by autoradiography.

Aliquots of each sample are analyzed by Western blotting to normalize autoradiograph signals to the abundance of NHE1 in the immunoprecipitated complex. We have used this approach to determine increases in NHE1 phosphorylation by RhoA- and Rac-mediated signaling pathways. As shown in Fig 7, activation of lysophosphatidic acid receptors, which are coupled to the activation of RhoA, increases NHE1 phosphorylation *in vivo*, which is blocked by the p160ROCK inhibitor Y-27632. Additionally, we have used phosphoamino acid analysis to confirm that similar residues of NHE1 are phosphorylated by p160ROCK *in vitro* and *in vivo*.[7]

Acknowledgment

Work discussed in this article was supported by National Institutes of Health Grants GM47413 and GM58642. D.L.B. is an established investigator for the American Heart Association.

[28] C. Sardet, L. Counillon, A. Franchi, and J. Pouyssegur, *Science* **247**, 723 (1990).

[31] Actin Filament Assembly in Permeabilized Platelets

By JOHN H. HARTWIG and KURT L. BARKALOW

Introduction

Platelets respond to soluble and tethered ligands by changing their shape and aggregating together into plugs that cover damaged vascular surfaces. Shape change results from a rapid and massive actin assembly that doubles the cellular actin filament concentration within 60 sec to reorganize the platelet cytoskeleton. Remodeling of the cytoskeleton involves the construction of two distinct actin-based structures: long cylindrical filopodia and flat lamellipodia. Because of its small cytoplasmic volume (7 fl), each resting platelet contains only 2000–15,000 actin filaments, allowing the organization of the entire cytoskeleton to be appreciated in the electron microscope and correlated with biochemical behavior. Like all other cells studied, actin assembly begins when actin filament barbed ends are exposed[1,2] either by uncapping of the barbed end of preexisting filaments[3,4] or by the *de novo* nucleation of actin as described for the Arp2/3 protein complex and its regulatory proteins.[5–13] We have developed a cell-free system to study the mechanism of actin filament barbed end exposure in platelets. Using octylglucopyranoside (OG) to permeabilize platelets, we have reconstituted PAR-1 receptor-mediated actin filament barbed end exposure.[14] This article describes the methodology used to permeabilize

[1] M. Carson, A. Weber, and S. Zigmond, *J. Cell Biol.* **103**, 2707 (1986).
[2] M. Cano, D. Lauffenburger, and S. Zigmond, *J. Cell Biol.* **115**, 677 (1991).
[3] J. Hartwig, *J. Cell Biol.* **118**, 1421 (1992).
[4] T. Azuma, W. Witke, T. Stossel, J. Hartwig, and D. Kwiatkowski, *EMBO J.* **17**, 1362 (1998).
[5] L. Machesky, R. Mullins, H. Higgs, D. Kaiser, L. Blanchoin, R. May, M. Hall, and T. Pollard, *Proc. Natl. Acad. Sci. U.S.A.* **96** (1999).
[6] L. Ma, R. Rohatgi, and M. Kirschner, *Proc. Natl. Acad. Sci. U.S.A.* **95**, 15362 (1998).
[7] R. Rohatgi, L. Ma, H. Miki, M. Lopez, T. Kirchhausen, T. Takenawa, and M. Kirschner, *Cell* **97**, 221 (1999).
[8] T. Svitkina and G. Borisy, *J. Cell Biol.* **145**, 1009 (1999).
[9] M. Welch, J. Rosenblatt, J. Skoble, D. Portnoy, and T. Mitchison, *Science* **281**, 105 (1998).
[10] D. Winter, T. Lechler, and R. Li, *Curr. Biol.* **9**, 501 (1999).
[11] R. Kolluri, K. Tolias, C. Carpenter, F. Rosen, and T. Kirchhausen, *Proc. Natl. Acad. Sci. U.S.A.* **93**, 5615 (1996).
[12] R. Mullins, J. Heuser, and T. Pollard, *Proc. Natl. Acad. Sci. U.S.A.* **95**, 6181 (1998).
[13] T. Loisel, R. Boujemaa, D. Pataloni, and M.-F. Carlier, *Nature* **401**, 613 (1999).
[14] J. Hartwig, G. Bokoch, C. Carpenter, P. Janmey, L. Taylor, A. Toker, and T. Stossel, *Cell* **82**, 643 (1995).

cells and the assay used to detect actin filament barbed end exposure. This model system affords an opportunity to dissect the signaling cascade from membrane receptors to the machinery of the cell responsible for cell shape changes and locomotion.

Materials

Solutions

10× PHEM buffer stock[15]: 600 mM HEPES, 250 mM PIPES, 100 mM EGTA, and 2 mM MgCl$_2$, pH 6.9

Actin assembly buffer: 100 mM KCl, 0.2 mM EGTA, 10 mM HEPES, 0.2 mM MgCl$_2$, 0.5 mM dithiothreitol (DTT), and 0.5 mM ATP, pH 7.5

Monomeric actin buffer[16]: 2 mM Tris, 0.2 mM MgCl$_2$, 0.5 mM DTT, 0.5 mM ATP, and 0.2 mM CaCl$_2$

Reagents

10% (w/v) solution of *n*-octyl-β-D-glucopyranoside (Sigma Chemical Company, St. Louis, MO) in water (stored at $-20°$)

10% Triton X-100 in water

20 μM pyrene-labeled monomeric rabbit skeletal muscle actin[17]

1 mM solution of phallacidin in water

10 mM cytochalasin B dissolved in dimethyl sulfoxide (DMSO)

Antiprotease cocktail (100×, stored $-20°$): 1 mM benzamidine, 5.2 mM leupeptin, and 12.3 mM aprotinin

Procedures

Permeabilization of Platelets

Platelets are isolated from human blood as described by Azim *et al.*[18] Purified platelets are permeabilized with OG in buffers containing EGTA to inactivate calcium-dependent proteins, particularly metalloproteases, and calcium-activated actin regulatory proteins. Phallacidin is also included to

---

[15] M. Schliwa, J. van Blerkom, and K. Porter, *Proc. Natl. Acad. Sci. U.S.A.* **78**, 4329 (1981).

[16] J. Spudich and S. Watt, *J. Biol. Chem.* **246**, 4866 (1971).

[17] T. Kouyama and K. Mihashi, *Eur. J. Biochem.* **114**, 33 (1981).

[18] A. C. Azim, K. L. Barkalow, and J. H. Hartwig, *Methods in Enzymology* **325** [22] (2000) (this volume).

stabilize endogenous cytoskeletal actin filaments. The benefit of OG as a detergent is its high critical micelle concentration (cmc), which allows the duration of permeabilization to be terminated rapidly by simple dilution below the cmc with buffers lacking OG. Originally, 0.4–0.45% OG was optimal for platelets suspended in buffers containing 0.3% bovine serum albumin (BSA) (Fig. 1, based on Hartwig et al.[14]). However, the concentration of OG required to permeabilize platelets and other cells shifts downward (to 0.25%) if lipid-free BSA is used in the lysis buffer, suggesting that proteins and lipids added to the buffer can bind OG and thereby alter this value. Therefore, it is best to first titrate a given batch of OG versus the cell suspension to ascertain the minimal detergent concentration required for adequate permeabilization. Direct comparison with Triton X-100-permeabilized cells provides values for the maximal amount of cell extraction. The methodology to determine the appropriate OG concentration for permeabilization is detailed for platelets, but the approach is applicable for all cells in suspension.

Assay System

The assay has a final volume of 0.3 ml. Routinely, 90 μl of platelets at a cell density of 2 × 10⁸ platelets/ml (~10–30 μg cell protein in 90 μl) is mixed with 10 μl of 10× OG (2–5%) in 10× PHEM buffer containing 1 μM

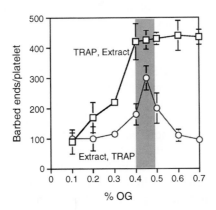

Fig. 1. OG permeabilizes the plasma membrane of platelets in a dose-dependent fashion. Treatment with 0.30–0.45% concentrations of OG (in fatty acid containing BSA buffers) places increasingly larger holes in the plasma membrane of resting cells but leaves large domains of membrane intact. The effect of receptor ligation on actin filament end exposure in intact cells (squares) and permeabilized resting cells (circles) is compared. The detergent concentration useful for permeabilized signaling studies is highlighted. When fatty acid-depleted BSA is used the shaded area shifts to the left to between 0.2 and 0.3%.

phallacidin, and a protease inhibitory cocktail of 10 nM benzamidine, 52 nM leupeptin, and 123 nM aprotinin (PHEM buffer). The maximum number of barbed ends is achieved in platelets activated with the PAR-1 activation peptide TRAP, at 25 μM, for 30 sec before permeabilization. Platelets are permeabilized for 1 min and then diluted with 185 μl of the actin assembly buffer. The assay starts when 15 μl of 20 μM pyrene actin is added; the solution is mixed and placed in the fluorimeter (see later). By varying the final OG concentration, the minimum amount of detergent required to detect filament ends in the actin assembly assay (see later) is determined, and at high detergent concentrations the maximal actin assembly rate values for cells extracted to completion with OG are ascertained. These values are compared to those from unpermeabilized cells (control for the background rate of actin assembly) and for cells permeabilized using a final concentration of 0.1% Triton X-100 instead of OG, which yields maximal cell permeabilization and barbed end numbers.

Assay for Filament Ends in Permeabilized Cells and Cell Lysates

Free barbed and pointed actin filament ends (as defined by the stereospecific binding of myosin subfragment 1 to actin) are detected by adding pyrene-labeled actin monomers to the lysates at concentrations sufficient to drive monomer addition at the two filament ends but not to self-nucleate within the time frame of the assay.[19] In this assay, a final concentration of 1 μM monomeric pyrene-labeled rabbit skeletal muscle actin is added to the platelet lysate.[3] Monomeric actin is maintained as a 20 μM stock in a low salt buffer, and endogenous nuclei (actin oligomers) are removed by ultracentrifugation at 400,000g for 15 min at 4°. The rate of actin assembly in the absence of added nuclei is extremely slow at low actin concentrations; permeabilized cells thus provide the nuclei to initiate actin assembly. Actin assembly is followed in a fluorimeter, excitation 366 nm, emission 386. The initial fluorescence and the final (12 hr) fluorescence when actin assembly/disassembly are at steady state are recorded, as well as the initial rate at which the fluorescence of the sample increases (filament assembly). The relative amounts of pointed versus barbed filament ends are determined by assaying the permeant cells in the presence and absence of 1 μM cytochalasin B. Cytochalasin inhibits pyrene–actin assembly at the barbed end in this assay, allowing the contribution of assembly from the pointed end to be determined.

[19] T. Pollard, *J. Cell Biol.* **103**, 2747 (1986).

Calculation of the Rate of Addition of Monomers to Filament Ends

Fluorescence contributed per actin monomer:

Final fluorescence − initial fluorescence = total fluorescence/actin monomers added (3×10^{-10} mol $\times$ 6.023×10^{23} = 18.07×10^{13} monomers per assay)

Rate actin monomers add to filaments:

Assembly rate (change in rate/sec) $\times$ fluorescence per monomer = monomers assembled into filaments per sec

Contribution of actin filament pointed ends to assembly rate:

Assembly rate in presence of cytochalasin = pointed end rate/1 monomer per μM actin per sec/2×10^7 platelets = free pointed ends

Number of actin filament barbed ends:

Assembly rate in absence of cytochalasin − assembly rate in presence = barbed end number/10 monomers per μM actin per sec/2×10^7 platelets = free barbed end number

Probing Signal Cascades

Once the minimal OG detergent concentration that maintains signaling from receptors to actin has been established empirically, experiments that explore signaling to actin in the permeant system can be initiated. It is important to note that at the minimal OG concentration, the cell suspension should remain opaque. Ninety microliters of cell suspension is extracted with 10 μl of 2.5% OG in the PHEM buffer defined earlier (in the absence of detergent-binding material) for 30 to 60 sec. Cells other than platelets may require significant changes in both the amount of OG and the duration of extraction. Permeabilization is terminated by dilution of the suspension with 185 μl of the actin assembly buffer.

Following dilution, permeant platelets retain signaling systems from receptors to actin. First, permeant cells spread on activating surfaces (Fig. 2). Second, PAR-1 signaling to barbed ends can be initiated by ligation of the receptor with 25–50 μM TRAP for 1 min before addition of the monomeric actin. Third, protein components can be added to the permeant cells and shown to stimulate, inhibit, and/or modify the actin assembly response. Table I lists agents that have been added to the permeant cell and their effect on barbed end exposure.

Platelets permeabilized with OG retain certain responses in addition

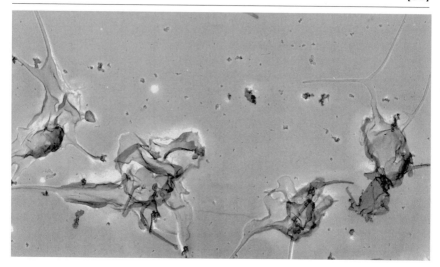

FIG. 2. OG-permeabilized cells retain their ability to spread when attached to glass surfaces. Glass is a potent activator of human platelets, which bind rapidly to the surface and spread.

TABLE I
AGENTS TESTED IN PERMEANT PLATELETS

Agent added	Stimulate (S+)/inhibit (I−)	Effect on PAR-1 signaling	Concentration required
TRAP	S++		
Phosphoinositides ($PI_{4,5}P_2$, $PI_{3,4}P_2$, $PI_{3,4,5}P_3$)	S++		20–100 μM
Phosphoinositide-binding peptides (gelsolin 160–169, 150–169)		I−−	10–25 μM
GTP–Rac, V12–Rac	S++		5–25 nM
N17Rac		I−	
GTP–cDc42	S+		50–100 nM
GTP	S++		
GDPβS		I−−	100 μM
V12–RacK186E		I−−	
PI 5-kinase 1a	S++		1–5 nM
PI-3-kinase	S++		1–5 nM
PI-5-kinase 1aD227A		I−−	

++, strong activation.
+, activation.
−, inhibition.
−−, strong inhibition.

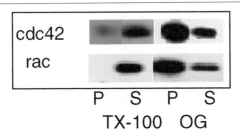

P   S   P   S
TX-100   OG

FIG. 3. Triton-soluble proteins remain in these permeabilized cells and are necessary for signal transduction from TRAP. Distribution of Rac and Cdc42 in cytoskeletal pellet (P) or supernatant (S) from platelets permeabilized with buffers containing 0.1% Triton X-100 or 0.5% OG. Triton X-100 permeabilization of resting platelets releases all of the Rac protein and ~90% of Cdc42. OG permeabilization, however, releases only a dramatically smaller fraction of Rac and Cdc42.

to end exposure. First, and not surprisingly, the small GTPases Rac and Cdc42 remain in the OG-permeant platelets (Fig. 3). In resting cells, ~60–70% of Rac and Cdc42 are bound to the permeant cell residue.[20] Second, cells remain responsive to ligation of receptors other than the PAR-1 receptor. As discussed earlier (Fig. 2), platelets permeabilized with OG as described retain all the cytoskeletal components necessary for them to adhere and activate on glass surfaces. This implies, in particular, that sufficient monomeric actin remains in the permeant cells to assemble filaments that form the resultant protrusions. Third, D4 and D5 containing polyphosphoinositides are synthesized. Fourth, we have tested other signals to actin in these cells. Cooling, a response that causes resting discoid platelets to change shape, increases their actin filament content and exposes barbed filament ends. Additionally, brief OG exposure allows human neutrophils to maintain their response to the N-formyl-L-methionyl-L-leucyl-L-phenyl-alanine receptor while providing intracellular access of reagents into them.

Conclusion

In summary, the OG permeabilization procedure is a simple technique that can be widely applied to cells and used to dissect pathways that regulate actin assembly. Many proteins released by Triton X-100 detergents remain associated with the OG-treated platelet. The permeabilization of platelets may be straightforward, in part, because platelet granules lack concentrated proteases and because of the high receptor density in the plasma membrane of platelets. Possible pitfalls of the permeabilization approach include the

[20] A. Azim, K. Barkalow, J. Chou, and J. H. Hartwig, *Blood* **95,** 959 (2000).

disruption of certain signaling cascades that require intact membrane or OG-solubilized components.

Acknowledgments

This work was supported by NIH Grants HL HL56252 and HL56949 and a Scientist Development Grant (K.L.B.) from the American Heart Association.

[32] Rho GTPases: Secretion and Actin Dynamics in Permeabilized Mast Cells

By MARK R. HOLT and ANNA KOFFER

Introduction

Permeabilized cells retain architectural integrity and respond to stimulation by calcium and/or GTPγS as well as to the activation of cell surface receptors.[1,2] After permeabilization, cytosolic components leak out of the cells: the rate and extent of leakage depend on pore size and solubility of cellular components.[3-7] Responsiveness of permeabilized cells decreases with increases in interval between permeabilization and cell activation ("rundown"). This is due to leakage and inactivation of various proteins and is affected greatly by ionic conditions and the permeabilization method. Responses to triggers that bypass cell surface receptors run down much slower than those to receptor activation. Recovery of cell activity, by controlling the intracellular environment and/or by addition of specific cytosolic proteins, forms the basis for mechanistic studies of various cell functions (e.g., exocytosis, endocytosis, phospholipid metabolism, and cytoskeletal responses). This article describes the use of primary mast cells, permeabilized by streptolysin O (SLO), for studying Rho family GTPases.

[1] T. W. Howell and B. D. Gomperts, *Biochim. Biophys. Acta* **927**, 177 (1987).
[2] E. Cunningham, G. M. H. Thomas, A. Ball, I. Hiles, and S. Cockcroft, *Curr. Biol.* **5**, 775 (1995).
[3] P. E. R. Tatham and B. D. Gomperts, in "Peptide Hormones: A Practical Approach" (K. Siddle and J. C. Hutton, eds.), p. 257. IRL Press, Oxford, 1990.
[4] T. Sarafian, D. Aunis, and M. Bader, *J. Biol. Chem.* **262**, 16671 (1987).
[5] Y. Churcher and B. D. Gomperts, *Cell Regul.* **1**, 337 (1990).
[6] A. Koffer and B. D. Gomperts, *J. Cell Sci.* **94**, 585 (1989).
[7] A. Koffer and Y. Churcher, *Biochim. Biophys. Acta* **1176**, 222 (1993).

Streptolysin O is a bacterial exotoxin, produced by *Streptococcus pyogenes*. As a monomer, it binds to plasma membrane cholesterol and oligomerizes to form large arc- and ring-shaped pores.[8–10] The diameter of these pores is up to 30 nm, allowing passage of proteins of molecular size of up to 400 kDa. The optimal concentration of SLO, exposure time needed for permeabilization, and pore size vary for different cells; adherent cells usually require higher concentrations of SLO and/or longer exposure times than suspended cells. Conditions for permeabilization and subsequent activation influence the behavior of the cells greatly. For example, mast cells permeabilized in chloride buffer respond differently to calcium and guanine nucleotides than cells permeabilized in glutamate (or aspartate)-based buffers.[5,11] Moreover, the former run down much faster than the latter (unpublished observation, 1993). Cytoskeletal structures retain their integrity better in glutamate buffer than chloride buffer.

In the initial experiments, SLO was added together with Ca^{2+} and guanine nucleotides, i.e., cells were triggered at the time of permeabilization.[12] This protocol can be modified: binding of SLO to membranes is temperature independent, whereas oligomerization of SLO is temperature dependent.[13] Therefore, excess SLO (plus additives) can be removed by washing cells in cold buffer after they have been exposed to SLO on ice ("prebind" method). Cells are then triggered at the time of permeabilization by the addition of warm triggering solutions. Another protocol involves washing cells after permeabilization to ensure the removal of freely soluble ions, nucleotides, proteins, and excess SLO. Thus, nucleotides and ions leaking out from cells do not come into contact with those applied exogenously during stimulation. This allows better control over experimental conditions. The latter method is described in this article.

The first section of this article outlines the preparation of recombinant GTPases, an assay for GTP-binding activity, and subsequent loading of recombinant proteins with GTP. The second section describes procedures for introducing GTPases into mast cells and evaluation of their effects on the cytoskeleton and exocytosis.

[8] L. Buckingham and J. L. Duncan, *Biochim. Biophys. Acta* **729**, 115 (1983).
[9] S. Bhakdi, U. Weller, I. Walev, E. Martin, D. Jonas, and M. Palmer, *Med. Microbiol. Immunol. Berl.* **182**, 167 (1993).
[10] M. Palmer, R. Harris, C. Freytag, M. Kehoe, J. Tranum Jensen, and S. Bhakdi, *EMBO J.* **17**, 1598 (1998).
[11] A. Koffer, *Biochim. Biophys. Acta* **1176**, 236 (1993).
[12] B. D. Gomperts and P. E. R. Tatham, *Methods Enzymol.* **219**, 178 (1992).
[13] K. Sekiya, H. Danbara, K. Yase, and Y. Futaesaku, *J. Bacteriol.* **178**, 6998 (1996).

GST Fusion Protein Purification, Assay of GTP-Binding Activity, and GTP Loading

Bacterial expression of recombinant GST-GTPase fusion proteins is followed by an affinity purification after which recombinant GTPases are liberated by thrombin cleavage.[14] A GTP-binding assay is used to determine the amount of active protein as opposed to total protein, as different preparations result in distinct GTP-binding activities; storage also results in decreased activity. A similar procedure is applied to load protein with GTP for experimental purposes. This loading step is essential for work with permeabilized cells but not for microinjection where proteins are activated by endogenous nucleotide exchange factors.[15] For both procedures, freshly thawed protein is maintained (by removal of Mg^{2+} using EDTA) in a conformation where the nucleotide-binding site is accessible to added guanine nucleotide (tritiated for the binding assay). The nucleotide-binding site is then closed by the addition of excess Mg^{2+}. It is important that the binding assay and the loading protocol are performed at this stage, i.e., immediately prior to usage, rather than before freezing/storage. Freeze-thawing affects GTP-binding activity greatly. So far, the following proteins have been tested: constitutively active GTPases: V12Cdc42, V12Rac1, V14RhoA; dominant inhibitory GTPases: N17Cdc42, N17Rac1; and effector mutant GTPases: V12A35Rac1, V14A37RhoA. The RhoA inhibitor, C3 transferase, has also been purified. However, it must be noted that the N17 mutants of Rac and Cdc42 do not load efficiently with nucleotide. The reason for this is unclear. It is probable that the proteins are unstable in the nucleotide-free state and denature. Alternatively, the proteins may have nucleotide bound already (during bacterial expression), but are unable to release it. In either case, we do not load these proteins and effects are defined in terms of total protein added rather than active protein.

Bacterial Expression of Rho Family GTPases and C3 Transferase[14]

Briefly, *Escherichia coli*, transformed with pGEX-2T plasmids containing the appropriate cDNA, are used to inoculate medium (containing ampicillin) and grown overnight. Cultures are then diluted 10-fold into fresh medium and grown for another hour. After this, protein expression is induced with isopropylthiogalactoside (IPTG) for 3 hr. Cells are harvested by centrifugation, resuspended, and sonicated on ice. The suspension is then centrifuged, and the supernatant is retained and added to glutathione-

---

[14] A. J. Self and A. Hall, *Methods Enzymol.* **256**, 10 (1995).
[15] J. C. Norman, L. S. Price, A. J. Ridley, A. Hall, and A. Koffer, *J. Cell Biol.* **126**, 1005 (1994).

Sepharose beads to allow glutathione S-transferase (GST) fusion protein to bind. Beads are washed, and the polypeptide is liberated from GST using thrombin. After thrombin removal, polypeptides are dialyzed and concentrated. The concentrate is then frozen in 10-μl aliquots in liquid nitrogen and stored at $-80°$ (or in liquid nitrogen for long-term storage).

GTP-Binding Assay

This assay measures the guanine nucleotide-binding activity of purified GTPases using radiolabeled GDP, GTP (not labeled in the γ-phosphate position), or their analogs. We use [^{3}H]GDP, but the assay could be adapted for other radiolabels. The assay is optimal for measuring between 10 and 100 ng of active protein, depending on the activity of the radiolabeled nucleotide.

Reagents and Materials

Assay buffer: 50 mM Tris–Cl, pH 7.5; add 1 mM dithiothreitol (DTT) immediately before use
Wash buffer: 50 mM Tris–Cl, pH 7.5, 50 mM NaCl, 5 mM MgCl$_2$ 50 mM EDTA, pH 8.0
[^{3}H]GDP or [^{3}H]GTP: 1 μCi/μl (~12 Ci/mmol)
0.2-μm nitrocellulose filters and a vacuum manifold filtration block
Scintillation fluid and vials

Procedure

One microliter of freshly thawed protein is added to 19 μl ice-cold assay buffer. Depending on total protein concentration, 1–10 μl of this mixture is then diluted further in ice-cold assay buffer to give a final volume of 36 μl (containing 1 μg total protein). This is kept on ice, and 4 μl 50 mM EDTA, pH 8.0, is added. EDTA causes a conformational change in the GTP-binding protein that enables nucleotide exchange. [^{3}H]GDP (or [^{3}H]GTP) is diluted 1:10 in assay buffer on ice, and 2 μl of the diluted radiolabeled nucleotide is added to the protein and incubated at 30° for 10 min. The tube is then transferred to ice, and nucleotide exchange is quenched by the addition of 1 ml ice-cold wash buffer (containing Mg^{2+}). While the nucleotide-bound GTPase is on ice, a nitrocellulose filter is placed onto a filtration block and soaked with 1 ml wash buffer. The sample is added to the filter, filtered, and washed with 10 ml ice-cold buffer. The filter (with nucleotide-bound protein) is finally air-dried on a paper towel for at least 15 min. When completely dry, the filter is transferred to a scintillation vial containing 10 ml scintillation fluid, and the number of disintegrations per minute is determined. From this it is possible to determine how much radiolabeled guanine nucleotide is bound to protein and, therefore, the proportion of the active GTP-binding protein (assuming nucleotide and protein are bound in a 1:1 stoichiometry).

GTP-Loading Protocol

GTP loading should be performed using the aliquot of the same protein used for the GTP-binding assay. Furthermore, it is desirable to load the protein and test its activity at the same time. This allows a more accurate estimate of the amount of GTP-bound protein used in the experiment. The principles are the same as those for the GTP-binding assay. Again, guanine nucleotide analogs can be used if appropriate, e.g., GTPγS.

Reagents and Materials

Exchange buffer: 50 mM Tris–Cl, pH 7.5, 100 mM NaCl; add 1 mM DTT immediately before use

Quenching buffer: 100 mM MgCl$_2$, 50 mM Tris–Cl, pH 7.5, 100 mM NaCl

Glutamate buffer (GB): 20 mM PIPES–acetate, pH 6.8, 137 mM sodium glutamate, 2 mM MgCl$_2$, 1 mg/ml bovine serum albumin (BSA); filter with 0.2-μm filters

10 mM GTP/GDP (Boehringer, Mannheim, Germany)

500 mM EDTA, pH 8.0

Speedy desalting columns (Pierce, Rockford, IL)

Centricon columns, 10 kDa cutoff (Amicon, Danvers, MA)

Procedure

An aliquot of protein (9 μl) is diluted into exchange buffer to a final volume of 1.5 ml. GTP (or GDP) is then added (10 μM final concentration) along with EDTA (20 mM final concentration). The mix is incubated at 4° for 40 min. The exchange reaction is terminated by the addition of 0.4 ml ice-cold quenching buffer (5 mM final Mg^{2+} concentration). The mix is passed through a desalting column equilibrated with glutamate buffer (containing 1 mM DTT). The eluate is transferred to a 10-kDa Centricon column and centrifuged at 5000g, 4° until the volume is about 100 μl. This is then diluted with 3 ml of glutamate buffer (with DTT) and concentrated as before to remove any remaining traces of GTP/GDP. This step is repeated. The final filtrate is retained and used as a control buffer (see Comments). After assaying the protein concentration, the protein is ready to use. The protein is stored on ice or at 4° and should be used within 2 days.

Comments

Permeabilized mast cells can be stimulated by trace quantities of GTP (presumably by exchange onto endogenous GTP-binding proteins). It is thus an obvious requirement that unbound GTP is removed from the GTP-loaded proteins either by dialysis or by several concentration steps on a Centricon column. It is also important to control for the possible release of GTP from loaded proteins either by spontaneous release or by protein denaturation. For optimal control, mutants should be used that bind GTP (without hydrolyzing it) in a constitutively active conformation but cannot bind effectors (effector domain mutants). Such mutants are V14A37RhoA, V12A35Rac1, and V12A35Cdc42 (the latter is not available yet). We also use the final filtrate from the Centricon column, as the only GTP present would be that from denatured protein.

Isolation, Permeabilization, and Stimulation of Rat
Peritoneal Mast Cells

Purification of mast cells, preparation of calcium buffers, and assay of the released hexosaminidase have been described previously.[12] Only brief versions of these procedures are given here. We focus on the processing of cells after permeabilization.

Reagents and Materials

Chloride buffer (CB): 20 mM PIPES–Cl, pH 7.2, 137 mM NaCl, 2.7 mM KCl, 2 mM MgCl$_2$, 5.6 mM glucose, 1 mg/ml BSA, 1.8 mM CaCl$_2$; filter using 0.2-μm filters

87.8% (v/v) Percoll solution: 175.6 ml of Percoll (Pharmacia, Piscataway, NJ; usual density ~1.13 mg/ml), 20 ml 100 mM phosphate buffer (containing 1.5 M NaCl), pH 7.0, 4.4 ml H$_2$O (freeze in 2-ml aliquots at $-20°$)

Glutamate buffer: as described previously

SLO: Purchased from Murex Diagnostics Ltd, Dartford, UK (distributed by Corgenix, Peterborough, UK); each bottle contains freeze-dried powder consisting of SLO together with additives. This is reconstituted with water to give SLO activity of 20 IU/ml. The additives consist of reducing agent (1.4% cysteine), 0.1% albumin, and isosmotic salts (after reconstitution). This solution is divided into 200-μl aliquots, freeze dried, and stored at 4°. Just before use, 200 μl of H$_2$O is added (and warmed and mixed) to make a stock solution (20 IU/ml). SLO is used at the final concentration of 0.4 IU/ml. The stock solution will keep for ~2 weeks at 4° but must be warmed up before use. SLO can also be obtained from VWR Scientific (CN) (manufactured by DIFCO, Piscataway, NJ) and from Sigma (St. Louis, MO), but in our hands, the Murex reagent has given the most reproducible results.

100 mM MgATP: 1 g Na$_2$ATP (trihydrate: molecular weight 605.2, Boehringer), 6.61 ml 0.5 M Tris, 8.25 ml 20 mM MgPIPES; add H$_2$O to a final volume of 16.5 ml

100 mM GTPγS: This stock solution is purchased from Boehringer

Ca^{2+}/EGTA buffer system: Free Ca^{2+} concentration is buffered by a 3 mM Ca^{2+}/EGTA buffer system, pH 6.8; preparation of stock solutions has been described previously[3,12]

Fluorogenic hexosaminidase substrate: 1 mM 4-methylumbelliferyl-N-acetyl-β-D-glucosaminide (Sigma), 0.1% dimethyl sulfoxide (DMSO), 0.01% Triton X-100, 200 mM citrate, pH 4.5. Dissolve reagent in DMSO first. Triton reduces surface tension and thus artifacts. Filter using Whatman (Clifton, NJ) filter paper No. 1 and store at −20°.

Cell fixation: 3% (w/v) paraformaldehyde, GB, pH 6.8 (without BSA), 3 mM EGTA, 4% (w/v) polyethylene glycol (PEG, molecular weight 6000)

F-actin staining: 0.5 μM TRITC–phalloidin with or without 80 μg/ml of lysophosphatidylcholine (Sigma), GB, pH 6.8, 3 mM EGTA (stocks: 100 mM TRITC–phalloidin; 40 mg/ml lysophosphatidylcholine. Both in ethanol at −20°. LPC is warmed and vortexed before use)

8-well slides ("Multitest" 8-well slides, ICN Biomedicals, Aurora, OH)

96-well plates, transparent

96-well plates, black

8-channel pipettes

Procedures

Isolation of Rat Peritoneal Mast Cells

Rat peritoneal mast cells are obtained by peritoneal lavage of rats (Sprague-Dawley) with "chloride buffer" (CB) as described.[12] Cells are pelleted by centrifugation (5 min at 250g, room temperature) and resuspended in 10 ml of CB. The suspension is filtered through a nylon mesh and overlayed onto a 2-ml cushion of Percoll solution. Centrifugation (5 min at 250g, room temperature) allows the dense mast cells to pellet, whereas neutrophils, macrophages, and red blood cells remain at the buffer:Percoll interface. The buffer and Percoll are removed, the pellet is collected, and cells are resuspended in 10 ml CB. Cells are centrifuged again (5 min at 250g at room temperature) and resuspended in chloride or glutamate buffer (as required) to remove any residual Percoll. About 1×10^6 cells (>90% purity) are obtained from one rat. This is sufficient for at least one 96-well plate of secretion assays, 10 flow cytometry assays, or five 8-well slides.

In some experiments, it is desirable to deplete endogenous ATP from intact cells with metabolic inhibitors. In this case, intact cells are incubated with CB where glucose has been omitted and 10 μM antimycin A with 6 mM 2-deoxyglucose included. More than 90% depletion of ATP is achieved in 20 min at 30°.[7]

Permeabilization

Suspended Cells. One milliliter of suspended intact cells (in CB) is washed with 9 ml of GB, pelleted, and resuspended in 0.5 ml GB (at 30°). The same volume of 2× permeabilizing solution (i.e., 0.8 IU SLO/ml, 6 mM EGTA, GB, pH 6.8) is then added. The time of SLO addition corresponds to the time of permeabilization; the period between this and cell triggering must be constant. After a 2-min incubation at 30°, cells are washed with 10 ml GB, pelleted, and resuspended in the required volume of GB (at room temperature).

A "prebind" method is useful for studies of proteins that leak out of permeabilized cells: excess SLO does not interfere with protein samples during electrophoretic analysis. Cells are incubated in 1 ml cold CB containing 0.8 IU SLO/ml on ice with shaking. After 5 min, 9 ml of cold GB (*without* BSA) is added and cells are centrifuged (5 min at 250g, 4°). Cells are then resuspended in a small volume of GB containing 100 μM EGTA. Permeabilization is initiated by transferring this suspension to a 30° water bath. Aliquots are removed at specific intervals, centrifuged (1 min at 12,000g), and processed for SDS–PAGE analysis.

Glass-Attached Cells. Rat peritoneal mast cells (~1×10^6 cell/ml CB) are plated onto 8-well slides, 25 μl per well, placed into a humidity chamber, and left to attach for 1 hr at room temperature. Cells are then washed twice with 25 μl GB and exposed for 8 min at room temperature to 25 μl of SLO at 0.4 IU/ml in GB and 3 mM EGTA. Cells are then washed (2 × 25 μl GB) to remove soluble components and excess SLO.

When starting with a new system, cell permeabilization should be tested using ethidium bromide (50 μM in GB). This normally impermeant agent moves into the nuclei of permeabilized cells and binds to DNA; this causes an increase in its (red) fluorescence.[3]

Pretreatment and Stimulation of Permeabilized Cells

Suspended cells are usually stimulated in transparent 96-well plates; the order of additions depends on the type of experiment. To introduce protein

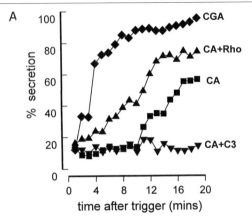

Fig. 1. Time course of calcium-induced secretion. (A) Permeabilized mast cells were pretreated for 5 min on ice with control buffer, 1 μg/ml C3/0.5 mM NAD$^+$, or 8 μg/ml (active) V14RhoA and then triggered with calcium (pCa 5)/3 mM ATP with (CGA) or without (CA) 30 μM GTPγS. Cells were incubated at 30°, and aliquots were removed at the indicated time intervals for hexosaminidase assay. V14RhoA was loaded with GTP and dialyzed. The final dialysate was used for the control. The response of permeabilized mast cells to calcium/ATP exhibits a characteristic delay in the onset of secretion (CA).[11] V14RhoA (like GTPγS, i.e., CGA) eliminates the delay and enhances hexosaminidase release (CA + Rho). C3, however, is inhibitory (CA + C3).[19] (B) The 8-well block used for time course experiments.

(e.g., GTPase) into cells before activation, 30 μl of protein (in GB-50 μM EGTA) is pipetted into the wells, followed by the addition of a 30-μl permeabilized cell suspension (in GB). Pretreatment with C3 transferase requires the presence of 0.5 mM NAD$^+$ to enable ADP-ribosylation (this is not required when microinjecting). The time of preincubation can vary from 1 to 30 min (5 min at room temperature is usually sufficient). "Rundown" occurs during this preincubation; therefore, it is important to match the interval between permeabilization and triggering exactly.

Various combinations of calcium, MgATP, and GTPγS (or GTP) are used for cell stimulation. The following four conditions (final concentrations) are used most frequently: EA (control): 3 mM EGTA/3 mM MgATP; EGA: 3 mM EGTA/50 μM GTPγS/3 mM MgATP; CA: 3 mM calcium/ EGTA buffer at pCa 5/3 mM MgATP; and CGA: 3 mM calcium/EGTA buffer at pCa 5/50 μM GTPγS/3 mM MgATP. In the example described previously, 30 μl of appropriate triggering solution, containing three times the required concentration of each component in GB, is added to 60 μl of pretreated cells. After 20 min at 30°, cells are quenched with 100 μl of ice-cold GB per well. The 96-well plates are then centrifuged (5 min at 250g,

B

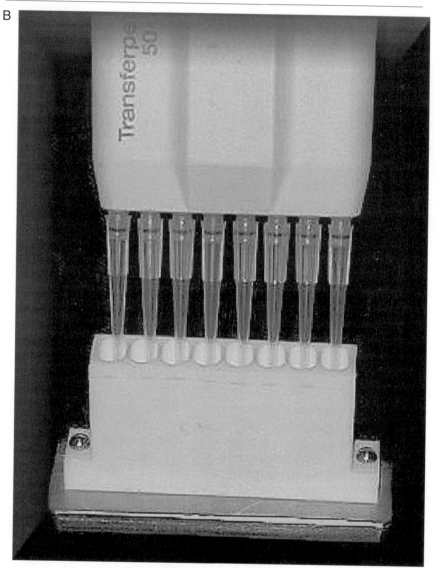

Fig. 1. (*continued*)

4°), and 50 μl of each supernatant is transferred to black 96-well plates containing 50 μl of hexosaminidase substrate in each well and the reaction is allowed to develop for 1 hr at 37°. To stop the reaction, 100 μl of unbuffered 0.5 M Tris is added to each well. A fluorescence plate reader is used to measure the fluorescence of the reaction product (4-methylumbelliferone).[12] The pelleted cells can be used for flow cytometry (see later).

For time course experiments (e.g., Fig. 1A), the just-described procedure is performed in an 8-well block (shown in Fig. 1B). Triggers are added to pretreated cells, and at required intervals, 90-μl aliquots are removed (using an 8-channel pipette) and placed into a 96-well plate containing 100 μl of ice-cold GB per well.

Glass-attached cells are pretreated (after permeabilization and washing) with 15 μl of protein (in GB-50 μM EGTA) for 5 min at room temperature, in a humidity chamber. Fifteen microliters of 2× triggering solution is then added. After 20 min at 30°, 15 μl of the supernatant is removed from each well and is transferred to a transparent 96-well plate (on ice) containing 85 μl ice-cold GB in each well. The plate is then centrifuged (5 min at 250g at 4°) to remove any detached cells. Fifty microliters of each supernatant is then transferred to black 96-well plates to assay for the release of hexosaminidase. The remaining attached cells are then fixed with 3% paraformaldehyde (in GB, 3 mM EGTA, 4% PEG) and processed for microscopy.

Analysis of F-actin Content and Morphology

Suspended Cells: Flow Cytometry. Morphological changes associated with secretion are reflected by pronounced changes in light scatter characteristics of secreting cells. Thus, degranulating mast cells can be distinguished from basal cells. After staining cells with rhodamine–phalloidin (RP), information about the total F-actin content of responding and nonresponding cells is obtained by combining light scatter data (i.e., forward angle, 1.5–19°, and 90° light scatter) with fluorescence data.[16] Freehand "gates" are drawn around subpopulations on two parameter light scatter distributions, using the Elite software. The fluorescence of cells falling within these gates is plotted on one parameter histograms. Staining with RP is performed either after (postlabeling) or before (prelabeling) cell activation. The former provides data about total F-actin content as a net result of cortical disassembly and *de novo* polymerization; the latter reports the loss of existing F-actin.

POSTLABELING. Staining with RP is performed after cell triggering and quenching but before fixation in order to avoid breakage of fragile fixed cells during washing. Cell pellets, obtained after centrifugation of cells in

[16] J. C. Norman, L. S. Price, A. J. Ridley, and A. Koffer, *Mol. Biol. Cell* **7**, 1429 (1996).

96-well plates, are resuspended in 50 μl of cold 0.5 μM RP in GB (with 3 mM EGTA) and stained for 10 min on ice. Cells (in 96-well plates) are then washed three times with 200 μl GB and fixed by resuspending gently in 50 μl 3% paraformaldehyde. Fixed cells are then transferred to tubes containing 250 μl of GB lacking BSA (which blocks the jet of the cytometer); the final cell density should be $\sim$5 $\times$ 10^5/ml.

PRELABELING. After permeabilization/washing, cells are resuspended in cold GB containing 100 μM EGTA and 0.1 μM RP and incubated for 10 min on ice. This low concentration of RP does not interfere with cytoskeletal responses.[15] They are then washed with a large volume of GB, pelleted, resuspended in GB, and used for pretreatment and activation as described previously. Cell pellets, obtained after centrifugation of 96-well plates, are resuspended in 50 μl 3% paraformaldehyde and processed as described earlier.

Different protocols of RP staining provide different data. Table I shows how V14RhoA affects the calcium-induced loss of F-actin and secretion. Using RP prelabeled cells, V14RhoA-induced enhancement of cortical F-actin disassembly can be seen. Pretreatment with 2 μM unlabeled phalloidin stabilizes actin filaments against disassembly; subsequent postlabeling with RP reveals an increase in F-actin due to V14RhoA-induced *de novo* actin polymerization. V14RhoA-induced enhancement of secretion is apparent with both protocols.

Attached Cells: Fluorescence Microscopy and Image Analysis. RP staining of attached cells is usually performed after cell activation and fixation, although all the variations described earlier are possible. Fixed cells are

TABLE I
F-ACTIN CONTENT AND SECRETION[a]

| | Relative F-actin content | | % Secretion | |
| | Cortical F-actin disassembly (A) | De novo F-actin polymerization (B) | | |
Treatment			A	B
Control (EA)	100.0	100.0	7	3
CA	58.4 ± 1.2	104.8 ± 1.1	35	38
CA/V14RhoA	43.1 ± 1.9	136.1 ± 2.4	43	46

[a] Permeabilized cells were pretreated with either 0.1 μM RP(A) to measure cortical disassembly or 2 μM unlabeled phalloidin (B) to prevent disassembly, followed by further pretreatment with or without V14RhoA (8μg/ml). Cells were then stimulated with calcium (pCa 5)/3 mM MgATP (CA). The latter cells were additionally postlabeled with 0.5 μM RP to measure *de novo* F-actin polymerization. The fluorescence intensity of control cells (exposed to EGTA/ATP, EA) was taken as 100%. Means from triplicate samples, each containing 5000–10,000 cells, are shown for F-actin; secretion values are averages of duplicate assays for released hexosaminidase.

EGTA / ATP   Ca / ATP

Con

Rho

C3

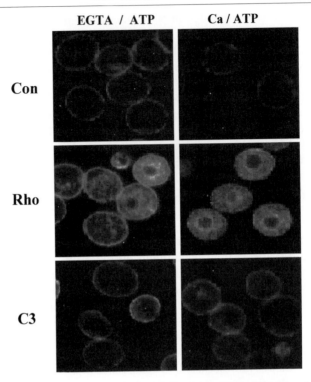

FIG. 2. F-actin morphology. Permeabilized cells were pretreated (5 min on ice) with GB (Con), V14RhoA (8 μg/ml) (Rho), or 1 μg/ml C3/0.5 mM NAD+ (C3). Cells were then stimulated with 3 mM EGTA/3 mM ATP (EA) or calcium (pCa 5)/3 mM ATP (CA). After 20 min at 30°, cells were fixed and postlabeled with 0.5 μM RP. Pretreatment of cells with V14RhoA induced an increase in F-actin content in the cell interior and, in the presence of calcium, enhanced disassembly of the cortex. C3 blocked both of these.[19]

labeled with 0.5 μM RP in GB-100 μM EGTA containing 80 μg/ml lyso-phosphatidylcholine for 20 min at room temperature. The detergent improves the access of RP to the cytoskeleton and the images obtained are better than those seen after Triton X-100 (0.1%) treatment. Figure 2 shows RP-postlabeled cells after pretreatment with V14Rho or C3. Note the increase in interior F-actin structures in the presence of V14RhoA. Similar structures appear when cells are stimulated with GTPγS and these are completely blocked by C3 transferase.

Concluding Comments

Permeabilized mast cells proved to be a valuable tool for studying the function of Rho-related GTPases. Rho, Rac, and Cdc42 were shown to be crucial for regulation of the late steps of exocytosis.[17,18] As expected, these GTPases also regulate the mast cell cytoskeleton in profoundly different ways.[15] So far, evidence indicates that this occurs independently of secretion.[16,19] This system will be extremely useful for defining further the signal transduction pathways regulating the cytoskeleton and exocytosis.

Acknowledgments

This work was supported by grants from the National Asthma Campaign and the Medical Research Council.

[17] L. S. Price, J. C. Norman, A. J. Ridley, and A. Koffer, *Curr. Biol.* **5**, 68 (1995).
[18] A. M. Brown, A. J. O'Sullivan, and B. D. Gomperts, *Mol. Biol. Cell* **9**, 1053 (1998).
[19] R. Sullivan, L. S. Price, and A. Koffer, *J. Biol. Chem.* **274**, 38140 (1999).

[33] Isolation and *in Vitro* Contraction of Stress Fibers

By Kazuo Katoh, Yumiko Kano, and Keigi Fujiwara

Introduction

When cultured cells spread well on a substrate surface are stained with antiactin or other specific probes for actin filaments, an array of needle-shaped structures, called stress fibers, is revealed in the cell. Since their first description in cultured cells,[1] this cytoskeletal structure has intrigued many cell biologists. There is a long list of suggested physiological roles of stress fibers, including cell adhesion, cell spreading, cell shape determination, cell locomotion, production of isometric tension within cells, organization of cell surface domains, and cell transformation.[2] More recently, their role in mechanosignal transduction has been proposed.[3-7]

[1] W. H. Lewis and M. R. Lewis, *in* "General Cytology" (E. V. Cowdry, ed.), p. 385. University of Chicago Press, Chicago, 1924.
[2] H. R. Byers, G. E. White, and K. Fujiwara, *Cell Muscle Motil.* **5**, 83 (1984).
[3] K. Katoh, M. Masuda, Y. Kano, Y. Jinguji, and K. Fujiwara, *Cell Motil. Cytoskel.* **31**, 177 (1995).
[4] K. Burridge and M. Chrzanowska-Wodnicka, *Annu. Rev. Cell Dev. Biol.* **12**, 463 (1996).
[5] P. F. Davies, K. A. Barbee, M. V. Volin, A. Robotewskyj, J. Chen, L. Joseph, M. L. Griem, M. N. Wernick, E. Jacobs, D. C. Polacek, N. dePaola, and A. I. Barakat, *Annu. Rev. Physiol.* **59**, 527 (1997).
[6] D. Ingber, *FASEB J.* **13**, S3 (1999).
[7] Y. Kano, K. Katoh, and K. Fujiwara, *Circ. Res.* **86**, 425 (2000).

Most studies on stress fibers are either phenomenological or morphological. There are many reasons for this, but two major advances in stress fiber cell biology have been made. One is the discovery that RhoA and Rho kinase are involved in stress fiber formation.[8-10] These studies have identified signaling molecules that play roles in the formation of stress fibers. This discovery has opened a new avenue to study the signaling pathway(s) for the organization of this cytoskeleton. Another is our demonstration that stress fibers can be isolated.[11] This has made it possible to study the basic properties of stress fibers. Using isolated stress fibers, we have directly shown the contractile nature of stress fibers. Isolated stress fibers are useful for studying their properties and functions, but they are also useful as a nonmuscle contraction model system for investigating the regulatory mechanism for the actomyosin-based contractility in nonmuscle cells. This article describes in detail how to isolate stress fibers en masse from cultured cells. It also describes how to observe contraction of stress fibers. The isolation procedure is simple and provides stress fibers that are pure enough for biochemical (i.e., enzymological) as well as structural and other studies. A similar approach may be useful for isolating other contractile apparatuses from both nonmuscle and muscle cells and tissues.

Isolation of Stress Fibers

Materials

Cells

Culture cells in five round (150 mm in diameter) culture dishes (Greiner, Frickenhausen, Germany) for 3–4 days. Cells should be subconfluent and well spread. Needless to say, they should contain richly developed stress fibers at the end of culture.

Solutions

Stock Solutions. Prepare the following stock solutions and store them at appropriate temperatures until use. As all of the working solutions must be at 0°–4°, the temperature of these stock solutions should be equilibrated to 4° prior to use. In addition to making these stock solutions, a tank of deionized water should be cooled to 4° in a cold room.

---

[8] A. J. Ridley and A. Hall, *Cell* **70**, 389 (1992).
[9] A. J. Ridley and A. Hall, *EMBO J.* **13**, 2600 (1994).
[10] M. Amano, K. Chihara, K. Kimura, Y. Fukata, N. Nakamura, Y. Matsuura, and K. Kaibuchi, *Science* **275**, 1308 (1997).
[11] K. Katoh, Y. Kano, M. Masuda, H. Onishi, and K. Fujiwara, *Mol. Biol. Cell* **9**, 1919 (1998).

4× phosphate-buffered saline (PBS): Use 19.2 g of Dulbecco's PBS powder (Nissui, Tokyo, Japan) to make 500 ml. Store at 4°.

250 mM TEA (triethanolamine) (Wako, Osaka, Japan): Store at 4°.

Trasylol (Bayer, Leverkusen, Germany): As they come (5-ml vials). Store at 4°.

1 mg/ml leupeptin (Peptide Institute, Osaka, Japan): Dissolve in deionized water and store at −20° in 1-ml quantity in small vials.

1 mg/ml pepstatin (Peptide Institute): Dissolve in dimethyl sulfoxide (DMSO) and store at −20° in 1-ml quantity in small vials.

20% (v/v) Nonidet P-40 (NP-40) (BDH, Poole, UK): In deionized water. Store at 4°.

20% (v/v) Triton X-100 (Wako, Osaka, Japan): In deionized water. Store at 4°.

Working Solutions. Prepare fresh working solutions for each session using stock solutions, already cooled to 4°. The volume indicated for each working solution is for isolating stress fibers from five large (150 mm in diameter) round culture dishes.

PBS (500 ml): Dilute the PBS stock four times with cold deionized water. Trasylol–PBS (500 ml): 20 μg/ml Trasylol, 1 μg/ml leupeptin, 1 μg/ml pepstatin, and 125 ml 4× PBS. Add deionized water to 500 ml (pH 7.2).

Low ionic strength extraction solution (5 liter): 2.5 mM TEA, 20 μg/ml Trasylol, 1 μg/ml leupeptin, and 1 μg/ml pepstatin; pH ~8.2 (no adjustment).

Extraction buffer I (300 ml): 0.05% NP-40, 20 μg/ml Trasylol, 1 μg/ml leupeptin, 1 μg/ml pepstatin, and 75 ml 4× PBS. Add deionized water to 300 ml (pH 7.2).

Extraction buffer II (300 ml): 0.5% Triton X-100, 20 μg/ml Trasylol, 1 μg/ml leupeptin, 1 μg/ml pepstatin, and 75 ml 4× PBS. Add deionized water to 300 ml (pH 7.2).

Sucrose gradient: Make 1.0, 1.2, and 1.5 M sucrose solutions in Trasylol–PBS.

Other Supplies

Z-shaped injection needle (Fig. 1): A 23-gauge injection needle is carefully bent in the form of a letter Z using forceps. Care should be exercised so that the channel of the needle remains open.

Methods

1. Prepare all of the working solutions fresh, and their temperature equilibrated to 4°. All of the following procedures should be carried out in a cold room or on ice at a regular laboratory bench.

FIG. 1. An injection needle bent in the form of a letter Z.

2. Aspirate culture medium and wash cells briefly with ~30 ml Trasylol–PBS. Extract cells with ~30 ml low ionic strength extraction solution for 20–40 min while shaking the culture dishes gently on a flat shaking table. Change the extraction solution 10–15 times during this extraction procedure. Soluble components in the cell are extracted and cells appear translucent with blebs on the surface.

3. Extract cells with ~30 ml extraction buffer I for 5 min. Looking under a phase-contrast microscope (total magnification × 100) one can see that the dorsal side of many cells has broken away and is floating freely. Many nuclei are also freed. Some "intact" cells may be broken by applying a gentle stream of buffer using a pipette. At this stage, only the basal portion of cells is still attached to the culture dish.

4. Extract what is attached on the culture dish with ~30 ml extraction buffer II for 5 min (with two changes) with gentle agitation.

5. With gentle shaking, wash the culture dish with ~30 ml Trasylol–PBS for 15 min with two changes. This wash completely removes Triton X-100 in which stress fibers are unstable. Although stress fibers are still attached to the substrate surface, this preparation is a good model system for showing their contractility. To completely isolate and obtain a purified fraction of stress fibers, continue following the procedure.

6. Using a rubber policeman, scrape off stress fibers from the culture dish and suspend in a small volume (less than 5 ml) of Trasylol–PBS. The

detached stress fibers are still embedded in sheet-like structures, presumably the cell cortex.

7. Collect detached stress fibers into a 10-ml syringe and forcefully extrude twice through a Z-shaped needle (Fig. 1). This step frees individual stress fibers from the cell cortex. The Z-shaped needle creates turbulent flow inside the needle and makes it more efficient to obtain free-floating stress fibers.

8. Centrifuge the stress fiber suspension at 1000g for 5 min to remove large cell debris. Isolated stress fibers in the supernatant can be collected by centrifugation at 100,000g for 1 hr. The pellet contains highly enriched isolated stress fibers. This preparation can be used for reactivation studies, as well as morphological and certain biochemical analyses.

9. The last bit of contaminating cell debris may be removed by running the pelleted material through a density step gradient consisting, from the bottom, of 1.5, 1.2, and 1.0 M sucrose in Trasylol–PBS. Resuspend the pellet into 2 ml Trasylol–PBS and centrifuge through the sucrose density gradient at 100,000g for 1 hr. Stress fibers can be recovered in the 1.2 M sucrose fraction as well as at the 1.2–1.5 M interface.

10. Collect the stress fiber containing fraction, dilute it with Trasylol–PBS, and centrifuge at 100,000g for 1 hr.

Notes

1. Although the suppliers of various compounds are indicated, we do not intend to imply exclusivity.

2. The five dishes are treated simultaneously in order to obtain a pellet at the end of step 10.

3. Because stress fibers tend to break apart in solutions containing Triton X-100, avoid exposing them to this detergent more than is necessary.

4. It is critical to use a Z-shaped needle for isolating stress fibers. Extruding stress fibers through a straight needle will not produce the desired effect. Stress fibers break into small fragments when this process is repeated several times.

5. For buffers, PIPES (10 mM) or HEPES (10 mM) can be used in place of phosphate in Trasylol–PBS. For isolating stress fibers for electron microscopy, phosphate in PIPES or HEPES buffer should be used throughout the isolation procedure. It is also acceptable to change the buffer just before fixation to prevent aggregation of uranyl acetate. Tris buffer should not be used because it significantly reduces the final volume of isolated stress fibers.

Application

Using the procedure just described, we have isolated stress fibers from human foreskin fibroblasts (FS-133) and bovine carotid arterial endothelial cells.[11] Figure 2a shows intact cells fixed without extraction and stained with fluorescent phalloidin, which stains actin filaments in cells. In addition to stress fibers, diffuse staining comes from actin filaments not associated with stress fibers. Once cells are extracted and their dorsal sides removed, the diffuse staining is no longer present, revealing a spectacular view of stress fibers (Fig. 2b). When the entire isolation procedure is completed, a pellet of stress fibers is obtained. Figure 3a is such a pellet stained with fluorescent phalloidin, showing clumped stress fibers of various sizes. Individual isolated stress fibers are stained with several antibodies, demonstrating that polypeptides known to be components of stress fibers in the cell are still associated with isolated stress fibers (Figs. 3b–3e). Their ultrastructure is essentially the same as stress fibers inside cells (Fig. 4).[3]

Our rough calculation indicates 1/20–1/30 of the total protein of the cell is recovered in the final isolated stress fiber fraction. The actin content in cultured cells such as fibroblasts is estimated to be roughly 10% of the total cellular protein, and myosin is thought to be of the order of 1–2%.[12] Because stress fiber is the most prominent actomyosin-containing structure in these cells, it is not surprising that this cytoskeletal structure consists of 3–5% of the total protein of the cell. Figure 5 shows SDS gels of a crude cell extract (lane 1) and the final pellet of isolated stress fibers (lane 2). There are about a dozen major bands in the isolated stress fiber, and we have identified six of them by immunoblotting.[11] Others need to be identified.

In Vitro Contraction of Stress Fibers

Materials

Solutions

Wash solution: 100 mM KCl, 2 mM EGTA [O,O'-bis(2-aminoethyl) ethylene glycol-N,N,N',N'-tetraacetic acid] (Wako), 10 mM imidazole (Wako), pH 7.2.
Reactivation solution: 75 mM KCl, 3 mM MgCl$_2$, 1 mM CaCl$_2$, 1 mM EGTA, 0.1 mM ATP, and 20 mM imidazole (pH 7.2).

---

[12] K. Fujiwara and T. D. Pollard, *J. Cell Biol.* **71,** 848 (1976).

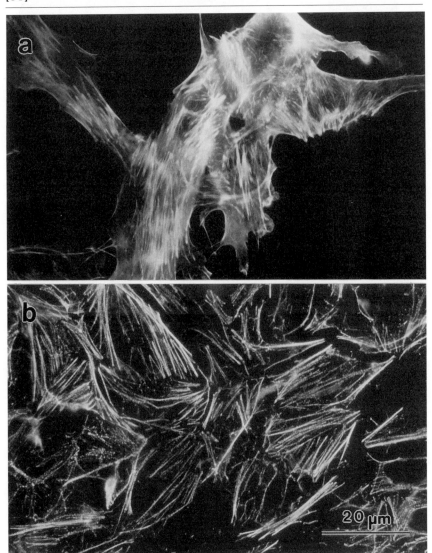

FIG. 2. Fluorescence micrographs showing stress fibers in endothelial cells before extraction (a) and those after the treatment with extraction buffer II (b). These specimens were stained with fluorescently labeled phalloidin, which specifically binds to actin filaments. Stress fiber preparations similar to that shown in (b) are used for reactivation experiments.

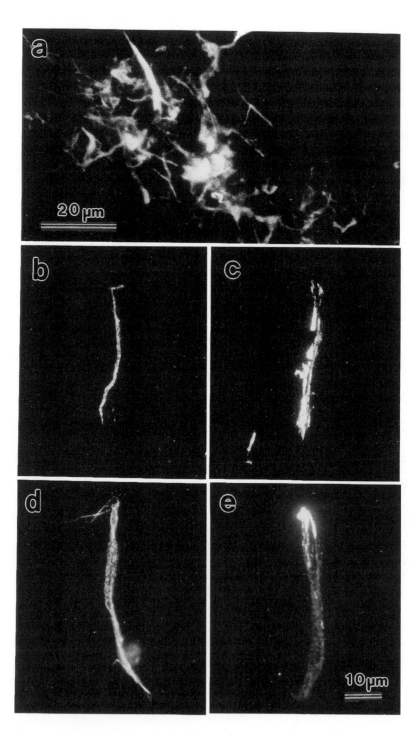

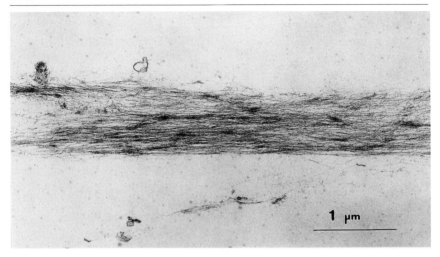

Fig. 4. A transmission electron micrograph of an isolated stress fiber of fibroblasts. Note that the ultrastructural features of stress fibers in the cell are preserved in this specimen.

Equipment

A high-resolution phase-contrast microscope is equipped with a high-resolution charge-coupled device (CCD) TV camera (Hamamatsu Photonics, Hamamatsu, Japan) and a digital image processor (Image Sigma-II, Nippon Avionics, Tokyo, Japan). Images are recorded using a high-resolution laser disk recorder (TEAC, Tokyo, Japan). This optical setup is necessary to observe stress fiber models and isolated stress fibers, which have very low contrast.

Methods

1. Culture cells on 18 × 18-mm coverslips, and follow the procedures described earlier up to step 5. Stress fibers are attached to the coverslip.

2. Rinse coverslips thoroughly with the wash solution for 15 min with three changes to remove Triton X-100.

3. Make a perfusion chamber. Two narrow parallel strips of Scotch tape separated by an appropriate distance are attached to a 2.5 × 7.5-cm slide

Fig. 3. Isolated stress fibers of fibroblast after sucrose gradient centrifugation were stained with various probes. Stress fiber pellet stained with fluorescent phalloidin (a) and highly magnified single isolated stress fibers stained with fluorescent phalloidin (b), antimyosin (c), anti-α-actinin (d), and antivinculin (e).

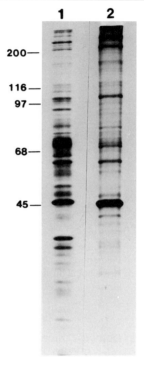

Fɪɢ. 5. SDS–gel electrophoretic patterns of fibroblast extract (lane 1) and isolated stress fiber pellet (b). The numbers on the left indicate molecular weights.

glass. A coverslip with stress fibers is placed upside down on the tape, making sure that the gap between the two glass surface is filled with the wash solution. The coverslip is glued to the slide glass using nail polish. The surface of the coverslip is cleaned carefully for optical clarity.

4. Mount the slide glass on a phase-contrast microscope stage and focus on stress fibers.

5. From one side of the perfusion chamber, 50 μl of the reactivation solution is applied while drawing solution from the other side using a piece of filter paper. Stress fibers begin to contract within several seconds.

Notes

Observing contraction of free stress fibers is difficult. We have recorded the contraction of free stress fibers that are attached to a poly-(D-lysine)-coated glass surface. However, such stress fibers go in and out of focus and

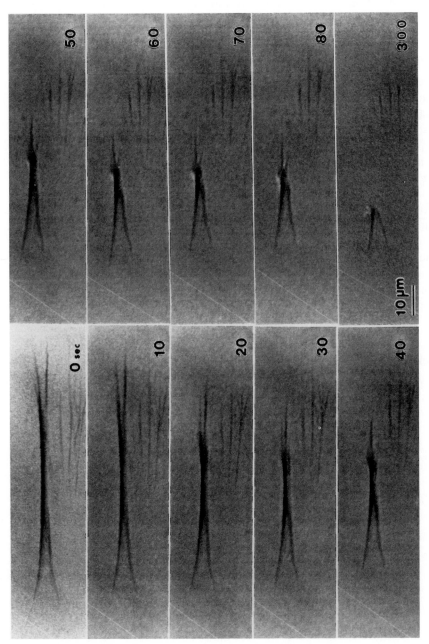

FIG. 6. Contraction of a fibroblast stress fiber model followed under a polarization microscope. Stress fibers are in dark contrast. At time 0, stress fibers were exposed to the contraction solution. With time, the stress fibers shortened.

the contraction is not smooth, presumably depending on how they are attached to the glass surface.

Application

The contraction of stress fibers is best observed by using those still attached to the substrate surface. Figure 6 shows stress fibers observed under a polarization microscope with the CCD camera, image processor, and recording system described earlier. At time 0, the reactivation solution is perfused and images are presented every 10 sec. Stress fibers are in dark contrast. We have characterized this contraction and concluded that it is an actomyosin-based, ATP-driven contraction.[11]

We have found that Rho kinase is involved in stress fiber contraction. Both RhoA and Rho kinase are present in isolated stress fibers. We have shown that the stress fiber model can contract in the absence of Ca^{2+} and that this contraction is inhibited by a Rho kinase inhibitor. In this case, the myosin regulatory light chain is phosphorylated, not by myosin light chain kinase, but by Rho kinase. Our studies indicate that stress fiber contraction is regulated by two independent systems: one by the Ca^{2+}-dependent myosin light chain kinase system and the other by the Ca^{2+}-independent Rho kinase system.

Acknowledgments

This work was supported in part by grants-in-aid for Scientific Research from the Ministry of Education, Science, and Culture of Japan; grants from the Ministry of Health and Welfare of Japan; and a grant from the Organization for Pharmaceutical Safety and Research.

[34] Inhibition of Rho GTPases Using Protein Geranylgeranyltransferase I Inhibitors

By Saïd M. Sebti and Andrew D. Hamilton

GTPases, Signal Transduction, and Cancer

Extracellular signals that induce cells to divide, differentiate, or die trigger a complex network of cellular signaling pathways.[1-6] Most often this process relies on the transfer of biological information from cell surface receptors to the nucleus,[3,4] and GTPases belonging to the Ras superfamily play a pivotal role in this transfer.[7] Several GTPase subfamilies have been studied in this context, including Ras proteins (H-, N-, and K-Ras), Rho proteins (RhoA, Rac1, and Cdc42), and R-Ras proteins [R-Ras and R-Ras2 (TC-21)].[7-9] One thoroughly studied area where these GTPases are intimately involved is that of peptide growth factor signal transduction. Here the GTPases use their GTP/GDP cycle as a molecular switch to transduce growth factor signals from receptor tyrosine kinases to the nucleus.[4,7,10] These growth signals are turned off by hydrolysis of the bound GTP to produce the inactive GDP-bound state.[7-9] Mutations in Ras inhibit GTPase activity, lock Ras in its GTP-bound form, and result in a growth factor-independent, constitutively activated signal that leads to uncontrolled growth and malignant transformation.[11-13] This Ras-dependent un-

[1] N. Dhanasekaran, *Oncogene* **17**, 1329 (1998).

[2] T. Dragovich, C. M. Rudin, and C. L. Thompson, *Oncogene* **17**, 3207 (1998).

[3] J. S. Gutkind, *Oncogene* **17**, 1331 (1998).

[4] A. C. Porter and R. R. Vaillancourt, *Oncogene* **17**, 1343 (1998).

[5] N. Dhanasekaran, S. T. Tsim, J. M. Dermott, and D. Onesime, *Oncogene* **17**, 1383 (1998).

[6] M. B. Jarpe, C. Widmann, C. Knall, T. K. Schlesinger, S. Gibson, T. Yujiri, G. R. Fanger, E. W. Gelfand, and G. L. Johnson, *Oncogene* **17**, 1475 (1998).

[7] S. L. Campbell, R. Khosravi-Far, K. L. Rossman, G. J. Clark, and C. J. Der, *Oncogene* **17**, 1395 (1998).

[8] D. J. Mackay, F. Esch, H. Furthmayr, and A. Hall, *in* "Cytoskeleton and G-proteins in the Regulation of Cancer" (N. Kuzumaki, ed.), p. 103. Kokoku, Japan, 1998.

[9] I. M. Zohn, S. L. Campbell, R. Khosravi-Far, K. L. Rossman, and C. J. Der, *Oncogene* **17**, 1415 (1998).

[10] F. McCormick, *Nature* **363**, 15 (1993).

[11] J. B. Gibbs, I. S. Sigal, M. Poe, and E. M. Scolnick, *Proc. Natl. Acad. Sci. U.S.A.* **81**, 5704 (1984).

[12] J. P. McGrath, D. J. Capon, D. V. Goeddel, and A. D. Levinson, *Nature* **310**, 644 (1984).

controlled growth is believed to be directly implicated in a large number of cancers, as 30% of human tumors express *ras* oncogenes containing such mutations.[14,15] Similar mutations in RhoA, Rac1, Cdc42, R-Ras, and TC-21 result in GTP-locked proteins that can induce tumorigenesis, invasion, and/or metastasis in experimental models, although such mutations have not been identified so far in human tumors.[8,9,16–19] However, ample evidence shows that Rho proteins are required for transformation by Ras and other oncoproteins,[8,9] and Ras, Rho and R-Ras proteins represent ideal targets for rational anticancer drug discovery.

PFTase and PGGTase I

One approach that has been taken to interfere with oncogenic signaling of the Ras superfamily of proteins is to design compounds that block their prenylation, a lipid posttranslational modification that is required for the transforming activity of these GTPases.[20–22] Two types of prenylations exist: farnesylation and geranylgeranylation.[23] Protein farnesyltransferase (PFTase) and protein geranylgeranyltransferase I (PGGTase I) catalyze the covalent attachment of the lipids farnesyl and geranylgeranyl, respectively, to the cysteine sulfhydryl at the carboxyl terminus of proteins that end in the consensus sequence CAAX, where C is cysteine, A is an aliphatic residue, and X is any amino acid.[23] PGGTase I prefers proteins where X is leucine, whereas PFTase prefers those where X is methionine or serine. H-, N-, and K-Ras are farnesylated, whereas RhoA, Rac1, Cdc42, R-Ras, and TC-21 are geranylgeranylated.[23]

PGGTase and PFTase are $\alpha\beta$ heterodimers that share the α subunit (48 kDa). Although the β subunit of PGGTase I (43 kDa) is distinct from

[13] R. W. Sweet, S. Yokoyama, T. Kamata, J. R. Feramisco, M. Rosenberg, and M. Gross, *Nature* **311**, 273 (1984).

[14] M. Barbacid, *Annu. Rev. Biochem.* **56**, 779 (1987).

[15] M. Barbacid, *in* "Important Advances in Oncology" (V. T. Devita, S. Hellman, and S. A. Rosenberg, eds.), p. 3. Lippincott, Philadelphia, 1986.

[16] A. M. Chan, T. Miki, K. A. Meyers, and S. A. Aaronson, *Proc. Natl. Acad. Sci. U.S.A.* **91**, 7558 (1994).

[17] A. D. Cox, T. R. Brtva, D. G. Lowe, and C. J. Der, *Oncogene* **9**, 3281 (1994).

[18] S. M. Graham, A. D. Cox, G. Drivas, M. G. Rush, P. D'Eustachio, and C. J. Der, *Mol. Cell. Biol.* **14**, 4108 (1994).

[19] R. Saez, A. M. Chan, T. Miki, and S. A. Aaronson, *Oncogene* **9**, 2977 (1994).

[20] S. M. Sebti and A. D. Hamilton, *Pharmacol. Ther.* **74**, 103 (1997).

[21] A. D. Cox and C. J. Der, *Biochim. Biophys. Acta* **1333**, F51 (1997).

[22] J. B. Gibbs and A. Oliff, *Annu. Rev. Pharmacol. Toxicol.* **37**, 143 (1997).

[23] F. L. Zhang and P. J. Casey, *Annu. Rev. Biochem.* **65**, 241 (1996).

that of PFTase (46 kDa), the β subunits have 35% amino acid sequence identity.[24,25] PFTase and PGGTase I do not require full-length substrates to prenylate the cysteine,[26,27] and small peptides as short as 4 amino acids, such as CVLS, can be farnesylated with kinetic parameters similar to those of full-length H-Ras.[26,27] This indicates that the structural requirements for enzyme recognition are contained within "CAAX."

Because a large number of PFTase and PGGTase I substrates such as H-, K-, and N-Ras, RhoA, Rac1, Cdc42, TC-21, and R-Ras are either oncogenes or are required for the transforming activity of these oncogenes,[7,9,14–19,28,29] both PFTase and PGGTase I have been suggested as novel molecular targets for anticancer drug discovery. However, the focus has been on PFTase with little attention given to PGGTase I as a target.[20–22] We have turned our attention to PGGTase I as a novel anticancer therapy target. One of the main reasons is that, as discussed earlier, many substrates of PGGTase I, such as Rac1, RhoA, Cdc42, R-Ras, and TC-21 can promote uncontrolled growth, tumorigenesis, and/or metastasis.[7–9,15,28] Furthermore, some of the PGGTase I substrates, such as RhoA, Rac1, and Cdc42, are required for transformation by Ras and other oncoproteins.[9] Finally, inhibition of PFTase in human cancer cells results in K-Ras and N-Ras becoming substrates for PGGTase I.[30]

PGGTase I and PFTase Inhibitors

Because CAAM tetrapeptides are potent inhibitors of PFTase (IC_{50} values ranging from 20 nM for CVFM to 200 nM for CVIM), we have used the CAAM scaffold as a starting point to design peptidomimetics that are cell permeable and that are less susceptible to protease degradation.[20] Our initial strategy was to replace the central aliphatic dipeptide "AA" by a

[24] F. L. Zhang, R. E. Diehl, N. E. Kohl, J. B. Gibbs, B. Giros, P. J. Casey, and C. A. Omer, *J. Biol. Chem.* **269**, 3175 (1994).

[25] J. F. Moomaw and P. J. Casey, *J. Biol. Chem.* **267**, 17438 (1992).

[26] Y. Reiss, S. J. Stradley, L. M. Gierasch, M. S. Brown, and J. L. Goldstein, *Proc. Natl. Acad. Sci. U.S.A.* **88**, 732 (1991).

[27] J. L. Goldstein, M. S. Brown, S. J. Stradley, Y. Reiss, and L. M. Gierasch, *J. Biol. Chem.* **24**, 15575 (1991).

[28] Y. Huang, R. Saez, L. Chao, E. Santos, S. A. Aaronson, and A. M. Chan, *Oncogene* **11**, 1255 (1995).

[29] G. J. Clark, M. S. Kinch, T. M. Gilmer, K. Burridge, and C. J. Der, *Oncogene* **12**, 169 (1996).

[30] D. B. Whyte, P. Kirschmeier, T. N. Hockenberry, I. Nunez-Oliva L. James, J. J. Catino, W. R. Bishop, and J. K. Pai, *J. Biol. Chem.* **272**, 14459 (1997).

moiety that is hydrophobic in nature and that lacks peptidic features.[31-37] FTI-276 is a third-generation CAAM peptidomimetic, in which reduced cysteine is linked to methionine by an aromatic spacer, 2-phenyl-4 amino-benzoic acid[35] (Fig. 1). Using a similar strategy, we have also made several PGGTase I inhibitors. GGTI-297 is a CAAL peptidomimetic in which reduced cysteine is linked to leucine by 2-naphthyl-4 aminobenzoic acid[38,39] (Fig. 1). To eliminate more peptidic features from these molecules, we replaced cysteine with several different groups and found that imidazole derivatives give highly potent and selective inhibitors. For example, FTI-2148 and GGTI-2154 (Fig. 1) have higher potency and selectivity than their corresponding cysteine-containing analogs (see Table I).[40]

Preparation of Partially Purified PGGTase I and PFTase from Daudi Burkitt's Lymphoma Cell Line

Human Daudi cells (ATCC, Rockville, MD) are grown in RPMI containing 5% (v/v) iron-supplemented calf serum, 1% (v/v) penicillin/streptomycin at 37°, and 10% CO_2 atmosphere. The cells are homogenized in lysis buffer [10 mM HEPES (pH 7.4), 1 mM $MgCl_2$, 1 mM EDTA, 0.1 mM dithiothreitol (DTT), and 0.1 mM phenylmethylsulfonyl fluoride (PMSF)] and centrifuged at 4° for 30 min at 12,000g, followed by 60,000g centrifugation at 4° for 60 min of the supernatant. PGGTase I and PFTase from 60,000g supernatant fractions are then partially purified by DEAE-Sephacel chromatography where they are eluted with 50 mM Tris (pH 7.5) containing 0.3 and 0.4 M NaCl, respectively. PGGTase I and PFTase specific activities from these fractions are compared to and found to be identical to those

[31] M. Nigam, C. M. Seong, Y. Qian, A. D. Hamilton, and S. M. Sebti, *J. Biol. Chem.* **268,** 20695 (1993).

[32] Y. Qian, M. Blaskovich, M. Saleem, S. Wathen, A. Hamilton, and S. M. Sebti, *J. Biol. Chem.* **269,** 12410 (1994).

[33] A. Vogt, Y. Qian, M. Blaskovich, R. Fossum, A. Hamilton, and S. M. Sebti, *J. Biol. Chem.* **270,** 660 (1995).

[34] E. Lerner, Y. Qian, A. D. Hamilton, and S. M. Sebti, *J. Biol. Chem.* **270,** 26770 (1995).

[35] E. Lerner, Y. Qian, M. A. Blaskovich, R. Fossum, A. Vogt, A. Cox, C. Der, A. D. Hamilton, and S. M. Sebti, *J. Biol. Chem.* **270,** 26802 (1995).

[36] Y. Qian, M. A. Blaskovich, C.-M. Seong, A. Vogt, A. D. Hamilton, and S. M. Sebti, *Biorg. Med. Chem. Lett.* **4,** 2579 (1994).

[37] Y. Qian, A. Vogt, S. M. Sebti, and A. D. Hamilton, *J. Med. Chem.* **39,** 217 (1996).

[38] A. Vogt, Y. Qian, A. D. Hamilton, and S. M. Sebti, *Oncogene* **13,** 1991 (1996).

[39] T. McGuire, Y. Qian, A. D. Hamilton, and S. M. Sebti, *J. Biol. Chem.* **271,** 27402 (1996).

[40] J. Sun, M. A. Blaskovich, D. Knowles, Y. Qian, J. Ohkanda, R. D. Bailey, A. D. Hamilton, and S. M. Sebti, *Cancer Res.* **59,** 4919 (1999).

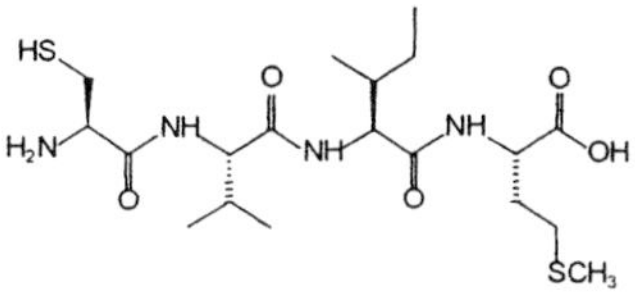

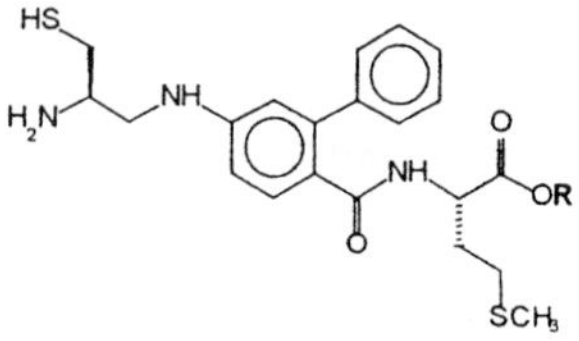

FTI-276 R = H
FTI-277 R = CH₃

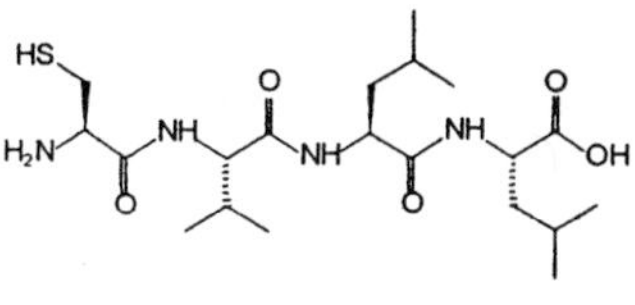

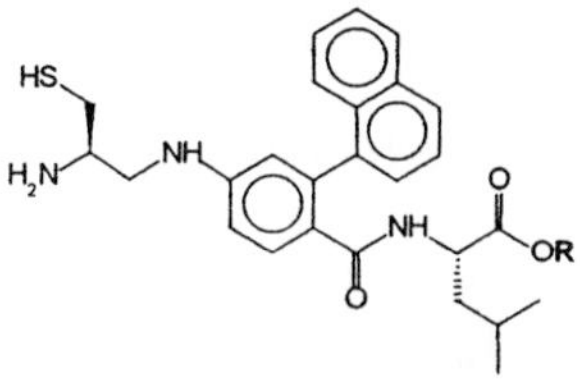

GGTI-297 R = H
GGTI-298 R = CH₃

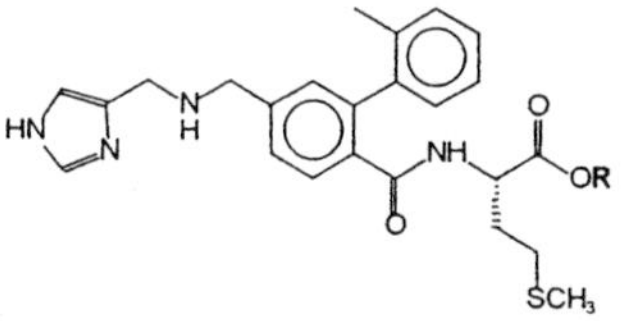

FTI-2148 R = H
FTI-2153 R = CH₃

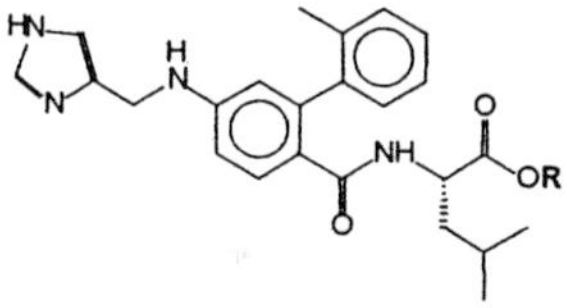

GGTI-2154 R = H
GGTI-2166 R = CH₃

FIG. 1. Design of CAAX peptidomimetics as FTIs and GGTIs.

TABLE I
INHIBITION IC$_{50}$ VALUES FOR FTIs AND GGTIs

Compound/methyl ester	In vitro activity			Whole cell processing (methyl esters)		
	PFTase (nM)	PGGTase I (nM)	Selectivity (-fold)	Ras (μM)	Rap1 (μM)	Selectivity (-fold)
FTI-276/FTI-277	0.5	50	100	0.3	>30	>100
FTI-2148/FTI-2153	1.4	1700	1200	0.03	>30	>1000
GGTI-297/GGTI-298	203	55	4	>15	5	>3
GGTI-2154/GGTI-2166	5600	21	266	>30	0.3	>100

from the 60,000g supernatant fractions. Furthermore, the ability of FTIs and GGTIs to inhibit PFTase and PGGTase I, respectively, is also the same when either preparation is used. Therefore, partially purified PFTase and PGGTase I, either from 60,000g supernatant fractions or from DEAE-Sephacel fractions, can be used (see later).

PGGTase I and PFTase Activity Assays: Inhibition by GGTIs and FTIs

PGGTase I and PFTase Activity Assays

PGGTase I and PFTase activities from 60,000g supernatants of human Burkitt lymphoma (Daudi) cells (ATCC) are assayed by measuring the amount of [^{3}H]GGPP and [^{3}H]FPP transferred to recombinant p21 H-ras CVLL and p21-H-Ras CVLS, respectively. The supernatant fractions (40 μg) are incubated in 50 mM Tris, pH 7.5, 50 μM ZnCl$_2$, 20 mM KCl, 3 mM MgCl$_2$, and 1 mM DTT. The reaction is incubated at 37° for 30 min with recombinant Ha-Ras-CVLS (11 μM) and [^{3}H]FPP (625 nM; 16.3 Ci/mmol) for PFTase and with recombinant Ha-Ras-CVLL (5 μM) and [^{3}H]geranylgeranyl pyrophosphate (525 nM; 19.0 Ci/mmol) for PGGTase I. For inhibition studies, the peptidomimetics are premixed for 5 min at 4° with PFTase and PGGTase I before adding the reaction mixture. Recombinant Ha-Ras-CVLS is prepared as described previously[31] from *Escherichia coli* strain PRI expressing H-Ras-CVLS (Dr. Robert Crowl, Hoffman LaRoche Inc., Nutley, NJ).[41] Recombinant Ha-Ras-CVLL is prepared from PR13-Q bacteria obtained from Drs. C. J. Der and A. D. Cox (University of North Carolina, Durham, NC) as described previously.[42] After incubation for 30 min at 37°, the reaction is stopped and filtered on glass fiber filters as described previously[43,44] and as follows: To each sample is added 500 μl 4% sodium dodecyl sulfate (SDS) and then 500 μl 30% trichloroacetic acid (TCA) to precipitate proteins. Samples are vortexed and then incubated on ice for 45 min. After incubation, 2 ml of 2% SDS/6% TCA (w/v) is added to each sample. Finally, samples are filtered onto Whatman (Clifton, NJ) GF/B glass fiber filters using a Millipore (Bedford, MA) 12-sample filtering manifold. Each vial is washed twice with 2 ml 2% SDS/6% TCA. Each filter is washed five times with 2 ml 6% TCA. Finally, each filter is placed in 5 ml scintillation fluid, and the amount of labeled Ras protein bound to each filter is counted by scintillation counting.

[41] J. C. Lacal, E. Santos, V. Notario, M. Barbacid, S. Yamazaki, H. F. Kung, C. Seamans, S. McAndrew, and R. Crowl, *Proc. Natl. Acad. Sci. U.S.A.* **81**, 5305 (1984).
[42] A. D. Cox, M. M. Hisaka, J. E. Buss, and C. J. Der, *Mol. Cell. Biol.* **12**, 2606 (1992).
[43] Y. Reiss, J. L. Goldstein, M. C. Seabra, P. A. Casey, and M. S. Brown, *Cell* **62**, 81 (1990).
[44] Y. Reiss, M. C. Seabra, J. L. Goldstein, and M. S. Brown, *Methods* **1**, 241 (1990).

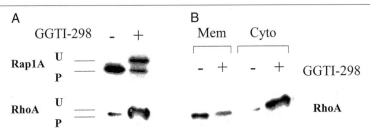

FIG. 2. Inhibition of RhoA and Rap1A processing in intact cells.

Inhibition of RhoA and Rap1A Processing in Cultured Cells by
PGGTase I Inhibitors

*Detection of Inhibition of RhoA and Rap1A Processing by Band Shift
Assay on SDS–PAGE*

Human Ha-Ras oncogene-transformed NIH 3T3 cells[35] or human lung
adenocarcinoma Calu-1 cells[45] are treated twice every 24 hr for 2 days with
GGTI-298 (15 μM) or vehicle [10 μM DTT in 0.1% (v/v) dimethyl sulfoxide
(DMSO)]. Cells are lysed in lysis buffer [30 mM HEPES, pH 7.5, 10 mM
NaCl, 5 mM MgCl$_2$, 25 mM NaF, 1 mM EGTA, 1% Triton X-100, 10%
(v/v) glycerol, 2 mM PMSF, 25 μg/ml leupeptin, 10μg/ml aprotinin, 10
μg/ml soybean trypsin inhibitor] for 30 min on ice. Cleared lysates (10–100
μg) are electrophoresed onto a 12.5% SDS–PAGE, transferred onto nitrocellulose, and immunoblotted with antibodies against RhoA or Rap1/Krev1
(Santa Cruz Biotechnology, Santa Cruz, CA). Positive antibody reactions
are visualized using peroxidase-conjugated goat antirabbit immunoglobulin
G (IgG) and goat antimouse IgG, respectively, and the enhanced chemiluminescence detection system (ECL; Amersham Corp., Chicago, IL).

Figure 2A shows that Calu-1 cells treated with vehicle contain only fully
processed RhoA and Rap1A. Treatment with GGTI-298 for 48 hr results
in inhibition of the geranylgeranylation of both RhoA and Rap1A, as
evident from the appearance of the corresponding slower migrating unprocessed forms of RhoA and Rap1A (Fig. 2A).

GGTI-298: Decreasing Membrane-Bound and Increasing Cytosolic RhoA

Unlike Rap1A, processed (P) and unprocessed (U) RhoA migrate very
closely, and it is difficult to detect a band shift (Fig. 2A). Therefore, the
band shift assay just described may not be appropriate for evaluating the

---

[45] J. Sun, Y. Qian, Z. Chen, J. Marfurt, A. D. Hamilton, and S. M. Sebti, *J. Biol. Chem.* **274,**
6930 (1999).

effects of PGGTase I inhibitors on the processing of some geranylgeranylated proteins such as RhoA. An alternative approach in assessing the effect of PGGTase I inhibitors on the processing of Rho proteins is to determine their effects on membrane-bound and cytosolic RhoA. This can be accomplished by isolating membrane and cytosolic fractions from vehicle and GGTI-298 treated cells prior to SDS–PAGE protein separation and immunoblotting. To this end, Calu-1 cells are plated on day 1 and treated with either vehicle or GGTI-298 (15 μM) on day 2 and treatment repeated on day 3 with fresh media containing either vehicle or GGTI-298(15 μM.) On day 4, cells are collected and pellets are resuspended in ice-cold hypotonic buffer (10 mM Tris, pH 7.5, 5 mM MgCl$_2$, 1 mM DTT, 1 mM PMSF) and sonicated (using Branson Sonifier 450 at 70% duty cycle; 2 × 5 sec) to break up the cell pellet. The cell suspension is then centrifuged at 2000 rpm for 10 min to clear debris after which the supernatant is spun for 60 min at 100,000g (4°) to separate membrane and cytosol fractions. The supernatant represents cytosolic fractions, whereas the pellet resuspended in the buffer described earlier represents membrane fractions. Aliquots of each fraction are added to equal volumes of 2× the HEPES lysis buffer as described earlier and put on ice for 30 min. Equal amounts of lysates from cytosolic and membrane fractions are electrophoresed on 12.5% SDS–PAGE gels and immunoblotted with the anti-RhoA antibody as described earlier for the band shift assay. Figure 2B shows that vehicle-treated Calu-1 cells contain only membrane-bound RhoA with barely detectable amounts of cytosolic RhoA. Treatment with GGTI-298 (15 μM) for 48 hr results in a large decrease in membrane-bound RhoA, which is paralleled with a large increase in cytosolic RhoA. Thus, GGTI-298 treatment of Calu-1 cells inhibits PGGTase I, resulting in inhibition of the geranylgeranylation of RhoA and, hence, its accumulation in the cytosol.

A third approach to determining the effects of PGGTase I inhibitors on the processing of Rho proteins is to first label cells with radiolabeled [³H]mevalonic acid, [³H]geranylgeraniol or [³H]geranylgeranyl pyrophosphate prior to immunoprecipitation of the geranylgeranylated Rho protein of interest and running SDS–PAGE. This method can be used with either the band shift or cytosolic vs membrane fractionation assays. This radiolabeling method has been described in detail previously.[46]

Acknowledgment

This work was supported by the NIH (CA67771).

[46] G. L. James, J. L. Goldstein, M. S. Brown, T. E. Rawson, T. C. Somers, R. S. McDowell, C. W. Crowley, B. K. Lucas, A. D. Levinson, and J. C. Marsters, Jr., Science 260, 193 (1993).

[35] Imaging Spatiotemporal Dynamics of Rac Activation in Vivo with FLAIR

By Chester E. Chamberlain, Vadim S. Kraynov, and Klaus M. Hahn

Introduction

FLAIR (*FL*uorescent *A*ctivation *I*ndicator for *R*ho GTPases) is a biosensor system that maps the spatial and temporal dynamics of Rac activation in living cells. The approach is based on microinjection of a fluorescently labeled domain from p21-activated kinase into cells expressing green fluorescent protein (GFP)–Rac. The injected domain (called PBD, for p21-binding domain) binds only to Rac–GTP, and not to Rac–GDP.[1,2] Within living cells, PBD binds to the GTP–Rac wherever it has bound GTP, bringing the Alexa 546 dye (Molecular Probes, Eugene, OR) on the PBD near the GFP on the Rac to produce fluorescence resonance energy transfer (FRET). Thus the FRET signal marks subcellular locations where Rac is activated. This can be quantified to follow the changing levels and locations of Rac activation or to trace the kinetics of total Rac activation on an individual cell basis.

The labeling of PBD with Alexa and mammalian expression vectors for the expression of Rac–GFP are described in detail elsewhere.[3] This article describes a detailed protocol for the production of pure PBD, as this required considerable optimization. It then discusses how to generate cell images suitable for quantitative analysis of Rac activation and finally gives procedures and caveats for generating two types of data: images showing the spatial distribution of Rac activation within cells and curves showing the kinetics of Rac activation in single cells.

PBD Expression and Purification

PBD is expressed in the form of C-terminal hexahistidine (His_6) fusion from the prokaryotic expression vector pET23 (Novagen, Madison, WI). It was determined experimentally that the highest levels of expression are observed when a vector containing plain T7 promoter (not T7*lac*) is used

[1] G. Thompson, D. Owen, P. Chalk, and P. Lowe, *Biochemistry* **37**, 7885 (1998).
[2] V. Benard, B. P. Bohl, and G. M. Bokoch, *J. Biol. Chem.* **274**, 13198 (1999).
[3] V. S. Kraynov, C. E. Chamberlain, G. Bokoch, M. Schwartz, S. Slabaugh, and K. M. Hahn, submitted for publication.

in combination with a BL21(DE3) strain [not the more stringent BL21(DE3)pLysS] of *Escherichia coli.* This system allows for "leaky" protein expression (Novagen). While the His$_6$ tag can be cleaved from the purified protein with thrombin, it is not necessary, as the tag does not have any significant effect on probe functionality.

Competent BL21(DE3) cells (Stratagene, La Jolla, CA) are transformed with pET23-PBD according to standard protocols[4] and plated on an LB plate containing carbenicillin. Cells do not degrade carbenicillin as quickly as ampicillin, so a higher percentage of cells retain the vector at the culture density appropriate for induction (Novagen). Five milliliters of LB media with 100 μg/ml carbenicillin is inoculated with a single colony of cells and grown in the shaker at 37° for 6–8 hr (until dense). Two milliliters of this is then used to inoculate 50 ml of LB*carb.* The rest of the culture is diluted 1:1 with glycerol and frozen for long-term storage at −80°. The 50-ml culture is incubated in the shaker overnight at 37°. The next morning, 1–2 liter of LB*carb* is inoculated with the overnight culture (15- to 20-ml culture/500 ml media) and grown in the shaker (37°) to OD$_{600}$ of 0.8–0.9 (about 2–3 hr). The cultures are then chilled briefly on ice to 30°–32° and put back in the shaking incubator turned down to 30°–32°. The protein is expressed at a lowered temperature to increase the portion of the correctly folded, soluble PBD. Isopropylthiogalactoside (IPTG) is added to a final concentration of 0.4–0.5 m*M,* and the cultures are allowed to grow for another 4–5 hr at 30°–32° (shaker). The cells are collected by centrifugation (8000g, 4 min) and stored as a pellet at −20° until use. Approximately 2.5–3 g of cells is usually obtained from each liter of culture.

Purification of PBD-His$_6$ is performed essentially as described in the Talon affinity resin manual (Clontech, Palo Alto, CA). The cells (3–5 g) are thawed in 20–30 ml of the lysis buffer [30 m*M* Tris–HCl, pH 7.8, 250 m*M* NaCl, 10% (v/v) glycerol, 5 m*M* MgCl$_2$, 2 m*M* 2-mercaptoethanol (2-ME), 1 m*M* phenylmethylsulfonyl fluoride (PMSF)], homogenized with a spatula, and sonicated (four pulses, 10–15 sec each). T4 lysozyme and DNase are added in catalytic amounts (approximately 100 μg/ml lysozyme and 500 U DNase) to help the lysis, and the suspension is incubated on ice with periodic mixing for 30 min. The cells are then centrifuged at 12,500 rpm for 30 min, and the supernatant containing PBD is carefully transferred into a 50-ml Falcon tube.

The Talon resin (1.5–2 ml) (Clontech) is washed twice with 10 volumes of the lysis buffer in a 15-ml Falcon tube, centrifuging in the swinging bucket centrifuge at low speed in between to separate the resin. The cell

---

[4] J. Sambrook, E. F. Fritsch, and T. Maniatis, *in* "Molecular Cloning: A Laboratory Manual" (C. Nolan, ed.), p. 18. Cold Spring Harbor Laboratory Press, Cold Spring Harbor, NY, 1989.

lysate is added to the 1.5 ml of washed Talon resin in a 50-ml Falcon tube and inverted or agitated gently (i.e., with an orbit shaker) for 20–30 min at room temperature.

The resin is then separated by centrifugation in a swinging bucket centrifuge. The supernatant containing the unbound fraction is removed and saved for SDS–PAGE analysis. The resin is then transferred into a new 15-ml Falcon tube and washed twice (10–15 min each, room temperature, orbit shaker) with 12 ml of the lysis buffer, without PMSF and 2-ME. The third wash is performed with lysis buffer + 10 mM imidazole (add 1 M stock in water, kept at −20°). After the final separation, the resin is resuspended in 2–3 ml of lysis buffer with 10 mM imidazole and pipetted into a column (0.5 cm in diameter). The resin is allowed to sediment by gravity flow until the fluid above the resin bed is almost gone, and then another 3–5 ml of lysis buffer with 10 mM imidazole is added to wash the column. Elution is performed using lysis buffer with 60 mM imidazole, and ca. 500-μl fractions are collected. PBD usually elutes in fractions 5–13 (total volume about 3–4 ml). An aliquot of each fraction is run on a 12% SDS–PAGE, and fractions containing pure PBD are combined and dialyzed against 1 liter of 25 mM NaP buffer (pH 7.3). A dialysis bag (SpectraPor 7) or dialysis cassette (Pierce, Rockford, IL) with a molecular weight cutoff value of 3,500 can be used.

After 2–3 hr of dialysis, the bag is wiped with a KimWipe and buried in Aquacide powder (Calbiochem) for 15–45 min at 4°, depending on the volume of the sample in the bag. This concentration process should be monitored carefully as complete drying may occur if the bag is left in the Aquacide for too long. The powder is scraped gently from the bag every 10–15 min to facilitate water absorption. When the sample reaches 0.5–1.5 ml in volume (3- to 10-fold concentration), the Aquacide is cleaned from the bag and the sample is removed carefully. The sample is centrifuged briefly (14,000 rpm, 2 min) to separate it from the precipitated material and transferred into a new dialysis bag or cassette. After the second dialysis step, the concentration of PBD is measured by taking a small aliquot (5–10 μl) and diluting into 50 mM Tris–HCl (pH 7.5–8.0) or other appropriate buffer. The extinction coefficient of PBD at 280 nm is 8250 (estimated from the primary sequence). On average, 1.5–2 mg of PBD is obtained per liter of cell culture.

Other methods of concentrating PBD were found to be less effective. For instance, centrifugal concentrators require prolonged centrifugations and result in nonspecific adsorption of the small PBD protein to the membrane. It is essential to perform dialysis after concentration with Aquacide. This prevents the ionic strength of the resultant protein preparation from becoming too high before labeling. Low ionic strength conditions are preferable to avoid excessive precipitation of the protein during attachment of the hydrophobic dye. As mentioned earlier, PBD labeling is described elsewhere.[3]

Using FLAIR to Examine Rac Nucleotide State in Cells

Loading GFP–Rac and Alexa–PBD in Cells

Cells are first transfected with GFP–Rac through nuclear microinjection. We use the EGFP variant (Clontech, Palo Alto, CA), which produces significantly brighter cells than wild-type GFP.[5] For microinjection of DNA and of PBD–Alexa, glass pipettes with a 1.0-mm outer diameter and a 0.50-mm inner diameter (Sutter) are pulled using a micropipette puller (Sutter Model P-87, Novato, CA) to make microinjection needles with tips of approximately 0.5 μm diameter. Rac–GFP cDNA is injected into Swiss 3T3 fibroblasts at 200 ng/μl using a constant needle pressure of approximately 100 hPa. DNA can be centrifuged prior to injection (20,000g for 15 min) to prevent clogging the needle.

Cells expressing GFP–Rac are microinjected with Alexa–PBD using a microscope with optics and illumination capable of revealing the GFP fluorescence (detection sensitivity is typically improved by using higher NA objectives and brighter light sources, such as a 100-W Hg arc lamp). Thus only GFP-expressing cells need to be injected. To reduce background fluorescence during injection and the following experiment, cells are placed in 1 ml of prewarmed Dulbecco's phosphate buffer solution (DPBS) containing 1000 mg/liter D-glucose and 36 mg/liter sodium pyruvate and supplemented with 0.2% bovine serum albumin (BSA), 1% L-glutamine, and 1% (v/w) penicillin–streptomycin. During injection and the following experiment, cells are mounted in a Dvorak live cell chamber (Nicholson, Gaithersburg, MD) preheated to 37° and maintained at 37° by a heated stage (20/20 Technology, Wilmington, NC). The microscope can be equipped with a motorized stage and shutter controls (Ludl, Hawthorne, NY) to monitor multiple stage positions in one experiment.

Cells that are barely expressing or expressing too much GFP–Rac are ignored. The former produce FRET too weak for recording, and in the latter, overexpressed Rac affects the biology of the cell. We have shown that cells expressing less than 300 intensity units (IU) do not display Rac-induced ruffling and altered morphology. The precise value of this cutoff will depend on the sensitivity of the imaging system and should be determined by each laboratory for a relevant biological behavior. A 100 μM solution of Alexa–PBD is centrifuged at 20,000g for 1 hr prior to injection and is then

[5] R. Heim, R. Y. Tsien, *Curr. Biol.* **6,** 178 (1996).

injected into the cytoplasm of cells expressing GFP–Rac. Lowering the needle into the region just adjacent to the nucleus seems to produce the best combination of efficient injection and cell health. After the injection, cells are placed back into the 37° incubator for 5–10 min to recover. Alexa–PBD could potentially act as an inhibitor of Rac activity, so controls are carried out showing that, for our imaging system, up to 1000 IU of Alexa–PBD does not inhibit induction of ruffling.

Imaging Rac Activation

Imaging experiments were performed in our laboratory using a Photometrics KAF1400 cooled charge-coupled device (CCD) camera and Inovision (Raleigh, NC) ISEE software for image processing and microscope automation. Although filters are undergoing further optimization, the best success to date has been achieved with the following filters designed with sharp cutoffs specifically for this purpose by the Chroma corporation (Brattleboro, VT): GFP, HQ480/40, HQ535/50, Q505LP; FRET, D480/30, HQ610/75, 505DCLP; and Alexa, HQ545/30, HQ610/75, Q565LP.

The exact camera settings depend on the type of experiment being performed. When the total Rac activity within the cell is being determined, images are not generated, so spatial resolution can be sacrificed for increased sensitivity. Images are taken using 3×3 binning with exposure times of 0.1, 0.1, and 0.5 sec for GFP, Alexa, and FRET respectively. When images are required, i.e., to examine the changing spatial distribution of Rac activation, 1×1 binning is used with exposure times of 1, 1, and up to 5 sec for GFP, Alexa, and FRET respectively. These settings depend on the sensitivity of the imaging system used and the desired trade-off between sensitivity and spatial or temporal resolution. Settings should always be chosen not to exceed the dynamic range of the camera.[6] Another important consideration during imaging experiments are motion artifacts. For fast-moving phenomena, features of the cell may move appreciably during the time between acquisition of the FRET and GFP images. This results in artifacts when the image is corrected for bleed through, as described in more detail later. This can only be prevented by reducing the time between exposures or by more expensive solutions, such as using two cameras simultaneously.

When total Rac activation is being determined, a picture of GFP–Rac and FRET is taken at each time point. Only one image of Alexa–PBD, usually at the initial time point, will also be required (for bleed-through corrections as described later). In contrast, when generating images (i.e., to determine the distribution of rac activation), an Alexa–PBD image must

---

[6] K. Berland and K. Jacobson, *in* "Video Microscopy" (G. Sluder and D. Wolf, eds.), p. 33. Academic Press, San Diego, 1998.

be taken at each time point. The reasons for this are discussed in the following section. If the cells are to be treated with some type of stimulus, it is helpful to take a series of images prior to stimulation as controls for noise, bleaching, and other artifacts.

Image Processing

Image analysis as described here is performed to follow the kinetics of total Rac activity within individual cells or to generate images that show the subcellular location of Rac activation. It is easy to generate artefactual data during this process, so this section emphasizes the proper application of corrections essential for quantitative imaging. We use common image processing operations whose precise implementation will depend on the software package used.[6] The correction factors must be applied rigorously when using the FLAIR system, as FRET signals will be low relative to other sources of fluorescence in the sample. The FRET signal must purposefully be kept low, as minimum quantities of fluorescent molecules should be used to prevent perturbation of cell behavior. Although the operations described are similar to those used for other fluorescent probes, there are important differences unique to application of FLAIR.

The following protocols assume use of a cooled CCD camera, which typically shows low levels of noise, linear response to light intensity, and little variation in response from pixel to pixel. It is valuable to use cameras and software with the greatest possible bit depth (allotment of computer memory to each pixel to maximize the number of possible intensity gradations). This is especially important for ratio operations, which are typically performed using 12-bit images or greater. Operations are described in the order in which they should be performed.

Registration

For corrections applied in later steps, it is important that each of the images taken using different excitation and emission wavelengths be registered so that the cells lie atop one another, with cell edges and internal features exactly coinciding. Different image processing software will accomplish this in different ways, but it usually involves manual translocation of one image so that it lines up with a second, fixed image. This is best accomplished by zooming in on cells and adjusting brightness and contrast to clearly see cell edges and internal features. Because the GFP–Rac signal is strongest in our experiments, it is used as the reference image. Errors in registration often become apparent as "shadow effects" when the bleed-through corrections described later are performed (see error illustrated in Fig. 1).

Background Subtraction

There are two methods commonly used for background subtraction. If the only intention is to follow the changing level of the FRET signal over time, and if the background (in the absence of cells) remains uniform across the field of view throughout the experiment, then it is sufficient to determine the background intensity in several regions of the image outside the cell. The average value of these intensities is then subtracted from each pixel in the image. This method can also be sufficient for following qualitative changes in the subcellular location of activation, but it must be used with caution. Subtle variations in background intensity across the cell could be large relative to the changes observed in FRET, producing significant artifacts. When quantifying the kinetics of total cell Rac activity over time

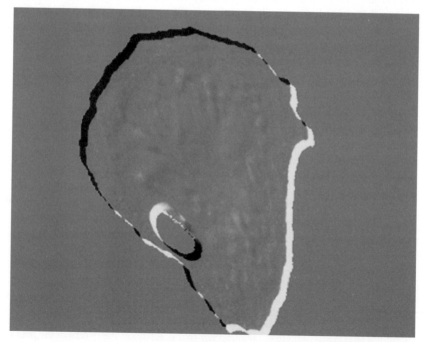

Fig. 1. Registration artifact. Before bleed-through correction, in which one image is subtracted from another, it is important to superimpose the two images precisely. Small differences in the microscope optics shift the position of one image relative to the other. If registration is performed incorrectly, a characteristic edge effect is seen, in which bleed-through correction results in artifactual bright and dark rims on exactly opposite sides of objects and cell edges.

(see following sections), it is better to take an image of a region of the coverslip containing no cells or fluorescent debris under the same conditions used for taking the real image. This background image is then subtracted from the real image prior to further analysis. A separate background image must be taken for each type of image (GFP, FRET, or Alexa) and at each time point when successive images are obtained. Artifacts that can be generated using the simpler background subtraction during quantitative analysis are illustrated in Fig. 2.

Masking

It is valuable to mask out regions surrounding the cells prior to further analysis. The edges of the cell are outlined in most software either manually or by eliminating all sections of the images below a certain intensity value (interactive threshholding). Regions outside the cell are thus identified and eliminated from further calculations. (The precise approach will depend on software. The mask is usually a binary image with all values within the cell equal to 1 and all outside equal to 0. The real image is multiplied by the mask.) The mask is best generated using the GFP image, which has

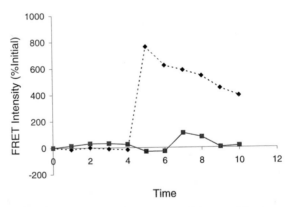

Time

FIG. 2. Background subtraction artifacts. If the efficiency of fluorescence detection is uneven across the field of view (i.e., uneven illumination, poorly corrected objectives, refractile objects), subtraction of a background image is superior to simply subtracting the intensity measured in an "empty" area of the image. In the example shown here, serum is added to a coverslip that has uneven illumination and poorly corrected optics. The serum produces a nonuniform increase in background fluorescence that is greater in the middle of the field of view. If the background is simply taken as intensity near the edge of the field, an artifactual increase in intensity will be observed after background subtraction (dashed line). When one obtains an actual image from a region of the coverslip that has no cells and subtracts this image, no artifactual intensity increase is seen (solid line). These effects are exaggerated when the real signal is not very large relative to the background.

the strongest signal and therefore the most clearly defined edges. When determining the total intensity within the cell, analysis should be carried out on the same pixels within the FRET, GFP, and Alexa images. Therefore, the same mask is applied to each image after registration, assuring that exactly the same pixels are analyzed.

Bleed-Through Correction

During FRET it is necessary to excite the donor fluorophore while monitoring emission from the acceptor fluorophore. It is extremely difficult to design FRET filters that see only FRET emission and block all GFP emission or block all light from Alexa excited directly rather than by FRET. To correct for "bleed-through" of such light into the FRET image, the fluorescence filters must be characterized by taking images of cells containing GFP–Rac or Alexa–PBD alone. For example, in bleed-through correction for GFP, cells are imaged using both the GFP and the FRET filter set. When observing GFP fluorescence through FRET filters, a fixed percentage of GFP emission will be seen. The total fluorescence intensity is determined for both GFP and FRET images from cells containing only GFP–Rac. A GFP bleed-through factor is computed for each cell by dividing the intensity through the FRET filters by that through the GFP filters (the "bleed-through factor" for GFP: FRET intensity/GFP intensity). This value is plotted against cell intensity for numerous cells, and a line is fit to these data to produce an accurate value of the bleed-through factor. It is important to use background-subtracted images. The process is repeated for Alexa–PBD. When the actual experiment is performed, Alexa–PBD, GFP–Rac, and FRET images are obtained. After background subtraction of all three images, the Alexa–PBD and GFP–Rac images are multiplied by the appropriate bleed-through factor and subtracted from the FRET image. This is an extremely important step, which must be applied carefully to prevent artifacts that appear to be regions of high FRET, especially as the magnitude of the FRET signal approaches that of the bleed-through. It is important not to use GFP–Rac or Alexa–PBD images that exceed the dynamic range of the camera ("overexposure") as they will not fully eliminate bleed-through. Motion artifacts (see earlier discussion) can also produce errors derived from bleed-through corrections, as illustrated in Fig. 3.

Production of Total Activation Curves (Bleaching Correction)

To determine how overall Rac activity within a cell changes over time, the total fluorescence intensity within a GFP–Rac and FRET image is determined at each time point. The intensity of the FRET image is divided

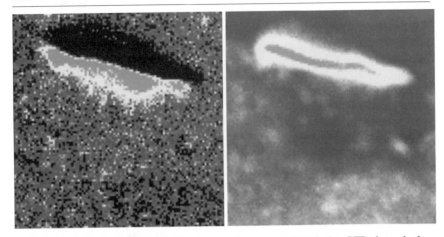

FIG. 3. Motion artifact. (Left) The FRET channel and (right) the GFP channel after corrections for bleed-through have been applied. In bleed-through correction, the GFP image is reduced in intensity by a precise percentage and is then subtracted from the FRET image to eliminate GFP emission being transmitted by the FRET microscope filters (see text). Because the object being imaged moved between acquisition of the GFP and FRET images, bleed-through subtraction produced an artifactual dark area in the FRET image and an artifactual localized increase in FRET.

by that of the GFP–Rac image. This ratio better reflects total Rac activity than FRET intensity alone. Division by GFP–Rac normalizes out errors due to bleaching of GFP, effects of uneven illumination, and other factors affecting both GFP and FRET signals. Because all FRET occurs through irradiation of GFP, bleaching of GFP will decrease both GFP and FRET emissions to the same extent. Therefore, the FRET/GFP ratio will be a measure of Rac activity that is not affected by bleaching. It is critical that each image be properly background subtracted and corrected for bleed-through. Bleed-through corrections are somewhat simplified when generating these curves, as they need not be carried out on actual images, but simply on the total intensity values derived from the images. The total intensity of the GFP–Rac image is multiplied by the GFP bleed-through factor, and the resulting value is subtracted from the FRET intensity. An analogous operation is performed for Alexa–PBD bleed-through. One need only obtain a single Alexa–PBD image (usually at the beginning of the time series) and use this for bleed-through correction of all images in the time series. If Alexa–PBD is not irradiated during the experiment, its bleaching will be negligible, and this single image will reflect the actual Alexa–PBD level throughout the entire time series.

Any ratio calculations are best performed using floating point operations. Large errors can be generated by software that truncates noninteger values into integers to display data as images. Such problems can be overcome by multiplying the images by a large scalar value prior to division. This value should be as high as possible without causing any pixel to exceed the bit depth of the image file. For example, a 12-bit image file can only hold values up to 4095 (2^{12}), so the constant selected must not cause the highest intensity value in the image to exceed 4095. When operations are performed on whole cell values rather than on images, as with production of the Rac activation curves described here, it is convenient to determine total cell intensities and then perform any division operations in a floating point spreadsheet program, such as Microsoft Excel.

Examining Changes in Localization of Rac Activation

The preceding sections described how to generate images showing the distribution of FRET within cells. A sequence of such images can be compared to show how localizations vary with time. While simple examination can suffice to show distribution changes, the overall intensity of FRET, and hence the perceived activation level, could become successively lower over time due to bleaching of the GFP. We have found that bleaching is not a serious issue for at least 20 images under the exposure conditions described here. The total GFP intensity at the beginning and end of the experiment should be examined to gauge bleaching effects. One can correct for bleaching by dividing each pixel in the image by the total intensity of GFP determined at the same time point.

Concluding Remarks

Many aspects of cytoskeletal control and signaling crosstalk depend on the localization of Rho family GTPase activation and may depend as well on the level and duration of activation. Signaling control by the precise dynamics of GTPase activation has been suggested by indirect experiments, but has been very difficult to quantify or study using previous methods. The FLAIR system can reveal Rac activation dynamics *in vivo*. We are currently working toward making FLAIR more readily accessible, either by using only genetically encoded fluorophores, such as GFP mutants, or by using peptide import sequences to simplify loading of the domain. The greatest benefits will be derived from accurate quantitation of Rac–GTP levels. The FLAIR system as described here can accurately report the changing activation levels within one cell, but the fluorescence response to the same level of Rac–GTP will vary from cell to cell. We are trying now to standardize the approach, eliminate this variability, and develop a system permitting quantitative comparisons of different individual cells or of different cell populations assayed simply in a fluorometer cuvette. It is hoped that FLAIR is an example of a generally applicable technology that can benefit researchers studying other protein families.

Acknowledgments

We are grateful to the Chroma Corporation for their outstanding help in designing filters optimized for FLAIR, to the Inovision corporation for their expert technical assistance and help in software design, and to Ella Blanc for expert secretarial assistance. We thank the National Institutes of Health (R01 GM-57464) and the Arthritis Foundation for financial support.

Section IV

Biological Assays of Rho GTPase Function

[36] Rho-like Gtpases in Tumor Cell Invasion

By EVA E. EVERS, ROB A. VAN DER KAMMEN, JEAN P. TEN KLOOSTER, and JOHN G. COLLARD

Invasion of T-Lymphoma Cells

Metastasizing tumor cells are characterized by their invasion into surrounding tissues. After entering the bloodstream, the malignant tumor cells extravasate at distant organs to establish metastases. In search for genes that can induce invasion and metastasis of T-lymphoma cells, we identified the *Tiam1* (T-lymphoma invasion and metastasis 1) gene in a retroviral insertional mutagenesis screen by selecting for invasive variants of the parental, noninvasive BW5147 T-lymphoma cell line[1] (see Method later). Activation of the *Tiam1* gene enabled BW5147 T-lymphoma cells to infiltrate into a monolayer of rat embryonic fibroblast (REF) cells as well as to metastasize after injection into the tail vein of syngeneic mice.[1] *Tiam1* encodes an activator or guanine-nucleotide exchange factor (GEF) of the Rho-like GTPase Rac1. Similar to Tiam1, constitutively active V12Rac1 induces an invasive phenotype in T-lymphoma cells.[1,2] V12Cdc42 is also able to induce invasion in BW5147 cells.[3] To dissect signaling pathways conferring invasion, we used the effector mutants of Rac and Cdc42 (kindly provided by C. Lamarche and A. Hall, MRC, London), which are defective in certain signaling pathways.[4] L61Rac1 induces invasion similar to V12Rac1. L61RacC40 neither binds to Pak kinase nor stimulates the Jun kinase (JKN) pathway, but still mediates cytoskeletal changes, such as membrane ruffling and the formation of focal complexes. L61RacA37 is able to activate Pak kinase and JNK, but fails to mediate cytoskeletal changes.[4] T-lymphoma cells retrovirally transduced[5] with the L61RacA37 effector mutant did not acquire an invasive phenotype, whereas L61RacC40 induced invasion of BW5147 cells. This suggests that signaling pathways

[1] G. G. M. Habets, E. H. M. Scholtes, D. Zuydgeest, R. A. van der Kammen, J. C. Stam, A. Berns, and J. G. Collard, *Cell* **77**, 537 (1994).

[2] F. Michiels, G. G. M. Habets, J. C. Stam, R. A. van der Kammen, and J. G. Collard, *Nature* **375**, 338 (1995).

[3] J. C. Stam, F. Michiels, R. A. van der Kammen, W. H. Moolenaar, and J. G. Collard, *EMBO J.* **17**, 4066 (1998).

[4] N. Lamarche, N. Tapon, L. Stowers, P. D. Burbelo, P. Aspenström, T. Bridges, J. Chant, and A. Hall, *Cell* **87**, 519 (1996).

[5] F. Michiels, R. A. van der Kammen, L. Janssen, G. Nolan, and J. G. Collard, *Methods Enzymol* **325** [26] (2000) (this volume).

mediating cytoskeletal remodeling, but not activation of Pak kinase or JNK, are required to confer invasiveness by Rac1. To dissect signaling pathways of Cdc42 that are able to induce invasion, we retrovirally transduced the effector mutants for Cdc42 in BW5147 cells. L61Cdc42A37 still activates Pak kinase and JNK, but it does not result in activation of Rac. This mutant still mediates the formation of filopodia. L61Cdc42C40 is defective in Pak kinase and JNK activation, but signals to Rac and induces the formation of filopodia.[4] Both effector mutants of Cdc42 were able to induce an invasive phenotype in T-lymphoma cells (F. Michiels and J. G. Collard, unpublished results, 1998). This suggests that invasion is dependent on a pathway shared by these two mutants, presumably involving alterations of the cytoskeleton. Furthermore, this is an indication that Cdc42-induced invasion can occur independently of Rac1. Indeed, expression of V12Cdc42 in these T-lymphoma cells did not activate Rac (S. van Delft and J. G. Collard, unpublished results, 1998). T-lymphoma cells expressing constitutively active V14RhoA did not acquire an invasive phenotype. However, inactivation of RhoA by treatment with C3 transferase inhibited invasion, indicating that RhoA activity is required but not sufficient to induce invasion. Coexpressed together with V12Rac or V12Cdc42, RhoA potentiated the invasive capacity of T-lymphoma cells.[3] Furthermore, invasion of T-lymphoma cells is dependent on the presence of serum. Serum can be replaced by the serum-borne lipids lysophosphatidic acid (LPA) and sphingosine-1-phosphate (S1P). These lipids act on distinct G-protein-linked receptors to activate RhoA and phospholipase C (PLC) signaling pathways. Invasion induced by LPA and S1P is preceded by the RhoA-dependent redistribution of the F-actin cytoskeleton, leading to the formation of actin-rich pseudopods. Cells coexpressing V12Rac1 and V14RhoA are still dependent on LPA/S1P to induce invasion, indicating that additional pathways independent of RhoA are required for the invasive phenotype.[3] Taken together, activation of Rho-like GTPases Rac1 or Cdc42 induces invasion of T-lymphoma cells, depending on pathways involved in cytoskeletal organization. Moreover, the invasive phenotype is dependent on LPA/S1P receptor-mediated RhoA and PLC signaling pathways mediating the formation of actin-rich pseudopods. The lipid-induced pseudopod formation also occurred in noninvasive T-lymphoma cells. Therefore, it is likely that enhanced Rac- and Cdc42-mediated cell spreading and adhesion of T-lymphoma cells (F. Michiels and J. G. Collard, unpublished results, 1998, and Ref. 6) to matrix components produced by the REF cells, as well as Rac1- and Cdc42-dependent cytoskeletal changes, are the major requirements for the induction of an invasive phenotype in lymphoid tumor cells.

[6] C. D'Souza-Schorey, B. Boettner, and L. van Aelst, *Mol. Cell. Biol.* **18**, 3936 (1998).

Fibroblast Invasion Assay

This *in vitro* invasion assay is based on the capacity of lymphoid cells to infiltrate into a monolayer of rat embryonic fibroblast (REF) cells. Noninvasive cells hardly infiltrate spontaneously into the REF monolayer (less than 2% invasion). Invasion can be induced by retroviral expression of Tiam1, V12Rac1, or V12Cdc42 in noninvasive BW5147 T-lymphoma cells.[5] Noninvasive cells that have not infiltrated into the REF monolayer can be removed by washing, and the remaining invasive variants appear as round, dark cells in between the REF cells (see Fig. 1). Note that this method can only be used to determine the invasion of lymphoid cells.

Method

1. Seed rat embryonic fibroblasts (REF 208F cells) in 24-well plates (4 × 10^4 cells/well) and culture them in Dulbecco's modified Eagle's medium (DMEM) containing heat-inactivated 10% (v/v) newborn calf serum until they are grown to confluency and have formed a monolayer (approximately 2–3 days).

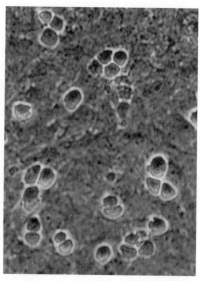

FIG. 1. Monolayer of REF 208F cells invaded by C1199Tiam1-expressing BW5147 T-lymphoma cells. (Left) A REF monolayer that has been incubated with noninvasive, parental BW5147 cells for 4 hr. Noninvasive floating T-lymphoma cells have been removed by washing. (Right) A monolayer incubated with invasive, C1199Tiam1-expressing BW5147 cells. Invasive cells have infiltrated in the REF monolayer and are seen as round, refractile cells in between REF cells.

2. Wash the REF monolayer once with RPMI 1640 medium containing 10% heat-inactivated fetal calf serum (FCS).

3. Harvest exponentially growing lymphoid cells by centrifugation at low speed (3 min), resuspend in a small volume RPMI 1640 medium supplemented with 10% FCS (5 to 10 ml), and count the cells.

4. Add $5-7 \times 10^5$ T-lymphoma cells contained in 1 ml fresh RPMI 1640 medium (supplemented with 10% FCS) to each well of REF cells and incubate for 4–16 hr, dependent on the invasive capacity.

5. Remove noninvaded and adherent cells by washing and mechanical agitation using phosphate-buffered saline (PBS) containing 1 mM CaCl$_2$ and 1 mM MgCl$_2$. For this, replace the medium in the wells with 0.5 ml PBS and carefully knock the side of the cell culture plate in a horizontal position against a hard surface (e.g., against the side of a table). Repeat the washing procedure four times until hardly any adherent cells are visible.

6. Fix the infiltrated REF monolayers for 15 min at room temperature in PBS containing 3.7% formaldehyde.

7. Infiltrated cells will appear as dark, highly refractile cells in between cells of the REF monolayer. Count infiltrated cells using a phase-contrast microscope (see Fig. 1).

8. Express the degree of invasion as the percentage of cells added. Less than 2% invasion, as is typically obtained with noninvasive lymphoid cells, is considered as background.

In our studies, most T-lymphoma cells expressing retrovirally transduced mutant Rho-like GTPases[5] were derived from noninvasive BW5147 cells.[1-3] Parental and retrovirally transduced cells are grown in suspension in RPMI 1640 medium containing 10% heat-inactivated FCS in cell densities varying between 2×10^4 and 2×10^6 cells/ml. In the invasion assay, 7×10^5 cells are incubated with the REF monolayer for 4 hr. For the determination of the invasive capacity of other lymphoid cells, such as normal activated T cells, Jurkat, HSB, CEM, and ESB T-lymphoma cells, different incubation times (4–16 hr) might be required, depending on the invasive capacity of the cell type tested.

Selection of Invasive Variants of Lymphoid Cells

Invasive variants that infiltrate into the REF monolayer infrequently can be isolated from the REF monolayer and be subjected to further selection cycles to enrich for the invasive variant (see Fig. 2). This powerful *in vitro* selection method allows the selection of invasive cells with high efficacy. One single invasive cell that is admixed to 1×10^6 noninvasive BW5147 cells can be isolated within three rounds of selection.[7]

[7] J. G. Collard, M. van de Poll, A. Scheffer, E. Roos, A. H. M. Hopman, A. H. M. Geurts van Kessel, and J. J. M. van Dongen, *Cancer Res.* **47**, 6666 (1987).

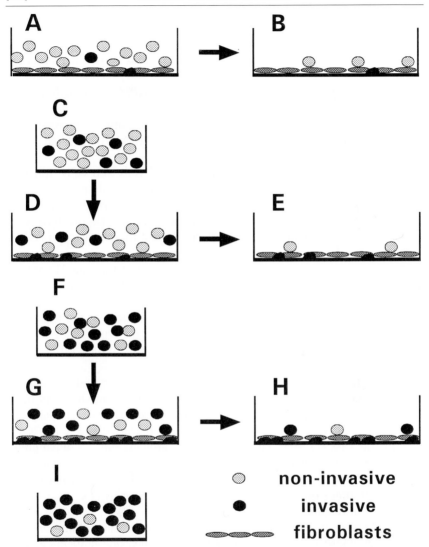

FIG. 2. Selection of invasive variants. In the first selection cycle (A–C), a heterogeneous population consisting of noninvasive and a few invasive variants of lymphoid cells is incubated with a REF monolayer. Floating and most adherent noninvasive variants that have not infiltrated into the monolayer are removed by washing (B). Invasive variants that have infiltrated in the REF monolayer are trypsinized and, together with REF cells, transferred to a larger cell culture flask. After attachment of REF cells, floating lymphoid cells are captured and expanded in a new flask (C). Subsequently, they can be used in additional selection cycles to further enrich for invasive cells (D–F and G–I).

Method

1. REF 208F cells (3×10^5) are seeded in DMEM containing heat-inactivated 10% newborn calf serum in a 25-cm² flask and grown until confluency.

2. The REF monolayer is washed once with RPMI 1640 medium containing 10% FCS.

3. Lymphoid cells (10×10^6) (containing invasive and noninvasive cells) are layered on the REF monolayer (in a 25-cm² flask) in 10 ml fresh RPMI medium supplemented with 10% FCS.

4. After 16 hr of incubation, noninfiltrated cells and cells adhering to the monolayer are removed by washing with PBS (containing 0.5 mM CaCl₂ and 1 mM MgCl₂) and mechanical agitation. For this, replace the medium with 4 ml PBS and carefully knock the side of the flask in a horizontal position against a hard surface. Repeat this washing procedure four times until hardly any adherent cells are visible.

5. Subsequently, the fibroblast monolayer cells with infiltrated lymphoid cells are trypsinized (2 min, in PBS containing 0.05% trypsin and 1 mM EDTA) and transferred to a larger 75-cm² flask.

6. When the fibroblasts have attached to the bottom of the flask (2–3 hr), the floating lymphoid cells are isolated and transferred to a new flask (75 cm²). Repeat this procedure one to two times to obtain a fibroblast-free culture of lymphoid cells.

7. The selected infiltrated lymphoid cells with contaminating noninfiltrating cells are collected in a 25-cm² flask. The flask is initially kept in an upright position for 1 to 2 days. Excess medium is removed after sedimentation of the cells. When the cells have reached a density of about 2.5×10^4 cells/ml, the flask is kept in a flat position. The growing cultures of lymphoid cells are expanded in regular RPMI 1640 medium. Expanded lymphoid cells can be used in a subsequent selection cycle. Selections are performed up to seven times or until invasive subpopulations are visible as assessed by phase-contrast microscopy.

Selection for invasive cell variants has been used in combination with the retroviral insertional mutagenesis to identify invasion-inducing genes.[1] For this, separate populations of noninvasive BW5147 T-lymphoma cells (each 10×10^6 cells) were infected by cocultivation for 24 hr with exponentially growing Moloney murine leukemia virus (MMLV)-producing NIH 3T3 cells (3×10^6 cells/75-cm² flask) in RPMI 1640 medium (supplemented with 10% FCS) with 8 μg/ml Polybrene (Sigma, St. Louis, MO) for efficient infection.[1] The infected lymphoid cells were separated from the adherent virus-producing cells and cultured for another 24 hr in fresh medium. Subsequently, cells were used in the selection protocol as described earlier to select for invasive variants induced by retroviral insertions. Independently generated invasive variants were subsequently analyzed for a proviral insertion cluster to identify the invasion-inducing gene.[1]

Function of Rho-like GTPases in Epithelial Cell–Cell Adhesion and Migration

In epithelial carcinoma cells, invasion and metastasis are often associated with reduced E-cadherin-mediated cell–cell adhesion.[8–10] Loss of or mutations in E-cadherin or the catenin proteins constituting the adhesion complex are often observed in these carcinomas. Moreover, reexpression of E-cadherin in tumors lacking this molecule restores E-cadherin-mediated cell–cell adhesion and inhibits invasion of these tumors, indicating an invasion suppressor function for E-cadherin.[8,9] Loss of E-cadherin has been found to be causal for the transition from adenomas to carcinomas.[11] In epithelial cells, family members of Rho-like GTPases have been shown to mediate E-cadherin-dependent cell–cell adhesions, which affect their invasive and migratory behavior. In Madin–Darby canine kidney (MDCK) cells and human keratinocytes, inactivation of Rac or Rho signaling interferes with the formation of E-cadherin-mediated cell–cell adhesions and results in dislocation of the cadherin complex members from adherens junctions.[12–15] The Rac activator Tiam1 localizes to adherens junctions in MDCK cells, and ectopic expression of Tiam1 or constitutively active V12Rac1 inhibits hepatocyte growth factor (HGF)-induced cell scattering by increasing E-cadherin-mediated cell–cell adhesion.[16] Moreover, Tiam1-mediated activation of Rac restores E-cadherin-mediated cell–cell adhesion in Ras-transformed, fibroblastoid MDCK-f3 cells, thereby reestablishing the epithelial morphology. Consequently, motility and invasion of these cells into three-dimensional collagen gels are inhibited by Tiam1/Rac acti-

[8] K. Vleminckx, L. Vakaet, M. Mareel, W. Fiers, and F. van Roy, *Cell* **66**, 107 (1991).

[9] U. H. Frixen, J. Behrens, M. Sachs, G. Eberle, B. Voss, A. Warda, D. Lochner, and W. Birchmeier, *J. Cell Biol.* **113**, 173 (1992).

[10] W. Birchmeier and J. Behrens, *Biochim. Biophys. Acta* **1198**, 11 (1994).

[11] A.-K. Perl, P. Wilgenbus, U. Dahl, H. Semb, and G. Christoferi, *Nature* **392**, 190 (1998).

[12] V. M. M. Braga, L. M. Machesky, A. Hall, and N. A. Hotchin, *J. Cell Biol.* **137**, 1421 (1997).

[13] K. Takaishi, T. Sasaki, H. Kotani, H. Nishioka, and Y. Takai, *J. Cell Biol.* **139**, 1047 (1997).

[14] C. Zhong, M. S. Kinch, and K. Burridge, *Mol. Biol. Cell* **8**, 2329 (1997).

[15] T.-S. Jou and J. W. Nelson, *J. Cell Biol.* **142**, 85 (1998).

[16] P. L. Hordijk, J. P. ten Klooster, R. A. van der Kammen, F. Michiels, L. C. J. M. Oomen, and J. G. Collard, *Science* **278**, 1464 (1997).

vation[16,17] (see Method). These findings identify the Rho-like GTPases, particularly Rac1, in processes involved in the maintenance of epithelial organization.[18]

In addition to mediating cell–cell adhesion, Rho-like GTPases play a pivotal role in the regulation of epithelial migration processes. An increased activity of RhoA is associated with Ras-induced cell transformation and cell motility.[14,19a] Ras-transformed mammary epithelial cells exhibit increased phosphorylation of the myosin light chain and show more stress fibers and focal adhesions due to elevated RhoA activity.[14] Similarly, Rac1 has been implicated in both cell–cell adhesion and motile responses of epithelial cells. In T47D mammary carcinoma cells, constitutively active V12Rac and V12Cdc42 have been shown to stimulate motility on collagen, thereby requiring the $\alpha_2\beta_1$ integrin and phosphoinositide 3-kinase (PI3-kinase).[19] PI3-kinase is identified as a downstream target of $\alpha_6\beta_4$ integrin signaling, leading to increased motility and invasion of colon carcinoma cells.[20] Rac1 activity is required downstream of PI3-kinase.[20] Similarly, Tiam1- but not V12Rac-mediated migration and invasion of Ras-transformed MDCK-f3 cells depend on PI3-kinase activity, required for localization and/or activation of the GEF.[17]

The cellular response toward activation of Rac, i.e., cell–cell adhesion or cell migration, can be regulated by the composition of the extracellular matrix. In Ras-transformed MDCK-f3 cells, Tiam1-mediated activation of Rac restores epithelial morphology and consequently inhibits migration on a fibronectin or laminin1 substrate. In contrast, on collagen substrates, Tiam1-mediated Rac activation promotes migration of these cells under conditions that prevent the formation of E-cadherin adhesions.[17] In not migrating cells on fibronectin or laminin1, Tiam1 localizes to adherens junctions, whereas it is present in the lamellae of migrating cells. Similar levels of Tiam1-mediated Rac activation are found on a fibronectin and collagen substrate, as determined by binding to a GST–Pak protein,[17] suggesting that the cellular localization and/or composition of the Rac signaling complex is regulated by the cell substrate. A similar substrate-dependent migratory behavior has been observed in an epithelial bladder carcinoma cell line, where collagens but not fibronectin or laminin substrates promoted

[17] E. E. Sander, S. van Delft, J. P. ten Klooster, T. Reid, R. A. van der Kammen, F. Michiels, and J. G. Collard, *J. Cell Biol.* **143**, 1385 (1998).

[18] E. E. Sander and J. G. Collard, *Eur. J. Cancer* **35**, 1302 (1999).

[19] P. J. Keely, J. K. Westwick, I. P. Whitehead, C. J. Der, and L. V. Parise, *Nature* **390**, 632 (1997).

[19a] G. C. Zondag, E. E. Evers, J. P. ten Klooster, L. Janssen, R. A. van der Kammen, and J. G. Collard, *J. Cell Biol.* **149**, 775 (2000).

[20] L. M. Shaw, I. Rabinovitz, H. H.-F. Wang, A. Toker, and A. M. Mercurio, *Cell* **91**, 949 (1997).

migration mediated by the integrin $\alpha_2\beta_1$.[21,22] Different integrin receptor(s) can initiate motile responses, dependent on the repertoire of matrix receptors present on the cell type studied.[19,20,22]

Both Rac1 and RhoA have been found to be activated downstream of the integrin $\beta1$ and to induce cell scattering, as demonstrated by reexpression of the $\beta1$ integrin subunit in $\beta1$ knockout neuroepithelial cells.[23] Similarly, the HGF-induced scatter response of epithelial MDCK cells is preceded by the transient activation of both Rac1 and RhoA.[19a] It thus appears that regulated activation of Rho-like GTPases by integrins and growth factor receptors is required to mediate epithelial cell migration. The reciprocal balance of Rac and Rho GTPase activities seems to be important for the initiation of a motile response. In NIH 3T3 fibroblasts, Tiam1-mediated activation of Rac leads to immotile cells with an epithelioid morphology and increased cadherin-based adhesions. This phenotype is characterized by the downregulation of RhoA activity by Rac1 and Cdc42, as determined by binding to a GST–Rhotekin fusion protein.[24] Rac-mediated inactivation of Rho occurs at the GTP level of Rho and is independent of Rac-mediated cytoskeletal changes.[19a,24] Expression of V14Rho in Tiam1-expressing fibroblasts leads to a more contractile fibroblastoid phenotype with loss of cell–cell contacts. Antagonistic effects of the Rac and Rho pathways have also been shown to occur between the downstream pathways of both GTPases.[25-27] It thus appears that the antagonism between Rac and Rho is regulated at different points of the respective signaling pathways, indicating that cells require a complex regulation pattern to balance Rac and Rho activities to elicit motile responses.

Invasion into Three-Dimensional Collagen Gels

Invasion into three-dimensional collagen gels has been used to determine the inhibition of invasion of Tiam1-expressing Ras-transformed MDCK-f3 cells.[16]

[21] G. C. Tucker, B. Boyer, J. Gavrilovic, H. Emonard, and J. P. Thiery, *Cancer Res.* **50**, 129 (1990).

[22] A. M. Vallés, B. Boyer, G. Tarone, and J. P. Thiery, *Cell Adh. Res. Commun.* **4**, 187 (1996).

[23] C. Gimond, A. van der Flier, S. van Delft, C. Brakebusch, I. Kuikman, J. G. Collard, R. Fässler, and A. Sonnenberg, *J. Cell Biol.* **147**, 1325 (1999).

[24] E. E. Sander, J. P. ten Klooster, S. van Delft, R. A. van der Kammen, and J. G. Collard, *J. Cell Biol.* **147**, 1009 (1999).

[25] L. C. Sanders, F. Matsamura, G. M. Bokoch, and P. de Lanerolle, *Science* **283**, 2083 (1999).

[26] K. Rottner, A. Hall, and J. V. Small, *Curr. Biol.* **9**, 640 (1999).

[27] M. A. Sells, J. T. Boyd, and J. Chernoff, *J. Cell Biol.* **145**, 837 (1999).

Method

This protocol yields 5 ml of 0.2% (v/v) collagen gels.

1. Solve 10 mg lyophilized collagen type I (from rat tail) in 2 ml sterile, ice-cold PBS containing 0.1% (v/v) acidic acid. Keep the mixture on ice, as otherwise gelation will occur. Alternatively, a mixture of collagen type I and collagen type IV can be used and will sometimes allow better detection of the embedded cells.

2. Add 1 ml DMEM containing 10% FCS and mix thoroughly.

3. Adjust the pH value with 5 and 1 M NaOH. In order to keep the collagen solution sterile, small samples are tested using pH paper strips.

4. Add 2 ml DMEM containing 20% FCS and mix thoroughly.

5. Mix the collagen gel with the appropriate number of trypsinized cells (densities will vary between cell types) and seed the mixture in 12- or 24-well plates. For MDCK cells (cultured in DMEM containing 10% FCS), we use between 5×10^4 and 10^3 cells/ml collagen gel. Typically, we add 1.5×10^4 MDCK cells to 300 μl collagen gel and seed the mixture in 24-well plates.

6. Immediately transfer the cells to a humidified incubator [37°, 5% (v/v) CO_2] for about 30 min to allow rapid gelation of the collagen/cell mixture.

7. Add 1.5 ml DMEM containing 10% FCS on top and incubate for 5–8 days. Refresh the cell culture medium every 2 to 3 days.

8. Invasion can be monitored by using a phase-contrast microscope using low magnification. Noninvasive cells appear as multicellular, spherical cysts. Invasive cells are seen as single, fibroblastoid cells dispersed throughout the gel matrix[16] (Fig. 3).

Three-dimensional collagen gels are also used to study epithelial differentiation processes (tubulogenesis).[28] For MDCK cells, the medium layered on top of the gels is supplemented with 10 ng/ml hepatocyte growth factor.[29] Alternatively, cells can be seeded on top of collagen gels to monitor invasion into the gel from the surface on.[8] However, in our hands, this method was only useful when using highly invasive cells.

Transwell Migration Assay

The migratory behavior of epithelial cells, including Tiam1- or V12Rac-expressing MDCK-f3 cells, is critically dependent on the experimental conditions applied. Inhibition of migration in Tiam1-expressing cells seeded

[28] S. A. Santoro, M. M. Zutter, J. E. Wu, W. D. Staatz, E. U. M. Saelman, and P. J. Keely, *Methods Enzymol.* **245,** 147 (1994).
[29] R. Montesano, K. Matsumoto, N. Nakamura, and L. Orci, *Cell* **67,** 901 (1991).

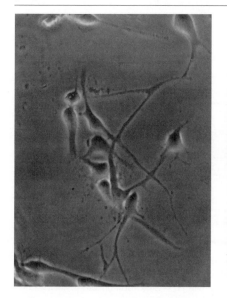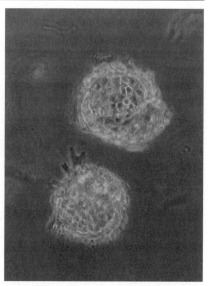

FIG. 3. Invasion into three-dimensional collagen gels. (Left) V12Ras-transformed MDCKf3 cells expressing nonfunctional Tiam1 protein (C580Tiam1) that invade as single fibroblastoid cells into the gel. (Right) C1199Tiam1-expressing MDCK-f3 cells that are inhibited in their migratory behavior by increased cell–cell adhesions and form multicellular, spherical cysts in the gel matrix.

on fibronectin and laminin1, and to a certain degree on collagen substrates, depends on the ability to establish cell–cell adhesions. High cell densities in the assay system will facilitate the formation of cell–cell adhesions and consequently favor reduced migratory behavior.[17] In order to detect inhibited migration of Tiam1-expressing cells, trypsinization of the cells is kept to a minimum to preserve cell–surface cadherin adhesion receptors. In contrast, Tiam1/Rac-mediated stimulation of migration is detected on collagen substrates when low cell densities are applied in the assay system (see Fig. 4), which makes the formation of functional cell–cell adhesions more difficult.

Method

1. Coat the membrane of a Transwell migration chamber (see Fig. 4) (diameter 6.5 mm; pore size 8 μm; Costar Corp., Cambridge, MA) on both sides with a suitable cell substrate. For MDCK-f3 cells, we solve 10 μg collagen type I, fibronectin, or laminin per milliliter PBS. For coating, 300 μl of the substrate solution is added to the lower chamber and 150 μl in the top chamber onto the upper side of the filter.

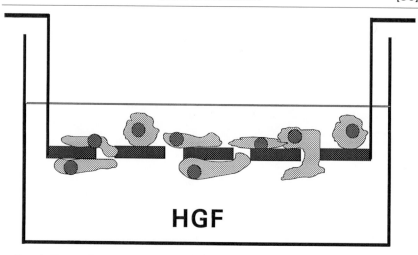

Fig. 4. Transwell migration assay. Upper and lower compartments are separated by a porous membrane that allows migration of cells contained in the upper well to the lower side of the filter toward a gradient of HGF. The filter can be coated with various cell substrates, such as collagens, fibronectin, or laminin1, to facilitate migration.

2. Coating can be done either overnight at 4° or for 2 hr at 37°. Rinse the filter membrane once with an excess of PBS and place it into the lower chamber containing 300 μl DMEM supplemented either with or without 10 ng/ml recombinant hepatocyte growth factor.

3. Parental or Tiam1-transduced MDCK-f3 cells (1.5 to 0.15 × 10⁵),[5] contained in 150 μl DMEM supplemented with 10% FCS, are added carefully into the upper chamber and distributed evenly by gentle agitation. Occasionally occurring air bubbles need to be removed.

4. The cells are incubated [37°, 5% (v/v) CO_2] for 4 to 5 hr and allowed to migrate to the underside of the filter.

5. Remove the nonmigratory cells on the upper membrane with a cotton swab and rinse carefully in PBS containing 0.5 mM $CaCl_2$ and 1 mM $MgCl_2$.

6. Fix the migratory cells at the bottom surface of the membrane for 10 min at room temperature in methanol.

7. Stain the fixed cells for 30 min in 0.2% (v/v) crystal violet (a 20 times stock solution is solved in ethanol).

8. Rinse shortly in PBS and count the migrated cells using a phase-contrast microscope.

Despite the motility-promoting effect of collagen in Tiam1-expressing MDCK-f3 cells at low cell densities in Transwell migration assays, we found reduced migration on collagen at high cell densities due to the formation of cell–cell adhesions.[17] This reduced migratory behavior on two-dimensional collagen at high cell densities is reflected in the inhibition of invasion into three-dimensional collagen gels of Tiam1-expressing cells.[16] Presumably, the Tiam1-induced formation of E-cadherin-mediated cell–cell adhesions is facilitated by the reduced movement of cells embedded in the collagen gel.

Acknowledgments

This work is supported by grants from the Dutch Cancer Society and the Netherlands Scientific Research Organization to J. G. Collard. E. E. Evers is supported by a fellowship from the Deutsche Forschungsgemeinschaft.

[37] Mouse Embryo Fibroblasts: A Genetic Model System for Studying Rho- and Ras-Dependent Cell Cycle Progression

By Giovanna M. D'Abaco and Michael F. Olson

Introduction

Immortalized cell lines have been used extensively to study the roles of Ras and Rho GTPase family proteins in the regulation of cell cycle progression. Studies in which NIH 3T3 fibroblasts were microinjected with recombinant Ras protein,[1] or the neutralizing anti-Ras monoclonal antibody Y13-259,[2] revealed the central role played by Ras in the regulation of mitogenesis. Similarly, experiments in which Swiss 3T3 fibroblasts were microinjected with recombinant active RhoA, Rac1, and Cdc42 protein[3] or with plasmids encoding dominant-negative versions of Rac1 or Cdc42,[3] or the Rho-specific inhibitor *Clostridium botulinum* C3 toxin,[3,4] showed that the Rho GTPase family also contributes to regulating the cell cycle.

Studies on Ras and Rho function have been extended to primary cells. One distinct advantage in using primary cells to examine Ras and Rho signal transduction pathways is that the possible influence of spontaneous undefined genetic alterations, which occur during the process of immortalization, is removed. Normal mammalian cell cultures have a finite capacity

[1] D. W. Stacey and H. F. Kung, *Nature* **310**, 508 (1984).
[2] L. S. Mulcahy, M. R. Smith, and D. W. Stacey, *Nature* **313**, 241 (1985).
[3] M. F. Olson, A. Ashworth, and A. Hall, *Science* **269**, 1270 (1995).
[4] M. Yamamoto, N. Marui, T. Sakai, N. Morii, S. Kozaki, K. Ikai, S. Imamura, and S. Narumiya, *Oncogene* **8**, 1449 (1993).

to proliferate *in vitro*.[5] After a limited but variable number of cell divisions, known as the Hayflick limit, growth ceases, the cells enter a crisis phase, and then senesce, a state marked by the inability to initiate DNA synthesis.[6] In some instances, a subpopulation of cells escapes the crisis phase and may eventually give rise to an immortalized cell line.[5] Although immortalized cell lines typically are relatively homogeneous and stable, a mutation(s) that facilitates immortalization may compromise conclusions drawn from their use in studies on cell cycle regulation. For example, the widely used NIH 3T3 murine fibroblast cell line is deleted at the INK4A locus,[7] which results in the loss of both the p16 cyclin-dependent kinase inhibitor (CDKI) and p19^{ARF}, a modulator of mdm2 and p53 function.

The turn toward primary cells has been due largely to the generation of genetically modified mouse models that express a transgene of interest or carry a targeted disruption of a specific gene. In addition, by cross-breeding mouse models, combinations of mutations or transgenes can be created easily. Primary cells derived from these mice maintain the same genotype, therefore, the influence of defined genetic alterations on Ras and Rho protein functions can be determined experimentally. As a consequence, the tools of genetic analysis, previously accessible principally to those working with genetically tractable organisms such as *Saccharomyces cerevisiae, Drosophila melanogaster,* or *Caenorhabditis elegans,* are now in the hands of mammalian cell biologists.

A number of studies have used genetically modified primary cell cultures to analyze Ras and Rho function in cell cycle regulation. Mouse embryo fibroblasts (MEFs) lacking the retinoblastoma gene product Rb or the CDKI p16, but not the CDKI p21$^{Waf1/Cip1}$, were found to have a reduced requirement for Ras function, particularly in G_1 to S phase cell cycle progression.[8,9] In addition, high-intensity signaling through the Ras/Raf/MEK/ERK pathway was found to induce cell cycle arrest via the induction of the p21$^{Waf1/Cip1}$ CDKI.[10-12] Primary fibroblasts[11] and hepatocytes[13] homozygous

---

[5] L. Hayflick and P. S. Moorehead, *Exp. Cell. Res.* **25,** 585 (1961).

[6] J. R. Smith and L. Hayflick, *J. Cell. Biol.* **62,** 48 (1974).

[7] S. Linardopoulos, A. J. Street, D. E. Quelle, D. Parry, G. Peters, C. J. Sherr, and A. Balmain, *Cancer Res.* **55,** 5168 (1995).

[8] D. S. Peeper, T. M. Upton, M. H. Ladha, E. Neuman, J. Zalvide, R. Bernards, J. A. DeCaprio, and M. E. Ewen, *Nature* **386,** 177 (1997).

[9] S. Mittnacht, H. Paterson, M. F. Olson, and C. J. Marshall, *Curr. Biol.* **7,** 219 (1997).

[10] A. Sewing, B. Wiseman, A. C. Lloyd, and H. Land, *Mol. Cell Biol.* **17,** 5588 (1997).

[11] D. Woods, D. Parry, H. Cherwinski, E. Bosch, E. Lees, and M. McMahon, *Mol. Cell Biol.* **17,** 5598 (1997).

[12] A. C. Lloyd, F. Obermuller, S. Staddon, C. F. Barth, M. McMahon, and H. Land, *Genes Dev.* **11,** 663 (1997).

[13] K. L. Auer, J. S. Park, P. Seth, R. J. Coffey, G. Darlington, A. Abo, M. McMahon, R. A. Depinho, P. B. Fisher, and P. Dent, *Biochem. J.* **336,** 551 (1998).

null for p21$^{Wafl/Cip1}$ did not arrest in response to high-intensity Raf signals, rather they were stimulated to proliferate. Interestingly, the induction of p21$^{Wafl/Cip1}$ by Raf was reported to be p53 independent in primary fibroblasts[11] but dependent on functional p53 in primary Schwann cells.[12] Rho was found to cooperate with Ras in stimulating primary fibroblasts to leave quiescence and synthesize new DNA by repressing the induction of p21$^{Wafl/Cip}$; fibroblasts lacking p21$^{Wafl/Cip1}$ did not require functional Rho for Ras-induced cell cycle progression through to S phase.[14] Through the use of primary cells, these studies have utilized the power of genetics to analyze Ras and Rho function in cell cycle regulation and have highlighted the possibility that signaling through Ras and Rho may give rise to tissue-specific responses.

This section describes methods for the preparation of mouse embryo fibroblasts followed by a description of how these cells are prepared for microinjection experiments.

Genotyping of Mice

When establishing a transgenic or knockout mouse colony, it is essential to verify the genotype of the founder mice. Genomic DNA prepared from a sample of mouse tail can be used to genotype the mice by polymerase chain reaction (PCR)-based methods or Southern blotting. Before obtaining a tail sample, each mouse is assigned an identification number. Mice can be identified by ear marking or a numbered tag can be used. Heterozygous mice are used to maintain a colony, and embryos obtained from crossing heterozygotes may be used to establish primary cell cultures. In most cases, primary cell lines can be established from homozygous wild-type, heterozygous, and homozygous mutant embryos. The genotype of the primary cell lines is determined by analyzing genomic DNA prepared from the head of each embryo. Some gene disruptions, e.g., Raf-1,[15] severely compromise the ability of homozygous mutant MEFs to proliferate; consequently, it may not be possible to establish cell lines from these homozygous mutant embryos.

Isolation of Genomic DNA from Mouse Tissue

Tissue from tail (0.5 cm is sufficient) or head is placed in an Eppendorf tube with 750 μl DNA Lysis buffer [50 mM Tris–HCl, pH 8, 100 mM NaCl, 100 mM EDTA, 1% sodium dodecyl sulfate (SDS) and 0.5 μg/ml proteinase K] and incubated at 55° overnight. The following day, the samples are

[14] M. F. Olson, H. F. Paterson, and C. J. Marshall, *Nature* **394**, 295 (1998).
[15] L. Wojnowski, L. F. Stancato, A. M. Zimmer, H. Hahn, T. W. Beck, A. C. Larner, U. R. Rapp, and A. Zimmer, *Mech. Dev.* **76**, 141 (1998).

vortexed for 5 min, and 310 μl of 5 M NaCl is then added to each sample followed by mixing for a further 5 min. The samples are spun at 13,000 rpm for 15 min at 4°, 800 μl of the supernatant is transferred to a new Eppendorf tube, and 500 μl 2-propanol is added to each tube to precipitate DNA. The samples are spun at 13,000 rpm at 4° for 15 min, washed once with 70% (v/v) ethanol at 4°, and resuspended in 100–200 μl TE (10 mM Tris–HCl, 1 mM EDTA, pH 7.5) with 4 μg/ml RNase (DNase free). Incubate the pellet at 37° for at least 2 hr to allow the DNA to dissolve and contaminating RNA to be digested. Each preparation yields approximately 500 μg of genomic DNA. For genotyping by PCR, 100–200 ng of DNA is sufficient.

Determination of Genotype by PCR

The precise strategy for determining the genotype of a mouse strain depends on how it has been modified genetically. In general terms, PCR oligonucleotides should be designed to give different sized products reflecting the presence or absence of the specific modification (Fig. 1A). An example of the PCR products obtained from homozygous wild-type, heterozygous, and homozygous mutant embryos at the p27^{Kip1} locus is shown in Fig. 1B.

Preparation of Mouse Embryo Fibroblasts

Breeding

Males reach maturity at approximately 6–8 weeks of age and may be used as studs for breeding until 8–10 months of age. Females may be used for breeding from 6 weeks to 4 months of age. Once a cross has been set up, female mice are monitored daily for a copulation plug. This consists of coagulated proteins from the seminal fluid and, in most strains, can be seen easily. In some cases, the plug is small and lies deep in the vagina and is only visible using a probe. The vaginal plug usually dissolves approximately 12–14 hr after mating. Each pregnant female will contain between 7 and 13 fertilized eggs, although this may vary depending on the mouse strain. The point at which a plug is observed is designated as 0.5 days postcoitus (dpc); embryos are typically isolated at 12.5–14.5 dpc, although this may vary according to the mouse model under study.

Embryo Dissection

Pregnant mice are culled by CO_2 treatment and cervical dissection. Swab the mouse with ethanol, cut the skin to expose the abdominal wall,

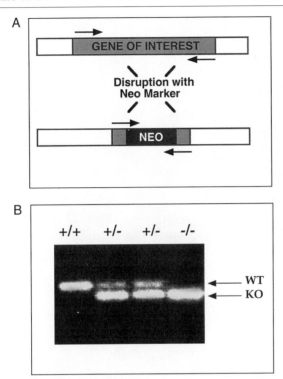

FIG. 1. Diagnostic PCR analysis of knockout mice. (A) Schematic representation of a representative PCR strategy for determining the genotype of a knockout mouse strain. Oligonucleotides for PCR amplification are shown as arrows. The unmodified gene gives a larger PCR product than the modified gene, allowing mice to be genotyped easily. (B) Typical result showing PCR products from p27^{Kip1} wild-type $(+/+)$, heterozygous $(+/-)$, and knockout $(-/-)$ mice.

and cut the viscera. Dissect the uterus and transfer it to a petri dish on ice containing cold sterile phosphate-buffered saline (PBS). From this point onward, the procedure should be continued in a laminar flow hood. Tear the uterus using two sterile forceps so as to free the embryo from the placenta and transfer to a fresh petri dish on ice. Each embryo is allowed to bleed fully and then is transferred to a fresh petri dish. Remove the liver (dark red) and then dissect the head (Fig. 2). The liver is removed to avoid contamination of fibroblast culture with other cell types of hemopoietic lineages. The head is retained for the preparation of genomic DNA, as outlined earlier, which can be used to determine the genotype of the resultant cell lines. The remaining carcass is cut into small pieces using two scalpel blades and transferred to a 15-ml screw-cap tube containing 5 ml

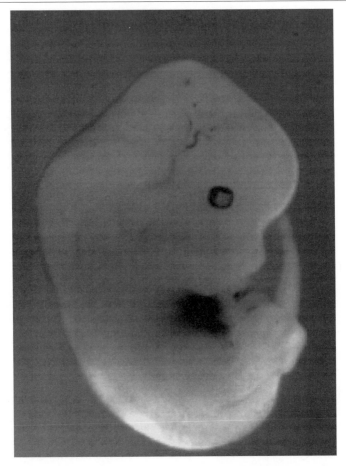

Fig. 2. Day 12.5 mouse embryo. After removing individual embryos from the uterus, primary fibroblasts are prepared as described in the text, taking special care to first remove the fetal liver (dark red spot in abdomen) in order to avoid contamination of the cultures with hemopoietic cells. Photo courtesy of A. Swain, Institute of Cancer Research, London.

of cold PBS. Between dissecting each sample, rinse forceps in 70% ethanol and use fresh scalpel blades for each dissection.

Cell Culture

After completing the dissection, allow the isolated tissue preparation to settle to the bottom of the tube, remove the PBS, and wash three times with fresh cold PBS. After removing the final PBS wash, add 2 ml of cold 0.25% (w/v) trypsin–EDTA to the sample. Incubate at 4° for 6 to 18 hr and then remove the trypsin. Place each tube at 37° for 20–30 min and then add 4 ml prewarmed DMEM containing 10% heat-inactivated fetal calf serum (FCS). Gently pipette up and down until cells are completely dispersed and suspended. Seed one embryo sample per 175-cm² tissue culture flask and add 20 ml DMEM containing 10% FCS and incubate at 37° in an atmosphere of 10% (v/v) CO_2. Change media 24 hr after seeding and continue to incubate until the cells form a confluent monolayer (approximately 48 hr).

While the cells are being incubated at this stage, it is important to minimize the handling of each tissue culture flask so as to allow MEFs to attach and begin dividing. Once the cells reach confluence, MEFs are split at 1 : 4 into 175-cm² flasks. Although cell types in addition to fibroblasts may be seen, such as cells from nerve and cartilage tissues, very few will survive this initial split. These MEFs are defined as passage 1. Once these cells again reach confluence they should be frozen in medium containing 20% (v/v) FCS and 10% (v/v) dimethyl sulfoxide at 3×10^6 cells per ampoule and placed in liquid nitrogen for long-term storage.

When recovering cells from liquid nitrogen, seed into a 25-cm² dish. Once the cells are confluent, split at 1 : 4 dilution of DMEM + 10% FCS (v/v) and continue to passage twice weekly at 1 : 4. We have found that it is important to batch test and select serum for optimal growth as there may be considerable differences in MEF proliferation between batches.

With continuous passages the growth characteristics of MEFs will change. The growth rate of the cells slows and it becomes increasingly more difficult to consistently synchronize populations of cells. In parallel, we have observed that the levels of the $p21^{Waf1/Cip1}$ CDK inhibitor (Fig. 3A) increase with additional passages, while others have observed a similar pattern with the p16 CDK inhibitor.[16] For these reasons, MEFs are not kept in culture beyond six passages.

Microinjection of Mouse Embryo Fibroblasts and Analysis by Immunolabeling

For microinjection experiments, cells are seeded at 2500 cells/ml in DMEM containing 10% FCS. Cells can be plated onto 60-mm-diameter tissue culture dishes or round 13- to 16-mm glass coverslips in appropriately sized wells. The following day, cells are washed once in DMEM with 0.5% FCS and insulin–transferrin–sodium selenite media supplement (Sigma, St. Louis, MO) and then maintained in this medium for 24 hr. We have

[16] I. Palmero, B. McConnell, D. Parry, S. Brookes, E. Hara, S. Bates, P. Jat, and G. Peters, *Oncogene* **15**, 495 (1997).

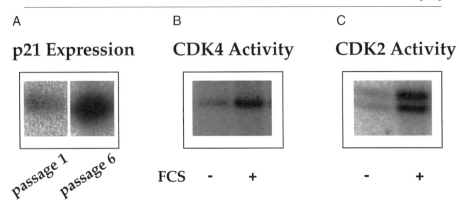

FIG. 3. Biochemical characterization of primary MEFs. (A) Expression of the p21[Wafl/Cip1] CDK inhibitor increases from passage 1 to 6. (B) CDK4 and (C) CDK2 activities after 20 hr of serum starvation (−) or restimulation with serum overnight (+). CDK4 and CDK2 were immunoprecipitated from MEF cell lysates and assayed for *in vitro* kinase activity using histone HI as substrate.

found that this treatment is sufficient to induce quiescence (Fig. 4A), and serum stimulation (Fig. 4B), as assayed by bromodeoxyuridine (BrdU) incorporation in most MEF strains. In addition, this serum-starvation protocol correlates with cyclin-dependent kinase activity (Fig. 4). For example, this treatment is sufficient to reduce cyclin-dependent kinase 4 (CDK4) (Fig. 3B) and CDK2 (Fig. 3C) while allowing their activities to be restimulated by serum. However, some MEF strains may require more extensive washes and longer incubation periods in medium with reduced serum to induce quiescence. For example, we found that p21[Wafl/Cip1−/−] MEFs required two washes in DMEM without serum and incubation in serum-free medium for 48 hr to achieve reproducibly low levels of BrdU incorporation. Therefore, it is important that a serum-starvation regimen that gives reproducible results be determined for each MEF cell line.

Following serum starvation, cells can be microinjected with plasmid DNA, antibodies, or recombinant protein as predicated by the experimental design. For example, Ras signaling to the Raf/MEK/ERK pathway is inhibited by microinjection with the Ras-neutralizing Y13-259 monoclonal antibody,[17] whereas Rho signaling is blocked following microinjection with *C. botulinum* C3 toxin.[18] In addition, it is essential to be able to detect the

---

[17] R. Marais, Y. Light, C. Mason, H. Paterson, M. F. Olson, and C. J. Marshall, *Science* **280,** 109 (1998).

[18] H. F. Paterson, A. J. Self, M. D. Garrett, I. Just, K. Aktories, and A. Hall, *J. Cell Biol.* **111,** 1001 (1990).

A. Serum-starved
MEFs

B. Serum-stimulated
MEFs

Phalloidin
stained

Anti-BrdU
stained

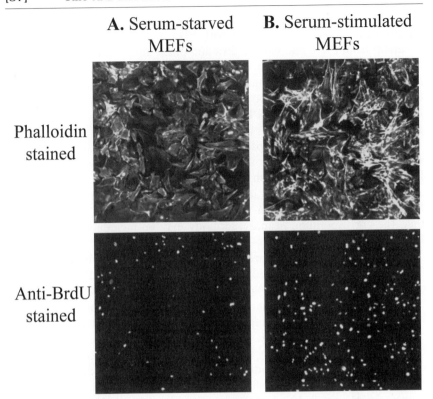

FIG. 4. Regulation of proliferation assayed by BrdU incorporation. (A) Incubation of primary MEFs in serum-free medium for 20 hr is sufficient to induce quiescence and reduce BrdU incorporation into newly synthesized DNA. (B) Stimulation of quiescent MEFs with medium containing serum for 20 hr increases BrdU incorporation significantly.

microinjected cells, therefore, it may be useful to include an injection marker such as a nonspecific IgG or high molecular weight fluorescent dextran. To monitor effects on proliferation, the cells are incubated with BrdU for 20 hr after microinjection. Cells are fixed and stained for immuno-fluorescence as detailed.

Fixation and Immunofluorescence Labeling of MEFs

1. Rinse cells once in phosphate-buffered saline containing 0.1% bovine serum albumin (PBS/BSA). Fix cells in 4% (w/v) paraformaldehyde for 10–15 min at room temperature. Rinse twice in PBS/BSA.

2. Permeabilize cells in 0.5% (v/v) Triton X-100 for 10 min at room temperature. Rinse several times in PBS without BSA.

3. For staining cells plated on glass coverslips, place a small strip of Parafilm onto filter paper wetted with PBS in a petri dish. Remove the coverslip from the tissue culture dish and place onto the Parafilm with the cells facing up. If cells have been plated onto 60-mm-diameter tissue culture dishes, then draw a 10- to 15-mm circle on the bottom of the dish and blot away excess liquid outside the circle. Spot 25–30 μl of RNase-free DNase I diluted to 1–2 units/μl in PBS onto the cells and incubate for 60 min at 37°.

4. If using glass coverslips, transfer back into tissue culture dish for washes. After washing three times in PBS/BSA, place coverslips onto a Parafilm strip in the petri dish. Incubate cells with anti-BrdU monoclonal antibody (diluted to 5 μg/ml in PBS/BSA) for 1 hr at room temperature. If protein expression from a microinjected plasmid is to be monitored, then the corresponding primary antibody may be included with the anti-BrdU antibody.

5. After washing cells three times in PBS/BSA, incubate cells with fluorescent secondary antibody to label incorporated BrdU. When it is appropriate, incubate cells with secondary antibody for detection of protein expression or the injection marker. Dilute secondary antibodies 1 : 200 and incubate for 1 hr at room temperature in a light-proof container.

6. Rinse cells three times in PBS and once in water, and then mount coverslips onto glass slides by inverting onto 5–10 μl of Moviol mountant containing 0.1% p-phenylenediamine as an antiquenching agent.[19] If cells have been plated on 60-mm dishes, blot away excess water outside of the drawn circle, spot 5–10 μl of mountant with antiquench onto cells, and place a coverslip on top.

Additional Comments

In order to extend the power of our microinjection experiments, we have made use of MEFs with a variety of genetic modifications. This approach has enabled us to begin to determine how signaling through Ras and Rho GTPases impinges on cell cycle regulation. Microinjection of primary cells as an experimental methodology has a number of advantages: (1) DNA, proteins, or antibodies can all be introduced into cells. (2) Time course data can be followed precisely from the point of microinjection. (3) The experimental microinjected cells can be compared to control microinjected cells and noninjected cells within the same field of view.

---

[19] G. V. Heimer and C. E. Taylor, *J. Clin. Pathol.* **27**, 254 (1974).

Microinjection does suffer from the limitation that biochemical analysis, although possible, is difficult. However, improvements in gene expression systems have made it possible to design biochemical experiments using primary cell lines. In particular, adenoviral gene delivery systems are able to infect a wide variety of cell types, including MEFs. Adenoviral gene transfer is not dependent on active cell division, allowing for the genetic manipulation of postmitotic cells, such as neurons. Additionally, it is possible to infect a large percentage of the target cells, thereby eliminating the need for lengthy drug selection protocols to obtain an enriched population of transduced cells.

Acknowledgments

This work was supported by a University Research Fellowship from The Royal Society (M.F.O.) and the Cancer Research Campaign (G.M.D. and M.F.O.)

[38] Analyses of Transforming Activity of Rho Family Activators

By Patricia A. Solski, Karon Abe, and Channing J. Der

Introduction

Members of the Rho family of small GTPases are regulators of actin cytoskeletal organization, cell cycle progression, and activators of transcription factors that target growth-promoting or antiapoptotic genes.[1–4] Therefore, it is not surprising that the aberrant activation of some Rho family proteins can promote many aspects of malignant transformation, including uncontrolled proliferation, invasion, and metastasis. Furthermore, multiple Rho family proteins have been implicated as important mediators of oncogenic Ras transformation. Rho family proteins have also been found to be essential for the transforming activity of other oncoproteins, such as members of the Dbl family of Rho guanine-nucleotide exchange factors (GEFs), G-protein-coupled receptors, and protein tyrosine kinase oncoproteins. NIH 3T3 mouse fibroblasts have

[1] L. Van Aelst and C. D'Souza-Schorey, *Genes Dev.* **11**, 2295 (1997).
[2] A. Hall, *Science* **279**, 509 (1998).
[3] R. Treisman, *Curr. Opin. Cell Biol.* **8**, 205 (1996).
[4] I. M. Zohn, S. L. Campbell, R. Khosravi-Far, K. L. Rossman, and C. J. Der, *Oncogene* **17**, 1415 (1998).

been a useful model cell system to evaluate transformation caused by aberrant Rho family protein function. The purpose of this article is to describe assays that are useful in assessing the transforming functions of oncoproteins that cause transformation, in part, by activating Rho family proteins. Two general applications of these assays are described. First, these assays can be used to implicate Rho family proteins in the transforming actions of oncoproteins. Second, these assays can also be used to delineate the structural and biochemical properties of a particu-

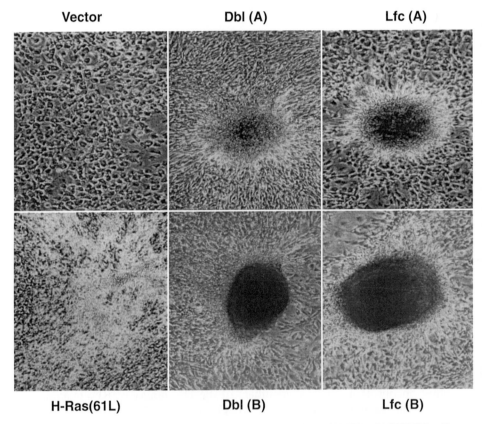

FIG. 1. Differences in the appearance of transformed foci in NIH 3T3 cells. NIH 3T3 cells were transfected with 20 ng pZIP-H-*ras*(61L), 100 ng pAX142-*dbl* HA1, or 100 ng pAX142-*lfc* D6HA plasmid DNAs. Foci of transformed cells caused by activated H-Ras(61L) are swirled in appearance and consist of highly refractile, spindle-shaped cells. In contrast, foci of transformed cells caused by activated Dbl or Lfc are much smaller in size and contain densely packed nonrefractile cells (A). Some foci begin to form mounds of cells that can detach from the dish (B). Foci were photographed at day 14 posttransfection. Magnification: 40×.

lar Rho activator that are important for promoting its transforming actions.

Background

Oncogene-transformed NIH 3T3 cells can exhibit many characteristics of a cancer cell. For example, Ras-transformed cells show an altered morphology, a reduced requirement for serum growth factors, loss of density-dependent growth inhibition, anchorage-independent growth, and the ability to form tumors in nude mice. The fact that NIH 3T3 cells are transfected easily, can undergo one-hit transformation by a wide variety of oncoproteins, and can provide both qualitative and quantitative information on transforming activity have led to the acceptance of this cell system as an important tool in the study of growth transformation.

·NIH 3T3 cells are sensitive to the transforming activities of a wide spectrum of functionally diverse oncoproteins. When a culture of NIH 3T3 cells is transfected with plasmid DNA that encodes the expression of an oncoprotein, the DNA will become stably integrated into the genome of only a small fraction of the cells (approximately 1 in 10^3). The oncoprotein-induced loss of density-dependent growth control and alteration in cellular morphology causes the outgrowth of a focus of proliferating, multilayered population of cells that can be distinguished readily from the background of confluent, well-attached, and nonrefractile untransformed cells.

Different oncogenes have been shown to induce foci of transformed cells that are distinct in their morphology. Activated mutants of Ras (G12V and Q61L) and serine/threonine protein kinases (e.g., Raf and Mos) cause the appearance of swirled-shaped foci of transformed cells that consist of highly refractile, spindle-shaped cells (Fig. 1). Activated tyrosine protein kinases cause the appearance of transformed foci that vary in focus shape and also contain highly refractile, rounded, poorly adherent cells (e.g., Src and Neu) (not shown). In contrast, activated mutants of RhoA (G14V or Q63L) or Rac1 (N115I) cause the appearance of transformed foci that are compact in shape and consist of densely packed and nonrefractile cells.[5–8]

---

[5] R. Khosravi-Far, P. A. Solski, M. S. Kinch, K. Burridge, and C. J. Der, *Mol. Cell. Biol.* **15**, 6443 (1995).

[6] R.-G. Qiu, J. Chen, F. McCormick, and M. Symons, *Proc. Natl. Acad. Sci. U.S.A.* **92**, 11781 (1995).

[7] R.-G. Qiu, J. Chen, D. Kirn, F. McCormick, and M. Symons, *Nature* **374**, 457 (1995).

[8] G. C. Prendergast, R. Khosravi-Far, P. A. Solski, H. Kurzawa, P. F. Lebowitz, and C. J. Der, *Oncogene* **10**, 2289 (1995).

Consistent with their function as GEFs that activate specific Rho family proteins,[9,10] Dbl family oncoproteins also cause the appearance of "Rho-like" foci of transformed cells. Figure 1 shows a primary focus assay in which activated versions of Dbl and Lfc induce transformed foci. Dbl is a GEF for Cdc42 and RhoA, whereas Lfc is a GEF for RhoA and Rac1. Thus, we have found that when a particular oncoprotein causes the appearance of Rho-like transformed foci, it has been a fairly reliable indicator that its transforming activity is mediated, in part, by the activation of specific members of the Rho family of proteins.

In contrast to Ras, GTPase-deficient mutants (analogous to the G12V or Q61L mutants of Ras) of Rho family proteins typically display very low (RhoA and Rac1) or no (Cdc42) focus-forming activity when assayed in primary focus-formation assays. In some cases, GDP/GTP exchange mutants (e.g., analogous to the F28L mutant of Ras) may exhibit more potent focus-forming activities than constitutively GTP-bound mutants.[11,12] For example, we have found that the N115I mutant of Rac1 causes greater focus formation than G12V or Q61L mutants in primary focus-formation assays.[5]

Rho activators, such as Dbl family proteins, vary widely in their focus-forming potencies that reflect whether they exhibit narrow or broad GEF target specificity.[13] Fgd1 is a specific activator of Cdc42; and a truncated variant of Fgd1 (pAX-*fgd1*), like activated Cdc42(12V), did not cause detectable focus-forming activity in primary focus-formation assays in NIH 3T3 cells.[14] Lsc, an activator of RhoA, displayed low focus-forming activity (~60 foci per pmol pAX-*lsc* D7HA DNA).[13] Lfc, an activator of Rho and Rac, displayed greater focus-forming activity (~200 foci per pmol pAX-*lfc* D6HA DNA). Dbl and Dbs are activators of Cdc42 and RhoA and both display potent focus-forming activity (>7000 foci/pmol pAX-*dbs* HA6 or pAX-*dbl* HA1). Vav is an activator of RhoA, Rac1, and Cdc42, and an NH$_2$-terminal truncated variant (ΔN-186 Vav) causes >4000 foci/pmol pAX-*vav*.[15] Interestingly, the focus-forming activities of these Dbl family proteins greatly exceed the activities seen with their corresponding GTPase targets. This may be due to Dbl family protein induction of increased GDP/GTP cycling, as well as increased levels of GTP-bound protein, or to their

[9] R. A. Cerione and Y. Zheng, *Curr. Opin. Cell Biol.* **8**, 216 (1996).

[10] I. P. Whitehead, S. Campbell, K. L. Rossman, and C. J. Der, *Biochim. Biophys. Acta* **1332**, F1 (1997).

[11] R. Lin, S. Bagrodia, R. Cerione, and D. Manor, *Curr. Biol.* **7**, 794 (1997).

[12] W. J. Wu, R. Lin, R. A. Cerione, and D. Manor, *J. Biol. Chem.* **273**, 16655 (1998).

[13] J. K. Westwick, R. J. Lee, Q. T. Lambert, M. Symons, R. G. Pestell, C. J. Der, and I. P. Whitehead, *J. Biol. Chem.* **273**, 16739 (1998).

[14] I. P. Whitehead, K. Abe, J. L. Gorski, and C. J. Der, *Mol. Cell. Biol.* **18**, 4689 (1998).

[15] K. Abe, I. P. Whitehead, J. P. O'Bryan, and C. J. Der, *J. Biol. Chem.* **214**, 30410 (2000).

activation of additional, as yet to be discovered, Rho family proteins. Another possibility is that Dbl family proteins not only activate Rho-dependent signaling pathways, but may also activate signals independent of their Rho targets.

NIH 3T3 Transformation Assays

Two general types of assays can be used to assess the transforming activity of oncoproteins that activate Rho family proteins. The first type of analyses involves primary focus-formation assays and they provide relatively quick and straightforward quantitation of transforming activity. The second type of assays involves the establishment of cells stably overexpressing activated Rho family proteins or their activators, which can be used for secondary focus-formation assays or to assess other aspects of cellular transformation. These include assays to determine the loss of density-dependent inhibition of growth on plastic, loss of anchorage-dependent growth, reduced requirement for serum growth factors, and tumorigenic growth potential in athymic nude mice.

Maintenance of Cell Stocks

The proper maintenance of stock cultures of NIH 3T3 cells is essential to avoid problems that can hinder reproducible, quantitative transformation assays. Although they possess properties of normal cells, NIH 3T3 cells have acquired genetic defects that render them highly sensitive to the one-hit transforming actions of many different oncoproteins. Consequently, they are also highly prone to spontaneous transformation if they are allowed to persist at confluent densities. Because some spontaneous foci of transformed cells are Rho-like in appearance, this is a particularly important concern if NIH 3T3 cells are to be used to assay for Rho-transforming activity. Another consideration in this assay is that not all sources of NIH 3T3 cells are alike. Different sources of NIH 3T3 cells may not behave similarly in the assay and may require different transfection protocols. It has also been found that some strains of NIH 3T3 cells are much less sensitive to transformation by activated Rho family proteins.[5]

Our NIH 3T3 cells (sometimes referred to as the UNC strain) are propagated in Dulbecco's modified Eagle's medium (DMEM) supplemented with 10% (v/v) calf serum, 100 U/ml penicillin, and 10 μg/ml streptomycin at 37° in a humidified 10% (v/v) CO_2 incubator. For focus formation assays, passage of the stock cells should be limited to three to four passages from each frozen stock ampule. It is prudent to create a large enough inventory of frozen stock ampules (20 to 40 ampules), which allows transfection assays for several months from the same stock cultures. Thus, assays done over several months can be performed with cells that have a common passage history, which enhances the reproducibility of the assays.

To establish a large frozen stock inventory, subconfluent NIH 3T3 cells are trypsinized (0.05% trypsin–EDTA) and seeded at 10^3 cells per 100-mm dish (and a total of 20 to 40 dishes). The cultures are fed every 3 to 4 days with fresh growth medium until the cells reach a confluency of approximately 70% confluency (approximately 13 to 15 days after plating). The cells are then removed by trypsin–EDTA treatment, washed in fresh growth medium, and resuspended in freezing medium [DMEM supplemented with 20% calf serum and 10% dimethyl sulfoxide (DMSO)]. Each freezing ampule should contain cells taken from one 100-mm dish. The ampules are frozen at $-80°$ and then transferred to liquid nitrogen for long-term storage.

Cells are reconstituted from frozen cell stocks by thawing quickly in a $37°$ water bath, suspended with 10 ml of growth medium, pelleted gently by low-speed centrifugation, and resuspended and transferred to three to six 100-mm dishes. The exact number of dishes used is based on the objective of having approximately 5×10^6 cells per dish after 3 days. A more detailed description of the maintenance of NIH 3T3 cells has been described previously.[16]

Reagents for Transfection

1.25 M CaCl$_2$: Autoclave to sterilize. Store at room temperature.
HEPES-buffered saline (HBS), pH 7.05: To 900 ml of distilled H$_2$O, add 8.0 g of NaCl, 0.37 g KCl, 0.19 g Na$_2$HPO $\cdot$ 7H$_2$O, 1.0 g glucose, and 5.0 g HEPES (free acid). Adjust pH to exactly pH 7.05 with 5 N NaOH. Adjust the volume to 1 liter, autoclave to sterilize, and store at room temperature. The correct pH is critical for good DNA precipitation. We use the tissue culture grade, free acid version of HEPES from Sigma (St. Louis, MO).
45% glycerol/HBS: Mix autoclaved glycerol and HBS at 45 : 55 (v/v). Store at room temperature.
Carrier DNA: High molecular weight carrier DNA is needed to promote the formation of calcium phosphate precipitate of the transfected plasmid DNA. This can be extracted from cell lines or animal tissue or purchased from a commercial supplier. We have found that calf thymus DNA from Boehringer Mannheim (Indianapolis, IN) yields the best results in focus-formation assays. Prepare a 1-mg/ml stock in sterile H$_2$O and store at $-20°$.

---

[16] G. J. Clark, A. D. Cox, S. M. Graham, and C. J. Der, *Methods Enzymol.* **255**, 395 (1995).

Of the reagents just described, HBS is the most critical for focus-formation assays. If the pH is off, the formation of an effective calcium phosphate DNA precipitate can be compromised and will result in lower transfection and transformation efficiencies. Another variable is the calf serum used to supplement the DMEM. We have found that the frequency of spontaneous background foci, as well as oncogenic Ras focus-forming activity, can vary widely between different lots of sera. We therefore test several lots of sera and select the serum that produces the maximum number of Ras-induced foci with the least number of background foci. Once we find one with these qualities, we purchase enough to perform transfections for 1 to 2 years. One source of serum that has given consistently good results from lot to lot is the Colorado Serum Co. (Denver, CO). Finally, do not use fetal calf serum; NIH 3T3 cells seem to grow better in calf serum.

Plasmid DNA

As stated earlier, activated Rho proteins exhibit several thousandfold less focus-forming activity when compared to activated forms of Ras (which typically cause $\sim5 \times 10^3$ foci/μg plasmid DNA). Hence, with suboptimal assay conditions, this can result in no foci being observed with activated Rho proteins. Several key factors can contribute to focus development with a weakly transforming protein. Careful maintenance of NIH 3T3 stock cultures is imperative for Rho focus-formation assays. A high background of spontaneous foci not only makes it difficult to see Rho-induced foci, but may inhibit their formation. Another key factor is the eukaryotic expression vector used to express the Rho protein. Our vectors of choice are ones that promote high levels of protein expression in stably transfected lines. We have found the pZIP-NeoSV(x)1[17] and pAX142[18] mammalian expression vectors to be the most useful for these analyses. Expression of the cDNA insert is regulated by the Moloney long terminal repeat (LTR) promoter or the EF1α promoter, respectively. Rho proteins expressed from cytomegalovirus (CMV) promoter-based vectors, while very good for transient expression signaling assays, usually exhibit much lower activity in primary focus-formation assays.

Transfection Protocol

We have described the procedures and amounts required for a 20-dish transfection assay, where quadruplicate dishes are used for each experimen-

---

[17] C. L. Cepko, B. Roberts, and R. C. Mulligan, *Cell* **37**, 1053 (1984).
[18] I. P. Whitehead, H. Kirk, C. Tognon, G. Trigo-Gonzalez, and R. Kay, *J. Biol. Chem.* **271**, 18388 (1995).

tal condition. This would involve one set of dishes to control for the appearance of spontaneous transformed foci and one set of dishes for a positive control. We usually use 20–50 ng per dish of pZIP-H-*ras*(61L) plasmid DNA. This would leave enough dishes for three sets of experimental conditions.

1. Plating of cells. Stock NIH 3T3 cells are plated the day before transfection at a density of 2.5×10^5 cells per dish. Add 5.0×10^6 total suspended cells to 100 ml growth medium that has been prechilled on ice. Pipette the medium to suspend the cells and add 5 ml per dish. Allow the cells to attach (approximately 20 min) in the tissue culture hood before transferring the dishes to a 37° incubator. This will minimize the concentration of the cells in the center of the dish due to the swirling of the growth medium when the dishes are transferred.

2. Preparation of DNA solution. Add 0.4 ml of the carrier DNA stock solution (1 mg/ml) to 8.6 ml HBS, vortex vigorously to mix, and aliquot 1.8 ml into five 15-ml polystyrene conical tubes. Add the desired amount of plasmid DNA to each tube. For the positive control, we usually add 20 ng of pZIP-H-*ras*(61L) plasmid DNA per dish. This should result in approximately 30–60 transformed foci per dish. For experimental dishes, the amount will depend on the expected potency (10 ng to 2 μg). The equivalent amount of empty vector plasmid DNA should be used for the negative control dishes. Vortex to mix in plasmid DNAs. Then add 0.2 ml 1.25 M $CaCl_2$ to each tube, mix initially by gentle pipeting, and vortex to complete the mixing for 20–40 sec. Allow precipitate to form for 20 min at room temperature. A fine white precipitate should be visible immediately on mixing and should increase with time. Volumes per 60-mm dish are 0.43 ml (HBS), 0.02 ml 1 mg/ml carrier DNA, and 0.05 ml 1.25 M $CaCl_2$, yielding a total volume of 0.50 ml.

3. Transfection. Add 0.5 ml per dish of the precipitated DNA solution dropwise directly onto the growth medium. Swirl gently to distribute the DNA. Incubate for 3 to 4 hr at 37° in a humidified 10% CO_2 incubator.

4. Glycerol shock. To increase the efficiency of DNA uptake, a glycerol shock of the transfected cultures is done immediately after the incubation with DNA. For a 20-dish transfection assay, prepare a 15% glycerol/ HBS solution by mixing together 7 ml of the 45% glycerol/HBS stock with 14 ml HBS. Aspirate off the DNA-containing growth medium, rinse the cells once with 3 ml of serum-free DMEM, and add 1 ml per dish of the 15% glycerol/HBS solution. Because the glycerol shock is rather toxic to the cells, the total time of exposure to the glycerol solution should be 4 to 5 min, from time of addition to time of rinsing with the addition of 5 ml growth medium. Return cultures to the 37° incubator.

Ras-induced transformed foci should be visible after 10 to 14 days, whereas RhoA- or Rho activator-induced foci may not be visible until 14 to 21 days. An inverted phase-contrast microscope is used to visualize and quantitate the appearance of transformed foci in live cultures. Alternatively, medium from the dishes can be removed and replaced with 2 ml per dish of fix (10%, v/v, acetic acid, 10%, v/v, methanol in H_2O) to fix the cultures to allow quantitation at a later time. This fixation procedure does not alter the morphology of the foci drastically and these fixed cultures can provide accurate measurements of focus-forming activity. Another less accurate and sensitive method of counting involves staining of the fixed cultures with crystal violet (0.4% in ethanol) to visualize the foci.[16]

Cooperation and Secondary Focus-Formation Assays

Because activated Rho family members display very weak or no focus-forming activity when assayed in primary transfection assays, two other focus-formation assays are used routinely in our laboratory that provide a more sensitive detection of transforming activity. The first assay is a cooperation, primary focus-formation assay. Ras-transforming activity is mediated by the cooperative action of multiple signaling pathways that include the Raf/MEK/ERK kinase cascade and those that lead to activation of Rho family proteins. Consequently, activated Raf cooperates with activated Rho family proteins and causes a synergistic enhancement of focus-forming activity (Fig. 2).[5-7] For these assays, we use the weakly activated Y340D mutant of human Raf-1.[19] When transfected alone, this Raf mutant displays little or no focus-forming activity. However, coexpression of Raf(Y340D), together with activated Rho GTPases or their activators (at concentrations that alone result in no focus-forming activity), will cause focus-forming activities that range from severalfold to more than 10- to 50-fold above their additive activity. For these assays, we typically combine 1 μg per dish of pZIP-raf(Y340D) along with 1 μg per dish of plasmid DNA encoding the Rho or Rho activator. However, trial assays with different amounts of each plasmid should be done to find the optimal ratio of plasmid DNAs to obtain the maximum level of cooperation. The appearance of transformed foci arising from these cooperation assays is distinct from those caused by Raf alone (Fig. 3). However, the appearance varies with the specific activated Rho family member and can range from typical Rho-like foci to those that exhibit a refractile appearance similar to that caused by activated Raf or Ras.

Other activated mutants of Raf-1 have been used in this assay, such

[19] J. R. Fabian, I. O. Daar, and D. K. Morrison, *Mol. Cell. Biol.* **13**, 7170 (1993).

A

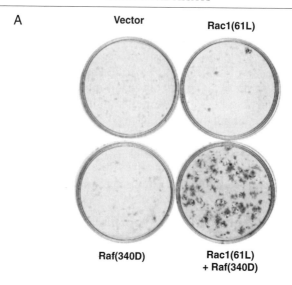

B

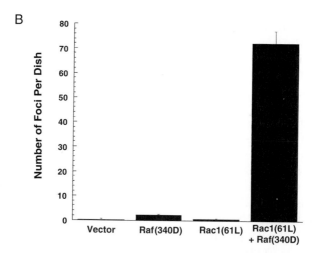

Fig. 2. Cooperation assay with Rac1 (61L) and Raf(340D). NIH 3T3 cells were cotransfected with pZIP-NeoSV(x)1 expression plasmids encoding the indicated proteins (1 μg DNA for each). At 18 days posttransfection, the plates were fixed and stained with 0.4% crystal violet to visualize the transformed foci (A) and to quantitate the number per dish (B). Under these assay conditions, cultures transfected with activated Rac1 or Raf-1 caused the appearance of little or no foci of transformed cells. However, coexpression of Rac1 and Raf caused focus-forming activities that were more than 30-fold above additive. Data shown are the average of triplicate dishes.

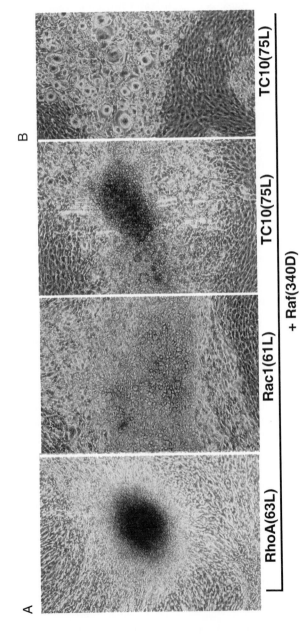

Fig. 3. Rho family members induce transformed foci with distinct appearances in a cooperation assay with Raf(340D). NIH 3T3 cells were cotransfected with pZIP-NeoSV(x)1 expression plasmids encoding the indicated proteins (1 μg DNA per dish). Transformed foci were photographed at day 18 posttransfection. Magnification: 40×. (A) RhoA(63L)/Raf(340D)-induced foci are also a densely packed clusters consisting of very small cells. Rac1(61L)/Raf(340D)-induced foci are also a densely packed cluster, but cells within the focus are larger than cells within the RhoA focus. TC10(75L)/Raf(340D) foci are similar to Rac-induced foci. (B) Very large multinucleated cells found in some Rac1- and TC10-induced foci. TC10(75L) contains a mutation analogous to the Q61L mutant of Ras.

as membrane-targeted Raf (Raf-CAAX) or N-terminally truncated (Raf-22W).[6,7,20] However, because these activated Raf-1 proteins possess focus-forming activities that are comparable to those of activated Ras mutants, the amount of DNA used is substantially less to minimize the level of Raf-induced transformed foci (10 to 100 ng DNA per dish). With these mutants, small changes in transfection efficiency or DNA concentrations cause the appearance of foci on Raf-transfected only dishes. It is therefore an advantage to use the Raf(Y340D) construct because it is easier to achieve Raf-transfected only dishes that contain no transformed foci.

The second type of focus formation assay is a secondary focus-formation assay that uses NIH 3T3 cells stably expressing the oncoprotein (see later). For the secondary focus assay, 10^3 stably expressing cells and 10^6 untransfected NIH 3T3 cells are mixed and then plated onto 60-mm dishes. The efficiency of the stably transfected cells to form multilayered colonies of cells is reflected by the appearance of foci of densely packed cells that can be visualized readily within the monolayer of untransformed, density-inhibited NIH 3T3 cells.[14] The cultures are fed every 2 days with growth medium, and the appearance of transformed foci is quantitated after 7 days.

Growth Transformation Assays

The second type of transformation assays involves analyses of NIH 3T3 cells stably expressing the desired oncoprotein. Briefly, NIH 3T3 cells are transfected as described earlier for focus-formation assays (50–100 ng plasmid DNA per dish). Three days after transfection, one-third of the transfected cultures are replated into a 100-mm dish containing growth medium supplemented with 400 μg/ml G418 (Geneticin; GIBCO, Grand Island, NY) or 400 μg/ml hygromycin. After approximately 2 weeks, the several hundred drug resistance colonies that have arisen are then trypsinized and pooled together, with one-tenth of the culture replated into another 100-mm dish containing growth medium supplemented with G418 or hygromycin. These mass populations of stably transfected cells are then used to assay for the acquisition of the various transformation phenotypes that are described in the following section.

Anchorage-Independent Growth

The acquisition of anchorage-independent growth potential represents one of the best *in vitro* correlates to *in vivo* tumorigenic growth potential.

[20] J. K. Westwick, Q. T. Lambert, G. J. Clark, M. Symons, L. Van Aelst, R. G. Pestell, and C. J. Der, *Mol. Cell. Biol.* **17**, 1324 (1997).

Colony formation in soft agar is the most widely used assay to evaluate anchorage independence of growth. NIH 3T3 cells need to adhere to a solid matrix in order to remain viable and proliferate. Transformed cells lose this requirement and therefore can form proliferating colonies of cells when suspended in a semisolid agar medium.

Reagents

A 1.8% Bacto-agar (w/v, in H_2O) stock is prepared by boiling the solution in a microwave oven to dissolve the agar. Fifty-milliliter aliquots are placed in 100-ml bottles and autoclaved to sterilize. Store at room temperature. A 2× DMEM stock solution is also required, along with the usual growth media components.

Soft Agar Assay

A similar protocol for this assay has been described in previously.[16] Briefly, a 0.6% agar (in growth medium) bottom layer is plated first and allowed to harden. The cultures are trypsinized to generate a single-cell suspension, counted, and resuspended in 0.4% agar (in growth medium) to form the top layer. For weakly transforming Rho proteins, it is best to use 10^4 to 10^5 cells per plate, as they cause a low efficiency of colony formation that ranges from 0.1 to 1%. For comparison, a greater than 10% colony-forming efficiency is seen with Ras-transformed cells. Rho-induced colonies also develop more slowly and are visible after 14 to 30 days, as compared to the 10 days it takes for Ras-transformed cells to develop colonies. It is important to include both positive (e.g., Ras-transformed cells) and negative (empty vector-transfected) control cell populations, as it is very easy to kill all the cells if the agar is too hot or if the agar was made incorrectly. The following section describes the procedure to make up a 14-dish assay using one batch of an agar/medium solution. In this case, the cells should be trypsinized and counted before mixing the agar and growth medium. The reason for this being that after the bottom layer is plated, the rest of the agar/medium is placed at 45°. Both the serum and medium will deteriorate if kept at this temperature for extended time periods. For a 14-dish assay where duplicates are used, one set would be used for the negative control and one set for the positive control, which would leave five sets of experimental conditions.

1. Preparing the reagents. Sterile 1.8% Bacto-agar stock solution is melted in a microwave oven and placed in a 60° bath to cool. The 1× and 2× DMEM solutions and the calf serum are placed in a 37° bath to prewarm. A 45° bath will also be needed.

2. *Preparing the cells.* Trypsinize each set of cells and transfer to a sterile 15-ml conical tube. The cells are spun down and resuspended in fresh growth medium. The cells are then counted to determine the number of cells per milliliter. It is important at this step to also confirm that the cells are a uniform single-cell suspension. To a 5-ml sterile culture tube add 10^4 cells to growth medium for a total volume of 0.5 ml.

3. *Preparing the bottom layer (0.6% agar).* For fourteen 60-mm plates, add 33 ml of 2× DMEM, 23 ml of 1× DMEM, 10 ml of calf serum, and 0.1 ml of 1000× penicillin/streptomycin to a sterile bottle. This recipe allows for 16 ml extra agar/medium. Warm this solution to just 45° and then add 33 ml of the melted agar. Quickly plate 5 ml onto 60-mm dishes and allow to solidify at room temperature in the tissue culture hood. Any bubbles should be removed gently with a pipette. The rest of the agar/medium is put back in the 45° bath.

4. *Preparing the top layer (0.4% agar).* Prewarm the culture tubes containing the cells to 37°. Add 1 ml of the agar/medium to the 0.5 ml of cells. Quickly, but gently, pour or pipette the cell/agar solution onto the solid bottom layer. Allow the top layer to solidify before transferring the plates to a 37° humidified 10% CO_2 incubator. The colonies will be visible after 2 weeks (if Ras-transformed cells are used as a positive control, these colonies will show up earlier).

One technical complication of this assay is that the agar-containing medium for the cell suspension may become solidified before the top layer is made. Another involves the loss of viability of the trypsinized cells prior to mixing with the agar-containing medium. To circumvent these potential complications, the bottom layer agar and the top layer can be made up separately. This allows for the bottom layer to be made up and plated before the cells are trypsinized. The cells can then be counted and resuspended. The top agar layer solution can then be made up immediately before plating such that the medium and serum will not be at 45° for a prolonged time period. The volumes per 60-mm dish are:

	Bottom layer	Top layer
Bacto-agar	1.65 ml	0.33 ml
2× DMEM	1.65 ml	0.33 ml
1× DMEM	1.15 ml	0.23 ml
Calf serum	0.5 ml	0.1 ml
1000× pen/strep (penicillin/streptomycin)	5.0 μl	1.0 μl
Total volume	5.0 ml	1.0 ml

Growth in Low Serum and Analysis of Growth Rate

Oncogene-transformed NIH 3T3 cells typically show an increased rate of growth, will grow to a higher saturation density, and have a lower requirement for serum in their growth medium. To establish a growth rate curve, drug-selected cells are trypsinized, counted, and plated onto 60-mm dishes at a concentration of 10^4 cells per dish. Time points are taken every 2 days, for a total of 10 days or five time points. Therefore 10 plates for each cell line are plated to produce duplicate sample numbers. The time points are taken by trypsinizing and counting the cells from each of the plates separately. For the early time points, cells should be resuspended in a low volume of medium after trypsinizing. The graph of the time points will establish both a growth rate and a saturation density for each of the cell lines.

For growth in low serum, three plates of 10^3 cells per 60-mm dish are plated. The cells are allowed to attach overnight in growth medium supplemented with 10% calf serum. The next day, the growth medium is replaced with DMEM supplemented with 0.5, 2, or 10% calf serum (duplicate dishes for each condition). In addition to cells expressing the oncoprotein under evaluation, parallel dishes are plated with either empty vector-transfected or Ras-transformed NIH 3T3 cells to provide negative and positive controls, respectively. The cells are then incubated for 6 to 10 days, with replacement with fresh media every 2 to 3 days. The plates are then fixed and stained with crystal violet. Whereas untransformed NIH 3T3 cells become quiescent when maintained in 0.5 to 2% serum-containing growth media, Ras-transformed cells continue to proliferate, at a somewhat reduced level, at these low serum concentrations. NIH 3T3 cells transformed by activated Rho or Dbl family proteins will also proliferate in low serum but at a reduced rate when compared to Ras-transformed cells.

Other Applications

The transformation assays described in this article can also be used in conjunction with inhibitors of Rho protein activity to determine the requirement of specific Rho family members for transformation by different oncoproteins. For example, the coexpression of dominant-negative mutants of Rac1, RhoA, RhoB, RhoG, Cdc42, and TC10 (analogous to the S17N mutant of Ras) has been used to demonstrate that the function of multiple Rho family proteins is required for full Ras transformation.[4]

Because there are potential concerns regarding the specificity of these dominant-negative proteins and because they target Dbl family GEFs to cause an indirect inhibition of small GTPase function, the use of other approaches to inhibit Rho family protein function is also recommended. Coexpression of p190 Rho GAP, which preferentially inhibits Rho, can block the transforming activity of Dbl family members.[21,22] Inhibitors of Rho effector binding or function have also been useful, providing evidence of Rho-mediated oncogenic signaling. The C3 botulinum toxin specifically inhibits RhoA, and not Rac1 or Cdc42, and we have used a C3 transferase expression plasmid to block RhoA-dependent cellular processes.[23] Another type of inhibitor Y-27632, which inhibits the Rho effector ROCK, has been utilized to show that this effector is required for both transformation and invasion of certain cell types.[24] Expression of peptide fragments that contain the GTPase-binding domain of downstream effectors can also be used. For example, we and others have used expression plasmids encoding the Cdc42–GTP-binding domain of WASP as a specific inhibitor of Cdc42 function.[25] When coupled with other assays to assess oncoprotein regulation of actin organization or upregulation of GTP-bound GTPase levels, these inhibition studies provide useful assays to implicate specific Rho family proteins in oncoprotein transformation.

Finally, transformation assays can also be helpful in determining the important structural and biochemical properties for the oncogenicity of certain proteins. Many studies have used mutants of the Dbl family members to dissect the important structural requirements of these proteins in transformation. Several studies using constructs that encode different domains of various Dbl family members in focus formation assays have shown that DH and PH domains are absolute requirements for Dbl transformation.[10]

Concluding Remarks

In summary, we have described NIH 3T3 assays for the analyses of transformation caused by aberrant activation of Rho family proteins. Similar assays using Rat-1 fibroblasts have also been described by Symons and colleagues.[6,7] Another useful cell system for assessing Rho transformation utilizes the T47D human breast carcinoma cell line. We showed previously that activated Rac1 and Cdc42, as well as activated Dbl family protein, can alter the differentiation potential of these cells and also promote their

[21] Y. Zheng, M. F. Olson, A. Hall, R. A. Cerione, and D. Toksoz, *J. Biol. Chem.* **270,** 9031 (1995).
[22] I. P. Whitehead, R. Khosravi-Far, H. Kirk, G. Trigo-Gonzalez, C. J. Der, and R. Kay, *J. Biol. Chem.* **271,** 18643 (1996).
[23] I. E. Zohn, T. Lugo, M. Chrzanowska-Wodnicka, M. Symons, C. J. Der, and R. Kay, *Oncogene,* in press (1999).
[24] E. Sahai, T. Ishizaki, S. Narumiya, and R. Treisman, *Curr. Biol.* **9,** 136 (1999).
[25] M. Symons, J. M. J. Derry, B. Karlak, S. Jiang, V. Lemahieu, F. McCormick, U. Francke, and A. Abo, *Cell* **84,** 723 (1996).

increased motility and invasion *in vitro.*[26] Details of the use of T47D cells for the analyses of Ras and Rho protein-mediated induction of motility and invasion are presented elsewhere.[27]

[26] P. J. Keely, J. K. Westwick, I. P. Whitehead, C. J. Der, and L. V. Parise, *Nature* **390,** 632 (1997).
[27] P. J. Keely, *Methods Enzymol.* **333,** [23], in preparation.

[39] Rho GTPases and Cell Migration–Fibroblast Wound Healing

By Catherine D. Nobes

Introduction

Cell migration is critical for many natural biological processes, including embryonic morphogenesis, wound healing, and host defense against infections.[1,2] Cell migrations plays a role also in abnormal biological processes such as tumor cell metastasis. Migrating cells in culture display three types of dynamic actin structure: lamellipodia and filopodia, which are sheet- and finger-like protrusions at the leading edge of the cell, and contractile actin–myosin filaments or stress fibers, which traverse the cell from front to back. At least part of the driving force for cell movement appears to be nucleation of actin monomers at the leading edge of the cell.[3] In addition to dynamically regulating actin structures, migrating cells need to attach to the underlying extracellular matrix via integrin adhesions. Equally important is their capacity to regulate the detachment of these adhesive structures at the back of the cell so as not to leave a trailing tail behind. Finally, cells are required to move their nucleus and organelles forward via a process described as traction.[4]

The Rho family of proteins plays a key role in regulating the assembly of actin structures within cells.[5] The three most well-characterized members of this family are Cdc42, Rac, and Rho. It has been shown in many cell types that Cdc42 regulates filopodia protrusion, Rac regulates lamellipodia

[1] B. L. Hogan, *Cell* **96,** 225 (1999).
[2] P. Martin, *Science* **276,** 75 (1997).
[3] T. J. Mitchison and L. P. Cramer, *Cell* **84,** 371 (1996).
[4] D. A. Lauffenburger and A. F. Horwitz, *Cell* **84,** 359 (1996).
[5] A. Hall, *Science* **279,** 509 (1998).

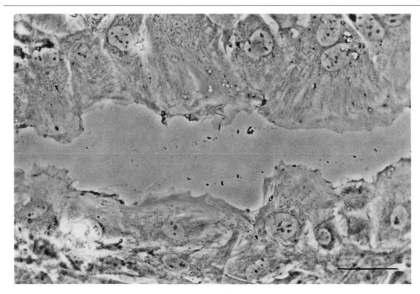

Fig. 1. Phase-contrast image of a REF "scrape" wound 3 hr after wounding showing polarized morphology of wound edge cells. Bar: 20 μm.

extension or membrane ruffling, and Rho triggers the assembly of actin stress fibers.[6-9] In addition, these three molecular switches regulate the assembly of integrin-containing adhesion complexes that link these filamentous actin structures to the underlying matrix.[8]

Wound healing in tissue culture cell monolayers is a commonly used model for analyzing the molecular mechanisms underlying cell migration. A strip of cells is removed from a confluent monolayer of cells that are contact inhibited for movement. The wound edge cells and those cells in the rows behind them migrate into the wound space and continue to move in an essentially unidirectional and synchronous fashion until they meet cells migrating toward them from the opposing wound edge (Fig. 1). Cell–cell contact then results in a cessation of forward movement. Cells used frequently in this wound assay are primary rat embryo fibroblasts (REFs) and mouse embryo fibroblasts (MEFs). REFs and MEFs are more motile than fibroblast cell lines, such as Swiss 3T3 cells, and exhibit a net translocation of around 0.5–1 μm/min. Because these cells display good contact-

[6] A. J. Ridley and A. Hall, *Cell* **70,** 389 (1992).
[7] A. J. Ridley, H. F. Paterson, C. L. Johnston, D. Diekmann, and A. Hall, *Cell* **70,** 401 (1992).
[8] C. D. Nobes and A. Hall, *Cell* **81,** 53 (1995).
[9] R. Kozma, S. Ahmed, A. Best, and L. Lim, *Mol. Cell. Biol.* **15,** 1942 (1995).

dependent inhibition of movement, wound-induced cell migration is directed into the wound space, and cells polarize toward the wound. MEFs prepared from knockout mice provide an opportunity to test the function of specific genes in regulating fibroblast migration.

This article describes methods to analyze the roles of the Rho GTPases in regulating distinct parameters of cell migration.

Materials

Recombinant Rho GTPases (constitutively active and dominant negative mutant versions) and C3 transferase, for microinjection, are expressed as glutathione S-transferase (GST) fusion proteins in *Escherichia coli,* purified on glutathione agarose beads, and cleaved proteolytically to remove GST as described.[10,11] Cdc42 activity can also be inhibited by microinjecting a fragment of its effector protein, Wiskott-Aldridge syndrome protein (WASp), comprising amino acids 201–321.[12] The WASp fragment is also prepared as a glutathione S-transferase fusion protein. The p160[ROCK] inhibitor, Y-27632, is a kind gift of Toshio Hamasaki (Yoshitomi Pharmaceutical Industries Ltd., Japan) and is prepared as a stock solution of 10 mM in H_2O.[13]

Preparation of Rat Embryo Fibroblast Cells

Primary REFs are prepared as described previously.[14] E14 rat embryos are dissected from the uterus and yolk sac, and their heads and viscerae are removed. The torso is placed in phosphate buffered saline (PBSA), and the tissue is minced into pieces smaller than 1 mm^3 with a scalpel and digested with 5 ml trypsin/EDTA for 20 min at 37°. Proteolysis is stopped by the addition of 5 ml Dulbecco's modified Eagle's medium (DMEM)/ 10% fetal calf serum (FCS). The suspension is taken up with a 10-ml pipette, the tissue is allowed to settle in the tip of the pipette, dripped into a 6-cm dish, and the remainder of the trypsin solution is discarded. Five milliliters DMEM/10% FCS is added and the tissue is disaggregated by forcing up and down through a 5-ml pipette. A trypan blue exclusion test for living cells is performed and cells are subsequently plated at a density of 3 × 10^7 cells in a 15-cm dish in 30 ml culture medium [DMEM/10% FCS supple-

[10] A. J. Self and A. Hall, *Methods Enzymol.* **256,** 3 (1995).
[11] A. J. Ridley, *Methods Enzymol.* **256,** 313 (1995).
[12] C. D. Nobes and A. Hall, *J. Cell Biol.* **144,** 1235 (1999).
[13] M. Hirose, T. Ishizaki, N. Watanabe, M. Uehata, O. Kranenberg, W. H. Moolenaar, F. Matsumura, M. Maekawa, H. Bito, and S. Narumiya, *J. Cell Biol.* **141,** 1625 (1998).
[14] H. Land, *Methods Enzymol.* **254,** 37 (1995).

mented with streptomycin (100 μg/ml) and penicillin (100 IU/ml)] at 37° and 10% (v/v) CO_2. Secondary cells can be grown to passage 7 after which the cells hit growth crisis and most die by apoptosis. Our wound-induced migration assays are performed on cells between passages 3 and 7.

Wounding REF Monolayers and Microinjection of Wound Edge Cells

REFs for wound assays are seeded on acid-washed, 13-mm glass coverslips (No. 1½, Chance Propper, England) in 4- or 24-well multidishes (Nunc, Denmark). Cell density at the time of wounding is critical for reproducible wound widths and reproducible wound closure. Cells are seeded in 1 ml culture medium at high density, 12×10^4 cells per coverslip, and are wounded 24 hr later when the cells have formed a confluent monolayer. The wound is made by scraping the cell monolayer. We make small wounds (100–150 μm diameter) using a microinjection needle broken at the shaft to a diameter of about 100 μm and flame polished. These wounds close after 5–6 hr, and the time course of wound closure is very reproducible. The wounds are wide enough to force wound closure to occur by cell migration and not simply by cell stretching.

Immediately after wounding, cells at the wound edge retract, round up, and are difficult to microinject. The cytoplasm of these cells can be microinjected with Rho proteins 30 min later when the cells have respread. Immediately prior to microinjection, coverslips are transferred to a 60-mm petri dish in 5 ml culture medium. Cell microinjection is performed as described previously.[11,15] A region of wound is randomly chosen, and a continuous stretch of 10–15 leading edge cells on opposing sides of the wound are microinjected. Since not only front row cells participate in wound closure, cells in rows 2, 3, and 4 behind these lead edge cells and on both sides of the wound are also microinjected (Fig. 2). Recombinant proteins (generally at concentrations between 0.5 and 1 mg/ml active protein, see Table I) are coinjected with a fixable, fluorescent injection marker, fluorescein isothiocyanate (FITC)- or Texas Red-conjugated dextran, molecular weight 10,000 (Molecular Probes, Eugene, OR) at 2–5 mg/ml.

Fixing, Staining, and Photographing Cells

We have stained wound edge cells for vinculin (VIN-11-5, Sigma, St. Louis, MO) to localize focal adhesions and complexes, for the centrosome apparatus using an antibody to pericentrin (M8–1, a kind gift from Elly Tanaka, Heidelberg, Germany), and for the Golgi apparatus using a rat

---

[15] H. F. Paterson, P. Adamson, and D. Robertson, *Methods Enzymol.* **256,** 162 (1995).

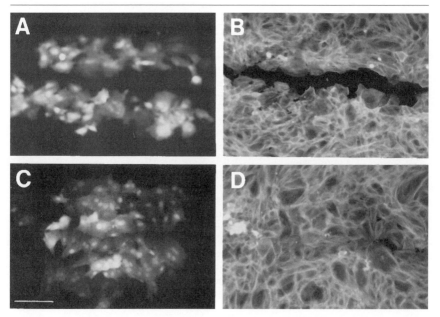

FIG. 2. REF wound assay from which measurements of wound closure are made. Thirty to 60 min after wounding, three rows of wound edge cells were coinjected with N17Cdc42 and fluorescent dextran (A and B) or fluorescent dextran alone as a control (C and D). Both control and Cdc42-inhibited wounds were fixed after 6 hr when the control wound had closed. Cells were stained to reveal filamentous actin structures with rhodamine-conjugated phalloidin (B and D). Injected fluorescent dextran is shown in A and C. Bar: 200 μm.

monoclonal antibody (23C), which recognizes β-COP (a kind gift from Colin Hopkins, University College, London, UK), and microtubules with a rat antitubulin antibody (YL1/2, Serotec Ltd., Oxford, UK). Fixation and immunofluorescent labeling of wounds are performed at room temperature according to the following protocol:

1. Fix with 4% paraformaldehyde (w/v)/PBSA for 10 min. (Rinse cells briefly with PBSA before fixation and between each incubation step.)
2. Permeabilize in 0.2% Triton X-100 (v/v)/PBSA for 5 min.
3. Quench residual formaldehyde with sodium borohydride (0.5 mg/ml) in PBSA for 10 min.
4. Incubate with primary antibodies diluted in PBSA, titrated to obtain optimal staining.
5. Incubate with appropriate FITC-conjugated secondary antibodies (Jackson ImmunoResearch Labs, Inc.) and with rhodamine-conjugated phalloidin (0.1 μg/ml, Sigma) diluted in PBSA.

TABLE I
EFFECTS OF ACTIVATING AND INHIBITING Rho GTPases ON
WOUND CLOSURE

Protein[a]	Concentration (mg/ml)[b]	Wound Closure (%)
V14RhoA	0.4	11
V12Rac1	1.0	90
V12Cdc42	1.0	95
C3 transferase	0.07	80
C3 transferase	0.3	20
N17Rac1	1.0	21
N17Cdc42	1.0	47
WASp[(201–321)]	0.5	45
Dextran	2.0	94

[a] Proteins were injected into wound edge REFs.
[b] The concentration of active protein in the case of GTPases was determined by a GTP-binding assay.[10] The concentration of C3 transferase and WASp fragment was determined using a Bio-Rad protein assay kit.

Cells processed for filamentous actin (w/v) staining alone or microtubule staining are fixed in 4% (w/v) paraformaldehyde/1% (v/v) gluteraldehyde/ PBSA and incubated with rhodamine-conjugated phalloidin at 0.2 μg/ml. Gluteraldehyde is necessary for preserving fine actin structures such as filopodia.

6. The underside of converslips is wiped dry and coverslips are mounted by inverting onto 5 μl Moviol mountant[16] containing p-phenylenediamine (a few grains/ml) as an antifade agent. Mounted coverslips are examined using a fluorescence microscope with 10 and 20× (air) objectives and 40 and 63× (oil immersion) objectives.

Wound Closure Measurements

Semiquantitative measurements are made of control and experimental wounds by measuring the distance across the wound. Wound width at the time of microinjection is calculated from control wounds that are fixed at the time of microinjection. Wounds with microinjected patches are fixed at a time when sister uninjected wound edges have just met. The state of wound closure is monitored by frequent observation using phase-contrast optics. Randomly chosen regions of wound, fixed and stained with rhoda-

[16] G. V. Heimer and C. E. Taylor, J. Clin. Pathol. 27, 254 (1974).

mine-conjugated phalloidin, are photographed using a 20× objective. Mean wound width in micrometers is determined (by averaging the width every 30 μm) and percentage wound closure is calculated. For microinjected wounds, wound width is measured as the distance between the leading edges of microinjected cells (on either side of the wound) as visualized by coinjected fluorescent dextran, as some uninjected cells crawl between injected cells into the wound space. In this case, measurements are made every 15 μm over a length of wound 300–400 μm. From the average values for each injection, percentage wound closure is calculated (see Table I).

Inhibiting Rac and Cdc42 activity in wound edge cells, by microinjection of dominant negative proteins, prevents wound closure (see Table I and shown for Cdc42 in Fig. 2). Rac-inhibited wound edge cells are unable to protrude lamellipodia and do not move into the wound space. This can be determined by staining with rhodamine-conjugated phalloidin (see earlier discussion) and by time-lapse cinematography (see later). Inhibiting Cdc42 activity does not block lamellipodia protrusion and only partially blocks cell motility, but Cdc42 is required for cells to polarize toward the wound (see later). Microinjection of C3 transferase (to inhibit Rho) at a concentration of 0.07 mg/ml is sufficient to induce a loss of Rho-dependent actin stress fibers, without affecting cell motility. However, injection of higher concentrations of C3 transferase (0.3 mg/ml) will cause cells to detach from the underlying matrix and, under these conditions, wound closure is inhibited significantly (Table I). Stress fibers and focal adhesions can be blocked more directly by inhibiting p160^{ROCK}, one of the downstream targets of Rho, mediating stress fiber assembly, by bathing in compound Y-27632 (20 μM).[13] Under these conditions, wound closure speed is enhanced by about 30%.[12] These data suggest that stress fibers and focal adhesions may hinder cell crawling.

Cell Polarity Determination

Migrating REFs develop a polarized morphology with a wide, ruffling lamella restricted to the front edge of the cell and a more narrow trailing tail (Fig. 3). Cell polarization at the wound edge also appears to involve the reorganization of the microtubular network, the positioning of the centrosome in front of the nucleus, and the realignment of the Golgi apparatus in the direction of movement.[17,18] We use Golgi apparatus position relative to the nucleus as a readout of cell polarization in response to

---

[17] A. Kupfer, D. Louvard, and S. J. Singer, *Proc. Natl. Acad. Sci. U.S.A* **79**, 2603 (1982).
[18] U. Euteneuer and M. Schliwa, *J. Cell Biol.* **116**, 1157 (1992).

wounding (Fig. 3). In most cells the Golgi apparatus is positioned juxtaposed to the nucleus and can be localized by immunolabeling, as described earlier, using a rat monoclonal antibody (23C) that recognizes β-COP. The position of the immunolabeled Golgi apparatus in each cell is recorded as described in Kupfer et al.[17] Wound edge cells are divided up into three × 120° sectors, one of which faces the edge of the wound (see Fig. 3B). Wounds are fixed at different times after wounding and the percentage of cells with their Golgi apparatus oriented in the direction of movement (i.e., in the front

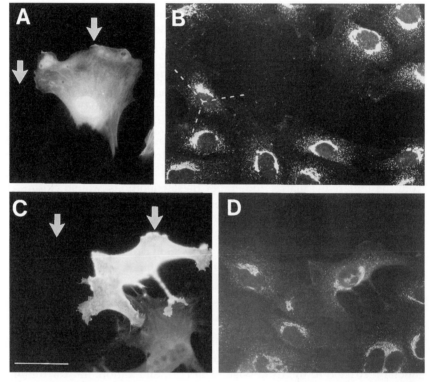

Fig. 3. Cell polarity is regulated by Cdc42. Leading edge REFs in control wounds develop a polarized morphology with lamellipodia and membrane ruffles restricted to the wound side (visualized in A by microinjection of fluorescent dextran). REFs migrating into the wound also reorient their Golgi apparatus forward of the nucleus in the direction of movement (B). Expression of myc-tagged WASp fragment (to block Cdc42) inhibits wound-induced cell polarization and the Golgi apparatus reorientation (C and D). Expressed myc-tagged WASp in visualized by indirect immunofluorescence and clearly leads to a loss of polarity with lamellipodia protruding from all sides of the cell (C) and the Golgi apparatus (costained in D) not oriented in the direction of the wound. The position of the wound edge in A and C is indicated by arrows. Bar: 20 μm.

segment) is calculated. The time required for wound-induced Golgi apparatus reorientation is different for different cell types and, in the case of REFs, appears to be about 5–6 hr.[12] The role of Rho GTPases in regulating Golgi apparatus reorientation can be examined by microinjecting mutant Rho GTPases into the wound edge cells.[12] These injection studies reveal that Cdc42 activity is required both for morphological polarization and for reorientation of the Golgi apparatus toward the leading edge of cells (Fig. 3).

Time-Lapse Imaging of Wound Edge Cells

Time-lapse video microscopy provides information about the dynamic behavior of migrating cells that cannot be obtained from "snapshot" observations of fixed cells. Cell movements are recorded using a charged coupled device (CCD) camera (Panasonic BL-22) and a time-lapse controller (EOS Electronics, South Glamorgan, UK) attached to a Zeiss inverted microscope. Cells are maintained at 37° and are placed in a slide flask (Nunc, Denmark) sealed to maintain a level of 10% CO_2. Microscope images are recorded at a rate of 10 frames every 60 sec on a Sony Betacam video recorder. After the time-lapse period, cells can be fixed and mounted to identify which have been injected.

Acknowledgments

Kate Nobes is a Lister Institute Research Fellow. Much of this work was done while a postdoctoral fellow in the laboratory of Alan Hall and was funded by the Cancer Research Campaign (UK).

[40] Rho GTPases and Cell Migration: Measurement of Macrophage Chemotaxis

By GARETH E. JONES, ANNE J. RIDLEY, and DANIEL ZICHA

Introduction

Cell migration is a complex process involving dynamic, coordinated changes in cell adhesion and in the cytoskeleton. Directed migration is stimulated by chemoattractants, and leukocytes are able to detect small differences in the concentration of these substances. The composition of the extracellular matrix also has a profound influence on the rate of cell migration. Thus the intracellular signaling pathways mediating cell migration have to respond to diverse signals from outside the cell and translate these into very precisely tuned changes within the cell. Of the signaling proteins implicated in cell migration response, the Rho family of GTPases appears to have the properties that would allow them to act concertedly as coordinators of cell migration. Initially identified as regulators of actin organization, they also regulate cell adhesion to the extracellular matrix and to other cells and, in the long term, they influence gene expression (see reviews by van Aelst and D'Souza-Schorey[1] and Hall[2]). The Rho proteins can be activated in response to cytokines and growth factors and to extracellular matrix proteins. They can interact with many different target proteins within cells, allowing them to activate a number of downstream signals simultaneously (reviewed in Aspenström[3]). Finally, they show extensive cross talk with each other,[2] which could allow them to coordinate different aspects of cell locomotion.

Of the mammalian Rho family proteins, the best characterized for their ability to regulate actin organization are Rho, Rac, and Cdc42. The roles of Rho, Rac, and Cdc42 in regulating actin organization were first characterized in detail in Swiss 3T3 fibroblasts. Constitutively active mutants of Rho and Rac induce the formation of stress fibres and lamellipodia, respectively, when microinjected into quiescent Swiss 3T3 cells.[4,5] Conversely, microinjection of C3 transferase to inhibit Rho or of a dominant-negative Rac mutant to inhibit Rac inhibits growth factor-induced formation of these structures. Activated Cdc42 protein induces the extension of filopodia, which are finger-like plasma membrane protrusions containing actin filament bundles.[6] Under appropriate conditions, Cdc42, Rac, and Rho can activate each other sequentially in a cascade: Cdc42 can induce Rac-mediated lamellipodium formation, and Rac can induce Rho-mediated stress fiber formation.

To determine how Rho, Rac, and Cdc42 regulate cell migration and chemotaxis, we have used a mouse macrophage cell line, BAC1.2F5, that resembles primary macrophages in being dependent on colony stimulating factor-1 (CSF-1) for survival and proliferation, as well as exhibiting many of the markers of normal activated macrophages.[7,8] Crucial to our interests,

[1] L. van Aelst and C. D'Souza-Schorey, *Genes Dev.* **11,** 2295 (1997).
[2] A. Hall, *Science* **279,** 509 (1998).
[3] P. Aspenström, *Curr. Opin. Cell Biol.* **11,** 95 (1999).
[4] A. J. Ridley and A. Hall, *Cell* **70,** 389 (1992).
[5] A. J. Ridley, H. F. Paterson, C. L. Johnson, D. Diekmann, and A. Hall, *Cell* **70,** 401 (1992).
[6] C. D. Nobes and A. Hall, *Cell* **81,** 53 (1995).
[7] C. Morgan, J. W. Pollard, and E. R. Stanley, *J. Cell. Physiol.* **130,** 420 (1987).
[8] C. A. Boocock, G. E. Jones, E. R. Stanley, and J. W. Pollard, *J. Cell Sci.* **93,** 447 (1989).

we have also shown that BAC1.2F5 cells show chemotactic migration to a source of CSF-1.[9] Thus we have a system in which we can use microinjection of Rho proteins to analyze their role in regulating the migration and chemotactic responses of cells.

The Boyden chamber is the most widely used method for assaying chemotaxis and is based on scoring cells that have migrated into or through a filter membrane toward a source of putative chemotactic factor. Two limitations of this type of approach are that the local concentration gradients of chemotactic factor in and around the pores of the filter membrane are variable and unknown and that the resulting cell behaviour is unobservable and can only be deduced from the final distribution of the cell population. Unequivocal confirmation of chemotaxis requires direct observation of the cells together with knowledge of the direction and magnitude of the concentration gradient.

The Dunn chamber is a development of the Zigmond chamber,[10] the only devices available that allow direct cell observation in a concentration gradient approaching a linear steady state. The Dunn chamber was developed to study long-term chemotaxis,[11] particularly in slowly moving cultured cells, and the principal design criterion was therefore an improved stability of gradient.[12]

Materials

The murine BAC1.2F5 macrophage cell line was kindly supplied by Dr. Richard Stanley (Albert Einstein College of Medicine, New York). Human recombinant macrophage colony-stimulating factor (CSF-1 is also called M-CSF) is obtained from R&D Systems (Abingdon, Oxon, UK), and unless otherwise stated, all tissue culture media and supplements are obtained from ICN-Flow Laboratories (High Wycombe, UK). Dunn chemotaxis chambers are obtained from Weber Scientific International (40 Udney Park Road, Teddington TW11 9BG, UK) and coverslips (18 × 18 mm, No. 3) are from Chance-Propper (Smethwick, UK). All recombinant proteins used for microinjection are purified from *Escherichia coli* containing pGEX-2T vectors that encode each glutathione *S*-transferase (GST) fusion protein as described previously.[13] GST is cleaved from all proteins prior to injection.

[9] S. E. Webb, J. W. Pollard, and G. E. Jones, *J. Cell Sci.* **109,** 793 (1996).
[10] S. H. Zigmond and J. G. Hirsch, *J. Exp. Med.* **137,** 387 (1973).
[11] D. Zicha, G. A. Dunn, and A. F. Brown, *J. Cell Sci.* **99,** 769 (1991).
[12] D. Zicha and G. A. Dunn, *Exp. Cell Res.* **221,** 526 (1995).
[13] A. J. Self and A. Hall, *Methods Enzymol.* **256,** 3 (1995).

Methods

Cell Culture

Stock cultures of BAC1.2F5 macrophages are maintained at 37° in a humidified atmosphere of 5% CO_2 and 95% air (v/v) in Dulbecco's modified Eagle's medium (DMEM) containing 10% heat-inactivated fetal calf serum (Globepharm, Esher, UK), 38 ng/ml CSF-1, 2 mM L-glutamine, 0.15 mM L-asparagine, 15 nM β-mecaptoethanol, 77.5 U/ml streptomycin, and 25 U/ml penicillin.

To obtain cells for microinjection and chemotaxis assays, 2×10^4 cells/well are seeded into 25-mm-diameter petri dishes containing sterile 18-mm^2 No. 3 glass coverslips. The coverslips have been marked previously with a diamond pen on the reverse side to facilitate the subsequent localization of microinjected cells and their positioning on the Dunn chamber chemotaxis bridge as described later. For analysis of cytokine-induced cell migration of microinjected cells, cells were maintained in full medium for 24 hr after seeding on coverslips, followed by thorough washing in medium without added CSF-1 and subsequent incubation in CSF-1-deficient medium for a further 24 hr. This regime brings the macrophages to a quiescent state in which the cells have lost membrane ruffles and lamellipodia and become rounded in morphology.

Microinjection of BAC-1.2F5 Macrophages

General Principles of Microinjection. Microinjection is a technique that introduces macromolecules into cells rapidly, allowing analysis of the immediate cellular changes occurring in response to the injected substance. The technique for microinjecting cultured cells was described in detail by Graessmann and Graessmann[14] and has been used to inject protein, RNA, or DNA. The substance to be injected is loaded into glass microcapillaries that have been pulled to a fine point at one end to less than 1 μm diameter. A micromanipulator is used to position the point of the glass pipette very close to the cells to be injected. The other end of the pipette is attached via tubing to a pressure regulator. Air pressure applied to this end of the pipette forces the protein solution out of the pointed end of the pipette. The pipette is manipulated so that it pierces the plasma membrane of a cell transiently, allowing the solution in the pipette to enter the cell. The pipette remains within the cell for only a very short period (<0.5 sec) and then is removed, allowing the membrane to reseal. The volume of solution

---

[14] M. Graessmann and A. Graessmann, *Methods Enzymol.* **101,** 482 (1983).

introduced into cells is between 5 and 10% of their total volume, or approximately 10^{-14} liter.

Microinjection of Recombinant Proteins

Recombinant proteins are stored in small aliquots in liquid nitrogen and thawed immediately prior to use on ice. They are diluted in 150 mM NaCl, 50 mM Tris, pH 7.5, 5 mM MgCl$_2$, on ice. The concentration of recombinant protein injected into cells has to be determined by testing different titrations. Routinely, however, constitutively active Cdc42, Rac1, and RhoA proteins are injected at between 200 and 300 μg/ml (as determined by [^{3}H]GDP binding). C3 transferase is injected at approximately 2 μg/ml, and dominant-negative Rac1 and Cdc42 are injected at about 0.75 mg/ml (total protein concentration, determined by Bradford assay). To identify microinjected cells, rabbit immunoglobulin G (IgG) (Sigma Chemical Co., St. Louis, MO) at 0.5 mg/ml is microinjected together with the recombinant proteins.

In our laboratory, cells are observed by phase-contrast microscopy on a Ziess IM35 inverted microscope. The petri dish is maintained at 37° by a heated stage (Zeiss, model TRZ 3700), and the pH of the medium is maintained by use of a flow of sterile 10% CO$_2$ over the apparatus. Initially, the marked area on the coverslip is located using a low-power objective, but cells are observed during microinjection with a 40× objective lens and 10× eyepieces.

Cells are injected with microcapillaries: 1.2-mm-bore glass pipettes (Clark Electroinstruments, Reading UK) are used to pull microcapillaries of approximately 0.4 μm tip diameter, with a programmable Pipette Puller (Campden Instruments, Model 773, Leicester, UK). Protein solution (1 μl) is loaded into a microcapillary using an Eppendorf (Cambridge, UK) microloader tip. Cells are injected in our laboratory using an Eppendorf microinjector (Transinjector Model 5246) and micromanipulator (Model 5171). All the cells residing within the marked area (approximately 15) are injected, and the coverslip is then returned to a humidified 5% CO$_2$ incubator at 37° for 20 min before mounting on the Dunn chemotaxis chamber. This latter detail proved to be a prerequisite for successful execution of the later procedures. Spread macrophages are quite amenable to microinjection, but the rounded quiescent cells we use here are likely to detach from the substratum unless given a short period in which to recover from the trauma of injection.

Dunn Direct Viewing Chemotaxis Chamber

The Dunn chamber is made by grinding two concentric wells into one face of a 1-mm-thick glass slide (Figs. 1 and 2). The two concentric

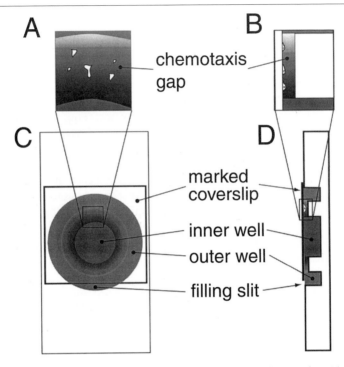

FIG. 1. The Dunn chemotaxis chamber shown from a top view (A and C) and from a side view in a cross section (B and D). Increasing intensity in the chemotaxis gap represents the increasing concentration of the chemotactic factor.

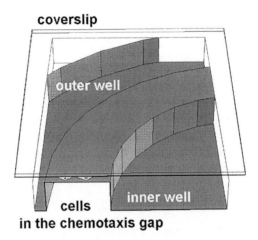

FIG. 2. A part of the Dunn chemotaxis chamber in a three-dimension representation.

wells are separated by an annular ridge 1 mm wide, which is polished by 20 μm under the level of the surrounding surface of the slide. The coverslip containing the previously microinjected cells is inverted over the chamber such that the marked area containing the microinjected cells overlies a section of the annular ridge. The inner well is blind and serves as a sink for the chemotactic factor, which is placed into the outer annular well (at 38 ng/ml CSF-1) by means of a hypodermic needle inserted into the filling slit. The polished annular ridge forms a narrow gap interconnecting the wells where the gradient develops by diffusion. Here the cells can be observed crawling on the undersurface of the coverslip. A linear gradient establishes in the gap of the assembled chamber within a short time, depending on the molecular weight of the chemotactic factor (see tabulation below). The difference between the concentrations of the chemoattractant in the wells determines the slope of the linear gradient. The diffusion of the chemoattractant produces a net flux and, as a result, the slope declines in time. The time taken by the concentration difference to halve is called "half-life" and characterizes the process.

Molecular weight	Gradient formation	Gradient half-life
10,000 to 20,000	30 min	30 hr
350 to 370	10 min	10 hr

Time-Lapse Imaging of Macrophage Motility in Dunn Chamber

The assembled direct viewing chemotaxis chamber is placed onto a stage of a microscope adapted for time-lapse recording. The observation of live cells requires a contrasting technique. Simple phase contrast usually gives good results. Differential interference contrast (Nomarski) can also be used. Reliable detection of chemotaxis requires large numbers of observed cells. Therefore a lower magnification objective lens (typically 10×) should be selected. This magnification encompasses nearly the full width of the annular ridge (and thus the CSF-1 gradient) in the chemotaxis gap. An inverted or upright microscope can be used in this situation. When detailed morphology of responding macrophages needs to be studied with higher power magnification, an upright microscope should be used.

In the past we used a variety of microscopes and simple charge-coupled device (CCD) cameras for our work. We now use a Zeiss Axiovert 135 TV microscope equipped with a CCD camera (Hamamatsu, Orca I), an automatic shutter (Uniblitz, Vincent Associates, Rochester, NY), and a shutter controller (Ludl Electronic Products Ltd., Hawthorne, NY). A Perspex box has been designed to contain the microscope stage and is fitted with a temperature-controlled heating system (custom made from controller

208 2739, T probe 219 4674, heater element 224 565, low noise a.c. fan 583 325, RS Components, UK). The camera and the shutter are driven by Acquisition Manager (AQM) software (Kinetic Imaging Ltd., Liverpool, UK). AQM allows a live image from the microscope to be observed on the computer screen. The live image is used for focusing and localizing a part of the chemotaxis gap suitable for recording. It is useful to align the high concentration end of the chemotaxis gradient with the top edge of the screen, as demonstrated in Fig. 1A, because then the chemotactic response can be detected as directional movement of cells upward. The live image is also useful for setting up the illumination intensity, the exposure time, brightness, and gain of the camera. In order to protect the cells from photo damage, the illumination intensity and the exposure time should be kept low. This requires fairly high camera gain and brightness. The quality of the camera determines what level of light exposure can be achieved. With a cooled, low-light CCD camera, such as Orca I, a very low illumination (e.g., minimum voltage setting with one interference filter and one green filter) can be used for a contrasting microscopic technique with a short exposure (e.g., 10 msec). The combination of interference and green filters protects cells from infrared photodamage. Once a satisfactory image has been achieved, the lapse interval and number of frames need to be selected. Because macrophages move fairly slowly (approximately 30 μm/hr), a lapse interval of 4 min is usually adequate. The full chemotactic response of macrophages does not last for more than 5 hr, and therefore around 75 frames should be sufficient for its detection. After the recording has been started, the AQM software allows a review of the section so far recorded, using another copy of the software, which can be started while the first recording is still in progress. This feature enables the abortion of an unsuccessful recording after a few hours and repeated attempts can be achieved easily. If required, the second copy of the AQM software can also be used to drive a second time-lapse recording system.

Identification of Microinjected Cells

After the Dunn chamber assay is finished, it is necessary to determine which of the cells were microinjected. The coverslip is lifted gently from the Dunn chamber and cells are fixed in phosphate-buffered saline (PBS) containing 4% formaldehyde and 3% sucrose (w/v) and then permeabilized with 0.5% Triton X-100 in PBS. They are then incubated for 30–40 min with a 1:400 dilution of fluorescein isothiocyanate (FITC)- or tetramethyl-rhodamine isothiocyanate (TRITC)-labeled goat anti-rabbit immunoglobulin antibody (Sigma). Coverslips are washed extensively with PBS, mounted on slides in Moviol, and viewed by epifluorescence microscopy. Comparison of stained cells with the final frame of the recorded sequence identifies injected versus uninjected cells easily. The latter set acts as an internal control for cell behavior, as these should always show significant chemotaxis toward CSF-1.

Analysis of Directional Migration

The recorded sequence of images can be analyzed in the image-processing program Lucida (also provided by Kinetic Imaging). The first part of the analysis consists of an interactive tracking of the identified cells, which is followed by analysis of the trajectories. In the cell-tracking module, use the mouse to select the cell-marking mode. When the mouse pointer is positioned in the center of a cell, the mouse button is used to mark the position and advance to the next frame. Individual cells are tracked in this way one by one throughout the sequence. The marked positions of cells can be saved into a file on request or at the point of exit. The file name with the cell positions will have an extension ".cel" by default and its individual lines will contain the following information: cell number, frame number, x coordinate, and y coordinate.

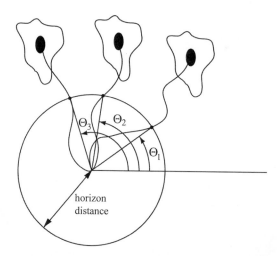

Fig. 3. Graphic representation of the directions derived from the trajectories of three cells. The trajectories were shifted with their starting point to the origin. The direction of a cell is the angle on which the cell reaches the horizon distance. The horizon distance can be set by the experimenter to any chosen distance prior to statistical analyses of the data; we normally choose arbitrary units such as multiples of cell diameter (approximately 10, 20, and 30 μm). A series of such horizon distances should be chosen when analyzing crucial experiments. Whereas fewer cells will reach horizons of greater magnitude, any unimodal pattern of cell migration will still be measured by the statistical analyses.

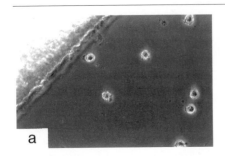

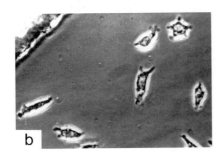

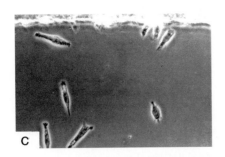

A further module in the Lucida program processes this file. Trajectories of individual cells become converted into directions. The direction of a trajectory is the angle on which the cell reaches an arbitrary horizon distance (Fig. 3). Nonmotile cells, which never reach the horizon distance, are automatically precluded from the analysis. This is a desirable feature because chemotaxis is defined for translocating cells. A conversion including n cells, translocating more than the horizon distance, generates a set of directions $\theta_1, \theta_2, \ldots, \theta_n$. The program for evaluation of the directionality then calculates the mean direction, D, from

$$\overline{S} = 1/n \sum \sin(\theta_i)$$
$$\overline{C} = 1/n \sum \cos(\theta_i)$$
$$\tan(\overline{D}) = \overline{S}/\overline{C}$$

The mean resultant length, $\overline{R}$, is then calculated from

$$\overline{R} = (\overline{C}^2 + \overline{S}^2)^{1/2}$$

The Rayleigh test for unimodal clustering of directions provides a p value using the mean resultant length, $\overline{R}$. If the null hypothesis about the uniform distribution is rejected by the Raleigh test ($p < 0.01$ or $p < 0.05$), then we can conclude that the cells are showing a significant directional response, and the 99 or 95% confidence interval of the mean direction, $\overline{D}$, can be determined on the basis of n and $\overline{R}$. In order to confirm that the response is chemotaxis to the known gradient, however, it should also be demonstrated that the direction of the gradient lies within the confidence interval of the mean cell direction $\overline{D}$. The program then generates a summary report.

These analytical protocols belong to a suite of software originally designed at The Randall Centre for our own use. However, the increasing interest in our measurement techniques has encouraged us to incorporate our routines into the commercial package available from Kinetic Imaging so that others may now purchase this software for their own use.

FIG. 4. Macrophage chemotaxis toward CSF-1. Cells maintained for 24 hr in medium lacking CSF-1 are rounded in morphology and remain so for many hours after assembly of the Dunn chamber. After 2 hr (a) no cell spreading can be observed and no cell movement is recorded outside a 10-μm horizon (approximately 1 cell diameter) on the circular histogram. On isotropic stimulation by CSF-1 (b), cells spread rapidly and then polarize,[9] and by 2 hr significant cell migration is seen, but the horizon plot analysis demonstrates that motile cells migrate in all possible directions. In a gradient of CSF-1 (c), cell polarization is significantly skewed along the direction of the gradient,[9] and macrophages preferentially migrate up the gradient, as can be seen from the horizon plot analysis.

Results

Macrophage Chemotaxis toward CSF-1

BAC1.2F5 cells show a strong chemotactic response to gradients of CSF-1. Figure 4 shows phase-contrast images of groups of macrophages situated on the annular bridge of a Dunn chamber. In Fig. 4a, cells have been previously deprived of cytokine for 24 hr prior to assembly of the chamber and monitoring for a further 2 hr in medium without added CSF-1. The rounded morphology characteristic of cytokine-deprived BAC1.2F5 cells is retained, and no significant cell movement is seen. The accompanying circular histogram shows that of 37 cells measured, none had migrated beyond a 10-μm horizon from its origin. The effects of adding CSF-1 are shown in Figs. 4b and 4c. In Fig. 4b, cytokine was added in equal concentrations to both inner and outer wells of the Dunn chamber so that the macrophages are bathed in an isotropic cytokine medium. Within 5 min, the cells have established an elongated morphology and begin to migrate.[9] This can be quantified as described, and the horizon plot demonstrates the migration of 36 cells beyond the 10-μm horizon. However, the plot also shows that application of the Rayleigh test fails to find any bias in the mean cell direction of movement. Thus CSF-1 can be regarded as a chemokinetic agent as it clearly stimulates cell migration. When cells are subjected to a gradient of CSF-1, as shown in Fig. 4c, cell polarization and cell migration are again established rapidly, with most cells being polarized along the axis of gradient formation. In this case, BAC1.2F5 macrophages show a strong positive chemotaxis toward the source of cytokine, which is demonstrated by application of the Rayleigh test to data for cell trajectories. The circular histogram shows the resulting mean vector of cell migration and the 95% confidence interval for 31 cell trajectories.

Effects of Rho, Rac, and Cdc42 on Macrophage Migration and Chemotaxis

Using the methodology described earlier, we measured the effects of microinjection of a range of dominant-negative and constitutively active Rho proteins on cell migration and chemotaxis. A full description of these experiments appears elsewhere (see Allen et al.[15]); here we show how the chemotaxis assay and analytical techniques we have developed can be applied to studies of cell migration. The insult of microinjection itself has the effect of slowing the rate of cell locomotion, but does not disturb the chemotactic behavior of BAC1.2F5 cells. The circular histogram plotted

[15] W. E. Allen, D. Zicha, A. J. Ridley, and G. E. Jones, J. Cell Biol. 141, 1147 (1998).

for cell trajectories of macrophages injected with an inactive protein (V12A35Rac 1) shows that 23 out of 30 microinjected cells migrated beyond a 10-μm horizon and that their migration was chemotactic, i.e., toward a source of CSF-1 (Fig. 5, control). In contrast, injection of either constitutively active (V12Rac1, data not shown) or dominant-negative (Fig. 5, N17Rac1) Rac inhibited cell migration. Only 9 out of 37 injected cells moved beyond a 10-μm horizon, and their trajectories were random with respect to the gradient of cytokine. This failure of migration was due to a failure to produce lamellipodia in the case of N17Rac1-injected cells and the continued radial spreading of cells and failure to polarize induced by V12Rac1. Thus we conclude that Rac regulates cell migration. Similarly, Rho is required for cell migration via its effects on cell contractility and Rho activity must be regulated precisely in order for cells to migrate effectively.[15]

The most significant finding in our study was that Cdc42 is required for BAC1.2F5 cell chemotaxis, but not for migration. Constitutively active (V12Cdc42) protein stimulates the production of filopodia around the cell margin, which correlates with a failure of the cells to polarize. Surprisingly, cells microinjected with the dominant inhibitory mutant of Cdc42 (N17Cdc42) showed a significant increase in migration speed when com-

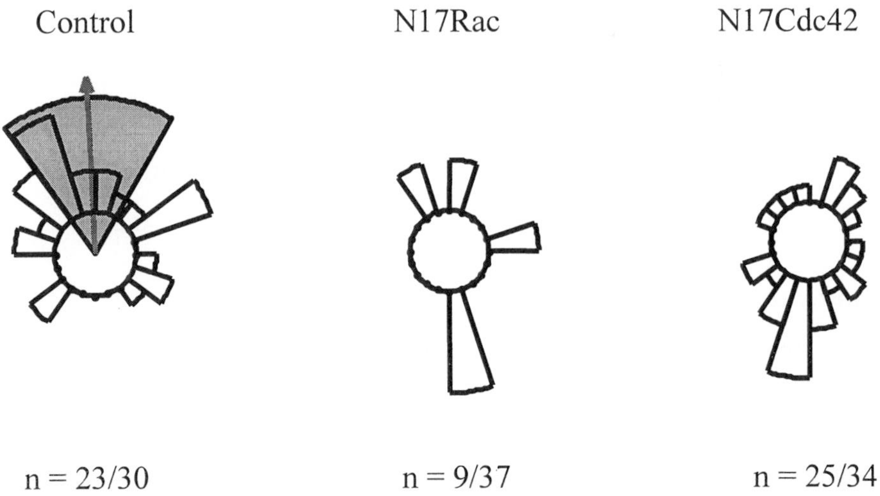

Control   N17Rac   N17Cdc42

n = 23/30   n = 9/37   n = 25/34

FIG. 5. Effects of Rac and Cdc42 on macrophage migration and chemotaxis. Horizon plots of cell migration of microinjected cells exposed to gradients of CSF-1 show that a control-injected population (V12A35Rac1) still migrates up a gradient of the cytokine. However, microinjection of the dominant-negative Rac1 (N17Rac) blocks cell movement. In contrast, microinjection of the dominant-negative Cdc42 (N17Cdc42) does not inhibit cell migration, but the motile population can no longer sense the gradient of CSF-1.

pared with control microinjected cells (33 versus 15 μm/hr, respectively).[15] We found that this increase was linked to the less polarized morphology of microinjected cells: the lamellae of N17Cdc42-injected cells showed great expansion with associated increase of ruffling activity compared to controls. However, despite their polarized shape and active migration, these macrophages did not sense the chemotactic gradient of CSF-1 (Fig. 5, N17Cdc42). Cell trajectories plotted on a circular histogram demonstrate that no unimodal clustering of cell paths could be discerned using the Rayleigh test. Thus, N17Cdc42-injected cells fail to polarize in response to a CSF-1 concentration gradient, but rather do so in a random, stochastic manner. Cdc42 is therefore not required for migration and actually exerts a restraint on the speed of locomotion, but it is essential for detection of the cytokine gradient.

Concluding Remarks

A combination of direct cell observation and rigorous statistical analysis makes the use of the Dunn chamber ideal in cases where small numbers of cells are examined for their migratory phenotype following experimental manipulation. Protein microinjection allows immediate, rapid analysis of the effects of specific signaling molecules on cell migration and has the advantage of avoiding the complications that can arise with transfection approaches, where long-term changes in cells induced by the expressed proteins cannot be excluded.

[41] Rho GTPases and Macrophage Phagocytosis

By Jayesh C. Patel, Alan Hall, and Emmanuelle Caron

Introduction

Phagocytosis is the conserved process whereby cells recognize and engulf large particles. It plays a crucial role in diverse functions ranging from feeding in *Dictyostelium* to removal of apoptotic cells and to host defense mechanisms in higher eukaryotes. In mammals, phagocytosis is primarily undertaken by specialized cells of the hematopoietic lineage, such as macrophages and neutrophils. These "professional" phagocytes bear multiple cell surface receptors, which orchestrate specific signal transduction mechanisms according to their target ligands.[1,2] Despite the variety of phagocytic

[1] K. Kwiatkowska and A. Sobota, *BioEssays* **21**, 422 (1999).
[2] A. Aderem and D. M. Underhill, *Annu. Rev. Immunol.* **17**, 593 (1999).

receptors, they all drive engulfment via the reorganization of the underlying cellular actin cytoskeleton.[3] This dependence on actin rearrangements strongly implicates the Rho family of small GTPases as promising candidates for regulating phagocytosis.[4,5] Indeed, accumulating evidence has firmly established that the three family members, Rho, Rac, and Cdc42, play crucial roles in uptake during different phagocytic mechanisms.[6,7]

Signal Transduction during Phagocytosis

Phagocytic signaling is best characterized for the Fc γ receptor (FcγR) and the integrin-based complement receptor CR3 (CD11b/CD18). These opsonic receptors preclude self-non-self-discrimination by binding targets coated with the immunoglobulin IgG and the complement component C3bi, respectively. There is compelling pharmacological and morphological evidence that distinct signaling mechanisms are used downstream of each receptor.[8,9] In particular, bacterial toxins have served as invaluable tools for dissecting the role of Rho GTPases within these signaling cascades. For example, toxin B from *Clostridium difficile,* which is a potent inhibitor of Rho, Rac, and Cdc42,[10] abolishes engulfment through both the FcγR and the CR3 receptor. In contrast, C3 transferase, under conditions where it specifically inactivates Rho,[11,12] prevents uptake through the CR3 receptor but not the FcγR.[7]

Such pharmacological techniques have long been popular. However, further analysis of signal transduction mechanisms requires specific inhibition of individual signaling components. This has traditionally entailed transfection of cells with dominant-negative (dn) or kinase dead constructs. However, most phagocytic cells are notoriously difficult to transfect and so new approaches, such as microinjection, are becoming more common. We have used microinjection of dnRho GTPases to demonstrate the distinct roles of these molecular switches in different phagocytic processes.[7] It is a

[3] S. Greenberg, K. Burridge, and S. C. Silverstein, *J. Exp. Med.* **172,** 1853 (1990).
[4] C. Nobes and A. Hall, *Cell* **81,** 53 (1995).
[5] A. Hall, *Science* **279,** 509 (1998).
[6] D. Cox, P. Chang, Q. Zhang, P. G. Reddy, G. M. Bokoch, and S. Greenberg, *J. Exp. Med.* **186,** 1487 (1997).
[7] E. Caron and A. Hall, *Science* **282,** 1717 (1998).
[8] S. L. Newman, L. K. Mikus, and M. A. Tucci, *J. Immunol.* **146,** 967 (1991).
[9] L. A. Allen and A. Aderem, *J. Exp. Med.* **184,** 627 (1996).
[10] I. Just, J. Selzer, M. Wilm, C. von Eichel Streiber, M. Mann, and K. Aktories, *Nature* **375,** 500 (1995).
[11] H. F. Paterson, A. J. Self, M. D. Garrett, I. Just, K. Aktories, and A. Hall, *J. Cell Biol.* **111,** 1001 (1990).
[12] A. J. Ridley and A. Hall, *Cell* **70,** 401 (1992).

very powerful technique to address the impact of known signaling molecules on different cellular processes.

General Methods for Study of Phagocytosis

Phagocytosis is a dynamic process that is influenced by a variety of environmental factors, such as time, temperature, pH, and osmolarity.[13] These conditions must be fixed as appropriate for particular investigations.

Choice of Targets

The phagocytic potential of cells may be assessed using a variety of different targets. However, size constraints dictate that particles should be >1 μm in diameter to ensure that uptake does not occur through pinocytosis.[14] Typical phagocytic targets include erythrocytes, latex beads, zymosan particles, live or dead microorganisms, and apoptotic cells. However, the majority of these can bind multiple phagocytic receptors, thereby complicating studies. To minimize this, targets can be coated with antibodies or ligands to direct their uptake to specific receptors. Fresh erythrocytes are particularly popular targets given their inability to bind phagocytes per se without prior opsonization.

Another criterion governing target choice is the technique used to identify internalized particles. For example, erythrocytes and some bacteria are susceptible to hypotonic lysis or enzymatic digestion (e.g., lysostaphin to remove extracellular *Staphylococcus aureus*.) These characteristics aid discrimination of targets that are bound to the phagocyte from those that are internalized.

Assessing Phagocytosis

A variety of techniques have been developed to determine the capacity of cells to phagocytose. The majority of these rely on the direct measurements of particle internalization rather than assessing associated phagocytic responses, such as the oxidative burst. In general, targets are fed to a monolayer or suspension of phagocytes for a specified time period under controlled conditions. Several experimental parameters, such as incubation time and target to phagocyte ratio, must be optimized prior to quantification.

---

[13] D. R. Absolom, *Methods Enzymol.* **132**, 95 (1986).
[14] M. Koval, K. Preiter, C. Adles, P. D. Stahl, and T. H. Steinberg, *Exp. Cell. Res.* **242**, 265 (1998).

The extent of phagocytosis is commonly expressed in two different ways: percentage phagocytosis ($\%P\phi$) or phagocytic index (PI). The former describes the ability of phagocytes to internalize ≥ 1 target and thus serves to indicate the "consistency" of the phagocytic response. In contrast the PI defines the number of targets engulfed by 100 phagocytes and so provides a general estimate of the phagocytic ability. Another parameter often quoted is the binding index (BI), which describes the number of targets bound by a population of phagocytes. It is an important criterion for assessing whether experimental conditions are adversly effecting receptor expression or the proximal steps of receptor clustering.

$$\frac{\% \text{ phagocytosis}}{(\%P\phi)} = \frac{\text{total phagocytes with} \geq 1 \text{ target inside}}{\text{total phagocytes}}$$

$$\frac{\text{Phagocytic index}}{(PI)} = \text{total targets internalized by 100 phagocytes}$$

$$\frac{\text{Binding index}}{(BI)} = \text{total targets bound by 100 phagocytes}$$

A variety of different protocols, most with considerable overlaps, have been developed to assess phagocytosis. Examples of the more common techniques are described later. In general, there is no consensus for one universally good method. Individuals should try to stick to one and cross-check its validity with at least one other method (Fig. 1a).

Microscopy Techniques. Analysis of target internalization by light or fluorescent microscopy is generally based on the manual counting of phagocytic events. Like most methods, this requires the ability to discriminate between internalized and surface-bound particles, but this is inherently difficult given the subtle changes that accompany target engulfment.

Hence, as mentioned previously, removal of attached particles by lysis or enzymatic digestion is an attractive option. However, conditions must be optimized to prevent loss of phagocytes and thus underestimation of the phagocytic capacity. For example, the plasma membrane of macrophages becomes more prone to hypotonic lysis on chelation of cytosolic Ca^{2+}.[15] Conversely, failure to eliminate all bound targets will increase the apparent phagocytic capacity. Moreover, an obvious limitation of this method is the inability to determine whether the experimental conditions alter the binding index adversely.

To address this problem, it is often possible to modify bound targets preferentially given their exposed nature. For example, surface-bound erythrocytes adopt a "small crenated" morphology when cells are fixed

[15] E. Picello, P. Pizzo, and F. Di Virgilio, *J. Biol. Chem.* **265,** 5635 (1990).

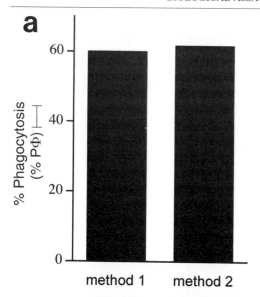

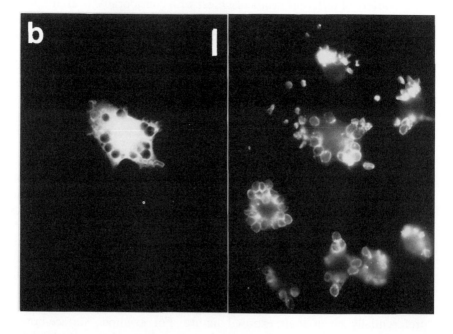

with formaldehyde (but not glutaraldehyde) at 4°.[16] In contrast, internalized erythocytes are protected within sealed phagosomes and so retain their spherical and swollen appearance. A variation on this theme is to employ fluorescently conjugated targets, such that those that are extracellular are amenable to quenching with a membrane-impermeable reagent.[17] Another approach is to fluorescently stain targets pre- and postphagocyte permeabilization with different fluorochromes. This results in double labeling of bound targets and single labeling of internalized particles.

Alternatively, the phagocytic cell may be stained with a cytosolic marker to negatively "highlight" the internalized particles within phagocytic vacuoles. The fluorescent dye, Acridine Orange, which stains the RNA-rich cytosol of macrophages, has been employed to good effect.[16] In this case, ingested targets appear as nonfluorescent "holes" in the cytoplasm. This technique is particularly reliable when coupled to the preferential modification of bound targets (see "Phagocytosis Assay").

Finally, although not routinely applicable, internalized particles can be assigned with little ambiguity following electron microscopy (EM). In this case, serial sectioning may reveal subtle differences between different phagocytic mechanisms. For example, EM data have shown that FcγR-mediated uptake occurs via the extension of pseudopods, whereas CR3 engulfment appears more passive, with particles sinking into cells.[9]

Target-Labeling Techniques. These methods generally avoid the time and tedium associated with manually counting individual phagocytic events. Instead they rely on feeding labeled targets and subsequently quantifying the extent of labeling for a population of phagocytes. Radioisotopes are frequently utilized to label inert or living particles.

New techniques exploiting flow cytometry technology have been developed. In particular, the use of labeled zymosan particles coupled with the digestion of bound particles has proved successful.[18] An alternative

[16] S. Greenberg, J. El Khoury, E. Kaplan, and S. C. Silverstein, *J. Immunol. Methods* **139**, 115 (1991).

[17] J. Hed, *Methods Enzymol.* **132**, 198 (1986).

[18] D. M. Underhill, J. Chen, L. A. Allen, and A. Aderem, *J. Biol. Chem.* **273**, 33619 (1998).

FIG. 1. (a) A comparison of two different techniques for assessing phagocytosis. Method 1 is as described in the phagocytosis assay. Method 2 involves labeling erythrocytes pre- and postphagocyte permeabilization with two different fluorescent markers. (b) Phagocytosis of IgG opsonized erythrocytes by J774.A1 macrophages. (Left) A cell microinjected with fluorescent dextran. Note that microinjection of cells per se does not perturb their ability to bind or internalize targets. (Right) Labeling of erythrocytes. Attached particles appear small and crenated, whereas those inside appear swollen and spherical, coincident within phagocytic vacuoles. Bar: 10 μm.

approach is to use fluorescein isothiocyanate (FITC)-labeled targets, which when extracellular can be preferentially "relabeled" to utilize double fluorescence activated cell sorting (FACS) assessment.[19] However, these techniques suffer from similar limitations to some of the microscopic methods, namely erroneous inclusion of bound targets as internalized. In addition the failure to visually observe the process denies one the ability to ensure the health of the phagocyte population under varying experimental conditions.

Target Viability Techniques. Some phagocytic assays use live microorganisms as targets. This may be exploited by determining the phagocytic index based on target viability postengulfment.[20] This is best illustrated when using bacterial targets. Following feeding, bacteria attached to the surface of the phagocyte can be destroyed using a membrane-impermeable antibiotic, such as gentamicin. The proportion of bacteria engulfed is then quantitated by lysing the phagocytes and measuring bacterial survival titers on solid media. However, this method requires that internalized bacteria are not killed within the experimental time frame. Moreover, adapting this method to the study of signal transduction during phagocytosis requires that the pathways studied do not interfere with bacterial survival within cells.

Role of Rho GTPases in Macrophage Phagocytosis

The assay described here involves feeding erythrocyte targets to a monolayer of macrophages. It details the use of microinjection to investigate the role of Rho GTPases in CR3- and FcγR-mediated phagocytosis. This single-cell assay benefits from its requirement for only two fluorescent channels, to label erythrocytes and injected constructs, thereby permitting one to stain for other proteins of interest.

Preparation of Ligand-Coated Erythrocytes

Materials

Sheep erythrocytes: Purchased from ICN (Costa Mesa, CA). Provided as a 10% suspension of erythrocytes in borate buffer, sodium acetate, and 0.1% thimerosal. Erythrocytes should not be used more than 1 month beyond their expiration date.

GVB^{2+} buffer (Sigma, St. Louis, MO): a gelatin-based buffer used for handling and preserving erythrocytes.

IgG anti-sheep erythrocytes (ICN): raised in rabbits. Used to opsonize erythrocytes for targeting to FcγR.

IgM anti-sheep E (ICN): raised in rabbits. Used to opsonize erythrocytes for fixation of C3b.

[19] E. C. de Boer, R. F. Bevers, K. H. Kurth, and D. H. Schamhart, *Cytometry* **25**, 381 (1996).
[20] P. Tang, V. Foubister, M. G. Pucciarelli, and B. B. Finlay, *J. Micro. Methods* **18**, 227 (1993).

C5-deficient serum (Sigma): used to fix C3b onto IgM-coated erythrocytes.

Procedure. To promote optimal phagocytosis, erythrocytes must be coated uniformly with a subagglutinating concentration of opsonin. Determination of this concentration is achieved easily by titrating the immunoglobulins to find the maximal concentration that does not induce visible agglutination of erythrocytes following incubation at 37°.

Preparation of IgG Erythrocytes

Approximately 0.5 μl fresh erythrocytes are opsonized for each coverslip of cells. Wash erythrocytes in 2 × 100 μl GVB^{2+} by centrifugation at 4000 rpm for 4 min in a microfuge. Resuspend the cells in 1 ml GVB^{2+} containing a subagglutinating concentration of rabbit anti-sheep erythrocytes IgG and rotate at room temperature for 20 min. Uniform opsonization can be verified by immunofluorescence, using conjugated anti-rabbit IgG. Finally, wash erythrocytes in 100 μl GVB^{2+} and resuspend in 10 mM HEPES-buffered Dulbecco's modified Eagle's medium (DMEM).

Preparation of C3bi Erythrocytes

As for IgG erythrocytes, except that cells are coated with IgM instead of IgG. These erythrocytes are washed 1 × 100 μl GVB^{2+} and further opsonized by incubation in 10% (v/v) C5-deficient serum for 20 min at 37°. Under these conditions, C3b is deposited rapidly onto the IgM-coated erythrocytes and is subsequently completely converted to C3bi,[21] the ligand for the CR3 receptor. Once again opsonization can be checked by immunofluorescent labeling with goat anti-C3 (Sigma) followed by conjugated anti-goat IgG.

Microinjection of dnGTPases

Materials

J774.A1 cells: a murine macrophage cell line (available at ATCC, Rockville, MD) used as phagocytes for the assay. The cells are cultured in DMEM (GIBCO, Grand Island, NY) supplemented with 10% heat-inactivated fetal calf serum (FCS) (Sigma), 5% (v/v) Pen/Strep (penicillin/streptomycin) at 37°, 10% CO_2.

myc-dnRho GTPases: constructs are cloned into the eukaryotic expression vector pRK5, with an N-terminal myc tag and prepared by CsCl purification.

Serum-free DMEM: 1.36% (w/v) DMEM (GIBCO), 100 mM pyruvate

[21] S. L. Newman and L. K. Mikus, *J. Exp. Med.* **161**, 1414 (1985).

(GIBCO) and 5% Pen/Strep (GIBCO) in doubly distilled H_2O. Filter sterilize.

Microinjector/manipulator (Eppendorf): see Nobes.[22]

Micropipette puller (Sutter Instruments, Novato, CA): see Nobes.[22]

Glass capillaries (Clark Electromedical Instruments, Reading, England): see Nobes.[22]

Procedure. Seed J774.A1 macrophages overnight onto acid-washed glass coverslips (13 mm diameter) in four-well plates (Nunc) at a density of 10^5/ml/well. Immediately prior to injection, transfer cells to 10 mM HEPES-buffered serum-free DMEM. This promotes a transient spreading of macrophages, thus allowing one to distinguish the nucleus more easily.

Microinjection is performed using pipettes pulled manually from borosilicate capillaries. It is worthwhile spending time and effort to optimize conditions for pulling micropipettes, as they are often the key to successful microinjections (see Nobes[22]). In general, capillaries should be made to have a tip diameter of <1 μm.

Prepare fresh solutions in PBS-A for microinjection by mixing CsCl-prepared myc-tagged cDNA constructs (0.1 mg/ml) with biotin dextran (2.5 mg/ml, molecular weight 10,000 lysine fixable, Molecular Probes, Eugene, OR) to mark injected cells. Microinject the nucleus of ~80 cells in a temperature (37°)- and CO_2 (10%)-controlled chamber using phase-contrast microscopy. Attempt to minimize the time cells are exposed to bright microscope light. Cells should be left in HEPES-buffered serum-free DMEM and returned to the incubator for ~3 hr for optimal expression.

Phagocytosis Assay

Procedure. Opsonized erythrocytes, resuspended in prewarmed HEPES-buffered DMEM, are added to microinjected cells at a ratio of 10:1, in a final volume of 500 μl/well. For phagocytosis of C3bi-opsonized erythrocytes, macrophages require preactivation[23] for 15 min at 37° with PMA (150 ng/ml) in buffered serum-free DMEM. Allow targets to interact for 20 min at 37°. This should be sufficient time for erythrocytes to sediment and initiate phagocytosis by macrophages. Engulfment is terminated by rinsing cells with serum-free DMEM and fixing with freshly prepared cold 4% PFA/PBS-A for 20 min at 4°. This fixation of cells with cold PFA at 4° is crucial for inducing morphological differences between bound and internalized erythrocytes.[16]

For immunofluorescent staining, permeabilize cells by incubation with 0.2% Triton X-100/PBS-A for 5 min at room temperature. Treat cells with

---

[22] C. D. Nobes, *Methods Enzymol.* **325** [39] (2000) (this volume).

[23] S. D. Wright and M. T. C. Jong, *J. Exp. Med.* **164,** 1876 (1986).

ammonium chloride (2 mg/ml) in PBS-A for 10 min to reduce free aldehyde groups formed during fixation. Wash coverslips 3×1 ml PBS-A and incubate cells in 0.5% BSA in PBS-A for 20 min to block nonspecific binding of antibodies. Finally, wash 3×1 ml PBS-A and process cells for immunofluorescence.

Cells can be triple stained to identify both erythrocytes and injected constructs, along with the injection marker. We routinely employ an AMCA-S-streptavidin conjugate (Molecular Probes) to visualize microinjected cells, 9E10 mAb followed by FITC-conjugated F(ab')$_2$ anti-mouse IgG (Jackson, West Grove, PA) to detect expression of myc-tagged constructs, and finally rhodamine-conjugated donkey anti-rabbit IgG (Jackson) to stain IgG as well as C3bi opsonised erythrocytes. When necessary, staining for the injection marker can be omitted, thereby freeing one fluorescent channel to study other markers, such as tyrosine-phosphorylated proteins or F-actin. To block nonspecific binding of antibodies to the FcγR on macrophages, antibody incubations should be performed in the presence of excess IgG (human or mouse, Sigma). All antibody incubations are carried out at room temperature, in the dark, for 45 min by inverting coverslips onto 100-μl antibody solutions. Rinse coverslips in between each staining step by sequentially dipping $3\times$ into PBS-A and each time draining excess PBS-A.

After the final PBS-A wash, rinse cells further by dipping coverslips $2\times$ into doubly distilled H$_2$O and draining. Mount coverslips by inverting them onto 5 μl Mowiol 4-88 (Calbiochem, La Jolla, CA) mounting agent containing 1 mg/ml p-phenylenediamine (Sigma) as an antibleach agent.

Examine the cells by immunofluorescence microscopy and quantitate phagocytosis using two principal criteria: (1) morphology of erythrocytes and (2) negative highlighting of phagocytic vacuoles. As mentioned previously, fixation conditions promote extracellular erythrocytes to adopt a "small and crenated" morphology, whereas internalized targets appear spherical and swollen (Fig. 1b). Moreover, internalized erythrocytes are within phagocytic vacuoles and thus appear as negatively stained "holes" in the cytosol following staining of injection marker and constructs.

Based on this classification, count the total number of erythrocytes bound and internalized per microinjected (AMCA-S positive) myc expressing (FITC-labeled) cell. From these, calculate the binding index (BI) and percentage phagocytosis ($P\phi$) for each construct during CR3- and FcγR-mediated phagocytosis. Results should be expressed as the average $\pm$ SEM of at least three independent experiments in which at least 40 expressing cells were retrieved.

We have used this assay to investigate the role of Rho, Rac, and Cdc42 during phagocytosis through the CR3 and FcγR. The function of endogenous Rho GTPases was individually blocked by injecting dominant-negative constructs (N17Cdc42, N17Rac) or C3 transferase (to block Rho). Our results indicate that uptake through the two receptors occurs via two distinct mechanisms. Type I, for the FcγR, requires Cdc42 and Rac function, whereas type II, for the CR3 receptor, requires only Rho function (Fig 2).[7]

Limitations

One particular limitation with the phagocytic assay described is the lack of synchronization between individual phagocytic events. This arises due to the uneven nature in which particles sediment onto the macrophages, prior to initiating engulfment. Consequently, it is difficult to investigate proximal and transient signaling events, such as tyrosine phosphorylation. However, the assay may be modified easily to reduce this problem and promote synchronous phagocytosis. One way is to add erythrocytes to macrophages at 4°. This permits binding of targets but not their internalization. Once adhered, the phagocytes can be warmed to 37° to induce uptake. However, this method is not applicable to all studies. In particular, the binding of C3bi erythrocytes to the CR3 receptor is reduced dramatically

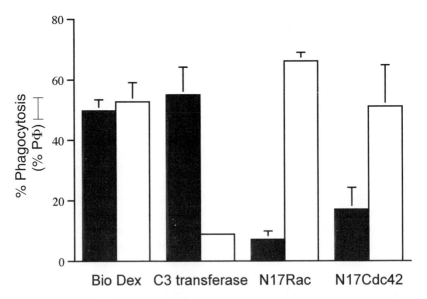

Fig. 2. Phagocytosis through the FcγR and the CR3 receptor occurs via two distinct mechanisms in J774.A1 macrophages. Macrophages were microinjected with C3 transferase, N17Rac, or N17Cdc42 to block endogenous Rho GTPase function and with Biotin Dextran (Bio Dex) as a control. Cells were challenged with IgG-E or C3bi-E to study the effect on FcγR (black bars) and CR3 receptor (open bars)-mediated phagocytosis.

at lower temperatures.[24,25] An alternative approach is therefore to promote uniform sedimentation of targets onto phagocytes by centrifugation.

The phagocytic assay is dependent on the availability of epitope-tagged, reliable dn constructs and the ability to coat targets with the appropriate opsonins. However, in principle, the same microinjection-based, single-cell approach can be taken to investigate many other signaling pathways relevant to phagocytosis or macrophage biology.

Acknowledgment

Supported by the Wellcome Trust. J. Patel is part of the MRC graduate program.

[24] J. A. G. Van Strijp, D. G. Russell, E. Tuomanen, E. J. Brown, and S. D. Wright, *J. Immunol.* **151,** 3324 (1993).
[25] S. D. Wright and S. C. Silverstein, *J. Exp. Med.* **156,** 1149 (1982).

[42] Rho GTPases and Axonal Growth Cone Collapse

By Alyson E. Fournier, Robert G. Kalb, and
Stephen M. Strittmatter

Introduction

The axonal growth cone is the specialized distal tip of the extending axon and consists of both lamellipodial and filopodial structures. Control of growth cone dynamics is critical for neurite outgrowth and guidance. These processes play key roles during development when neurites must navigate a complex array of guidance cues and during regeneration when the growth cone encounters myelin-associated inhibitory molecules that cause growth cone collapse.[1] The structure of the growth cone is supported by a microtubule- and actin-based cytoskeletal network, and the dynamics of this network control growth cone morphology.[2] For example, growth cone collapse is accompanied by a loss of F-actin in the peripheral region of the growth cone,[3] and localized treatment of the growth cone with factors that normally cause collapse can steer growth cones by inducing partial

[1] M. Li, A. Shibata, C. Li, P. E. Braun, L. McKerracher, J. Roder, S. B. Kater, and S. David, *J. Neurosci. Res.* **46,** 404 (1996).
[2] P. C. Letourneau, *Persp. Dev. Neurobiol.* **4,** 111 (1996).
[3] J. Fan, S. G. Mansfield, T. Redmond, P. R. Gordon-Weeks, and J. A. Raper, *J. Cell Biol.* **121,** 867 (1993).

asymmetric collapse.[4] In nonneuronal cells, the dynamics of structures such as lamellipodia, filopodia, and focal adhesions are under the control of Rho family GTPases.[5] Similarly, the growth cone consists of lamellipodia, filopodia, and point contacts. An understanding of the role of Rho GTPases in the control of growth cone dynamics will strengthen our understanding of growth cone guidance and collapse.[6] Previous studies have shown that modulating the activity state of Rho, Rac, and Cdc42 in neurons can modulate the growth cone response to collapsing agents.[7–9]

This article describes two methods to introduce Rho family GTPases into cultured neurons. These methods are designed to facilitate an analysis of the role of such GTPases in growth cone collapse by axonal guidance factors. One protocol incorporates the proteins by trituration and the other by infection with herpes simplex virus (HSV) vectors. Further, we describe a protocol to measure levels of activated Rac and Cdc42 in cultured neurons following the introduction of growth cone collapsing agents.

Preparation of GTPases

Monomeric human GTPases and *Clostridium botulinum* C3 transferase are produced in *Escherichia coli* as glutathione *S*-transferase (GST) fusion proteins as described previously.[10] To examine the effects of these proteins, wild-type (RhoA, Rac1, and Cdc42), constitutively active (V12Rac, V14RhoA, and V12Cdc42), and dominant-negative (N17Rac and N17Cdc42) forms of the GTPases are used. To study the affects of inactivating Rho, the C3 exoenzyme from *C. botulinum* is used to ADP ribosylate and inactivate Rho.[11,12] After purification on glutathione-Sepharose beads, the proteins of interest are cleaved from the GST moiety with thrombin, and thrombin is removed from the samples by absorption to *p*-aminobenzamidine-agarose. For trituration experiments, proteins are prepared by dialyzing overnight against trituration buffer. Proteins are concentrated to 5 mg protein/ml for Rho subfamily proteins and 0.1 mg protein/ml for C3

[4] J. Fan and J. A. Raper, *Neuron* **14,** 263 (1995).
[5] A. Hall, *Annu. Rev. Cell Biol.* **10,** 31 (1996).
[6] L. Luo, L. Y. Jan, and Y.-N. Jan, *Curr. Opin. Neurobiol.* **7,** 81 (1997).
[7] Z. Jin and S. M. Strittmatter, *J. Neurosci.* **17,** 6256 (1997).
[8] T. B. Kuhn, M. D. Brown, C. L. Wilcox, J. A. Raper, and J. R. Bamburg, *J. Neurosci.* **19,** 1965 (1999).
[9] M. Lehmann, A. Fournier, I. Selles-Navarro, P. Dergham, A. Sebok, N. Leclerc, G. Tigyi, and L. McKerracher, *J. Neurosci.* **19,** 7537 (1999).
[10] C. D. Nobes and A. Hall, *Cell* **81,** 53 (1995).
[11] S. T. Dillon and L. A. Feig, *Methods Enzymol.* **256,** 174 (1995).
[12] A. Hall, *Science* **249,** 635 (1990).

transferase using a Centricon concentrator (Amicon, Beverly, MA) and are stored in aliquots at $-80°$.

Trituration buffer: 25 mM Tris–HCl, pH 7.5, 150 mM NaCl, 5 mM MgCl$_2$, and 1 mM dithiothreitol (DTT)

Preparation of HSV GTPases

Generating Recombinant Viral Stock

A second strategy to introduce Rho family GTPases into primary neurons is to use replication-deficient HSV-1 viral vectors.[13,14] Viral preparation is based on inserting the gene of interest into an amplicon plasmid (pHSV-PrpUC) containing an HSV-1 origin of replication and an HSV-1 packaging site. This plasmid is transfected into a packaging cell line stably expressing the essential immediate early 2 (IE2) HSV gene. Cells are then infected with a defective HSV-1 helper virus lacking the IE2 gene to generate a population of HSV-1 particles with the plasmid packaged in the HSV-1 capsid. The resulting particles can be amplified to high titer with a minimal decrease in the amplicon-to-helper virus ratio.

The amplicon plasmid containing the gene of interest is transfected into the HSV packaging cell line 2-2 by Lipofectamine (Life Technologies, Grand Island, NY) -mediated transfection. Cells are left to recover at least 20 hr after transfection. Medium is removed from the plates and 5 ml Dulbecco's modified eagle's medium (DMEM) + 5% fetal bovine serum (FBS) is added. The culture is then infected with 30 μl of the replication-deficient HSV IE2 deletion mutant 5d/1.2 helper virus [5 × 10^7 plaque-forming units (pfu)/ml] and incubated overnight at 37°. One day after infection, cells should round up but remain adherent to the plate. The cells are then harvested by osmotic lysis. The medium is transferred to a 15-ml tube. Add 1 ml of 0.1× PBS to the cells on the plate and incubate for 2 min. Collect the cells with a Pasteur pipette and pass them through the pipette five times to burst the cells. Transfer the cells to a 15-ml tube with the medium. Add 100 μl of 10× PBS to restore ionic strength and centrifuge the cells for 10 min to pellet cell debris but not the virus particles. Transfer the supernatant to a new tube labeled P$_0$ and store at $-80°$. Determine the viral titer on PC12 cells and continue with anything above 10^4 pfu/ml. Methods for titering HSV have been described[13,14] and are facilitated greatly

---

[13] D. J. Fink, N. A. DeLuca, W. F. Goins, and J. C. Glorioso, *Annu. Rev. Neurosci.* **19**, 265 (1996).

[14] A. I. Geller, K. Keyomarsi, J. Bryan, and A. B. Pardee, *Proc. Natl. Acad. Sci. U.S.A.* **87**, 8950 (1990).

if the gene product of interest can be detected directly by immunohisto-logic methods.

Amplification of Viral Stock

Recombinant viral stocks are amplified further by three sequential rounds of infection.

P_1: 2-2 cells are plated at 4×10^5 cells per 6-cm dish in 5 ml DMEM + 10% FBS. Two days later the medium is replaced with 4 ml DMEM + 5% FBS, and 4 ml of P_0 supernatant is added. Following a 24-hr incubation, check the cells to see that they have rounded up and harvest by osmotic lysis.

P_2: 2-2 cells are plated at 1×10^6 cells per 10-cm dish (two dishes per sample) in 10 ml DMEM + 10% FBS. Two days later the medium is replaced with 6 ml DMEM + 5% FBS, and 4 ml of P_1 supernatant is added. Following a 24-hr incubation, check the cells to see that they have rounded up and harvest by osmotic lysis into 2 ml of 0.1× PBS per plate.

P_3: 2-2 cells are plated at 1×10^6 cells per 10-cm dish (four dishes per sample) in 10 ml DMEM + 10% FBS. Two days later the medium is replaced with 6 ml DMEM + 5% FBS, and 4 ml of P_2 supernatant is added. Following a 24-hr incubation, check the cells to see that they have rounded up and harvest by sonication. Cells are scraped into the media they are grown in and freeze/thawed three times using a dry ice/ethanol bath and a 37° water bath, being careful to minimize the time at 37°. Sonicate the cells to release the virus and pass through a 0.2-μm filter. Store the virus-containing supernatant in aliquots at −80°. Viral stocks are added to neu-ronal cultures at a concentration of approximately 10^6 pfu/ml.

Tissue Culture of Sensory, Sympathetic, and Retinal Ganglion Cell Neurons

This section describes the preparation of both dissociated and explant cultures of sensory neurons prepared from dorsal root ganglia (DRG) and sympathetic neurons prepared from sympathetic ganglia. We also describe the culture of retinal explants that provides for the study of retinal ganglion cell growth cones. It is possible to culture purified dissociated retinal gan-glion cells, but this requires immunopanning and is described in other studies.[15] E7–E8 embryonic chicks were used as the source for sensory, sympathetic, and retinal ganglion cell cultures. Both dissociated and explant cultures are prepared in four-well glass chamber slides (Fisher, Pitts-burgh, PA).

[15] A. Meyer-Franke, M. R. Kaplan, F. W. Pfrieger, and B. A. Barres, *Neuron* **15**, 805 (1995).

Precoating of Glass Chamber Slides

Chamber slide wells are coated sequentially with poly (L-lysine) (PLL) and laminin as follows: Incubate 0.5 ml of PLL (100 μg/ml in H_2O) in each well for 1 hr at room temperature. Wash the well three times with PBS. Coat the well with laminin (10 μg/ml in PBS) for 1 hr (sensory and sympathetic neurons) or overnight (retinal neurons) at 37°. Aspirate solution and add 0.5 ml of growth media per well.

Growth media: F12 (Ham) nutrient mixture, 10% fetal bovine serum (FBS), 2 mM L-glutamine, 100 U/ml penicillin, 100 μg/ml streptomycin, and 50 ng/ml 7 S-nerve growth factor (NGF)

Culture Conditions and Trituration Method

DRGs are dissected from E7–E8 chick embryos into a petri dish with PBS. The outer sheath and roots are removed from the DRGs, and 2-3 DRG explants are plated per chamber. For sympathetic explants, the entire chain of sympathetic ganglia is dissected from E8 chick embryos and placed in PBS. Using spring scissors, individual sympathetic ganglia are cut from the chain of ganglia, and 2-3 explants are plated per well. To avoid crushing the sympathetic explants, they should be transferred to the chamber slide with a Gilson P200 pipetter. To prepare retinal explants, individual retinae are dissected from the chick eye and placed in growth media. Using spring scissors, individual explants are cut from the central region of the retina. Strips can be cut through the central region of the retina and individual square explants can be cut from these strips. Three to four retinal explants are transferred to each well using a Gilson P1000 pipetter and are plated with the retinal ganglion cell layer in contact with the laminin substrate. Because extensive curling of the explants will impede their adherence, explants should be cut and transferred as quickly as possible, and explants should not be cut from extensively curled retinae.

For dissociated sympathetic and sensory neuron cultures, dissect DRGs or sympathetic ganglia chains, respectively, and transfer to a 1.5-ml microcentrifuge tube with PBS on ice. Spin in a bench top microcentrifuge at 3000 rpm for 1 min at 23° to pellet the tissue. Remove the PBS and add 300 μl of 0.05% trypsin and 0.53 mM EDTA (Life Technologies, Grand Island, NY) for 15 min at 37°, inverting the tube periodically. Add 1 ml growth media to the tube to inactivate the trypsin, and pellet the cells. Resuspend the pellet in 1 ml of growth media and dissociate the cells by passing them through a fire-polished Pasteur pipette approximately 10 times. Bring the cells up to a total volume of 5 ml with growth media and plate the cells twice for 30 min in 60-mm uncoated tissue culture plates. During these preplating steps, nonneuronal cells will adhere to the plate, and neuronal cells will be collected with the media. Collect the media and calculate cell concentration using a hemocytometer. Pellet the cells in a bench top microcentrifuge such that the desired number of cells for each treatment is aliquoted in individual tubes. For trituration experiments, neurons are then suspended in trituration buffer with Rho subfamily proteins at 5 mg protein/ml or with C3 transferase at 0.1 mg/ml. This high concentration of Rho subfamily proteins is critical for intracellular protein concentrations to reach levels capable of exerting biological effects. The cell suspension is then passed 50 times through a Gilson P200 pipette tip at a rate of one pass per second. The pipette tip should be kept below the surface of the cell suspension to prevent frothing and the introduction of air bubbles. Following trituration, neurons are plated in 25 volumes of growth media on PLL/laminin-coated slides. It is not necessary to remove the recombinant protein by pelleting and resuspending the cells. Plate approximately 25,000 cells per well.

Neurite Outgrowth and Growth Cone Collapse Assay

Trituration

For growth cone collapse assays following trituration, neurons are cultured for 4 hr until neurites have extended at least two cell body lengths and then test reagents are added directly to the media for 20–30 min. It is important to preequilibrate test reagents in the incubator before their addition to the culture. Growth cones are highly sensitive to changes in temperature and CO_2 concentration, and changes in these parameters may result in nonspecific growth cone collapse. Moreover, test reagents should be dialyzed against F12 medium when possible or added in small volumes to minimize buffer effects on the growth cones. Similarly, fixation conditions are critical to preserve growth cone morphology. Cultures are fixed by gently layering 750 μl of ice cold fixative directly into the well. The addition of sucrose to the fixative increases the density of the solution and allows it to displace the growth media to the top of the well.

Fixative: 4% paraformaldehyde (PFA), pH 7.2, 0.1 M PO_4, and 20% sucrose

After 20 min of fixation, cultures are washed three times with PBS and stained or coverslipped. The growth cone collapse assay for explant cultures has been described in detail.[16]

To examine the growth cone at intermediate stages of collapse, a time course of collapse must be determined empirically for the reagent of inter-

[16] J. A. Raper and J. P. Kapfhammer, *Neuron* **4**, 21 (1990).

est. For example, because collapse in response to 1 nM Sema3A occurs in approximately 10 min, cultures are fixed by directly adding PFA to the well 0, 2, 5, and 10 min following the addition of Sema3A. In contrast, DRG collapse in response to 100 μg myelin protein/ml occurs over a slower time course of approximately 30 min. The collapse assay can be performed on dissociated cultures following 3–4 hr of outgrowth or on explants following 12–16 hr of outgrowth.

In addition to measuring growth cone collapse, the length of neurite outgrowth may be an important parameter when measuring the response of the growth cone to various agents. Whereas growth cone collapse is often accompanied by neurite retraction in response to inhibitory agents, growth cones may also exhibit a collapsed morphology during periods of rapid outgrowth, as is the case with C3 treatment of DRG neurons.[7] For neurite outgrowth assays, triturated cells are plated for 1.5–2 hr. Agents to be tested are then added directly to the medium. After an additional 2- to 3-hr incubation, cells are fixed and total neurite length per neuron is measured for 75–150 neurons.

HSV Infection

To infect neuronal cultures with HSV virus, neurons are plated for 4–6 hr to allow the neurons to adhere to the slide. Viral stocks are then preequilibrated in the incubator and added to neuronal cultures at a concentration of approximately 10^6 pfu/ml. Cultures are grown for 24 hr and growth cone collapse assay is performed as described previously. Growth cone infection can be monitored by tagging the gene of interest with various reporters, such as green fluorescent protein (GFP) or LacZ. In the case of viral infection of dissociated cultures, an appropriate cell density must be determined that permits neurite outgrowth but allows for the visualization of isolated growth cones. In the case of dissociated DRGs, approximately 25,000 cells per well are appropriate for four-well chamber slides.

Because viral infection of explants requires a lengthy growth period (approximately 24 hr postinfection), neurites have a propensity to lift off the slide following the fixation step. To prevent this, we have used 3-aminopropyltriethoxysilane (Sigma, St. Louis, MO) treatment following paraformaldehyde fixation. Five hundred microliters of 3-aminopropyltriethoxysilane is diluted in 15 ml acetone. Following 20 min of paraformaldehyde fixation, PFA is aspirated from the wells, and 500 μl of the diluted 3-aminopropyltriethoxysilane is added to the well for 5 sec. The solution is then aspirated from the well and the culture is washed three times with PBS.

Growth Cone Staining

To facilitate growth cone collapse counts, and to examine the distribution of Rho family GTPases and F-actin at various stages of collapse, growth cones may be stained with rhodamine–phalloidin or antibodies against members of the Rho family of GTPases (see Fig. 1). For immunocytochemical staining for Rho family GTPases, explants are blocked for 1 hr at room temperature in 250 μl of blocking solution [3% (w/v) bovine serum albumin (BSA), 0.1% (v/v) Triton X-100], washed three times with PBS, and incubated 3–18 hr with 250 μl primary antibody in blocking solution. The following antibodies are used to examine the Rho family of GTPases: anti-Rac1 monoclonal antibody (Upstate Biotechnology, Lake Placid, NY; 4 μg/ml), anti-Rho monoclonal antibody (Santa Cruz Biotechnology, Santa Cruz, CA; 1 μg/ml), and anti-Cdc42 polyclonal antibody (Santa Cruz Biotechnology; 0.4 μg/ml). Secondary antibody incubations are for 1 hr at room temperature in blocking solution. To examine F-actin in growth cones, cultures are stained with 250 μl of rhodamine–phallodin (Molecular Probes, Eugene, OR; 0.33 μm) in 3% BSA and 0.1% Triton X-100 for 1 hr at room temperature. For double-staining experiments, rhodamine–phalloidin is added with the secondary antibody.

Rac Activation Assay

This section describes an assay to measure Rac activation state in dissociated DRG neurons following treatment with growth cone collapsing factors. The assay is based on the ability of the serine/threonine kinase p65^{PAK} to

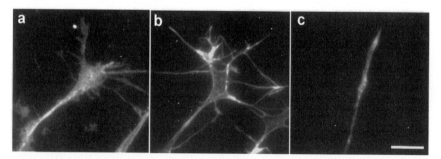

Fig. 1. Rac and F-actin localization in E7 chick DRG growth cones. Control (a and b), and Sema3A (c)-treated growth cones were stained with anti-Rac antibody (a) or phalloidin (b and c). In control growth cones, Rac staining is diffuse (a), whereas phalloidin-stained F-actin is concentrated in growth cone filopodia (b). Ten minutes following treatment with 1 nM Sema3A (c), the growth cone is completely collapsed. Bar: 20 μm.

bind Rac and Cdc42 (but not Rho) in a GTP-dependent manner.[17,18] The Cdc42/Rac interactive binding (CRIB) domain (amino acids 29–90 of Pak1) is expressed as a GST fusion protein in *Escherichia coli* and is utilized to isolate GTP-bound, activated Rac and Cdc42 by affinity chromatography. Following treatment of DRG neurons with the agent of interest, cell lysates are generated in the presence of high Mg^{2+} to prevent further guanine nucleotide exchange. The lysate is then incubated with GST–CRIB bound to Sepharose beads. Following several washes, proteins bound to the beads are eluted with SDS and are analyzed by SDS–PAGE and western blotting. Analysis with anti-Rac and anti-Cdc42 antibodies allows for a relative measure of activated Rac and Cdc42 in cell lysates.

Resin Generation

Inoculate 200 ml of LB medium containing 50 μg/ml ampicillin with 2 ml of a saturated culture of pGEX–CRIB-transformed *E. coli* and incubate for 12 hr at 23°. Induce expression of GST–CRIB by adding 1 mM IPTG and incubate for 3 hr at 23°. Pellet the bacteria and wash with 20 mM Na–HEPES, pH 7.4, 150 mM NaCl. Resuspend the pellet in 10 ml of glutathione-binding buffer [120 mM Na–HEPES, pH 7.4, 120 mM NaCl, 10% glycerol, 2 mM EDTA, and 1 mM phenylmethylsulfonyl fluoride (PMSF)]. Transfer to a fresh tube and sonicate extensively. Spin at 12,000 rpm for 30 min at 4°. Transfer the supernatant to a fresh tube and add Triton X-100 to a final concentration of 0.5%. Add 500 μl of glutathione bead slurry (i.e., 250 μl of beads) and mix by rotation for 1–2 hr at 4°. Wash five times with 500 μl of glutathione-binding buffer + 0.5% Triton X-100. Add 1 ml of glutathione-binding buffer to generate 1250 μl of resin (GST–CRIB resin). This resin can be stored at 4° and used at any time over the next 7 days.

Rac Activation Assay

Prepare dissociated DRGs as described earlier and plate 2×10^6 cells per 100-mm plate that has been sequentially precoated with PLL/laminin (see earlier discussion). After 24 hr of growth, cells are treated with the reagent of interest. Place the plate on ice and wash twice with ice-cold HBS. Add 250 μl of RIPA buffer with NaCl and $MgCl_2$. The addition of high salt is critical to dissociate large protein complexes and allow access to the GTPase. Inclusion of $MgCl_2$ stabilizes GTP binding. Wash with an

[17] P. D. Burbelo, D. Drechsel, and A. Hall, *J. Biol. Chem.* **270**, 29071 (1995).
[18] C. M. Waterman-Storer, R. A. Worthylake, B. P. Liu, K. Burridge, and E. D. Salmon, *Nature Cell Biol.* **1**, 45 (1999).

additional 250 μl of RIPA buffer with NaCl and MgCl$_2$ and pool the samples in a 1.5-ml microcentrifuge tube.

> RIPA + NaCl + MgCl$_2$: 500 mM NaCl, 1.0%, Triton X-100, 0.5% DOC, 0.1% SDS, 50 mM Tris–HCl, pH 8.0, 10 mM MgCl$_2$, and 1 mM PMSF

The assay should be done quickly and on ice to maximize the recovery of GTP-bound proteins before the GTP can dissociate or hydrolyze. Sonicate the samples and centrifuge at 15,000g for 10 min at 4°. Collect the supernatant and measure the protein concentration. After normalization for protein content, add lysates to fresh tubes with 60 μl of GST–CRIB resin (approximately 13 μl of settled beads) to each tube. Incubate for 30 min at 4°. Wash beads three times with HBS wash buffer (20 mM HEPES, 120 mM NaCl, 1.0% Triton X-100, and 10 mM MgCl$_2$). Add SDS sample buffer to each tube, boil, and subject to 12% SDS–PAGE. Transfer protein to a polyvinylidene difluoride (PVDF) membrane and probe with monoclonal anti-Rac (Upstate Biotechnology) or polyclonal anti-cdc42 antibodies (Santa Cruz Biotechnology). The intensity of the band reflects the level of Rac1 or Cdc42 activation.

Acknowledgments

The authors' work is supported by grants to S.M.S. from the National Institutes of Health and the American Paralysis Association. S.M.S. is an Investigator of the Patrick and Catherine Weldon Donaghue Medical Research Foundation. A.F. is supported by an F.C.A.R. fellowship.

[43] Study of *in Situ* Function of Cytoskeletal Proteins in Lamellipodia and Filopodia Using Microscale Chromophore-Assisted Laser Inactivation

By Eric V. Wong and Daniel G. Jay

Introduction

Understanding how cells move requires us to establish how the multitude of cytoskeletal proteins and their effectors act at the leading edge of the cell. The vast array of these cytoskeletal proteins, their multifunctional nature, and their complex interactions make fine study of their *in situ* functions difficult. Standard pharmacological interventions act globally, often causing catastrophic cellular alterations that make analysis of the precise roles of these proteins nearly impossible. For many proteins, there may be no specific pharmacological agents or function-blocking antibodies with which to interrupt normal functions. Genetic knockout strategies usually remove the protein chronically and globally. Many cytoskeletal proteins have essential roles so early in development that the embryos die before many tissues arise. Hence, it is difficult to use these methods to study the roles of proteins in processes that occur late in development or in adult tissue. Also, such approaches are limited to model systems for which genetic methods are well established and facile. The ectopic expression of dominant negative or constitutively active mutants has been very useful, particularly for the study of Rho family small GTPases.[1,2] These approaches have led to significant advances in the general roles of these proteins. However, these techniques will activate or inactivate all of the target proteins throughout the cell for an extended period of time. It has thus been more difficult to establish how these proteins act in real time at the leading edge of the cell. The highly dynamic nature of motile leading edge structures such as lamellipodia and filopodia requires a method to eliminate the function of cytoskeletal components such as Rho family members in a time- and space-controlled manner. To achieve such specific perturbation, we have developed microscale chromophore-assisted laser inactivation (micro-CALI). This article describes this technique, discusses its limitations, and explains how micro-CALI is set up and applied to studying *in situ* protein function. It concludes with examples of the application of micro-CALI to study cytoskeletal proteins in filopodia and lamellipodia and suggestions for future directions.

CALI and Micro-CALI

CALI was developed for the specific inactivation of protein function in cells and tissues.[3] CALI uses precisely aimed laser energy to inactivate proteins only within the area of laser irradiation. The principle of CALI is illustrated in Fig. 1. CALI inactivates only those proteins in the spot (~2 mm diameter) that have been targeted by nonfunction-blocking antibodies conjugated to the chromophore, malachite green (MG). Although most living tissue does not absorb light energy with a wavelength of 620 nm, MG absorbs 620-nm laser energy and generates short-lived hydroxyl radicals that attack the amino acid side chains of extremely close proteins, i.e., those specifically bound by the MG-labeled antibody.[4]

The damage caused by CALI is very spatially restricted,[5] with a half-

[1] A. J. Ridley and A. Hall, *Cell* **70,** 389 (1992).
[2] C. Nobes and A. Hall, *Curr. Opin. Genet. Dev.* **4,** 77 (1994).
[3] D. G. Jay, *Proc. Natl. Acad. Sci. U.S.A.* **85,** 5454 (1988).
[4] J. C. Liao, J. Roider, and D. G. Jay, *Proc. Natl. Acad. Sci. U.S.A.* **91,** 2659 (1994).
[5] K. G. Linden, J. C. Liao, and D. G. Jay, *Biophys. J.* **61,** 956 (1992).

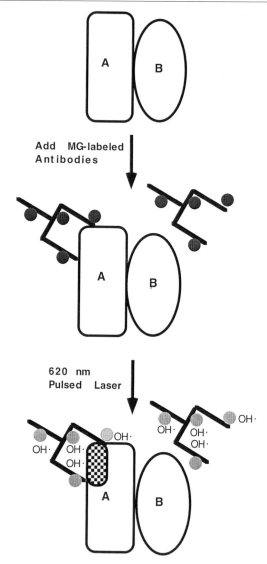

Fig. 1. The principle of CALI. Mixtures of proteins are incubated with MG-labeled antibodies that bind specifically to protein A but not protein B. Laser irradiation with 620-nm light excited the MG dye to generate hydroxyl radicals ($\cdot$ OH). These short-lived radicals cause damage over a very short distance and hence inactivate the bound protein A specifically without significantly affecting protein B.

maximal inactivation radius of 15 Å from each MG moiety. CALI has been demonstrated to inactivate specific subunits of protein complexes, such as the T-cell receptor, without significantly affecting the function of neighboring subunits.[6] CALI has also been able to inactivate part of a single large protein without affecting other domains within that protein. CALI of myosin V, using an antitail domain antibody, selectively inactivated the motor activity, while leaving its ATPase activity functional because the antibody-binding site and the ATPase domain are separated by more than 100 Å. CALI has been applied to a large variety of proteins in many biological systems (Table I). It has been shown to specifically inactivate over 90% of proteins treated with CALI, as assessed by the loss of activity and/or a change in physiological phenotype.

Of particular utility for the study of proteins required in cell motility is micro-CALI. Micro-CALI uses a laser beam focused to a 10-μm diameter.[7] Micro-CALI may be used to inactivate proteins within subregions of cells, such as the leading edge of motile cells.[8] Micro-CALI has been used extensively to address cytoskeletal proteins in the filopodia and lamellipodia of nerve growth cones. This approach has allowed us and others to inactivate a specific protein on a small region of a growth cone, such that the effects of a localized loss of function at the leading edge of the growth cone could be assessed. Combining micro-CALI with high-resolution video microscopy has provided unprecedented resolution in assessing the effect of the loss of individual cytoskeletal proteins on leading edge motility. In addition to myosin V, micro-CALI has also been used to examine the functions of myosin Iβ,[9] the actin-associated proteins talin,[10] vinculin,[10] ezrin,[8] and radixin,[11] and the microtubule-associated proteins tau[12] and MAP 1B. Micro-CALI of these and other growth cone proteins are summarized in Table II.

Limitations of CALI and Micro-CALI

An understanding of the limitations of CALI is required to interpret data obtained using this approach. CALI has best been applied to test the

[6] J. C. Liao, L. J. Berg, and D. G. Jay, *Photochem. Photobiol.* **62,** 923 (1995).

[7] P. Diamond, A. Mallavarupu, J. Schnipper, J. Booth, L. Park, T. P. O'Connor, and D. G. Jay, *Neuron* **11,** 409 (1993).

[8] R. F. Lamb, B. W. Ozanne, C. Roy, L. M. Garry, C. Stipp, P. Mangent, and D. G. Jay, *Curr. Biol.* **7,** 682 (1997).

[9] F. S. Wang, J. S. Wolenski, R. E. Cheney, M. S. Mooseker, and D. G. Jay, *Science* **273,** 660 (1996).

[10] A. M. Sydor, A. L. Su, F. S. Wang, A. Xu, and D. G. Jay, *J. Cell Biol.* **134,** 1197 (1996).

[11] L. Castelo and D. G. Jay, *Mol. Biol. Cell* **10,** 1511 (1999).

[12] C. W. Liu, G. Lee, and D. G. Jay, *Cell Motil. Cytoskel.* **43,** 232 (1999).

TABLE I
PROTEINS INACTIVATED BY CALI

Protein	System	Ref.
Enzymes		
β-Galactosidase	*In vitro*	3
Alkaline phosphatase	*In vitro*	3
Acetylcholinesterase	*In vitro*	3
Signal transduction molecules		
Calcineurin	*In vitro* and in cells	20
IP3 receptor	*In vitro* and in cells	21[a]
pp60c-src	*In vitro* and in cells	22
pp59fyn	*In vitro* and in cells	22
Surface proteins		
Grasshopper fasciclin I	*In vivo*	23
Grasshopper fasciclin II	*In vivo*	7
α chain of the T-cell receptor	In cells	6
β chain of the T-cell receptor	In cells	6
ε chain of the T-cell receptor	In cells	6
Drosophila patched protein	*In vivo*, mimics hypomorphic mutation	24
NCAM	*In vitro* and in cells	25
L1	*In vitro* and in cells	25
RGM	*In vitro*	26
FMRFamide receptor	*In vitro*	27[a]
Cytoskeletal proteins		
Myosin V	*In vitro* and in cells	9
Talin	In cells	10
Actin	*In vitro* and in cells	28
Kinesin	*In vitro*	14[a]
Tau	*In vitro* and in cells	12
Myosin 1β	In cells	9
Vinculin	*In vitro* and in cells	10
Ezrin	*In vitro* and in cells	8
Radixin	In cells	11
Transcription factors		
Drosophila even skipped	*In vivo*, mimics genetic loss of function	29
Tribolium even skipped	*In vivo*	30
Protein not inactivated by CALI		
Hexokinase		3
Glyceraldehyde 3-phosphate dehydrogenase		3

[a] Established by other groups independently.

TABLE II
GROWTH CONE PROTEINS INACTIVATED WITH MICRO-CALI

Growth cone molecule	Neurite outgrowth	Lamellipodia behavior	Filopodia behavior	Ref.
Myosin V	n.d.[a]	Unchanged	Net retraction	9
Myosin 1β	Unchanged	Expansion	Unchanged	9
Talin	Unchanged	n.d.	Stalling	21
Vinculin	Unchanged	n.d.	Increased bending and buckling	10 10
Radixin	Unchanged	Retraction and growth cone splitting	Unchanged	11
Zyxin	Unchanged	Unchanged	Collapse	32
Tau	Decreased	Retraction	Unchanged	12
MAP 1B	Retraction	Retraction	n.d.	33
L1CAM	Neurite retraction	Unchanged	Unchanged	25
NCAM	Unchanged	Retraction and growth cone turning	Retraction	25
Calcineurin	Unchanged	Retraction	Retraction	20
IP3R	Neurite retraction	n.d.	Unchanged	21
pp60 c-*src*	Increased	n.d.	Unchanged	22
pp59 fyn	Unchanged	n.d.	Unchanged	22

[a] Not determined.

role of a protein based on established *in vitro* biochemical and immunocyto-chemical data. As such, a simple hypothesis of function and a good cellular assay to test that hypothesis are required. The effects of CALI are not equivalent to the chronic and global loss of function generated by genetic knockout. Recovery of function depends on *de novo* protein synthesis that may occur within hours or days.[6] For micro-CALI, diffusion of unirradiated protein into the CALI-treated area will restore function. CALI can demonstrate a correlation between the acute loss of function and a change in cellular processes, but this link may be indirect. The negative result (i.e., no effect by CALI) is difficult to interpret for several reasons. CALI may not inactivate the protein if the functional domain is insensitive to free radical damage or if the dye-labeled antibody binds too far away from the domain required for function. CALI may not inactivate all of the initial activity, especially for naturally abundant proteins.[10] As such, the residual activity may be sufficient for normal physiological function. Although CALI is spatially restricted,[4-6] it is possible that for any specific antibody, the geometry of binding places an MG moiety in constant contact with a neighboring protein. Thus far, this has not been observed. It is also possible that the free radical damage caused by CALI could result in a constitutively

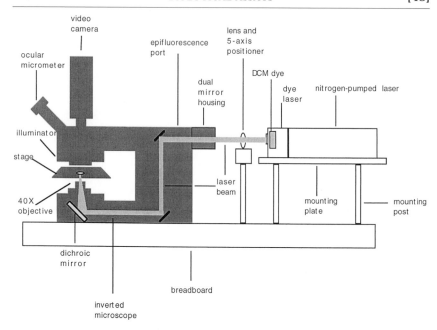

FIG. 2. The setup for micro-CALI. A 620-nm laser beam from a nitrogen-driven pulsed dye laser is directed into the epifluorescence optical path of an inverted microscope and focused through the microscope objective to a diameter of ~10 μm. A cell loaded with MG-labeled antibodies on the stage is brought into view and subjected to the focused laser beam to inactivate proteins within the cell that are specifically bound by the MG-labeled antibody.

active protein. This would confound interpretation, but it has also not yet been observed in the cases tested thus far.

Equipment and Setup of Micro-CALI

The setup of CALI has been described previously.[13] Unlike large-scale CALI, micro-CALI is readily accessible to researchers due to its relatively low cost and ease of setup. The first setup for micro-CALI used an upright microscope,[13] but recent models have used an inverted microscope and we have found this more useful for applying micro-CALI to cells in culture. The apparatus for this micro-CALI setup is shown in Fig. 2. A nitrogen-driven dye laser (VSL-337 and DLM-110, Laser Science Co., Acton, MA, or equivalent, using DCM fluorescent dye) generates approximately 30 μJ of laser energy per pulse. The laser is mounted on a breadboard or isolation

[13] A. E. Beermann and D. G. Jay, *Methods Cell Biol.* **44,** 715 (1994).

table (Newport Corp., Fountain Valley, CA) with the microscope so that the beam is directed to the center of the mirror reflecting the light into the microscope. This mirror is part of a dual-mirror lamp housing attachment available for most commercial research microscopes (consult your microscope dealer). This housing allows switching between laser light and a mercury arc lamp for use in fluorescence microscopy. A dichroic mirror that reflects light through the objective at wavelengths greater than 600 nm (Chroma Inc., Brattleboro, VT) is mounted in a vacant fluorescence filter set.

Alignment of the laser spot begins by placing a piece of paper over the dual-mirror housing and checking that the majority of the laser light is emitted in a single spot, although birefringence bands may be visible. Adjust the position of the laser and microscope so that the beam hits directly in the center of the mirror. Removing an objective lens from the nosepiece, coarse alignment of the beam is performed by placing a piece of lens paper on the stage and moving the microscope so that the laser beam coincides with the light coming through the illuminator (with aperture closed down to the minimum diameter). This procedure is then repeated with a low-power objective in place. If the beam is not parfocal with the stage a focusing lens placed in a five-axis positioner (Newport) should be interjected between the laser and the microscope. During the alignment procedure, neutral density filters should be employed to attenuate power, and care should be taken not to focus the beam on the face of the objective lenses.

The power of the beam is gauged before each use by emitting a single pulse onto a field of ink from a blue felt-tip permanent marker drawn on a slide. If the pulse can photobleach a spot, it is judged to provide sufficient power to perform micro-CALI. The size of the spot and location in the field of view are determined using an ocular micrometer and, when using video microscopy, marked on the video monitor. It should be noted that the laser spot is not circular, but elongated due to the laser beam profile of the laser. During an experiment, the area of interest is aligned with the laser spot, and the laser is turned on for continuous pulsed irradiation for 2 to 5 min.

Antibody Selection

The antibody selected to target the protein of interest is the primary determinant of success of a micro-CALI experiment. The antibodies of choice are generated against a specific part of the target protein, using synthetic, recombinant, or protease-generated polypeptides. They may be either monoclonal or polyclonal antibodies. Other reagents have been used, such as MG–streptavidin with biotinylated enzymes,[3] suggesting that other specific binding reagents may be used for CALI.

When using antibodies, the antibody must (1) specifically recognize the antigen in the tissue studied; (2) be transported to and retained at the cellular location of interest after loading into the cells; (3) bind with strong affinity to the protein under the experimental conditions; and (4) have no effects on cellular function without laser irradiation. It is useful to demonstrate by Western blot that only the target protein is recognized from an extract of the target cell type. For CALI of cytoskeletal proteins, the antibodies are loaded into living cells by a variety of methods discussed later in this article. Loading and specific binding of the antibodies are tested by immunocytochemistry. The loaded cells are fixed and permeabilized at a time point equivalent to the start of CALI, generally 2–6 hr after plating. Visualization of the antibodies is accomplished with the appropriate fluorescent secondary antibody. When possible, the MG-labeled antibodies are tested by CALI *in vitro,* particularly when assaying measurable enzymatic activities, e.g., motor activity of myosins or phosphorylation by kinases. To test for the effect of the antibody on cellular functions, we generally observe the cellular behavior of antibody-loaded cells. If these cells are indistinguishable from untreated cells, the antibody is deemed to not affect the cellular function that is being studied (e.g., cell motility). One should bear in mind that it is possible that other cellular functions that are not assayed are perturbed.

Malachite Green Labeling of Antibodies

Antibodies are labeled with malachite green isothiocyanate (MGITC) as described previously.[3,13] MGITC is stable when stored desiccated at $-20°$, but hydrolyzes quickly when exposed to moisture. A 10-mg/ml solution of MGITC (Molecular Probes Inc., Eugene, OR) is freshly prepared in dry dimethyl sulfoxide. A total weight:weight ratio of dye to IgG of 1:5 is used for labeling. MGITC solution is added to the antibody (in 0.5 M $NaHCO_3$, pH 9.5) in three equal aliquots every 5 min, with continuous shaking. After the last addition of dye, the mixture is incubated for an additional 15 min. The MG-labeled antibodies are separated from free dye by gel filtration through a prepacked PD-10 column (Amersham Pharmacia Biotech AB, Uppsala, Sweden) and eluted in the relevant buffer for experimental use, usually Hanks' buffered saline solution or phosphate-buffered saline. The efficacy of CALI is dependent on the labeling ratio, which can be found by measuring the optical density of the labeled solution at 620 nm (molar absorptivity of 150,000 M^{-1} cm^{-1}) to determine the dye concentration and dividing that value by the antibody concentration. We aim for a labeling ratio of four to eight dyes per antibody and found this to be optimal for CALI.[5] The MG-labeled antibody solution is usually concentrated using a Centricon filter (Amicon Corp., Danvers, MA) to ~1 mg/ml for loading into cells.

The labeling method is similar to other isothiocyanate-labeling reactions with free amino groups. However, because MG has a greater hydrophobicity and capability for π-stacking interactions, aggregation of hydrolyzed dye and trapping of protein onto insoluble precipitate may be problematic. Two competing reactions occur when the MGITC is added to the antibody. The isothiocyanate reacts with amino groups to form a stable thioester, and the isothiocyanate hydrolyzes. To compensate for loss by hydrolysis, a high concentration of dye is used in the reaction, and the reaction is carried out at high pH (9.5) to favor the reaction with free amino groups. A higher pH does not seem to enhance the reaction and risks antibody denaturation. A low protein concentration will also cause the hydrolysis reaction to predominate, so we generally aim for at least 0.4 mg/ml protein prior to the labeling reaction. This is generally limited only by the availability and/or cost of the antibodies. We occasionally add bovine serum albumin (BSA) to supplement the protein concentration in the labeling reaction. While BSA is also labeled, MG–BSA does not cause nonspecific damage in any cases tested. We have also found this reduces the nonspecific precipitation of the hydrolyzed product in the labeling reaction, perhaps because BSA may bind the hydrophobic hydrolysate to keep it in solution. The antibody solution must also not contain free amino groups other than the protein to be labeled (e.g., Tris and glycine buffers). These small molecules will react readily with the dye and lower antibody labeling. Furthermore, a purple precipitate that occurs when MGITC labels these small molecules may further trap protein.

Some of the MG label on the protein even after gel filtration is not covalently attached, but is associated hydrophobically on the surface of the protein. Over time (several weeks at $-20°$), the adhering dye may dissociate, limiting the useful shelf life of the labeled antibody. For this reason, each batch of labeled antibodies is immediately aliquoted and frozen and used for less than 6 months.

Studies have suggested that fluorescein and green fluorescent protein (GFP) may also be used for CALI using the appropriate wavelength of light.[14] It is clear that FITC-labeled antibodies to β-galactosidase are very efficient at inactivating this enzyme and this approach has been used to inactivate kinesin *in vitro*. There are two concerns when using fluorescein-

---

[14] T. Surrey, M. B. Elowitz, P. E. Wolf, F. Yang, F. Nedelec, K. Shokat, and S. Leibler, *Proc. Natl. Acad. Sci. U.S.A.* **95**, 4293 (1998).

or GFP-labeled probes for CALI *in vivo*. The wavelength of light used is somewhat phototoxic to cells and the radius of damage for excited fluorescein has not been shown to be sufficiently restricted to rule out damage to neighboring proteins. These issues will likely be resolved soon, and if CALI using these fluorophores were compatible with *in vivo* experiments, this would enhance the range of experiments that may be done using CALI.

Antibody Loading and Experimental Parameters

In order to perform micro-CALI of cytoskeletal effectors such as Rho, the antibody must be loaded into the cell. At present, we have successfully used trituration loading,[10] microinjection,[9] and electroporation to do so. Trituration loading of antibodies relies on transient shear-induced openings in the cell membrane to allow the influx of antibodies from the trituration medium into the cells. Although this process also shears any processes from the cell body, a significant population (~30%) survive the treatment and behave normally within an hour of plating. Greater than 80% of the surviving cells are loaded with antibody. Electroporation is useful in cases where the cells are too small or are otherwise recalcitrant to trituration loading, whereas microinjection is used primarily for very fine study of a small group of cells or their processes. To confirm loading of the antibody, fluorescein-conjugated nonimmune IgG (1 mg/ml) is mixed with the MG-labeled antibody. Based on fluorescence microscopy of the loaded cells by subsequent immunocytochemistry, we have determined that the antibodies are retained with an average half-life of 12 hr.

Several standard control experiments are performed to verify the selectivity of the inactivation. CALI has been used on a variety of cultured cells as well as living embryos without observable damage. The 620-nm wavelength light is not absorbed readily by animal cell components. However, it is still important to establish that 620-nm laser light has no effect on each cell type and for each particular assay in the absence of the MG–antibody. The MG-labeled antibody is also tested without laser to establish that it does not block function at the concentrations used. In fact, some function-blocking antibodies may be used if they require significantly higher concentrations for their effects on function. While other methods of inactivation are limited to comparing a treated cell with an untreated cell, micro-CALI also allows the comparison of an experimental side of the cell to the opposite, nonirradiated side. This allows the investigator to discern subtle functional changes in motile structures that may not be apparent when the target protein in a cell is inactivated globally.

Studying Lamellipodia and Filopodia with Micro-CALI and Video Microscopy

Combining micro-CALI with time-lapse video microscopy has allowed even subtle changes at the leading edge to be correlated with the loss of specific proteins and has thus implicated these proteins in precise cellular processes. The parameters we have measured include overall cell shape, filopodial length, bent or buckled filopodia, lamellipodial area, leading edge expansion or retraction, and the rates of change in these parameters. Video microscopy is done as described in Wang *et al.*[9] and images are taken every 15 sec. We have used both tube cameras and charge-coupled device cameras (both from Hamamatsu Photonics K. K., Hamamatsu City, Japan). Images are captured using NIH Image software or, more recently, commercially available software such as Openlab (Improvision, Coventry, UK). These software packages include morphometry programs that allow us to measure these parameters frame by frame.

Statistical tests are used to determine if the effect of CALI caused a significant effect within the region of irradiation. We generally compare several different data sets derived from cells subjected to CALI and control treatments. The data sets include observations before, during, and after laser irradiation, inside and outside the laser spot, and we also compare controls using cells loaded with MG-labeled nonimmune IgG or without dye-labeled reagents.

Micro-CALI of Actin-Associated Proteins during Motility

Because we are interested in cell and neuronal growth cone motility, we have applied micro-CALI to a variety of actin-associated proteins to study their functions in motility. Most of the work in our laboratory has focused on the nerve growth cone, but several experiments have now been done using fibroblasts.

We have investigated the effects of micro-CALI on two members of the ERM (ezrin–radixin–moesin) family of actin-binding proteins. These proteins are highly homologous and have binding domains for F-actin as well as membrane proteins such as CD44, leading to speculation that these proteins may function as membrane to cytoskeleton linkers. Ezrin expression, phosphorylation, and association with the actin cytoskeleton are increased by Fos transformation of fibroblasts. Because it is highly abundant in leading edge structures, ezrin was inactivated by micro-CALI at the leading edge of transformed and normal fibroblasts.[8] Micro-CALI of ezrin in transformed fibroblasts caused a rapid inhibition of membrane ruffling and net retraction of pseudopodia. In untransformed fibroblasts, micro-

CALI of ezrin at the leading edge, but not at the cell body, caused a pronounced loss of shape, rounding of the cell, and shearing of the unirradiated cell periphery. The related protein, radixin, was investigated in the lamellipodia of neuronal growth cones. Radixin has been shown to play a role in the activation of Rho family members by interaction with Dbl, a stimulatory GTP/GDP exchange protein.[15] Micro-CALI of radixin decreased the lamellipodial area within the irradiated spot, and when the laser was directed to the middle of the leading edge, the growth cone split into two small growth cones.[11]

Another set of experiments examined the role of the actin-binding proteins talin and vinculin in the actin-rich filopodia of neuronal growth cones.[10] Interestingly, the two disruptions caused different outcomes. When micro-CALI of talin was directed at the filopodia, extension and retraction of the irradiated filopodia ceased, whereas nonirradiated neighbors continued to move. However, when vinculin was inactivated, there was no effect on extension or retraction. Instead, the filopodia appeared to bend and buckle more frequently than on the control side of the growth cone, suggesting a role in maintaining the structural integrity of the filopodia.

Micro-CALI has also been used to investigate molecular motor proteins.[9] The unconventional myosins, myosin V and myosin Iβ, are both found in the nerve growth cone. The inactivation of myosin V decreased the rate of filopodial elongation, without affecting their rate of retraction. However, inactivation of myosin Iβ had no effect on the filopodia. Instead, asymmetric application of micro-CALI against myosin Iβ induced lamellipodial expansion and turning toward the irradiated side. These experiments were thus able to determine distinct roles for myosin V and myosin Iβ and showed that they act in different components of the leading edge.

Motility at the leading edge is thought to depend on localized actin polymerization. One protein implicated in this localization is zyxin, which binds to the Arp 2/3 complex.[16] We applied micro-CALI to the nerve growth cone where zyxin is concentrated in filopodial tips. This caused filopodial collapse. When CALI of zyxin was applied to fibroblasts, these cells showed a loss of cell spreading and locomotion during wound healing also consistent with a loss of leading edge actin polymerization.

Although we have not yet inactivated Rho family GTPases, CALI and micro-CALI have been used to inactivate Gα_0 in vitro and in neuronal growth cones. This treatment eliminates over 90% of the in vitro GTPase activity (K. Vancura and D. Jay, unpublished results). However, there was

[15] K. Takahashi, T. Sasaki, A. Mammoto, I. Hotta, K. Takaishi, H. Imamura, K. Nakano, A. Kodama, and Y. Takai, Oncogene 16, 3279 (1998).
[16] M. C. Beckerle, Cell 95, 741 (1998).

little discernible effect on chick dorsal root ganglion growth cone behavior caused by micro-CALI of $G\alpha_0$. This reflects both a strength and a potential liability of micro-CALI. In its specificity, the functions of a particular isoform within a family of proteins may be determined. However, functional redundancy by related proteins may mask the effects of eliminating one particular protein. The use of pan-specific antibodies that recognize all isoforms for CALI provides a solution to this liability.

Future Work

Micro-CALI has been useful in ascribing *in situ* function to cytoskeletal proteins at the leading edge of motile cells and growth cones. A major challenge for us is to address how these proteins act together to affect motility. This will require combining micro-CALI with other complementary approaches to test hypotheses of pathways of interaction. A key issue in these interactions is how these proteins act on the cytoskeleton to affect their roles and this remains to be established for most of the proteins we have studied. The next stage of analysis will be examining cytoskeletal dynamics in real time after micro-CALI of specific proteins thought to act on the cytoskeleton. This may be done by combining micro-CALI with methods to visualize F-actin[17] and microtubules[18] in living cells. This analysis will be particularly useful in assessing the functional roles of the Rho family GTPases as they have been ascribed to roles in filopodia, lamellipodia, and stress fibers in serum-starved Swiss 3T3 cells, but their precise roles in these structures are not clear in these cells. Moreover, their roles and functions in other cell types are complex[19] and will require further study. As antibodies or other specific probes become available to the Rho family members, CALI may prove to be a useful tool for understanding their cellular functions.

[17] J. A. Theriot and T. J. Mitchison, *J. Cell Biol.* **119,** 367 (1992).
[18] E. Tanaka and M. W. Kirschner, *J. Cell Biol.* **128,** 127 (1995).
[19] T. B. Kuhn, M. D. Brown, and J. R. Bamburg, *J. Neurobiol.* **37,** 524 (1998).
[20] H. Y. Chang, K. Takei, A. M. Sydor, T. Born, F. Rusnak, and D. G. Jay, *Nature* **376,** 686 (1995).
[21] K. Takei, R. M. Shin, T. Inoue, K. Kato, and K. Mikoshiba, *Science* **282,** 1705 (1998).
[22] D. Hoffman-Kim, A. Chen, A. Xu, T. F. Wang, and D. G. Jay, submitted for publication.
[23] D. G. Jay and H. Keshishian, *Nature* **348,** 548 (1990).
[24] D. Schmucker, A. Su, A. Beerman, H. Jackle, and D. G. Jay, *Proc. Natl. Acad. Sci. U.S.A.* **91,** 2664 (1994).
[25] K. Takei, T. A. Chan, F. S. Wang, H. Deng, and U. Rutishauser, *J. Neurosci.* **19,** 9469 (1999).
[26] B. K. Muller, D. G. Jay, and F. B. Bonhoeffer, *Curr. Biol.* **6,** 1497 (1996).
[27] J. J. Feigenbaum, M. D. Choubal, K. Payza, J. R. Kanofsky, and D. S. Crumrine, *Peptides* **17,** 991 (1996).
[28] A. M. Sydor, A. L. Su, F. S. Wang, A. Xu, and D. G. Jay, *J. Cell Biol.* **134,** 1197 (1996).
[29] R. Schroder, D. G. Jay, and D. Tautz, *Mech. Dev.* **80,** 191 (1996).

Acknowledgments

This work was supported by grants from the NIH (DGJ) and an NRSA (EVW). The authors thank Elisabeth Pollerberg for providing a manuscript prior to publication.

[30] R. Schroder, D. G. Jay, and D. Tautz, *Mech. Dev.* **80**, 191 (1999)
[31] F. S. Wang, submitted for publication.
[32] B. Benson, F. S. Wang, D. G. Jay, and M. C. Beckerle, submitted for publication.
[33] T. G. Mack, M. P. Koester, and G. E. Pollerberg, *Mol. Cell. Neurosci.* **15**, 51 (2000).

[44] Modulation of Actin Cytoskeleton by *Salmonella* GTPase Activating Protein SptP

By Jorge E. Galán and Yixin Fu

Introduction

Salmonella species have evolved the capacity to engage host cells in very sophisticated interactions.[1,2] This is most likely a consequence of the millions of years of close association of these bacteria with vertebrate hosts. Such long-standing association has undoubtedly refined the interaction of *Salmonella* with host cells, ultimately resulting in a relatively well-balanced coexistence. This statement may come as a surprise to some cell biologists that have experienced a bout of food poisoning caused by these bacteria. However, it is important to point out that, like for many other microbial pathogens, overt disease is usually a rare outcome of the encounter. Indeed, in the overwhelming majority of cases, *Salmonella* infections in healthy people go either unnoticed or are completely self-limiting. This is important to keep in mind to fully appreciate the sophistication of the battery of tools (so-called "virulence factors") that *Salmonella* has assembled to carry out its life cycle within the host. This is of particular significance to some *Salmonella* subspecies, as the process of host adaptation has limited their ecological niche to a single host (e.g., humans for *Salmonella typhi* or chickens for *Salmonella gallinarum*).

An essential aspect of the *Salmonella* life cycle is its ability to gain access to cells that are normally nonphagocytic, such as those of the intestinal epithelium. The internalization process is the result of a carefully orchestrated modulation of the actin cytoskeleton.[1,2] *Salmonella* accomplishes this

[1] J. E. Galán, *Proc. Natl. Acad. Sci. U.S.A.* **95**, 14006 (1998).
[2] J. E. Galán, *Curr. Opin. Microbiol.* **2**, 46 (1999).

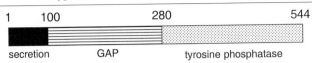

FIG. 1. Domain organization of SptP.

objective using a sophisticated organelle, the type III secretion system, which has evolved to inject bacterial proteins into the host cell.[3] Among this battery of *Salmonella* proteins, there is a subgroup that has evolved to modulate the actin cytoskeleton. One of these proteins is SopE (see Ref. 3a), which is an exchange factor for the Rho GTPases Cdc42 and Rac.[4] Delivery of SopE by *Salmonella* results in profuse actin cytoskeleton reorganization, membrane ruffling, macropinocytosis, and subsequent uptake of the bacteria into spacious phagosomes.[4] The efficiency of the uptake process is enhanced by another *Salmonella* protein, SipA, which exerts its function by binding actin.[5] SipA binding to actin results in a very significant reduction in the critical concentration for actin polymerization and a marked stabilization of actin filaments, as well as an enhancement of the bundling activity of the actin-binding protein T-plastin.[5,6] Remarkably, the rather dramatic reorganization of the actin cytoskeleton induced by *Salmonella* is short-lived, i.e., shortly after bacterial infection, the cell regains its normal architecture.[7,8] The reversion of the changes in the actin cytoskeleton is mediated by SptP, a *Salmonella* protein that is also delivered to the host cell by the type III secretion apparatus.[9–11] SptP is composed of two effector modules (Fig. 1): a carboxy-terminal domain that shares sequence similarity to the *Yersinia* YopH protein[12,13] and other protein tyrosine phosphatases and an amino-terminal domain that exhibits sequence similarity to the bacterial toxins YopE from *Yersinia*[14] and ExoS from *Pseudomo-*

[3] J. E. Galán and A. Collmer, *Science* **284**, 322 (1999).
[3a] A. Triebel and W.-D. Hardt, *Methods Enzymol.* **325** [8] (2000) (this volume).
[4] W.-D. Hardt, L.-M. Chen, K. E. Schuebel, X. R. Bustelo, and J. E. Galán, *Cell* **93**, 815 (1998).
[5] D. Zhou, M. Mooseker, and J. E. Galán, *Science* **283**, 2092 (1999).
[6] D. Zhou, M. S. Mooseker, and J. E. Galan, *Proc. Natl. Acad. Sci. U.S.A.* **96**, 10176 (1999).
[7] A. Takeuchi, *Am. J. Pathol.* **50**, 109 (1967).
[8] C. Ginocchio, J. Pace, and J. E. Galán, *Proc. Natl. Acad. Sci. U.S.A.* **89**, 5976 (1992).
[9] K. Kaniga, J. Uralil, J. B. Bliska, and J. E. Galán, *Mol. Microbiol.* **21**, 633 (1996).
[10] Y. Fu and J. E. Galán, *Mol. Microbiol.* **27**, 359 (1998).
[11] Y. Fu and J. E. Galan, *Nature* **401**, 293 (1999).
[12] I. Bolin and H. Wolf-Watz, *Mol. Microbiol.* **2**, 237 (1988).
[13] T. Michiels and G. Cornelis, *Microb. Pathog.* **5**, 449 (1988).
[14] A. Forsberg and H. Wolf-Watz, *J. Bacteriol.* **172**, 1547 (1990).

nas species.[15] Consistent with its sequence similarity, purified SptP protein exhibits potent tyrosine phosphatase activity.[9] Studies in our laboratory have established that the reversion of the *Salmonella*-induced actin cytoskeleton rearrangements mediated by SptP does not require its tyrosine phosphatase domain.[11] Rather, this modulation of the actin cytoskeleton is mediated by its amino-terminal domain, which exerts its function by acting as a GTPase-activating protein for Cdc42 and Rac. Consistent with its function, the amino-terminal domain of SptP exhibits a sequence motif with an invariant arginine ("arginine finger") that is present in all Rho GTPases and is essential for efficient catalysis.[11,16] Mutation of this arginine into an alanine results in the total abrogation of the SptP GAP activity and, consequently, a loss of its ability to mediate the recovery of the normal architecture of the actin cytoskeleton after bacterial infection.[11]

Thus, *Salmonella* has evolved the ability to alternatively stimulate and downmodulate the activity of Rho GTPases as a strategy to gain a foothold inside the cell without causing irreparable damage to the host cell. This not only constitutes a remarkable piece of evolution, but also provides cell biologists with valuable reagents to probe Rho GTPase function. This article reviews the methodology to isolate SptP as well as to probe its GAP activity. Procedures to isolate and assay the exchange factor SopE are described in this volume.[3a]

Expression and Purification of SptP

SptP is composed of at least three functionally recognizable domains (Fig. 1)[9,10,17]: (1) a secretion and translocation domain composed of the first ~100 residues that mediates its export and subsequent delivery into the host cell; (2) a GAP domain that is located approximately between residues 70 and 280; and (3) a tyrosine phosphatase domain that is located approximately between residues 280 and the carboxy-terminal end. The actual boundaries of these domains have not been mapped precisely and it is possible that there is an overlap between some of these domains. Like most type III-secreted proteins, however, the presence of the secretion and translocation domain severely affects the solubility of SptP. Because this domain is totally dispensable for function inside the host cell, the first 70 amino acids of SptP are removed to create a glutathione *S*-transferase (GST)–SptP$_{70-544}$ fusion protein that is more soluble and therefore more

[15] S. M. Kulich, T. L. Yahr, L. M. Mende-Mueller, J. T. Barbieri, and D. W. Frank, *J. Biol. Chem.* **269**, 10431 (1994).

[16] K. Scheffzek, M. R. Ahmadian, and A. Wittinghofer, *Trends Biochem.* **23**, 7257 (1998).

[17] Y. Fu and J. E. Galán, *J. Bacteriol.* **180**, 3393 (1998).

amenable to standard purification procedures.[11] For certain experiments it may be desirable to use a derivative of SptP devoid of tyrosine phosphatase activity. Two options are available: (1) the use of a catalytically defective mutant of SptP carrying a substitution at residue 481 that effectively abolishes tyrosine phosphatase activity (GST–SptP$_{C481S}$)[9] or (2) the use of a truncated form of SptP that lacks the tyrosine phosphatase domain all together (GST–SptP$_{70-480}$). The solubility of both these derivatives is good and therefore purification offers no difficulties. Purification of the different GST–SptP fusion proteins is carried out as per standard procedures. Briefly, *Escherichia coli* strains carrying the different plasmids are grown in 200 ml of L broth at 37° to an OD$_{600}$ of 0.5 when 0.1 mM isopropyl-B-D-thiogalactoside (IPTG) is added to induce expression of the fusion constructs. Cultures are grown for an additional 3 hr at 30° under heavy shaking and bacterial cells collected by centrifugation at 10,000g for 20 min at 4°. The cell pellet is washed once in phosphate-buffered saline (PBS) and resuspended in 10 ml of PBS containing 2 mM EDTA, 0.1% 2-mercaptoethanol, 1% Triton X-100, 0.2 mM phenylmethylsulfonyl fluoride (PMSF), and 5 mM benzamidine. Bacterial cells are disrupted in a French pressure cell, and bacterial lysates are centrifuged at 10,000g for 30 min at 4° to eliminate unbroken cells and cell debris. The clear bacterial lysate is then applied to 1 ml of 50% slurry of glutathione-Sepharose 4B beads (Pharmacia Biotechnology, Piscataway, NJ). The mixture is incubated at 4° for 45 min under gentle rocking, and the beads are washed three times with 50 volumes of cold PBS containing 1% Triton X-100. The bound protein is eluted by the addition of 10 mM glutathione in 50 mM Tris–HCl, pH 8. The elution step is repeated twice and the elutes are pooled. The eluted protein is concentrated to approximately 2 mg/ml by filtration through a Centricon filter (Amicon Bioseparations, Danvers, MA).

Binding of SptP to Rho GTPases

The binding of SptP to Rho GTPases is assayed by a GST pull-down protocol. Briefly, GST–Rac1 or GST–Rac1$_{L61}$ (or equivalent forms of Cdc42) (1–10 μg) purified as per standard protocols is loaded onto 25 μl of glutathione Sepharose 4B beads (50% slurry, Pharmacia Biotechnology) in 100 μl of PBS and incubated at 4° for 30 min with gentle rocking. The unbound protein is removed by several washes in binding buffer [20 mM phosphate buffer, pH 7.4, 1% Triton X-100, 1 mM dithiothreitol (DTT)], and the beads are resuspended in a final volume of 100 μl of the same buffer. Beads are added to purified SptP (10 μg/ml) that has been cleaved off the GST moiety by thrombin digestion in a total volume of 300 μl of binding buffer. Alternatively, beads are added to a lysate of *E. coli* expressing *Salmonella typhimurium* SptP under the control of a strong p*araBAD* promoter that is prepared by sonication in binding buffer. The binding reaction is carried out at 4° for 1 hr with end-over-end rotation. Beads are washed three times with binding buffer, and bound proteins are released by boiling in Laemmli sample buffer. Proteins are separated by SDS–PAGE and analyzed by Western immunoblot with a monoclonal antibody directed to SptP. SptP binds preferentially to GTP-loaded Rac and Cdc42. Typical results of this binding reaction are shown in Fig. 2.

GTPase Assay

Recombinant wild-type Rho GTPases are produced and purified as GST fusion proteins according to standard procedures. Rho GTPases (200 nM) are loaded with 50 nM of GTP (1:30 ratio of [α-^{32}P]GTP to cold GTP) in 20 μl of GTP-loading buffer (20 mM Tris, pH 8.0, 2 mM EDTA, 1 mM DTT) at room temperature for 10 min. The loading reaction is terminated by adding 3 μl of 100 mM MgCl$_2$ and placed on ice. Three microliters of loaded GTPases is added to 26 μl of hydrolysis reaction buffer (20 mM Tris, pH 7.5, 0.2 mM DTT, 1 mM GTP, and 1 mg/ml BSA) containing 1 μl (usually 10 nM) of the SptP or control proteins. After different incubation times, aliquots (4 μl) are removed and added to 2 μl of elution buffer (0.2% SDS, 5 mM EDTA, 5 mM DTT, 5 mM GTP, and 5 mM GDP) to terminate the reaction. Nucleotides are released by heating the samples at 65° for 5 min, and the eluted GTP and GDP are separated by thin-layer chromatography on polyethyleneimine-cellulose sheets (VWR Scientific, S. Plainfield, NJ) with 0.75 M KH$_2$PO$_4$, pH 3.5, as the developing solvent, followed by autoradiography and PhosphorImager quantitation (Molecular Dynamics, Sunnyvale, CA).

Alternatively, GAP activity can be measured by a filter-binding assay. Briefly, Rho GTPases are loaded as described earlier but using [γ-^{32}P]GTP instead and the loading reaction is terminated by the addition of 3 M MgCl$_2$ (10 mM final concentration). Reactions are initiated as described, and at different time points, aliquots (4 μl) are removed and added to 1 ml of ice-cold filter-wash buffer [20 mM Tris (pH 8.0), 100 mM NaCl, 5 mM

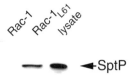

FIG. 2. Binding of SptP to GTP-loaded Rac. Bacterial extracts containing SptP were incubated with glutathione-agarose beads loaded with GST–Rac or GST–Rac$_{L61}$, and bound proteins were examined by a Western immunoblot using a monoclonal antibody against SptP.

MgCl$_2$]. Samples are applied onto nitrocellulose filters (0.45 μM, Millipore) that had been placed onto a vacuum manifold (Millipore, Bedford MA). Filters are washed three times with cold wash buffer, and the bound radioactivity is counted on a scintillation counter. In a standard reaction, a Rho GTPase : SptP ratio of 30 : 1 is usually used. Higher concentrations of SptP result in very rapid GTP hydrolysis and therefore an inability to measure the kinetics of the reaction. Typical results of these reactions are shown in Fig. 3. SptP exhibits potent GAP activity toward Rac and Cdc42 and 100-fold less activity toward Rho.[11] SptP exhibits no detectable GAP activity toward Ras.[11]

Microinjection of SptP into Ref52 Cells

Microinjection of SptP into Ref52 cells results in the total disruption of stress fibers and the normal architecture of the actin cytoskeleton.[10] However, comicroinjection of SptP and SopE (the *Salmonella*-encoded Rho GTPase exchange factor) results in no overt disruption of the actin cytoskeleton as these two proteins antagonize each others function.[11] Disruption of the actin cytoskeleton is presumably a consequence of the potent GAP activity of SptP toward Rac and Cdc42, as this activity is retained by

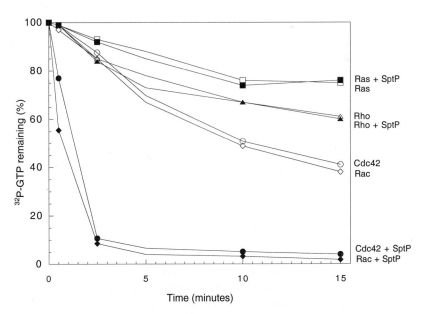

FIG. 3. GAP activity of SptP. The GAP activity of SptP toward Rac1, RhoA, Cdc42, and H-Ras was examined by a filter-binding assay.

a SptP$_{C481S}$, which is devoid of tyrosine phosphatase activity.[10] Ref52 cells are used to evaluate SptP activity because of their well-developed actin cytoskeleton. Ref52 cells are grown in Dulbecco's Modified Eagle Medium (DMEM) supplemented with 10% fetal calf serum on gridded coverslips to 70% confluency and serum-starved for 3 hr before microinjection. We have observed that not all batches of serum are adequate to sustain normal growth of these cells and the formation of well-developed stress fibers. Therefore, different batches of serum must be evaluated before their use. Purified GST–SptP proteins (GST–SptP$_{70-544}$ or GST–SptP$_{70-480}$) are microinjected into Ref52 cells usually at a concentration of 0.3–05 mg/ml in microinjection buffer (20 mM Tris–HCl buffer, pH 7.4, 20 mM NaCl, 1 mM MgCl$_2$, 0.1 mM EDTA, 5 mM 2-mercaptoethanol), although concentrations as low as 0.1 mg/ml retain a detectable effect. At different times after microinjection, cells are fixed in 3.7% formaldehyde for 1 hr and permeabilized by treatment with 0.2% Triton X-100 for 3 min. Cells are washed and stained with an antibody directed to GST or SptP to visualize microinjected cells and rhodamine-labeled phalloidin to visualized the actin cytoskeleton. Usually, disruption of the cytoskeleton can be seen as early as 15 min after microinjection, although the time is dependent on the amount of SptP microinjected into the cell. Microinjection of SptP at concentrations as low as 0.2 mg/ml results in marked disruption of the actin cytoskeleton (Fig. 4).

Use of *Salmonella* Effector Proteins for Study of Rho GTPase Function

Salmonella typhimurium SopE and SptP proteins are useful reagents to study Rho GTPase function. SopE can efficiently and directly activate members of the Rho GTPase family of small G-proteins, including Rac-1, Rac-2, Cdc42, RhoG, and, to a lesser extent, RhoA.[4] Therefore, introduction of SopE into cells results in the direct activation of Rho GTPases without the stimulation of any upstream pathways, which would result, for example, after their activation by extracellular stimuli. This allows the study of downstream pathways without the potential interference from the stimulation of other unrelated signaling pathways. Activation by SopE offers advantages over the introduction of constitutively active mutant forms of Rho GTPases as agonist-mediated stimulation of small G-protein may better mimic responses induced by physiological activation of these regulatory proteins. SptP is a unique reagent to specifically target Cdc42 and Rac. Transient expression or microinjection of the amino-terminal GAP domain of SptP (SptP$_{70-480}$) results in the inactivation of both of these Rho GTPases and therefore this reagent can be used to investigate their potential involvement in a given biological process. SptP offers an advan-

phalloidin   anti-GST

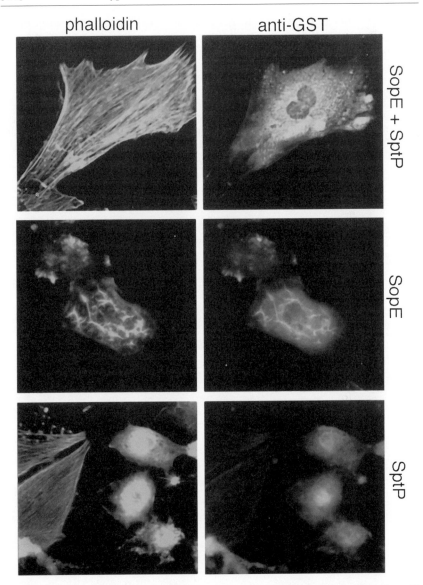

FIG. 4. Effect of GST–SptP microinjection into Ref52 cells alone or in combination with GST–SopE. The concentration of bacterial proteins was 0.4 mg/ml. The actin cytoskeleton was visualized by rhodamine–phalloidin staining 45 min after microinjection. From Fu and Galán.[11]

tage over other reagents that target these Rho GTPases as the *Salmonella* protein, unlike other bacterial toxins, does not irreversibly modify its targets.

Acknowledgments

We thank Sumati Murli for critical review of this manuscript. This work was supported by Public Health Service Grants AI30492 and GM52543 from the National Institutes of Health to J.E.G.

Author Index

Numbers in parentheses are footnote reference numbers and indicate that an author's work is referred to although the name is not cited in the text.

A

Aaronson, S. A., 31, 33, 34(25), 39, 141(11), 142, 382, 383, 383(16, 19)
Abe, J., 345
Abe, K., 27, 28(17), 31(17), 35(17), 36(17), 37, 42, 425, 428
Abe, M., 303, 304(1)
Aber, S., 326
Abo, A., 9, 58, 60(30), 141, 157, 166, 214, 217, 257, 333, 416, 440
Abramson, S. B., 104, 238
Absolom, D. R., 464
Adami, N., 321
Adams, T. E., 126
Adamson, P., 444
Aderem, A., 462, 463, 467, 467(9)
Adles, C., 464
Adra, C. N., 93
Aebersold, R., 200
Aebi, U., 222
Aelst, L. V., 93
Aepfelbacher, M., 278(27, 34), 282, 284
Aerbersold, R., 345
Aghazadeh, B., 26(12), 27
Ahmadian, M. R., 140, 498
Ahmed, S., 442
Ahn, N. G., 335
Akedo, H., 279(38), 284
Aktories, K., 17, 88, 115, 116(5), 121, 122, 125, 126, 127, 129(3), 133(15), 136, 264, 278(42), 284, 422, 463
Alape Giron, A., 116
Alberts, A. S., 273
Alessandra, E., 92
Algrain, M., 92
Allard, W. J., 8
Allen, K. M., 324

Allen, L. A., 463, 467, 467(9)
Allen, W. E., 238, 460, 462(15)
Allet, E., 321
Alnemri, E. S., 156, 166(11), 321
Altshuller, Y. A., 178
Altshuller, Y. M., 179
Amano, M., 149, 149(8–10), 150, 152(6), 153, 155, 179, 273, 274, 277, 277(12), 370
Amara, J. F., 286
Ambrose, D. M., 315, 316, 324(11)
Ames, B. N., 7
Anderson, J. M., 53
Anderson, R. A., 190, 191, 191(3), 192(3), 195(6)
Ando, S., 15, 16(10), 72
Angstrom, J., 116
Anraku, Y., 191
Antonny, B., 121
Anzai, K., 208
Aoki, J., 150
Araki, N., 330
Araki, S., 15, 93
Armstrong, S. A., 16
Arpin, M., 92
Arthur, J. M., 168
Arvidsson, A. K., 28, 53
Asada, M., 15, 16(10), 72
Asano, T., 153
Ashby, M. N., 101
Ashworth, A., 9, 264, 284
Aspenstrom, P., 10, 257, 403, 404(4), 450
Atsuchi, N., 303, 304(1)
Auer, K. L., 416
Auger, K. R., 197, 199
Aunis, D., 356
Avila, J., 201
Azim, A. C., 257, 258, 350, 355
Azuma, T., 349

505

G

Glaven, J. A., 26, 27, 28(17), 31(17), 33(7), 35(17), 36(17), 176
Gleeson, J. G., 324
Glomset, J. A., 168
Glorioso, J. C., 475
Goeddel, D. V., 381
Goehring, U.-M., 127, 133(15), 136
Goffinet, A. M., 201
Goins, W. F., 475
Goldschmidt-Clermont, P., 216, 295, 303, 304(4, 6)
Goldstein, J. L., 16, 383, 386, 388
Golla, R., 222
Gomperts, B. D., 356, 357, 357(5), 361(12), 362(3, 12), 369
Gomperts, S. N., 303, 304(3)
Gong, M. C., 278(45), 284
Gonzalez-Garcia, M., 322
Goody, R. S., 82, 83, 90(6), 140, 141(5), 142
Gordon-Weeks, P. R., 473
Gorman, S., 53
Gorski, J. L., 26, 33(9), 49
Gotoh, Y., 322
Gouin, E., 286, 289(10), 290(10), 291(10)
Gould, K., 257
Graessmann, A., 452
Graessmann, M., 452
Graham, F. L., 305, 306
Graham, R., 331
Graham, S. M., 382, 383(18), 430, 433(16)
Grannass, A., 122
Graves, J. D., 265
Gray, T., 115
Greenberg, M. E., 322
Greenberg, S., 238, 463, 467, 470(16)
Greene, L. A., 322
Grinstein, S., 334, 336, 345, 348(27)
Groblewski, G. E., 208
Groffen, J., 315, 320(2)
Grogan, A., 216
Gross, M., 381(13), 382
Grunwald, J., 92
GST Gene Fusion System, 341
Guarente, L., 17
Guillemot, J. C., 93, 286, 289(10), 290(10), 291(10), 293
Gupta, R., 156
Gutkind, J. S., 9, 39, 61, 62(5), 68(5), 381
Gutowski, S., 178

H

Habets, G. G., 26, 51, 52, 54(5), 61(1), 69, 403, 406(1, 2), 408(1), 409(1)
Haddara, W., 306
Hadding, U., 116
Hahn, C. S., 122
Hahn, H., 417
Hahn, K. M., 389
Haines, J. L., 28
Haines, K. A., 104
Halenbeck, R., 156, 166(10), 315, 321(6)
Hall, A., 9, 10, 16, 17, 24, 25, 30, 49, 56, 62, 66, 71(10), 72, 85, 91, 121, 140, 142, 156, 161, 175, 176(18), 190, 192, 257, 263(7), 264, 278(28), 279(43), 282, 284, 285, 319, 330, 333, 339, 358, 367(15), 369(15), 370, 381, 382(8), 383(8), 403, 404(4), 409, 411, 422, 425, 427(7), 440, 441, 442, 443, 446(10), 447(12), 450, 451, 462, 463, 474, 481, 483
Hall, C., 11
Hall, M. E., 214, 217(6), 257, 349
Hall, P., 156, 316, 324
Halvarsson, M. O., 116
Hama, C., 53
Hamajima, Y., 273
Hamanaka, I., 279(29), 282
Hamel, E., 79
Hamilton, A. D., 381, 382, 383(20), 384, 386(31), 387, 387(35)
Hammond, S. M., 177, 178, 179(5), 181(12), 188(5), 189(5)
Hampton, L. L., 208
Han, J., 26, 28, 37(6, 19), 38, 38(19), 39, 40(5), 41(4), 43(5), 44(5), 49(4), 339
Hancock, J. F., 15, 16, 16(3), 55, 56, 66, 72, 73(9)
Hannah, V. C., 16
Hannapel, E., 240
Hanser, H., 326
Hara, E., 421
Harada, H., 322
Harada, M., 279(29), 282
Harada, T., 279(46), 284
Harden, N., 324
Hardt, W.-D., 82, 83, 85(4), 87(2, 4), 88(4), 89(4), 90(2, 4), 91, 91(2, 4), 497, 498(3a), 502(4)
Hardy, S., 303, 304(3)

L

Subject Index

H

ISBN 0-12-182226-5

90038

9 780121 822262